REINFORCED CONCRETE FUNDAMENTALS

REINFORCED CONCRETE FUNDAMENTALS,

SI VERSION
FOURTH EDITION

PHIL M. FERGUSON
PROFESSOR EMERITUS OF CIVIL ENGINEERING
THE UNIVERSITY OF TEXAS AT AUSTIN

SI Conversion by
Henry J. Cowan
PROFESSOR OF ARCHITECTURAL SCIENCE
UNIVERSITY OF SYDNEY, AUSTRALIA

John Wiley & Sons
NEW YORK   CHICHESTER   BRISBANE   TORONTO

Library of Congress Cataloging in Publication Data

Ferguson, Phil Moss, 1899–
  Reinforced concrete fundamentals, SI version.

    Includes index.
    1. Reinforced concrete construction. I. Cowan,
Henry J. II. Title.
TA683.2.F4 1981   624.1′834   80-24409
ISBN 0-471-05897-1

Printed in the United States of America

10 9 8 7 6 5 4 3 2 1

PREFACE TO THE SI VERSION

At the time of writing there is no SI (metric) version of the American Concrete Institute's *Building Code Requirements for Reinforced Concrete (ACI 318–77)* on which the fourth edition of Professor Ferguson's book is based. The Canadian Standards Association's *Code for the Design of Concrete Structures for Buildings (CAN3-A23.3-M77)*, modeled on the ACI Code, is metric, but it is uncertain whether the forthcoming metric ACI Code will follow the Canadian practice for converting the formulas; for example, in one equation the English-unit term $\sqrt{f_c'}$ becomes $0.083\sqrt{f_c'}$ in metric units. To convert $\sqrt{f_c'}$ from psi to MPa exactly requires multiplication by $0.083\,035$; but to use this degree of precision in an empirical formula confers a quite unwarranted importance on the choice of 1 as a factor for multiplying $\sqrt{f_c'}$. In this case 1 indicates a number greater than 0.9 and smaller than 1.1, or perhaps greater than $\frac{3}{4}$ and smaller than $1\frac{1}{4}$. There is a proposal to use $\frac{1}{12}$ in the SI-metric ACI Code. This conveys a clearer impression of the degree of accuracy inherent in the number, and it is easier to memorize. I have therefore taken a calculated risk in assuming that the SI-metric ACI Code will use simple conversion figures. The difference between 0.083 and $\frac{1}{12}$ is only 0.4%, which does not alter the dimension of the structure, but the formula *looks* different.

The SI-metric system recognizes only conversion factors of 1 000, so that it is sometimes necessary to choose between the use of decimal fractions and very large numbers. This is particularly marked in calculations of the moments of inertia of the most common concrete sections which measure a few hundred millimeters in each direction. Since $1\,\text{m}^4 = 1 \times 10^{12}\,\text{mm}^4$, one can either express I in terms of $10^{-3}\,\text{m}^4$ or $10^9\,\text{mm}^4$. The choice depends to

v

some extent on the use to which the result will be put; for example, in an equation containing stresses (usually given in MPa, which is MN/m^2) the use of $10^{-3} m^4$ is more convenient.

Loads are generally specified in kilonewtons (kN) or kilopascals (kPa), that is, in thousands; but stresses are stated in megapascals (MPa), that is, in millions. Long dimensions, such as spans, are frequently in meters, but cross-sectional dimensions are in mm. Provided the equation is homogeneous, the units need not be converted; for example, a stress in MPa multiplied by one dimension in millimeters and divided by another dimension in millimeters remains in MPa. However, if units are mixed, it is best to convert all of them to one set of basic units. I have chosen meters and meganewtons as the basic units. Quantities in m, MN, and MPa remain unchanged; mm become $m \times 10^{-3}$ and kN become $MN \times 10^{-3}$. In the earlier part of the book the 10^{-3} is stated in each equation, but in the later problems it is assumed that the reader knows 400 mm = 0.4 m and 23 kN = 0.023 MN.

The metric bar sizes introduced in Canada, which will almost certainly be adopted in the United States, are more convenient than those in English units, because they are graded in $100 mm^2$ increments of cross-sectional area. However, there are only 8 American metric bar sizes, compared with twelve American bar sizes in English units, 13 British metric bar sizes, and 10 Australian metric bar sizes. Consequently the amount of reinforcement required in metric calculations is sometimes higher because the choice of bars is more restricted. On the other hand, the reduction in the number of bar sizes will presumably reduce construction cost.

No reference has been made to codes other than the ACI Code. The Canadian Code is based on the ACI Code, the Australian Code is different in several important sections, and the British Code differs greatly from the ACI Code. To refer to these codes, even though all are in SI-metric units, would have greatly added to the length of an already long book.

Three additional appendices give the cross-sectional areas of metric bars, a list of the symbols used in the book, and a list of the conversion figures used in the metrication.

Henry J. Cowan
Sydney, Australia
December, 1980

PREFACE

The following major changes have been made in this fourth edition.

1. Modernization to conform to the technical changes in the 1977 ACI Building Code.
2. Recognition of very significant changes in Code notation and format. The new symbols M_n, V_n, P_n, and so on replace the $\overline{M}$, $\overline{V}$, $\overline{P}$ of earlier editions; and all 1977 Code equations are in these terms, none in terms of unit shears.
3. The total use of SI units in many elementary examples in flexure, in a deflection example, and in a square footing design; related examples are in English units. This is a small move to encourage students to be mathematically bilingual. Chapter 1 briefly reviews the SI system used.
4. A major reorganization of material to simplify beginning class assignments. A few chapters have been internally rearranged to group the simpler and more usable material before adding the more complex coverage. More frequently, chapters have been split, with one now on shear in one-way members and another on two-way systems and torque problems. In slabs the first chapter on two-way slabs is now Chapter 12 for slabs on stiff beams, representing the many cases *not* eligible for the methods of Code Chapter 13; this presentation follows a method from the 1963 Code, one still valid. Chapters 13 and 14 on yield line and strip methods are basic material, but usually crowded out of a beginning course. The Code chapter on slab systems approach is presented in two parts: Chapter 15 on flat plates and flat slabs, representing the simpler cases, and Chapter 16, providing the general slab case including the interaction of two-way slabs with beams and columns. Treatment of columns has been separated into one chapter on "short" columns and a second on long columns.

The behavior of reinforced concrete is still given a prominent place in this textbook, with many pictures of members at failure. Because reinforced concrete is largely semiempirical, design engineers who understand how reinforced concrete behaves in approaching ultimate resistance have an advantage in assessing the many situations that they face day to day.

Some limits in the scope of this book have been necessary. Only a simple introduction to prestressed concrete is included, composite member coverage is kept brief, shearheads and brackets are outlined rather than covered in depth, joints have only limited treatment (but include two excellent references), and Vierendeel trusses and shear walls are omitted.

Service load analysis, as in the third edition, starts with transformed areas and deflections in the serviceability chapter and is summarized with the treatment of flexure in Appendix A, including a few design charts.

Although this book is now better arranged for the beginning student in reinforced concrete, it covers material adequate for a second semester of work. Many detailed examples appear in the text. Because it is primarily a book on basic philosophy, behavior, and theory, design is included chiefly as a teaching tool.

The author feels that an elementary course might well include most of Chapters 1 through 5, plus parts from Chapters 6, 7, 8, 10, 18, 20, and 21. I use Chapter 9 on retaining walls to pull early theory chapters together; other instructors prefer more from some of the partially covered chapters just listed. Students should have the ACI Building Code available and the Commentary would also be helpful.

I am indebted to many friends and colleagues. The repeated help from fellow faculty members is gratefully acknowledged. This book was greatly improved by comments or suggestions from Professors J. E. Breen, N. H. Burns, N. J. Carino, R. W. Furlong, and J. O. Jirsa.

Readers in the past have been most cooperative in reporting any errors they notice and these comments permit errors to be corrected at each reprinting. This is most helpful and appreciated.

Phil M. Ferguson
Austin, Texas
May, 1978

CONTENTS

1
INTRODUCTION

1.1 THE USE OF SI UNITS

As the preface has indicated, this edition differs from the last in several aspects. One of the more noticeable is the use of SI units and the total absence of English units. It is now assumed that nearly all engineering students at the junior year level have used SI units earlier and will find their use here creates no problem. Nevertheless, Sec. 1.4 summarizes SI units and prefixes and outlines a few general good practices for using them in calculations and on drawings. The aim is primarily to refresh the reader's memory in case he or she has not been using SI calculations recently or has not been using them in the area of forces and stresses, mass, weights, and energy.

It is hoped that the student will be able (or will learn) to think in SI units. Hence, no English equivalents are shown in the problems.

1.2 PROBLEMS IN THE USE OF SI UNITS

Reinforced concrete is not the simplest area in which to introduce SI units. In many respects it is an empirical field, because there is no mathematical basis that can be extended as a differential equation to many areas. It is nonlinear in behavior. Concrete is reinforced with steel because it is weak in tension and in most structural usage is uneconomical as plain concrete. To make reinforcement work economically, the concrete must crack in tension at service loads (fortunately, if well designed, with cracks too small to see without searching). But such physical responses mean there are many empirical constants from research studies that enter into designs.

In the United States and Canada most of these constants were

1

established using English units and the allowable values in many cases involve judgment decisions. For example, one of the unit shear values in psi is limited to $2\sqrt{f'_c}$, where f'_c is the concrete strength in compression in psi. If you transform this mathematically into megapascals (MPa), equal to Newtons per square millimeter (N/mm^2), you obtain something like 0.166 as the constant. But you know that the input was 2 and not 2.0 or 2.00. Hence, 0.166 seems unreasonable, somewhat like estimating a distance as 600 feet and having someone say you estimated it as 0.1136 miles (600/5280 = 0.113636 . . .). The 0.166 is actually no more accurate than the starting 2; hence a $\frac{1}{6}$ in SI usage would appear to be at least as accurate as the original input, even though it differs from 0.166 by several percent.

Specified concrete strengths will almost certainly be C20, C25, C30, and so on, where the number is the cylinder strength in megapascals. One pascal is one newton per square meter or N/m^2.

Likewise, new bar sizes are currently being written into a new ASTM standard. The smallest deformed bar would have a cross section of 100 square millimeters and be called a #10 bar (very roughly 10 mm diameter), the largest #55 with an area of 2500 mm^2. Grades of reinforcing bars will be Grade 300 and Grade 400 with the number the yield strength in megapascals (MPa). This book uses these as standards, on the assumption that these are now essentially fixed. These provide a reasonably sound basis for flexure calculations with SI units.

Totally outside the questions of concrete strength and reinforcement strengths and sizes, there are many other interlocking construction materials. Lumber sizes now nominally in full inches, such as 2 by 12 in. nominal size, $1\frac{5}{8}$ by $11\frac{1}{2}$ in. actual size, will probably differ in SI units; and lumber sizes often determine beam widths for economy. Joist forms will probably change size as they move to SI units. Loadings must change from pounds per square foot to newtons per square meter, very different quantities. For distributed loads kPa = kN/m^2 has been recommended as very convenient. Slabs now are typically designed for one foot strips because that width fits loads per square foot. The SI designer will use one meter widths.

1.3 THE STATUS OF SI UNITS

The United States is committed to change to SI units, but no calendar or schedule has been adopted. Hence, one should say "changing very slowly." Canada, on the other hand, is working toward a specific changeover date and a new code to match. England has been using an SI code for several years, but seems to have stopped short of using the term pascal (abbreviated as Pa). It uses N/mm^2 (newtons per square millimeter), which is an identity, but longer to write.

A number of countries do not want to give up kilograms for force even though it also means mass; the SI system is kilogram for mass *only*, newton for force, and pascal for unit stress. The SI usage discourages the use of centimeters, since either millimeters or meters can be written with less confusion from decimal places. Drawings for structures may be entirely in millimeters to avoid decimals, a fraction of a millimeter being negligible in most construction cases.

Australia, New Zealand, and South Africa have moved to SI units. In the United States ACI plans an SI version of the 1977 Code. Appendix D of the 1977 ACI Code lists SI equivalents for Code equations and Appendix E the MKS-metric equivalents.

1.4 SI UNITS FREQUENTLY APPLICABLE TO REINFORCED CONCRETE DESIGN

The SI standard has one (and only one) unit for each physical quantity, with these units subdivided into three classes of terms, which include the following related to structures:

Basic Units (out of a total of seven)		Symbol	Formula
length	meter	m	
mass	kilogram	kg	
time	second	s	
Supplementary Units (total of two)			
plane angle	radian	rad	
solid angle	steradian	sr	
Derived units (very large group)			
acceleration	meter per second squared		m/s^2
area	square meter		m^2
energy	joule	J	$N{\cdot}m*$
force	newton	N	$kg{\cdot}m/s^2$
moment of force	newton meter		$N{\cdot}m$
moment of inertia	kilogram meter squared		$kg{\cdot}m^2$
pressure, stress	pascal	Pa	N/m^2
elastic modulus	pascal	Pa	N/m^2
torque	newton meter		$N{\cdot}m$
velocity	meter per second		m/s
volume	cubic meter		m^3
work	joule	J	$N{\cdot}m$

*Note that a period at midheight is a multiplication sign.

Rather than a large number of digits, prefixes attached to units have been standardized. Multiples for 100, 10, 0.1, and 0.001 are available, but discouraged. (Centimeter usage is thus discouraged.) The prefixes most often useful in structural calculations are:

1 000 000* = 10^6	mega	M
1 000 = 10^3	kilo	k
0.001 = 10^{-3}	milli	m
0.000 001 = 10^{-6}	micro	μ

Preferred practice is to use the prefixes to keep isolated numbers in the range between 0.1 and 1000; but *not* in tabulations of data, where a given unit should be maintained even if the data scatter widely.

The units spelled out are kept in lower case, even though as abbreviations they may be capitalized, such as newton with the symbol N, pascal with the symbol Pa, and so on. The form newton/meter is discouraged for all cases where a unit is spelled out; use either newton per meter or N/m.

For more than three digits use no space for four places, but for more places skip a place to keep numbers in groups of three: 1213, but 12 130 and 1 121 300 and similarly in decimals 0.1213, but 0.012 13 and 0.012 130 4. Note that commas are not used, as we are accustomed to do in English units, because the comma in some areas means a decimal point.

SELECTED REFERENCES

1. American National Standards Institute, *ASTM/IEEE Standard Metric Practice*, ASTM E 380-76, IEEE Std 268-1976.
2. *Metric Practice Guide*, E 380-74, Amer. Soc. for Testing Materials, 1974.
3. *Recommended Practice for the Use of Metric (SI) Units in Building Design and Construction*, National Bureau of Standards, NBS Technical Note 938, 1977.

---

*Note that commas are *not* used in the open gaps. The gaps can be either full spaces or half spaces.

2
MATERIALS AND SPECIFICATIONS

2.1 CONCRETE MATERIALS AND PRODUCTION

Concrete for reinforced concrete consists of inert aggregate particles bound together by a paste made from portland cement and water. The paste fills the voids between aggregate particles and after the fresh concrete is placed it hardens as a result of exothermic chemical reactions between cement and water to form a solid and durable structural material. A typical cross section of hardened concrete is shown in Fig. 2.1a.

Although there are a number of standard portland cements, most concrete for buildings is made from Type I ordinary or standard cement (for concrete where the critical strength is needed in something like 28 days) or from Type III high-early strength cement (for concrete where strength is required in a few days). The heat generated by the different types of cements during the setting and hardening process varies widely, as indicated in Fig. 2.2. Where shrinkage and temperature stresses are important in the design, the volumetric change associated with these heat differences becomes significant. Air-entraining cement or admixtures for entraining air in the concrete are frequently used for greater workability or durability. Expansive cement to limit shrinkage is also on the market.

Aggregate consists of both fine and coarse aggregate, usually sand for the fine and gravel or crushed stone for the coarse aggregate. Lightweight aggregate made from expanded shale, slate, or clay has become increasingly important. Other aggregates, such as expanded slag, are also used. The size (and also the grading) of aggregate has an important influence on the amount of cement and water required to make a given unit of concrete

5

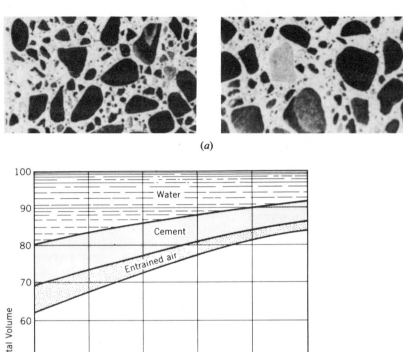

(a)

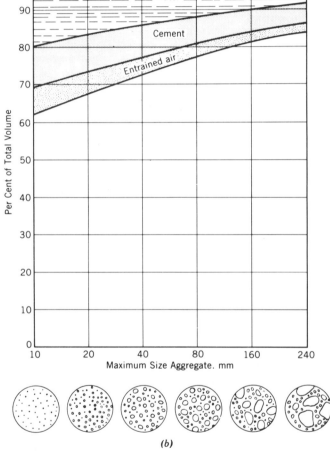

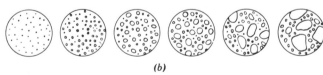

(b)

FIGURE 2.1 Components of a concrete mix. (*a*) Cross sections of concrete showing coarse and fine aggregate separated by cement paste. (Courtesy Bureau of Reclamation.) (*b*) Quantities of each material in 1 m³ of concrete. (After Reference 1, Bureau of Reclamation.)

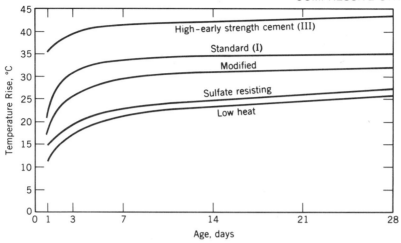

FIGURE 2.2   Temperature rise in concrete for various types of cement, when no heat is lost. Cement content $1:6\frac{1}{2}$ by volume. (After Reference 2 and 1, ACI and Bureau of Reclamation.)

of a given consistency (Fig. 2.1b). It also exerts a major influence on bleeding, ease of finishing, shrinkage and permeability.

The quantity of water relative to that of the cement is the most important item in determining concrete strength. The effect of the water-cement ratio on strength is indicated in Fig. 2.3. The water is sometimes controlled indirectly and approximately by specifying the cement content in terms of sacks per cubic yard of concrete.

It is important that concrete have a workability adequate to assure its consolidation in the forms without excessive voids. This property is usually indirectly measured in the field by the slump test (Fig. 2.4a) or the Kelly ball test (Fig. 2.4b). The necessary slump may be small when vibrators are used to consolidate the concrete. For methods of designing concrete mixes the student is referred to the American Concrete Institute's "Recommended Practice for Selecting Proportions of Normal and Heavyweight Concrete" (ACI 211.1-77),[4] "Recommended Practice for Selecting Proportions for Structural Lightweight Concrete (ACI 211.2-69),"[5] or the Portland Cement Association booklet, "Design and Control of Concrete Mixtures."[3]

Proper curing of concrete requires that the water in the mix not be allowed to evaporate from the concrete until the concrete has gained the desired strength. Figure 2.5 is representative of the variations in strength that can result from differences in curing. Note that air-dried concrete is able to gain strength if it is moistened at a later time. Temperature is also an important element in the rate at which concrete gains strength, low

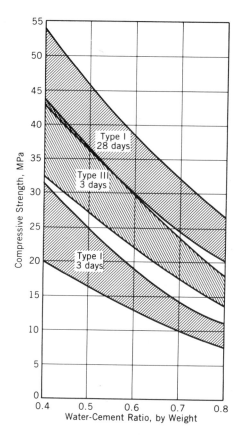

FIGURE 2.3 Effect of water-cement ratio on strength of non-air-entrained concrete at different ages. All specimens moist cured at 21°C. (Modified from Reference 3, Portland Cement Assn., earlier edition.)

temperatures slowing up the process but raising the potential strength if normal temperature is restored, as indicated in Fig. 2.6. Early high temperatures lead to rapid setting and some permanent loss of strength potential.

2.2 COMPRESSIVE STRENGTH

Depending on the mix (especially the water-cement ratio) and the time and quality of the curing, compressive strengths of concrete can be obtained up to 100 MPa or more. Commerical production of concrete with ordinary

(a)　　　　　　　　　　　　　(b)

FIGURE 2.4 Testing for workability of concrete. (<i>a</i>) Slump test. (Courtesy Portland Cement Assn.) (<i>b</i>) Kelly ball test. The "ball" penetration is read on the graduated shaft by the late Professor Kelly of the University of California. This mix is quite stiff.

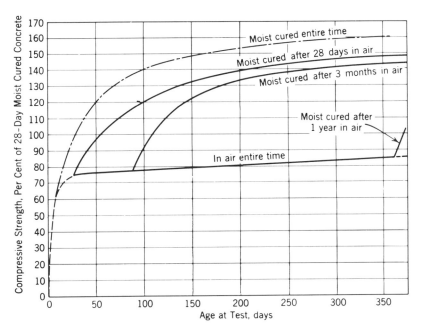

FIGURE 2.5 Effect of curing conditions on strength of concrete. (From earlier edition of Reference 3, Portland Cement Assn.)

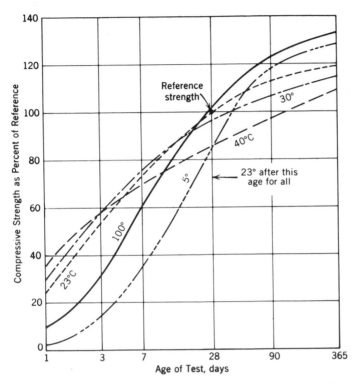

FIGURE 2.6 Compression strength attained at various ages and temperatures as percent of 28-day strength under curing at 23°C. Type I cement. (Modified from Reference 3, Portland Cement Assn.)

aggregates is usually in the 20 to 70 MPa range with the most common near 20 to 35 MPa. On the other hand, highway departments often expect strengths of 30 to 50 MPa. Because of the difference in aggregates, and to a lesser degree in cements, the same mix proportions result in substantially lower strengths in some sections of the country than in others. In these sections a lower water-cement ratio must be used.

Compressive strength f'_c is based on standard 150 mm by 300 mm cylinders cured under standard laboratory conditions and tested at a specified rate of loading at 28 days of age.* The designer should note that building concrete cured in place on the job will rarely develop as much strength as these standard cured cylinders. Separate cylinders should be made to check on

*Test procedure is covered in detail in ASTM C-39.

the quality of curing if this is desired. These same cylinders, however, are not suitable for checking the quality of the mix.

The ACI Code, discussed in Sec. 2.10 following, specifies the average of two cylinders from the same sample tested at the same age (usually 28 days) for a strength test. It specifies the frequency at which such tests shall be made and sets up this criterion:

4.8.2.3—Strength level of an individual class of concrete shall be considered satisfactory if both of the following requirements are met:
(a) The averages of all sets of three consecutive strength tests equal or exceed required f_c'.
(b) No individual strength test (average of two cylinders) falls below required f_c' by more than 500 psi (3 MPa).

Such a simple criterion can not always be perfect. The Code Commentary states that, even with the strength level and uniformity satisfactory, an occasional low strength may be indicated (at least once in 100 tests) and makes suggestions about allowance for these.

The average concrete strength for which a concrete mix must be designed must exceed f_c' by an amount which depends on the uniformity of plant production, that is, on how well the variations in operation are minimized. Where production records are available, the Code in 4.3 specifies the mix must be designed for an average strength that is 3 MPa above f_c' if the standard deviation has been held to less than 2 MPa (extremely good) with a larger increment for higher values. The requirement is 8 MPa above f_c' if the standard deviation is over 4 MPa or if suitable records are not available.

It must be emphasized that f_c' for design is *not* to be considered as the average strength of job cylinder tests. The design f_c' is nearer a minimum than an average, but it is not an absolute minimum. An individual test may be nearly 3 MPa low and still be acceptable if all averages of three consecutive tests are satisfactory.

With lightweight aggregates a mix design should definitely be made on the basis of trial batches. Many lightweight aggregates produce 20 MPa concrete and some easily give 40 MPa concrete under proper control.

A special treatment of concrete after it has set, which can lead to extremely high potential strengths, is currently in an interesting stage of development. Polymer impregnated concrete is made by impregnating ordinary concrete with a liquid organic monomer* which is subsequently polymerized by radiation or thermal catalytic methods. The result is greatly increased tension and compression strengths (up several hundred percent) accompanied by much improved impermeability, hardness, and durability.

---

*Such as methyl methacrylate.

2.3 TENSILE STRENGTH

The tensile strength of concrete is relatively low, about 10 to 15% of the compression strength, occasionally 20%. This strength is more difficult to measure and the results vary more specimen to specimen than those from compression cylinders. The modulus of rupture as measured from standard 150 mm square beams somewhat exceeds the real tensile strength. The value of $0.62\sqrt{f_c'}$ is often used for the modulus of rupture, in MPa when f_c' is in MPa. The cylinder splitting test, often called the Brazilian test, is the most respected test for tensile strength. A compression force applied full length to opposite elements splits the cylinder open along the connecting diameter.

Lightweight concrete in many, but not all cases, has a lower tensile strength than ordinary weight concrete. The ACI Code accepts either of the following approaches in computing shear and development of reinforcement. If the splitting tensile strength is specified and the concrete is proportioned under Code 4.2, the designer substitutes $1.8f_{ct} \lessgtr 1$ for $\sqrt{f_c'}$ in computing the allowable. If f_{ct} is not specified, the assumed shear resistance may simply be reduced by a factor of 0.75 for "all-lightweight" concrete and 0.85 for "sand-lightweight" concrete. The latter refers to lightweight concrete containing natural sand for the fine aggregate. Linear interpolation may be used for mixtures of natural sand and lightweight fine aggregate. Development length (Chapter 7) is modified in a similar way, with the ACI Code factors $0.56f_{ct}/\sqrt{f_c'}$, 1.33, and 1.18, since splitting strength enters here as a reciprocal.

2.4 SIGNIFICANCE OF LOW TENSILE STRENGTH

In a homogeneous elastic beam subjected to bending moment, one can calculate the bending stresses from $f = Mc/I$. The extreme fiber on one face carries compression, on the opposite face tension. If the beam is rectangular (or of any shape symmetrical about the centroidal axis), the maximum tensile stress equals the maximum compressive stress. In concrete construction, except in massive structures such as gravity dams and sometimes heavy footings, it is not economical to accept the low tensile strength of plain concrete as a limit on beam strength. It is generally more economical to design a beam such that compressive bending stresses are carried by concrete and tensile bending stresses are carried entirely by steel reinforcing bars. It is not possible for concrete to cooperate with steel in carrying these tensile stresses except at very low and uneconomical values of steel stress. Except in prestressed concrete, when the steel stress reaches about 40 MPa the tensile concrete starts to crack and the steel soon thereafter must pick up essentially all the tension necessary to provide for the applied moment. Hence, in ordinary reinforced concrete beams the tensile concrete is not assumed to assist in resisting the moment. Still, tension in the concrete does reduce the beam deflection considerably.

2.5 SHEAR STRENGTH

The shear strength of concrete is large, variously reported as from 35 to 80% of the compression strength. It is difficult to separate shear from other stresses in testing and this accounts for some of the variation reported. The lower values represent attempts to separate friction effects from true shears. The shear value is significant only in rare cases, since shear must ordinarily be limited to much lower values in order to protect the concrete against diagonal tension stresses (Chapter 5).

Diagonal tension stresses are often referred to as shear stresses, but this is actually a misnomer. If the student will keep in mind that true shear strength is rarely in question, it will not matter that the term "shear" is often loosely used for diagonal tension.

2.6 STRESS-STRAIN CURVE

Typical stress-strain curves for concrete cylinders on initial loadings are shown in Fig. 2.7. The first part of each curve is nearly a straight line, but there is some curvature at f_c equal to half the maximum value.* The maximum stress is designated as f'_c (but note the statistical nature of the f'_c value to be used in design, as given in Sec. 2.2). The curve for low strength concrete has a long and relatively flat top. For high strength concrete the peak is sharper. Special techniques are necessary to establish these curves on the steep downward sections beyond the peak stress. Otherwise the testing machine characteristics result in sudden instead of gradual failure. Figure 2.8b illustrates this difficulty with high strength concretes.

At strains beyond the peak value of stress, considerable strength still exists. It will be noted that the cylinder strain occurring near maximum stress is nearly the same for all strengths of concrete, being roughly 0.002 mm/mm.† In a cylinder a maximum strain of something like 0.0025 measures the useful limit for all concretes except those of low strength or those made with lightweight aggregate. In beams of ordinary reinforced concrete, the shape of the cross section is a factor and the steeper the strain gradient the greater is the usable edge strain for a given f'_c

---

*Nonlinearity results from the formation of microscopic internal cracks which lower the stiffness.

†Lightweight concrete with its lower modulus will give higher values.

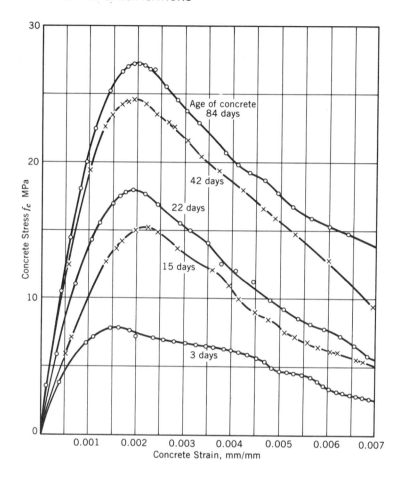

FIGURE 2.7 Concrete stress-strain curves from compression cylinders. (After Reference 6, Bureau of Reclamation.)

Observations show that unit strains of 0.0030 to 0.0045 normally occur before a beam fails; with f'_c over 40 MPa, the maximum observed strains are from 0.0025 to 0.0040. Tests[7] have proved conclusively that the stress-strain curve for the compression face of a beam is essentially identical with that for a standard test cylinder when stress is applied at the same rate, as shown in Fig. 2.8. However, confinement of the concrete by a spiral or by heavy ties can greatly increase the extreme strain value in compression, as can a steep strain gradient.

Strictly speaking, concrete on initial loading has no fixed ratio of f_c/ϵ which truly justifies the term "modulus of elasticity." The initial slope of

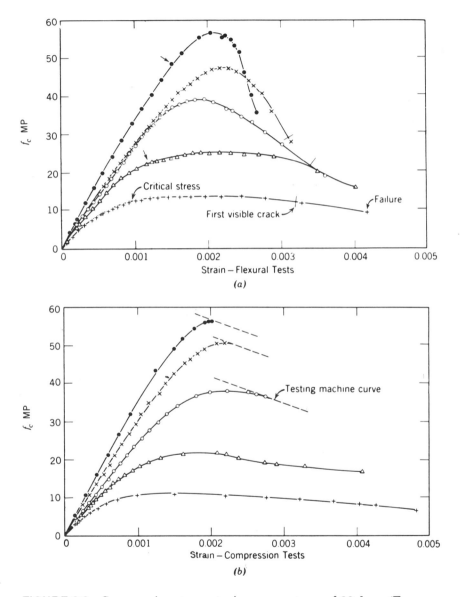

FIGURE 2.8 Compression stress-strain curves at age of 28 days. (From Reference 7, ACI.) (*a*) From flexural tests on 125 mm by 200 mm by 400 mm prisms. (*b*) From direct compression tests on 150 mm by 300 mm cylinders for the concretes shown in (*a*).

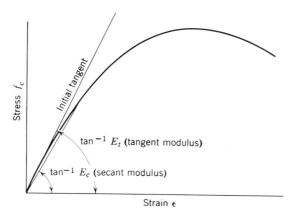

FIGURE 2.9 Tangent and secant modulus of elasticity determinations.

the stress-strain curve defines the initial or tangent modulus used with the parabolic stress method and occasionally elsewhere. The slope of the chord (up to about $0.5 f_c'$) determines the secant modulus of elasticity that is generally used in straight-line stress calculations (Fig. 2.9). When E or the term "modulus of elasticity" is used without further designation, it is usually the secant modulus which is intended.

The secant modulus in MPa is taken in the Code as

$$E_c = w^{1.5} 0.43 \sqrt{f_c'}$$

where w is the weight of concrete in the range between $1\,500\,\text{kg/m}^3$ and $2\,500\,\text{kg/m}^3$, and both f_c' and $\sqrt{f_c'}$ are in MPa. For normal weight concrete, E_c may be considered as $5\,000\,\sqrt{f_c'}$.

The use of a reduced modulus is one way to account for the time effects discussed in Sec. 2.7 and Sec. 2.8.

2.7 CREEP OF CONCRETE

The initial strain in concrete on first loading at low unit stresses is nearly elastic, but this strain increases with time even under constant load (Fig. 2.10a). This increased deformation with time is called creep and under ordinary conditions it may amount to more than the elastic deformation. Factors tending to increase creep include loading at an early age (while the concrete is still "green"), using concrete with a high water-cement ratio, and exposing the concrete to drying conditions. Concrete completely wet or

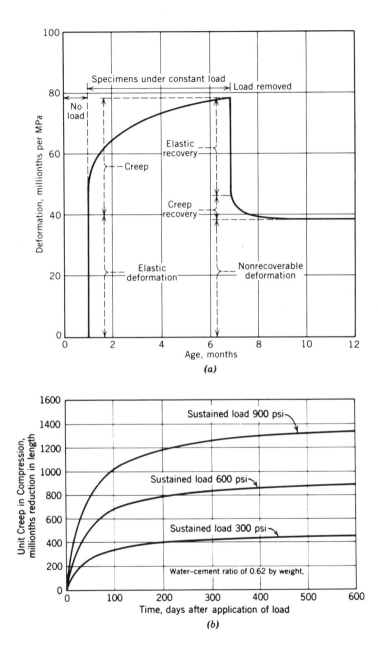

FIGURE 2.10. Creep of concrete (After Reference 1, Bureau of Reclamation.) (*a*) Creep and elastic deformation. (*b*) Effect of unit stress on unit creep for identical concretes. (*c*) Effect of age at loading. (After Reference 11, ACI).

17

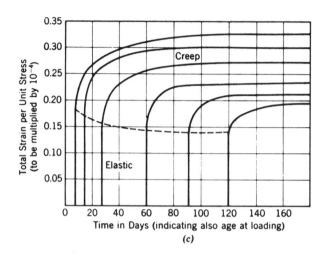

FIGURE 2.10   (cont.)

completely dried out creeps only a little, and in general creep decreases with the age of the concrete.

At stresses up to the usual service level, creep is directly proportional to the unit stress; hence elastic and creep deformations are essentially proportional in plain concrete members. Under overload conditions this proportionality no longer holds; and in reinforced concrete the constant modulus of the steel causes strain readjustments with time. While these are minor for tension steel, compression steel stresses may be more than doubled by creep.

The creep rate is more rapid when the load is first applied and decreases somewhat exponentially with time, as shown in Fig. 2.10b. Note that while unloading allows an immediate (but partial) elastic recovery (Fig. 2.10a), a further creep recovery occurs with time; but always this leaves a substantial residual deformation.

Creep is also much reduced by delaying the loading until the concrete is more mature, as indicated by Fig. 2.10c. This figure shows that both the initial elastic strain and the creep are reduced when the loading is delayed. However, temperatures from 50°C up substantially increase creep.

Creep is one common cause of deflections that increase with time. In reinforced concrete, without compressive steel, the final deflection will usually be from 2.5 to 3.0 times the initial deflection. It has been suggested that long-time deflections be calculated on the basis of a reduced modulus having a value one-third that of the instantaneous modulus, but see

Chapter 8 for better procedures. The Code requires (9.5.2.5)* that the additional long-time deflection for the sustained load be obtained by multiplying the immediate deflection by

$$(2 - 1.2\ A'_s/A_s) \geqslant 0.6$$

where A'_s/A_s is the ratio of compression reinforcement to tension reinforcement areas.

Creep deformation is not always harmful. Creep relaxes the stress effect of early deformation loadings; such stresses are so much reduced that creep in this situation is often given the special name of relaxation. For example, concrete stresses set up by differential settlement may be nearly eliminated where settlement occurs gradually but early in the life of the concrete, while it is still "green."

2.8   SHRINKAGE OF CONCRETE

As concrete loses moisture by evaporation, it shrinks. Since moisture is never uniformly withdrawn throughout the concrete, the differential moisture changes cause differential shrinkage tendencies and internal stresses. Stresses due to differential shrinkage can be quite large and this is one of the reasons for insisting on moist curing conditions. The larger the ratio of surface area to member cross section the larger will be the resulting shrinkage. Therefore, large specimens shrink much less than small ones.

In plain concrete completely unrestrained against contraction, a uniform shrinkage would cause no stress; but complete lack of restraint and uniform shrinkage are both theoretical terms, not ordinary conditions. With reinforced concrete even uniform shrinkage causes stresses, compression in the steel, tension in the concrete.

Expansive cement is sometimes used to minimize these shrinkage stresses. Because of expansive material in the cement, this concrete first expands a little. If it is partially restrained by embedded reinforcement, tension builds up in this steel and compression in the concrete. As the concrete later shrinks and cools it comes to equilibrium with less change from its initial length.

In ordinary concrete the amount of shrinkage will depend on the exposure and the concrete. Exposure to wind greatly increases the shrinkage rate. A humid atmosphere will reduce shrinkage; a low humidity will increase shrinkage. Shrinkage is usually expressed in terms of the

---

*The section number in the 1977 ACI Code is often listed in this book in this manner.

shrinkage coefficient s, which is the shortening per unit length. This coefficient varies greatly, with values commonly 0.0002 to 0.0006 and sometimes as much as 0.0010. An indication of how shrinkage varies with the water and cement content is given in Fig. 2.11a with shrinkage there

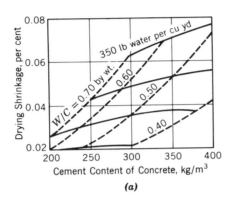

(a)

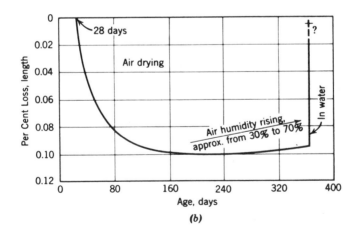

(b)

FIGURE 2.11 Shrinkage of concrete. (a) Relation of shrinkage to water content. (From Reference 2, ACI.) (b) Typical shrinkage-time curve starting with water-cured specimen (75 mm by 75 mm by 1 m) 28 days old. Note recovery when again placed in water; constant volume reached in about 24 hours of soaking. (Replotted and modified from Reference 8, The Engineering Foundation.)

expressed in percent. This figure can only show trends since the amount of shrinkage differs with materials and drying conditions. In lightweight concrete early shrinkage is noticeably reduced by the water contained in the pores of the lightweight aggregate.

Shrinkage is, to a considerable extent, a reversible phenomenon. If concrete is soaked after it has shrunk, it will expand to nearly its original size, as indicated in Fig. 2.11b. The recovery is now known not to be *total.*[10]

Shrinkage is one common cause of deflections that increase with time. Only symmetrical reinforcement can prevent curvature and deflection from shrinkage (Chapter 8).

2.9 REINFORCING STEEL

The yield strengths and ultimate strengths of the proposed grades for reinforcing bars under the SI system are shown in Table 2.1. The modulus of elasticity E_s is usually taken as 200 000 MPa. For design purposes it is usually assumed that the stress-strain curve of steel is elastic-plastic.

To increase the bond between concrete and steel, projections called deformations are rolled on the bar surface as shown in Fig. 2.12, with a pattern varying with the manufacturer. The deformations shown must satisfy ASTM Spec. A615-75 to be accepted as *deformed* bars. The deformed wire has indentations pressed *into* the wire or bar to serve as deformations.

Except for spirals, only deformed bars, deformed wire, and fabric made from cold drawn or deformed wire may be counted for strength in reinforced concrete under the ACI Code.

All standard bars are round bars, designated by size as #10 to #55; this number corresponding roughly to the bar diameter in millimeters. The bar weights and nominal areas, diameters, and circumferences or perimeters are tabulated in Table 2.2. Since the area is calculated from the weight, including that of the deformations, it is a nominal area as far as minimum cross section is concerned. The perimeter is the circumference of this nominal circular area.

TABLE 2.1   Reinforcement-Grades and Strengths

	Min Yield Point or Yield Strength, f_y	Ultimate Strength hf_u
Grade 300	300 MPa	500 MPa
Grade 400	400 MPa	600 MPa

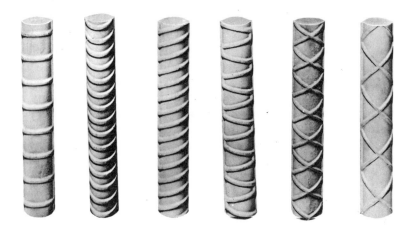

FIGURE 2.12 Various types of deformed bars. (Courtesy Concrete Reinforcing Steel Inst.)

The SI bar sizes are organized around convenient bar areas, but the bar numbering is a measure of the bar diameter (rather roughly).

Welded wire fabric is increasingly important for slab and pavement reinforcing. Fabric is made of cold-drawn wires running in two directions and welded together at intersections. A more recent development is deformed wire fabric and the use of larger rod sizes made up into fabric mats. Plain wire fabric depends largely on the welded crosswires for development. Deformed wire fabric usually has fewer crosswires with lighter welds and depends more, but not wholly, on the wire deformations for development.

TABLE 2.2 Weight, Area, and Perimeter of Individual Bars

Metric No.	Mass kg/m	Area, A_b mm^2	Diameter mm	Perimeter mm
10	0.785	100	11.3	35.5
15	1.570	200	16.0	50.1
20	2.355	300	19.5	61.3
25	3.925	500	25.2	79.2
30	5.495	700	29.9	93.9
35	7.850	1000	35.7	112.2
45	11.775	1500	43.7	137.3
55	19.625	2500	56.4	177.2

2.10 DESIGN CODES

A specification for reinforced concrete design may take the form of a code or a recommended practice. A code is written in the form of a law for enactment by public bodies such as city councils. It represents, usually, the minimum requirements necessary to protect the public from danger. It makes no attempt to specify the best practice, although it usually attempts to eliminate the most common mistakes, especially those involving safety. A recommended practice, on the other hand, attempts to define the best practice or at least to state satisfactory design assumptions and procedures. It may state reasons as well as methods.

The most significant reinforced concrete code in the United States is the "Building Code Requirement for Reinforced Concrete" (ACI 318-77). The Code is available from the American Concrete Institute,* Box 19150 Redford Station, Detroit, Mich., 48219. This code will be quoted frequently in this book, usually as the Code, or the ACI Code, or by a section reference number, as 4.3 was used in Sec. 2.2 of this chapter.

A Commentary on the 1977 Code is also available from the same address. The Commentary is not a legal part of the Code but it attempts to indicate some of the reasoning behind the various sections of the Code.

It might be noted that this Code uses some new notation that has been incorporated in this text, especially that discussed in Sec. 2.12g. All Code symbols are defined in each Code chapter and then summarized in Code Appendix B.

Two general codes have considerable usage: the Pacific Coast Building Officials Conference's "Uniform Building Code" and the Building Officials Conference of America "Basic Building Code," the latter using the ACI Code. In addition, many cities write their own codes, making some modifications in these more standard codes. Designers must at the beginning of any project determine the code under which they are required to operate.

Highway bridges are normally designed under specifications prepared by the American Association of State Highway and Transportation Officials (AASHTO).

In Europe, in addition to the various national codes, the CEB-FIP "International recommendations for the design and construction of concrete structures," dated June 1970, is highly regarded as a set of "Principles and Recommendations" valuable to code writing commissions. A "Model Code for Concrete Structures" using SI units and a more thorough coverage is in at least a third draft and nearing completion by Comité Euro-International du Béton.

---

*ACI members receive a nominal discount on the price.

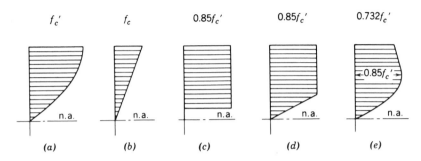

FIGURE 2.13 Stress distributions assumed on compression side of beam. (*a*) Simple parabola (ultimate). (*b*) Straight line (working stress). (*c*) Rectangular stress block (ultimate). (*d*) Trapezoidal stress block (ultimate). (*e*) Parabolic, with straight line (ultimate).

2.11 STRESS VERSUS STRENGTH CONCEPTS

The curved stress-strain diagram for concrete, with all strains increasing with time because of creep, complicates theoretical analyses. At low stress levels elastic analysis is reasonable, that is, stress varies almost linearly with the distance from the neutral axis. However, the reader should not confuse this with a feeling that such calculated stresses are therefore either elastic or exact. Shrinkage (usually nonuniform), creep, and cracking all complicate the stresses. Shrinkage tends to reduce compressive stress or increase tensile stress in typical cases. Creep increases the strains, whether tensile or compressive , in effect reducing the modulus of the concrete. Stresses computed elastically can only be index numbers, not real stresses.

Early investigators (before 1910) sensed this behavior and favored calculations based on ultimate strength, usually using the parabolic stress distribution of Fig. 2.13*a*. However, the apparent simplicity of the triangle of stress shown in Fig. 2.13*b* was strong enough to lead to the adoption of the so-called *working-stress design* (WSD).

In the 1930s field studies of strains in columns showed that the elastic concept of adjacent stresses in steel and concrete in proportion to the moduli values was not tenable; creep shifted much more of the initial load from concrete to steel. Tests on beams showed extreme face strains in compression, much in excess of the strain at peak stress, shown in Fig. 2.7 as approximately 0.002. It became clear that the descending part of the stress-strain curve for $\epsilon > 0.002$ was also important. In 1942 C. S. Whitney* presented a paper emphasizing this fact and showing how a probable

---

*Whitney was not the first to suggest the rectangular stress block, but he was the first in the United States to achieve some general acceptance.

stress-strain curve like those of Fig. 2.7 could, with reasonable accuracy,* be replaced with an artificial rectangular stress block simplification, as in Fig. 2.13c. With the rectangular stress block simplification, the 1956 ACI Code added an Appendix permitting *ultimate strength design* (USD) as an alternate to WSD. The 1963 Code gave the two methods equal standing.

Because of the nonlinear response of columns and prestressed concrete, engineers have been pushed toward ultimate strength design. Since 1971 the Code has been almost totally a *strength* code with strength meaning ultimate. However, an alternate method which for beams and slabs (not columns) was effectively a working stress method remained in the 1971 Code and constitutes Appendix B of the 1977 Code.

For strength design the Code permits any reasonable stress distribution to be assumed, if it agrees with tests, such as those in Figs. 2.13c to 2.13e or others. Practice has moved to the rectangular stress block because of its simplicity. Research often uses the shaded stress distribution of Fig. 2.13e, occasionally the dotted form. The use of the computer has also encouraged the development of a continuous function rather than this half parabola plus a straight line.

This book emphasizes the strength method and its advantages. Nevertheless, for two reasons it also includes working stress analysis briefly in Chapters 8, 17, and Appendix A. First, its assumption of linear stress distribution is appropriate for calculations of deflections under service load conditions and for crack investigations at service loads. Second, a minimum understanding of working stress design methods will improve communications between those trained in today's best methods and those more versed in older procedures.

2.12 SAFETY PROVISIONS

(a) Factor of Safety

A factor of safety is a concept of long standing, some reserve strength to care for unusual loads beyond those in the design. It has often been thought of as the ratio of yield stress to service load stress, although this concept is invalid where nonlinear responses exist. Correctly defined, the factor of safety is the ratio of the load that would cause collapse to the service or working load. Factor of safety is now such a misused term it almost requires a definition each time it is used.

*Accuracy with beams is excellent. With columns the results are more sensitive to the exact stress-strain curve assumed.

(b) Factors Determining Safety

Strength design as presented in the Code recognizes that a factor of safety is necessary for a number of reasons that are individually important but have no real interrelation. (1) Some are matters beyond the control of the engineer, such as the possibility of unforeseen future loads. (2) Others are in part under the engineer's control, such as: the quality of materials used, which the engineer can at least partially control by specifications and inspection; the usual dimensional tolerances in the sizing of beams and areas of bars and the placement of reinforcing, which can be closely inspected but have practical limitations inherent in manufacture and field construction. (3) Others relate to the importance of the member in maintaining the integrity of the structure; a local slab failure is not as serious as the collapse of a column. (4) Still other factors vary with the degree of safety factor that is justified, such as the hazards to life and limb from the collapse of a school building as compared to that of a simple shed for storage of equipment or material; these relative hazards are considered as already reasonably recognized in the size of the live load required by all general building codes.

(c) Partial Factors of Safety

Safety has always been clearer when thought of in terms of partial safety factors. In design these might well be in terms of loads, construction practices, the quality of the materials used, and the importance of the member or structure.

The loads specified in codes are usually not the average or mean loads, but more nearly the maximum loads. The European concept of characteristic loads is excellent. A characteristic load is one that has only a 5% probability of ever being exceeded. If one used such a load the strength above this load could be smaller than if one started from the mean or average load concept. The difficulty lies in identifying and specifying this characteristic load.

Construction practices introduce an element of uncertainty. If columns can only be built to a 12 mm tolerance (Code 7.5.2.1), it is well to note that a 288 mm square column is some 8% short of the area of a 300 mm square column.

Materials are also variables. The designer will not always find actually present at a critical spot the particular concrete strength that was assumed in design, in spite of good controls.

All the various overloads or deficiencies are less likely to occur together. Probability theory can predict the reduced probability of all bad aspects occurring together if the probability of each occurring separately were known. But here is always a basic problem; the input data cannot be clearly established in many cases.

(d) Code Safety Philosophy

The ACI Code separates safety provisions into two parts in the hope that ultimately (not in the present Code) different factors may be established for different qualities of specifications and inspection, and so on. The two factors now prescribed are load factors and ϕ factors.

Load factors (greater than unity) attempt to assess the possibility that prescribed service loads may be exceeded. Obviously, a specified live load is more apt to be exceeded than a dead load which is largely fixed by the weight of the construction. The ultimate strength of the members must care for the total of all service loads, each multiplied by its respective load factor; and the load factors are different in magnitude for dead load, live load, and wind or earthquake loading.

The ϕ factors (less than unity) are provided to allow for variations in materials, construction dimensions, and calculation approximations, that is, matters at least partially under the engineer's control. At present* ϕ varies only with the type of stress or member considered, recognizing that a column failure, for instance, would be more serious than a beam failure in most cases.

The general building code prescribes live loads that are assumed to cover adequately the relative hazards resulting from failure. For example, the prescribed live load for a school auditorium is properly oversafe compared to that for an office building. The general code also usually considers the probability of all areas being fully loaded at the same time.

(e) Load Factors

For dead and live loads the Code specifies that factored loads, factored shears, and factored moments be obtained from service loads by using the relation:

$$U = 1.4\,D + 1.7\,L \qquad \text{(Code Eq. 9.1)}$$

where U represents any of these required strengths for factored loads, D represents that for service dead load, and L that for service live load.

When wind W is included the total loading may be worse, but the chance of maximum wind occurring when an overload in both dead and

---

*At some future time the ϕ factor might have a range of values for different type specifications or inspection provisions.

live load exists is less than the chance of the overloads existing alone. Hence, the Code uses an 0.75 coefficient on the sum of all three:

$$U = 0.75(1.4\,D + 1.7\,L + 1.7\,W) \qquad \text{(Code Eq. 9.2)}$$

where L must be considered *both* as zero and as its maximum value. If L does act to reduce the total, this also suggests that D and W might act in an opposite sense. If so, an overlarge D is on the unsafe side and the third Code condition becomes necessary:

$$U = 0.9\,D + 1.3\,W \qquad \text{(Code Eq. 9.3)}$$

This equation underestimates D when it opposes W. (Note that $1.3\,W$ is approximately $0.75 \times 1.7\,W$, as in Eq. 9.2, with only two significant digits.)

The Code (9.2) also gives load factors for earthquake, lateral earth pressure, lateral liquid pressure, vertical liquid load, impact, and special effects such as settlement, creep, shrinkage or temperature change, but these special cases are not essential to the discussion here.

(f) Code ϕ Factors

The ϕ factor is multiplied into what is now called the nominal strength equation to obtain the reasonably dependable strength. The nominal strength equation might be said to give the "ideal" strength assuming materials are as strong as specified, sizes are as shown on the drawings, bars are of full weight, that calculations of load, shear, and moment are exact, and that the strength equation itself is scientifically correct. The practical dependable strength of a particular member will often be something less, probably sometimes very much less, since all these factors vary over some range. How low the strength may go can only be determined by probability theory; and the variables are not yet clearly enough defined to make this approach clearly better. Furthermore, the adequacy of theory varies, being lower for shear than for flexure. The Code provides for these variables by using these simple ϕ factors for the several cases listed (Code 9.3.2).

Flexure, with or without axial tension	0.90
Axial tension	0.90
Axial compression, with or without flexure:	
Members with spiral reinforcement conforming to Sec. 10.9.3	0.75*
Other reinforced members	0.70*
Shear and torsion	0.85

*Provisions are included for the gradual increase to 0.90 as the axial compression drops below ϕP_n of $0.10 f'_c A_g$ (Code Sec. 9.3.2c). See Sec. 18.7.

Bearing on concrete (See also Code 18.13 for prestress tendon
anchorage)   0.70
Flexure in plain concrete   0.65

   The ϕ is largest for flexure because the variability of steel is less than
that of concrete; and all flexural members are specified to be designed for
failure in tension, that is, in the steel. The ϕ values for columns are lowest
(favoring the toughness of spiral columns a little over tied columns).
Because columns fail in compression where concrete strength is critical,
there is some small danger that the analysis may miss the worst combination of axial load and moment; and the column is critical in the building.
For shear and torsion ϕ is intermediate since these depend on concrete
strength, but on $\sqrt{f'_c}$ rather than f'_c itself; the validity of shear and torsion
theory is also still questionable to some degree.

(g) Improved Code Notation

The 1977 Code clarifies the notation associated with member strengths.
The subscript u is now reserved for *required* strengths, that is, the M_u, V_u,
P_u, and so on, required for the factored loads. The theoretical strengths
provided are designated by the subscript n for nominal (or ideal or
theoretical) strengths and the usable portions thereof as ϕM_n, ϕV_n, ϕP_n, to
be compared to the required values:*

$$M_u \gtrless \phi M_n, \qquad V_u \gtrless \phi V_n, \qquad P_u \gtrless \phi P_n$$

(h) Author's Evaluation of Code Safety Provisions

The approximate factor of safety, which is given by the load factor divided
by ϕ, is essentially sound. Philosophically, the author feels that if the
same ratio had been held with each term a little lower, there would be
more encouragement to improve design practice, materials, and inspection.
Any of these would justify an increase in the ϕ value which is now
somewhat constrained by the very high 0.90 value of ϕ for flexure. By the
time allowance is made for necessary tolerances in construction dimensions and placement of reinforcing bars, underweight (but legal) bars, and
the like, the 10% margin between 0.90 and 1.00 looks none too large. In
fact, in the 1971 Code, to recognize improvement in design through new
Code provisions, better quality control of materials, and additional
research and experience, the ϕ was left unchanged and the load factors
lowered. This gave the intended reduction in the necessary factor of

---

*Those who have used earlier editions of this book will note that M_n, V_n, P_n replace $\bar{M}$, $\bar{V}$, $\bar{P}$;
also that now $M_u \gtrless \phi M_n$ replaces $M_u = \phi M$, and so on.

safety; only the philosophy that ϕ (not load factor) was to reflect such matters had to suffer.

At this time it appears that the ϕ of 0.85 for shear may be lowered (to 0.80?) in the near future to reflect re-analyses of research.

2.13 DUCTILITY IN REINFORCED CONCRETE CONSTRUCTION

Provision for ductility in reinforced concrete could, with considerable logic, be considered as part of the safety provisions, but probably it deserves the emphasis of separate consideration. Ductility in a structure or a member means the maintenance of strength while sizeable deformation or deflection occurs. The engineer seeks to build ductility into structures for several reasons. In typical indeterminate structures, ductility permits a heavily stressed portion to continue to carry its capacity load while deforming enough to bring lesser stressed neighboring portions more definitely into the resistance pattern. In slabs and beams ductility means a warning of overloads will be present in the form of excessive cracking and deflection. Where energy must be absorbed, as in blast and earthquake situations,* ductility is particularly important.

Although concrete in compression is far from being elastic at higher stresses, it is basically a brittle material that crushes in compression. In beams and slabs ductility is obtained by (1) using reinforcing steel, which is not brittle, to carry tension and then (2) limiting the amount of tensile steel to that which will yield before the concrete crushes in compression. This method of control is discussed in Chapter 3.

Reinforced concrete columns are basically brittle under vertical load, although spiral columns (Chapter 18) do introduce limited ductility. Under lateral loads (sideway) considerable ductility can be built into a column by proper detailing and the use of spirals. The detailing of reinforcement at the joints between beams and columns is critical in the development of ductility and the capacity to survive shock loadings.

Where ductility cannot be achieved in a detail or a design, added strength is necessary to insure that any potential failure be initiated at a more ductile section or in a more ductile manner. In the Code this concern shows, for example, in more conservative ϕ values for columns and in longer bar laps for tension splices. Chapter 23 based on Code Appendix A gives special provisions for seismic design details and is largely directed toward building ductility into the structural frame.

---

*An official study in the United States recommends that some earthquake provisions be included in design much more broadly than in the past, exempting only a portion of seven states, with none totally exempt.

2.14 LIMIT DESIGN

Present strength design is based on moments and shears calculated from elastic analysis of members and frames. Strength evaluations of the cross sections, nevertheless, involve inelastic action at the critical design sections. The elastic analysis gives moments and shears generally on the safe side, but their use involves at least a philosophical inconsistency. Although it is not usually computed, it should also be noted that the influence of the cracking of some of the members and the influence of the joints where beam and column have a common section can modify the usual* computed moments more than most designers realize.

In view of these problems, limit design concepts look very attractive. A limited arbitrary reassignment of moments is now permitted, but limit design as such is still considered in the development stage in the United States. A number of countries in Europe have moved further in this direction. Limit design is discussed further in Chapter 11.

2.15 CALCULATION ACCURACY

Reinforced concrete is still not a precise theory and, hence, judgment is always an important factor in the application of this theory.

A distinguished engineer and the chairman of the ACI committee responsible for the 1963 ACI Building Code, Raymond C. Reese, introduced his *CRSI Design Handbook* with these comments on the accuracy of calculations (slightly abbreviated here):

> If ... involved mathematical methods ... lead one to believe that the design of reinforced concrete structures requires a high degree of precision, the reverse is the case. Concrete is a job-made material, and control cylinders that do not vary more than 10% are remarkably good. Reinforcing bars are shop-made; yet variations in strength characteristics run 3% to 5%; rolled weights can vary $3\frac{1}{2}$%. Formwork is field-built; frequently a 2×8 or 2×10 (50 mm × 200 mm or 50 mm × 250 mm) (measuring, respectively, $7\frac{5}{8}$ in. and $9\frac{1}{2}$ in. that is, 194 mm and 241 mm) is used to form the soffit of an 8 in. (200 mm) or 10 in. (250 mm) beam. Bars that are held in place to an accuracy of between $\frac{1}{8}$ in. (3 mm) and $\frac{1}{4}$ in. (6 mm) are extremely well placed. Two-figure accuracy is sufficient for almost all problems in reinforced concrete design.
>
> ... the time and effort of the designer is best spent in recognizing and providing for ... tensions wherever they may exist, not in striving for a high degree of precision by carrying figures to an unmeaning number of significant places.
>
> On the other hand, ... when numbers are subtracted, significant figures are often lost. It is, therefore, recommended, more for control of the compu-

*Based on members with a constant EI, often based on gross concrete sections.

tations, for ready checking, and to keep the computer alert, rather than for any effect on the completed structure, that figures be carried to three significant places. . . . The following table is suggested.

RECORD VALUES TO THE FOLLOWING PRECISION

Loads to nearest 0.1 kPa = 100 Pa; 0.1 kN/m = 100 N/m; 1 kN concentration
Span lengths to 10 mm
Total loads and reactions to 1 kN
Moments to nearest 10 N·m
Individual bar areas to 10 mm²
Concrete sizes to 10 mm for small and 25 mm for large dimensions
Bar spacings in slabs to 25 mm
Effective beam depth to 1 mm

As a practical limit the designer should note the construction tolerances on effective depth d and clear cover specified under Code 7.5.2.1:

d of 200 mm or less   ± 10 mm   Min. cover − 10 mm
$d > 200$ mm   ± 12 mm   Min. cover − 12 mm

"except that tolerance for the clear distance to formed soffits shall be minus 6 mm and tolerance for cover shall not exceed minus one-third of the concrete cover required in the contract drawings or in the specifications."

2.16 HANDBOOKS

Since this book is written primarily as a textbook, it does not concern itself greatly with office practice and the use of design aids. Nevertheless, the student should be aware that curves and tables can speed up design considerably.

The *Ultimate Strength Design Handbook*, Vol. 1[13], published by ACI in 1973 as SP-17(73) has many useful tables and charts. It is currently under revision and will be available soon in revised form to agree with the 1977 Code. Volume 2, SP-17A(78) covering column design, was recently issued.

The *CRSI Handbook*[12], 2nd Ed., 1975, published by the Concrete Reinforcing Steel Institute has many tables of allowable design loads on various types of members, some with details of the reinforcing steel. This book will undoubtedly be revised to match the 1977 Code.

In the detailing of structures the *Manual of Standard Practice for Detailing Reinforced Concrete Structures* (ACI 315-74)[14] is most helpful as a guide to the clear presentation of the necessary details. It is a drafting rather than a design manual, although it calls attention to some important considerations in designing reinforced concrete.

SELECTED REFERENCES

1. *Concrete Manual*, U.S. Bureau of Reclamation, Denver, Colo, 7th ed., 1963.
2. *ACI Manual of Concrete Inspection*, ACI SP-2, Detroit, 1974.
3. "Design and Control of Concrete Mixtures," Portland Cement Association, Chicago, 11th ed., 1968.
4. ACI Committee 211, "Recommended Practice for Selecting Proportions for Normal Weight Concrete (ACI 211.1-74)," (Revised, 1975).
5. ACI Committee 211, "Recommended Practice for Selecting Proportions for Structural Lighweight Concrete (ACI 211.2-69)," *Jour. ACI*, *66*, No. 5, May 1969, p. 365.
6. David Ramaley and Douglas McHenry, "Stress-Strain Curves for Concrete Strained Beyond Ultimate Load," *Lab. Rep. No. Sp-12*, U.S. Bureau of Reclamation, Denver, Colo., 1947.
7. Eivind Hognestad, N. W. Hanson, and Douglas McHenry, "Concrete Stress Distribution in Ultimate Strength Design," *Jour, AÇI, 27*, Dec. 1955; *Proc., 52*, p. 455.
8. J. L. Savage, Ivan E. Houk, H. J. Gilkey, and Fredrik Vogt, *Arch Dam Investigation*, Vol. II, *Tests of Models of Arch Dams and Auxiliary Concrete Tests*, Engineering Foundation, New York, 1934.
9. "Standard Specification for Deformed and Plain Billet-Steel Bars for Concrete Reinforcement," *ASTM Spec. A615-76a*, ASTM, Philadelphia, 1976.
10. Richard A. Helmuth and Danica H. Turk, "The Reversible and Irreversible Drying Shrinkage of Hardened Portland Cement and Tricalcium Silicate Pastes," *Journal*, PCA Research and Development Laboratories, *9*, No. 2, May 1967.
11. Lev Zetlin, Chas. H. Thornton, and I. Paul Lew, "Internal Straining of Concrete," *Designing for Effect of Creep, Shrinkage, and Temperature in Concrete Structures*, SP-27, Amer. Concrete Inst., Detroit, 1970, p. 323.
12. *CRSI Handbook* 2nd ed., Concrete Reinforcing Steel Institute, Chicago, 1975.
13. *Ultimate Strength Design Handbook*, Vol. I, ACI, SP-17(73) Detroit, 1973.
14. *Manual of Standard Practice for Detailing Reinforced Concrete Structures (ACI 315-74)*, ACI, Detroit, 1974.

3
FLEXURAL ANALYSIS OR REVIEW OF BEAMS

3.1 THE RESISTING COUPLE IN A BEAM

Statics shows that an external bending moment on any beam must be resisted by internal stresses that can be indicated as a resultant tension N_t and a resultant compression N_c. Unless there is axial load, summation of horizontal forces (Fig. 3.1) indicates that N_t must equal N_c and that together they form a couple. In a reinforced concrete beam the reinforcing steel is assumed to carry all the tension N_t (Sec. 2.4) and thus N_t is located at the level of the steel. The compressive force N_c is the resultant of compression stresses over some depth of beam and thus its location is not established from statics alone. Let the arm or distance between these N_t and N_c forces be designated as z, as shown in Fig. 3.1. Evaluation of z is first presented as a portion of Sec. 3.5. The further depth below the steel is chiefly to fireproof the steel, to protect it from moisture, and to assist in bonding the reinforcing steel to the concrete; it does not influence present

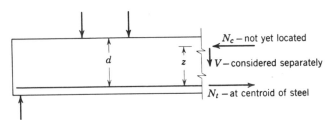

FIGURE 3.1   Internal resisting couple in a beam.

34

flexural strength calculations. However, early Code recognition of cover (and bar spacings) in the bond strength evaluation of beams is indicated by analyses made of recently published studies (Sec. 7.1).

The resisting couple idea can be applied to homogeneous beams. It is simpler for rectangular beams with straight-line distribution of stresses than is the use of $f = Mc/I$. For steel I-beams it is more awkward, but is useful for rough estimates based on neglect of the web area. For reinforced concrete beams it has the definite advantage of using the basic resistance pattern.

3.2 DISTRIBUTION OF COMPRESSIVE STRESS

In reinforced concrete beams the compressive stress varies from zero at the neutral axis to a maximum at or near the extreme fiber. How it actually varies and where the neutral axis actually lies depend both on the amount of the load and on the history of past loadings. This variation is the result of several factors: (1) The spacing and depth of tension cracks depend upon whether the beam has been loaded before, and how heavily. (2) Shrinkage stresses and creep of concrete are important factors in relation to stress distribution, factors very difficult to include in an analysis. (3) Most important, the stress-strain curve for concrete as indicated in Figs. 3.2 and 2.9 is not a straight line.

Experiments confirm that, even for reinforced concrete cracked on the tension side of the beam, unit strains vary as the distance from the neutral axis,* the maximum compressive strain increasing with the moment from ϵ_1 to ϵ_2, to ϵ_3, ... on to failure. The curve of Fig. 3.2a thus indicates that the compressive stress distribution on the first application of loading (without any shrinkage) would go through several stages. The neutral axis would shift location with the stress pattern to keep $N_t = N_c$. For example, the small shaded triangle $O\epsilon_1$ would represent the early stresses, as shown more conveniently in Fig. 3.2b. (Actually, at the very start of the first loading before the beam has cracked in tension the neutral axis would be lower, usually below middepth.) Similarly, ϵ_2 (Fig. 3.2a) would limit a larger nontriangular stress distribution as in (c). The maximum unit stress would be f'_c in (d), with strain at ϵ_3 in (a). The maximum total compression in (e) would occur at some strain value ϵ_4 which is defined later. A detailed analysis would show that as the higher strains are reached, the stiffness of the concrete decreases. When the tension is all carried in the elastic range by the steel, the neutral axis has to drop as indicated in Fig. 3.2b–e to keep N_c increasing as rapidly as N_t.

---

*Exactly at a crack this is not quite true, but over an average gage length it is statistically true.

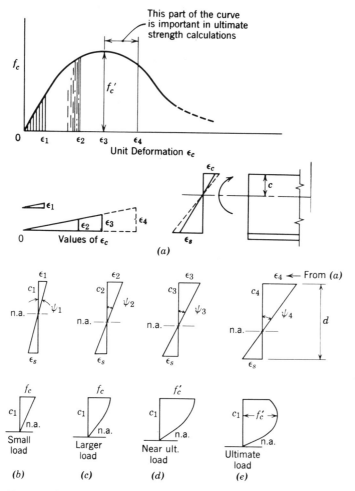

FIGURE 3.2 (*a*) Stress-strain curve for concrete of medium strength and possible strain-stress combinations. (*b–e*) Shifting strain and stress patterns with increasing moment up to compression failure ($f_s < f_y$). The triangles show the unit rotations ψ at each stress level.

Design for this compression type of failure is not permitted, because the ψ unit rotations here are restricted to those in the upper triangles of Fig. 3.2*b–e*. Member stiffness and deflection are directly related to the angle changes per unit length, and these control ductility (Sec. 3.3).

Ductility, measured simply as the rotation ψ per unit length of member, is important in reinforced concrete design and is discussed in the next section (3.3). In the usual continuous type member, ductility under overload conditions is evidenced by widening tension cracks and increased deflections that warn of approaching failure. At the same time the member is absorbing energy from the loading and its local reduced stiffness results in shifting the resisting moment to lesser strained sections. This ability to adjust to local overstress is a valuable asset for continuous reinforced concrete construction. The next section shows how much this ductility is influenced by the design Code.

Under service loads it probably is not feasible, and it may not even be possible, to do more than estimate rather crudely the approximate magnitude of real stresses. Certainly these are complicated by shrinkage and creep and depend on how much cracking has been induced by previous loadings. Specifications formerly emphasized stresses at working loads, but laboratory tests of reinforced concrete beams show that actual deformations and stresses at working loads only faintly resemble values conventionally calculated on the basis of straight-line stresses (Fig. 3.2b). Hence, design procedures have gradually shifted to ultimate strength methods, with checks at service load for deflection and cracking.

Fortunately the *ultimate* strength of reinforced concrete beams can be predicted or calculated with quite satisfactory accuracy. Design on the basis of (ultimate) strength is now standard in the 1977 ACI Code.

Reference 2 by Park and Paulay includes a more thorough coverage of the research background in this area.

3.3 TENSION AND COMPRESSION FAILURES DUE TO BENDING MOMENT; MEMBER DUCTILITY

Beams may fail from moment because of weakness in the tension steel or weakness in the compression concrete. Failure from stresses primarily related to shear will be considered later and separately, as in the case of steel construction, since it is only in very deep and relatively short beams that these influence bending strength.

Most beams are weaker in their reinforcing steel than in their compression concrete. Both the Code and economy require such design. Such beams fail under a load slightly larger than that which makes $N_t = f_y A_s$, where f_y is the yield-point strength of the steel and A_s is the area of the steel. Since $f_y A_s$ represents the nominal usable strength of steel, the further increase in moment resistance needs explanation. When the steel first reaches its yield point, the compressive stress distribution may be like that previously illustrated in Fig. 3.2c or d. A slight additional load at this stage causes the steel to stretch a considerable amount. The increasing steel deformation in turn causes the neutral axis to rise (when the tension is on the bottom) and the center of compression

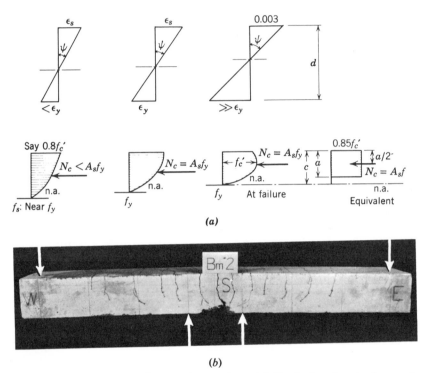

(a)

(b)

FIGURE 3.3 Flexural failure in tension. (a) Variation in strains and compressive stresses as beam approaches failure in tension. (b) Beam after failure under negative moment (tension on top).

N_c therefore moves upward, Fig. 3.3a. This increase in the arm z between N_c and N_t gives an increased resisting moment $N_t z$ even though N_t is essentially unchanged.* The rising of the neutral axis also reduces the area under compression and thereby increases the unit compressive stress required to develop the nearly constant value of N_c. This process continues until the reduced area fails in compression, as a secondary effect, at about the same ϵ_4 as indicated in Fig. 3.2. These changes (neglecting any tension in the concrete) are summarized in Fig. 3.3a. This type of failure (inverted because in a negative moment region) is shown in Fig. 3.3b.

Since the ultimate c to neutral axis is smaller in Fig. 3.3 than in Fig. 3.2 while the maximum ϵ_c is about the same, the unit rotation ψ at

---

*When the steel has no sharp yield point, N_t increases above the yield strength enough to make calculations based on a yielding steel conservative. The same problem arises when a short yield plateau introduces a strain hardening region.

ultimate is greater, usually by 35% to over 200%. Hence, such an *under-reinforced* beam shows greatly increased deflection after the steel reaches the yield point, giving adequate warning of approaching beam failure; the steel, being ductile, will not actually pull apart even at failure of the beam. The increase in z will rarely be more than a few percent, say 3 to 8%. If the concrete stress is very high before the steel stress f_s reaches the yield-point value of f_y, the increase in z may be very small.

If the concrete reaches its full compressive strength just as the steel reaches its yield-point stress, the beam is said to be a *balanced* beam (at failure). Such a beam requires very heavy steel, is rarely economical, and is not allowed by the Code.

Before a beam fails from weakness in compression, the top elements of the beam shorten considerably under the final increments of load, causing the neutral axis to move lower down in the beam. Such movement of the neutral axis increases the area of concrete carrying compression and importantly increases the total N_c that can be carried (Fig. 3.5). The greatly increased N_c is offset slightly by a reduced arm z. The concrete finally fails suddenly and often explosively in compression, and the steel

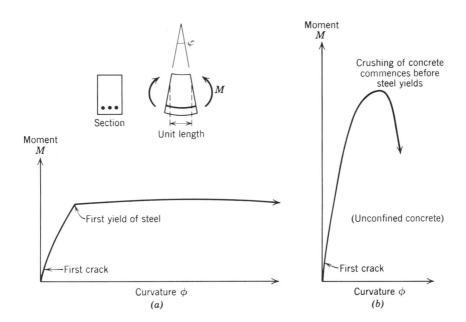

FIGURE 3.4 Moment-curvature relationships for singly reinforced beam sections. (*a*) Section failing in tension, $\rho < \rho_b$. (*b*) Section failing in compression, $\rho > \rho_b$. From Reference 2. (Reproduced by permission of the publisher.)

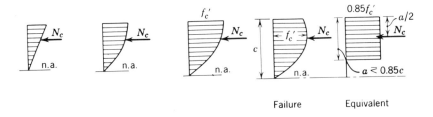

Failure   Equivalent

FIGURE 3.5 Variation in compressive stresses as beam approaches failure in compression. See text for values of a. (Corresponding strains are similar to those in Fig. 3.2b–e.)

stress remains below the yield point unless the beam is exactly balanced. These stress changes, again ignoring any tension in the concrete, are summarized in Fig. 3.5. Such a beam is called an *overreinforced* beam.

In a plot (Fig. 3.4) of resisting moment M against ψ for a given cross section, energy absorbed per unit length is proportional to the area under the curve. Such a curve for a balanced beam would be much like the curve on the right of the figure with the peak at a point where f'_c and f_y are reached simultaneously. The limited area under the curve, up to the peak value, measures the energy absorption capacity available before damage becomes severe. The curve shown at the left would be for a lower A_s, possibly 40% of balanced reinforcement.* The lower the A_s, the lower the M capacity and the longer the slightly rising flat portion of the curve created by the rising neutral axis (Fig. 3.3a), while the reinforcement continues to yield without damage, except for wider tension cracks. This "flat" extension stops at the ψ value causing the concrete to crush. The larger area under this curve indicates this member's greater ductility. The ratio of curvature at ultimate to that at yield is called the ductility factor. As A_s is increased, the higher the M at first yielding becomes and the shorter the near-horizontal length, with a lower ductility resulting. The Code assures some ductility by limiting the maximum A_s to 75% of the balanced A_s. Limitation of deflections (Chapter 8) or economy of design often demand larger members that need less A_s; 40 to 50% of balanced A_s is a very common range.

For design purposes real final stress distribution may be replaced adequately by an equivalent rectangle of stress (pioneered in this country by Whitney) of intensity $0.85\,f'_c$ and depth a as shown in the final sketches in Figs. 3.3a and 3.5. For rectangular beams, the shaded area of the rectangular stress block should equal that of the real stress block and their

---

*Also see Figs. 11.1 and 11.2.

centroids should be at the same level. The value of a given in the Code is intended to give this result and is recommended:

For $f_c' \le 30$ MPa (C30 concrete) $a = 0.85\ c$
For $f_c' > 30$ MPa, reduce the 0.85 factor* linearly at the rate of 0.08 per 10 MPa, but not to any value less than 0.65.

3.4 ANALYSIS OR REVIEW VERSUS DESIGN

In analysis, or review of a given design, whether for actual stresses or for permissible moments, the engineer deals with given beams, known for both dimensions and steel. The engineer has no control over the location of the neutral axis, which lies at a definite but initially undefined depth.

In design, loads and ultimate stresses are known and some or all the dimensions remain to be fixed. In this case designers have some control over the location of the neutral axis. They can shift it where they want it, to the extent that the change in dimensions changes the depth of rectangular stress block or magnitude of N_t.

The student should understand clearly this fundamental difference between analysis (or review) and design problems. Analysis for various types of beams is discussed first, then design.

3.5 THE BALANCED RECTANGULAR BEAM

The balanced beam in ultimate strength design is not a practical beam, but the concept is fundamental to the philosophy of the Code.
The Code 10.33 specifies:

> For flexural members ... the ratio of reinforcement ρ provided shall not exceed 0.75 of the ratio ρ_b that would produce balanced conditions for the section under flexure without axial load. For members with compression reinforcement, the portion of ρ_b balanced by compression reinforcement need not be reduced by the 0.75 factor.

The first use and discussion of the second Code sentence quoted above is in Sec. 3.14 on double-reinforced beams. The ratio $\rho = A_s/bd$, where A_s is the area of steel reinforcement and b and d are shown in Fig. 3.6a. The Code (10.2.3) requires that the concrete strain not exceed 0.003, as also shown in Fig. 3.6a for the balanced condition.

These regulations represent an attempt to assure the ductile failure produced by yielding of steel as compared to the brittle type of failure occurring when the failure is in compression. (See also Sec. 3.6.)

*This factor is designated β_1.

The analysis of the balanced beam starts from the strain triangles at failure as shown in **Fig. 3.6a**. The steel strain will be f_y/E_s and the maximum concrete strain 0.003, which is conservatively in agreement with test observations. The neutral axis can be located from the strain triangles, most simply by considering the compression strain triangle and the large dotted triangle.

$$c_b = \frac{0.003}{0.003 + f_y/E_s} d$$

For $E_s = 200\text{ GPa} = 200\,000\text{ MPa}$ and f_y in MPa:

$$c_b = \frac{600}{600 + f_y} d \tag{3.1}$$

Experimental work has established the necessary depth a of stress block to go with a given neutral axis, or conversely the neutral axis distance c to go with a given stress block depth a. The Code requirement for this has already been stated in the last paragraph of Sec. 3.3 above. With a_b established, the balanced reinforcement ratio $\rho_b = A_{sb}/bd$ can now be found from the resisting couple $N_{tnb} = N_{cnb}$*, as indicated in **Fig. 3.6c**. The n subscript on each term indicates that each is now a nominal (or ideal) value; the strength reduction factor ϕ must still be applied to N_n values or to M_n values to establish the *design* strength provided. Then $N_{tb} = \phi N_{tnb}$, $N_{cb} = \phi N_{cnb}$, and $M_{un} = \phi M_{nb}$. Usually only the nominal values

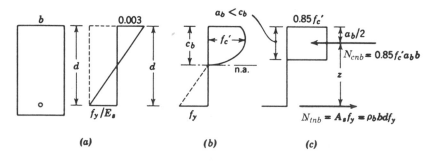

(a)   **(b)**   **(c)**

FIGURE 3.6 Balanced beam. (*a*) Strains. (*b*) Stresses. (*c*) Equivalent stresses and forces.

---

*Normally, triple subscripts can be avoided. For example, the *b* subscript is needed only to emphasize that a balanced beam is being considered. The subscript *n* emphasizes that these are nominal (or ideal) values that do *not* include the strength reduction value ϕ (equally applicable to both terms).

of N_c or N_b are needed, since the change to design strength can conveniently be applied to the moment term.

The following numerical approach is recommended to the student as presenting a clearer picture.

For $f'_c = 25$ MPa, $f_y = 300$ MPa, the closing comments in Sec. 3.3 show $a_b = 0.85\ c_b$. Then for a rectangular beam of width b:

$$c_b = \frac{600}{600 + 300}\ d = 0.667\ d$$

$$a_b = 0.85\ c_b = 0.85 \times 0.667\ d = 0.567\ d$$

$$N_{cnb} = 0.85\ f'_c a_b b = 0.85\ f'_c \times 0.567\ db = 0.481\ f'_c bd$$

If b and d are in m, and f'_c is in MPa, the force N_{cnb} is in MN (meganewtons). Since $N_{cnb} = N_{tnb} = A_{sb} f_y$,

$$0.481\ f'_c bd = (\rho_b bd) f_y$$

$$0.481 \times 25 = \rho_b \times 300, \qquad \rho_b = 0.040$$

This could be written algebraically, with β_1 representing a/c

$$\rho_b = 0.85\ \beta_1 \frac{f'_c}{f_y} \frac{600}{600 + f_y}$$

with all stresses in MPa. Although there is little practical use for the balanced moment in beam design, it is simple to establish from the nominal resisting couple.

$$z = d - a_b/2 = d - 0.567\ d/2 = 0.717\ d$$

If d is in meters, z is in meters.

$$M_{nb} = N_{tnb} z = (\rho_b bd) f_y z = (0.040\ bd)300 \times 0.717\ d = 8.60\ bd^2$$

or

$$M_{nb} = N_{cnb} z = (0.481\ f'_c bd)0.717\ d = 0.345 \times 25\ bd^2 = 8.60\ bd^2$$

Design strength $= \phi M_{nb} = 0.90 \times 8.60\ bd^2 = 7.74\ bd^2$

where ϕM_{nb} is in MN $\cdot$ m if b and d are in m and f'_c and f_y in MPa. This example has been followed through in detail because the methods also apply to nonbalanced beams when either c, N_{nt}, or a is known.

Other shapes of beams such as T-beams, double-reinforced beams, and irregular shapes, can likewise be evaluated as balanced beams. The neutral axis is at the same relative depth in every case, being established from strains rather than stresses. The procedures with varying widths and with compression reinforcement are discussed starting in Sec. 3.11.

3.6 TENSILE STEEL LIMITATION—FOR DUCTILITY*

Since a balanced beam will ultimately fail suddenly in compression, the Code, as quoted in Sec. 3.5, limits the tensile reinforcement to a maximum of $0.75\rho_b$ both for beams and for some lightly loaded columns (Code 10.3.3). In beams this is equivalent to requiring $N_t = 0.75\,N_{cb}$. Thus, when the reinforcement reaches yield, the concrete will still have roughly an added one-third reserve strength in compression. The usual ductile reinforcement assures member ductility, with wide cracking and more deflection as a warning of approaching capacity.

It is rarely economical to make beams so small that this rule becomes an operative limit. Both deflections and costs normally point toward larger beams with even smaller steel ratios.

3.7 CODE MAXIMUM MOMENT FOR RECTANGULAR BEAMS

The limitation to $0.75\ \rho_b$ as a Code maximum is equally well described by $N_t = 0.75\,N_{tb}$ or $N_c = 0.75\,N_{cb}$ and for the single case of a rectangular beam also by $c = 0.75\,c_b$ or $a = 0.75\,a_b$. For the rectangular beam the last is the most direct form. Then the Code maximum moment and reinforcement ratio for $f'_c = 25$ MPa and $f_y = 300$ MPa can be found as follows:

$$c_b = 0.667\,d, \qquad a_b = 0.567\,d, \text{ from Sec. 3.5}$$
$$\text{Max } a = 0.75 \times 0.567 = 0.425\,d$$
$$\text{Max } N_c = 0.85\,f'_c \times 0.425\,d \times b = 0.361\,f'_c bd$$
$$z = d - a/2 = d - 0.212\,d = 0.788\,d$$
$$\text{Nominal } M_n = N_c z = 0.361\,f'_c bd(0.788\,d) = 0.284\,f'_c bd^2$$
$$\text{Dependable moment} = \phi M_n = 0.9(0.284\,f'_c bd^2)$$
$$= 0.256\,f'_c bd^2 = 6.40\,bd^2 = k_m bd^2$$

Since $N_t = N_c$,
$$\rho bd f_y = N_c = 0.361\,f'_c bd$$
$$\text{Max } \rho = 0.361\,f'_c/f_y = 0.361 \times 25/300 = 0.0301$$
$$\text{Nominal } M_n = N_t z = (0.0301\,bd f_y)(0.788\,d) = 0.0237\,f_y bd^2$$
$$\text{Design moment} = \phi M_n = 0.9 \times 0.0237\,f_y bd^2 = 0.0213\,f_y bd^2 = 6.40\,bd^2$$

This dependable moment ϕM_n would, with more decimal places, be the same whether based on N_c or N_t since M_n represents a couple. For

---

*Reference 2 has a very extensive treatment of ductility.

TABLE 3.1 Limiting Constants for Rectangular Beams

Steel:	$f_y = 300$ MPa				$f_y = 400$ MPa			
Concrete Gr. f_c' MPa	k_m MPa	k_n MPa	100ρ	a/d	k_m MPa	k_n MPa	100ρ	a/d
C20 20	5.11	5.68	2.40	0.43	4.73	5.25	1.62	0.38
C25 25	6.40	7.11	3.01	0.43	5.92	6.58	2.03	0.38
C30 30	7.68	8.53	3.61	0.43	7.10	7.89	2.44	0.38
C35 35	8.65	9.61	4.01	0.41	7.98	8.87	2.71	0.36
C40 40	9.51	10.56	4.36	0.39	8.77	9.74	2.94	0.35
$\rho = 0.18\dfrac{f_c'}{f_y}$	$0.144\,f_c'$	$0.160\,f_c'$	$0.18\dfrac{f_c'}{f_y}$	0.21	$0.144\,f_c'$	$0.160\,f_c'$	$0.18\dfrac{f_c'}{f_y}$	0.21

convenience, designate the constant 6.40* as k_m^* and say design moment $\phi M_n = k_m bd^2$. Similarly, values of k_n are usable with $M_n = k_n bd^2$.

Table 3.1 lists a number of values of maximum k_m, k_n, ρ and a/d for easy reference. (This table is repeated as Table B.5 in Appendix B.)

The significance of entries in Table 3.1 for $\rho = 0.18\, f_c'/f_y$ is discussed near the end of Sec. 3.8.

The legal Code limit for ρ is definitely 0.75 ρ_b. However, a designer can be as logical in permitting a greater ρ (with moment capacity of the member limited to that for 0.75 ρ_b) as in permitting the use of an actual f_y of 350 MPa where Grade 300 stell is specified. Accordingly, the author limits the k_m value quite strictly and is less concerned (in nonseismic work) about extra steel which can be ignored for strength. In most design, economy calls for something like half of balanced reinforcement.

3.8 ANALYSIS OF RECTANGULAR BEAMS

Only underreinforced beams, as discussed in Sec. 3.6, are permitted by the Code and their analysis is simple. First establish whether the given ρ, which is simply A_s/bd, is less than (or equal to) 0.75 ρ_b, the maximum ρ, tabulated in Table 3.1. (Of course, the equivalent check can be made later, at some risk to earlier calculations, in terms of k_m of maximum a/d.) If $\rho < 0.75\ \rho_b$, the beam obviously fails primarily in tension. The resulting stretching of the steel will raise the neutral axis until the final secondary compression failure occurs at the compression strain ϵ_4 in Fig. 3.2, taken in the Code as 0.003. The usual analysis, however, avoids the necessity of using this strain directly for simple cases.

Since $N_t = N_c$, the ultimate moment is reached when the compressive area can just support a N_c equal to $A_s f_y$. By using the Code simplification as shown in Fig. 3.7, the compression N_c can be evaluated as an equivalent block of uniform stress, of intensity 0.85 f_c' and height a (Fig. 3.7b). Then $N_c = N_t$ becomes:

$$0.85\ f_c'ab = A_s f_y$$

$$a = \frac{A_s f_y}{0.85\ f_c' b} = \frac{\rho f_y d}{0.85\ f_c'}$$

where $\rho = A_s/bd$. The arm between N_c and N_t is $d - a/2$. Hence

$$M_n = N_t z = A_s f_y \left(d - \frac{a}{2} \right)$$

---

*This constant carries a unit of MPa for b and d in m, and moments in MN·m. The symbol k_m is in accord with the new ACI notation, but is not yet standardized; the notation K or K_w has also been used.

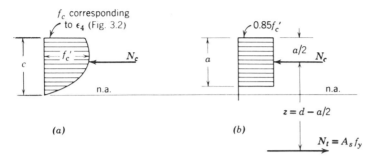

FIGURE 3.7   The Code (or Whitney) compressive stress block. (a) Probable stress distribution. (b) Simplified model for calculations.

where M_n is the nominal moment strength under ideal conditions and the value of a is found from the above relation. The design moment $= \phi M_n = 0.9\,M_n$.

An alternate form for the formula for ϕM_n in terms f/f' is:

$$\phi M_n = \phi f'_c bd^2 \omega (1 - 0.59\,\omega)$$

This is sometimes convenient and involves only an algebraic transformation of the above process. In terms of the steel, ϕM_n could also be written:

$$\phi M_n = \phi A_s f_y d \left(1 - \frac{0.59\,\rho f_y}{f'_c}\right)$$

The author recommends for students the forms first developed above (using A_s).

The Code places certain minimum steel requirements on beams.

To resist calculated moments the minimum reinforcement (Code 10.5.1) is $\rho_{\min} = 1.4/f_y$ unless *all* the flexural reinforcement for the member is at least 4/3 that required by analysis. This means $\rho_{\min}$ of 0.0047 for Grade 300 and 0.0035 for Grade 400 bars. For joists or T-beams with webs in tension $\rho_{\min}$ is based on web width rather than flange width. For slabs of uniform thickness the minimum drops back to temperature and shrinkage requirements (Code 7.12.2) of 0.0020 of *gross* slab area for Grade 300 bars and 0.0018 for Grade 400 bars, with maximum bar spacing of five times the slab thickness but not more than 500 mm.

The steel yield strength in calculations is limited to 550 MPa (Code 9.4). The Code specifies (3.5.3.3) that if the yield strength exceeds 400 MPa the f_y used shall be that corresponding to a strain of 0.0035 determined by tests on full size bars.

The ratio of a/c used is to be decreased 0.08 for each 10 MPa of f'_c above 30 MPa (Code 10.2.7c), as already noted at the close of Sec. 3.3, to reflect the more triangular or less curved nature of the stress-strain curve for these higher strength concretes (Fig. 2.8a).

Under service loads, deflection of members must be considered and deflection calculations are discussed in Chapter 8. When the length-depth ratios of Table 3.2 (a copy of ACI Code Table 9.5a*) are not satisfied, or when the member supports partitions that might be cracked, deflections must be calculated. This criterion does not necessarily separate members without deflection troubles from those with excessive deflections. It simply reflects, in a somewhat oversimplified fashion, a boundary where the designer must certainly become aware of deflection.

The 1963 Code required that deflections be checked where ρ exceeded $0.18 f'_c/f_y$. There then was a greater design experience with members designed with approximately this steel ratio under working stress

TABLE 3.2 (Code Table 9.5a). Minimum Thickness of Beams or One-way Slabs Unless Deflections Are Computed[a]

| Member | Minimum Thickness, h | | | |
	Simply Supported	One End Continuous	Both Ends Continuous	Cantilever
	Members not supporting or attached to partitions or other construction likely to be damaged by large deflections.			
Solid one-way slabs	$l/20$	$l/24$	$l/28$	$l/10$
Beams or ribbed one-way slabs	$l/16$	$l/18.5$	$l/21$	$l/8$

[a]The span length l is in m.

Values given shall be used directly for members with normal density concrete ($w = 2\,400$ kg/m^3) and Grade 400 reinforcement. For other conditions, the values shall be modified as follows:

(a) For structural lightweight concrete having unit weights in the range $1\,500$–$2\,000$ kg/m^3, the values in the table shall be multiplied by $1.65 - 0.0003\, w_c$ but not less than 1.09, where w_c is the unit weight in kg/m^3.

(b) For f_y other than 400 MPa, the values shall be multiplied by $0.4 + f_y/690$.

*The code is available from ACI as noted in Sec. 2.10.

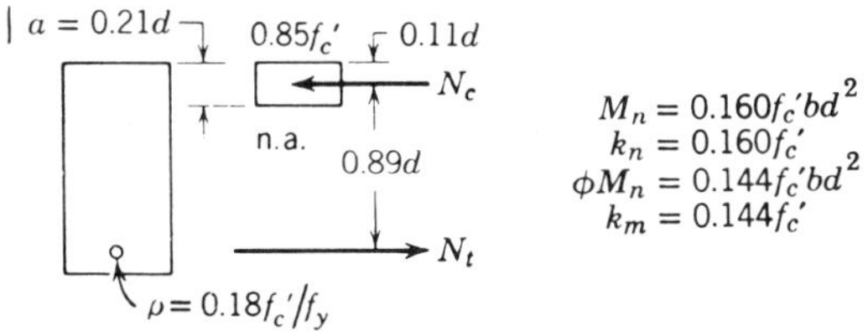

$$M_n = 0.160 f_c' b d^2$$
$$k_n = 0.160 f_c'$$
$$\phi M_n = 0.144 f_c' b d^2$$
$$k_m = 0.144 f_c'$$

FIGURE 3.8 Design constants for $\rho = 0.18\, f_c'/f_y$, usable with either English or SI units.

methods; often it is still an economical ratio to use. Larger reinforcement ratios imply smaller beam sizes and hence larger deflections. Short-span beams can use larger ratios without deflection problems; long-span beams can present deflection problems even when a smaller ratio is used. Long-span beams, cantilever construction, and shallow members always require that special attention be given to deflections. As a guidepost from past experience, the criterion still has value.

As a sample of the determination of beam constants for a given ρ, consider Fig. 3.8 and the following for $\rho = 0.18\, f_c'/f_y$:

$$N_t = A_s f_y = \rho b d f_y = 0.18\, f_c' b d$$

$$a = N_t/(0.85\, f_c' b) = (0.18\, f_c' b d) \div (0.85\, f_c' b) = 0.21\, d$$

$$z = d - 0.5 \times 0.21\, d = 0.89\, d$$

$$M_n = N_t z = 0.18\, f_c' b d \times 0.89\, d = 0.160\, f_c' b d^2$$

$$\text{Design moment} = \phi M_n = 0.9 \times 0.160\, f_c' b d^2 = 0.144\, f_c' b d^2$$

$$k_m = 0.144\, f_c'$$

In this form the constants are the same for both English and SI units.

3.9 RECTANGULAR BEAM EXAMPLES

(a) A rectangular beam has $b = 280$ mm $= 0.28$ m, $d = 500$ mm $= 0.5$ m, $A_s = 3$-#25 $= 1500$ mm$^2 = 1.5 \times 10^{-3}$ m^2, C25 concrete ($f_c' = 25$ MPa), Gr 400 steel ($f_y = 400$ MPa). Calculate design moment $= \phi M_n$.

Solution

$$\rho = A_s/bd = 1.5 \times 10^{-3}/0.28 \times 0.5 = 0.0107$$

$$\rho_{min} = 1.4/f_y = 1.4/400 = 0.0035 \qquad \text{O.K.}^*$$

$$\text{Ultimate } N_{tn} = A_s f_y = 1.5 \times 10^{-3} \times 400 = 0.6 \text{ MN} = 600 \text{ kN}$$

$$N_{cn} = 0.85 \, f_c' b a = 0.85 \times 25 \times 0.28 \, a = 5.95 \, a = N_{tn} = 0.6 \text{ MN} = 600 \text{ kN}$$

$$a = 0.101 \text{ m} = 101 \text{ mm} < \text{max. of } 0.38 \, d \text{ (Table 3.1)} = 0.38 \times 500 = 190 \text{ mm} \qquad \text{O.K.}$$

or

$$c = a/0.85 = 119 \text{ mm},$$

$$\epsilon_s = \epsilon_c (d - c)/c = 0.003(500 - 119)/119 = 0.0096 \gg \epsilon_y \qquad \text{O.K.}$$

$$z = d - a/2 = 500 - 101/2 = 450 \text{ mm} = 0.45 \text{ m}$$

$$M_n = N_{tn} z = 600 \text{ kN} \times 0.45 \text{ m} = 270 \text{ kN} \cdot \text{m}$$

$$\text{Design strength } M = \phi M_n = 0.9 \, M_n = 0.9 \times 270 = 243 \text{ kN} \cdot \text{m}$$

The fact that tension controls this maximum could have been established by two other methods: (1) by showing that $\rho < 0.0203$ (the maximum in Table 3.1); (2) by showing that $k_m = M_u/bd^2 < 5.92$ MPa. Since $a < 0.212 \, d = 106$ mm (Fig. 3.8), deflections are probably no problem. The length-depth ratio is not available for checking deflection.

(b) If in (a) the steel is changed to $4 - \#35 = 4000 \text{ mm}^2 = 4 \times 10^{-3} \text{ m}^2$, calculate the design moment capacity and corresponding f_s.

Solution

With the steel area increased to 4000 mm² both ρ and a from Example (a) are greatly increased and may exceed the permitted limits of Table 3.1. The simpler check is on ρ.

$$\rho = A_s/bd = 4000/(280 \times 500) = 0.0286 > 0.0203 \text{ of Table 3.1} \qquad \text{N.G.}$$

The beam does not satisfy the Code Sec. 10.3.3 requirement for ductility. If ductility is not a critical concern in this case, the author would accept this beam for a design moment based on limiting values defined by Table 3.1, as discussed at the close of Sec. 3.7. This would leave the steel much understressed at the design load. The easiest evaluation on this basis is from the k_m value of Table 3.1, 5.92 MPa.

$$\text{Design moment} = \phi M_n = k_m bd^2 = 5.92 \times 0.28 \times 0.5^2$$

$$= 414 \times 10^{-3} \text{ MN} \cdot \text{m} = 414 \text{ kN} \cdot \text{m}$$

The nominal f_s at this value of ϕM_n is easily found from the limiting a used here, which Table 3.1 shows to be $0.38 \, d = 190$ mm. Then $z = d - a/2 = 500 - 190/2 = 410$ mm $= 0.41$ m.

$$f_s = \phi M_n/(\phi A_s z) = 414 \times 10^{-3}/0.9 \times 4 \times 10^{-3} \times 0.41 = 280 \text{ MPa}$$

*Designers in the United States often use O.K. and N.G. (or OK and NG without the periods) to mean satisfactory for use or "no good" and to be discarded.

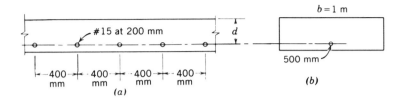

FIGURE 3.9 Design strip and effective steel area in slab analysis. (a) Slab cross section. (b) Design strip.

Since the steel operates at only 70 percent of f_y, it appears that $3 - \#10$ would be as good as the $4 - \#10$ assumed to be used.

Deflections may be large at this moment since $\rho \geqslant 0.18\ f_c'/f_y = 0.0112$, but the lower f_s would help some.

(c) How much steel can be effectively used in the beam of (a)? What is the design moment, ϕM_n, for this area of steel?

Solution

Table 3.1 shows the maximum percentage of reinforcement as 2.03 percent and the corresponding k_m as 5.92 MPa. Hence,

$$\text{Max. } A_s = 0.0203 \times 280 \times 500 = 2840 \text{ mm}^2 = 2.84 \times 10^{-3} \text{ m}^2$$

$$\text{Design moment} = \phi M_n = k_m bd^2 = 5.92 \times 0.28 \times 0.5^2$$
$$= 414 \times 10^{-3} \text{ MN} \cdot \text{m} = 414 \text{ kN} \cdot \text{m}$$

As an alternate, A_s and M_n could be evaluated from the maximum a/d, calculating successively a, N_{cn}, $A_s = N_{cn}/f_y$, and design moment $= \phi M_n = \phi N_{cn}(d - a/2)$. Deflection would also need investigation.

3.10 SLABS

A one-way slab is simply a wide, shallow, rectangular beam insofar as analysis is concerned. The reinforcing steel is usually spaced uniformly over its width. For convenience, a 1-m width is generally taken for analysis or design, since loads are frequently specified in terms of load per square meter; on a 1-m strip, this unit load becomes the load per linear meter. Since the slab can average the effect of steel over some width, the effective A_s may well correspond to a fractional number of bars in the 1-m width. For example, $\#15$ bars at 400 mm spacing, as in Fig. 3.9a, give $A_s = 200 \times 1000/400 = 500 \text{ mm}^2/\text{m}$ for the 1 m design strip, Fig. 3.9b.

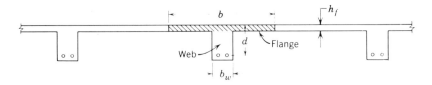

FIGURE 3.10   T-beam as part of a floor system.

3.11 ANALYSIS OF T-BEAMS

Because slabs and beams are ordinarily cast together as shown in Fig. 3.10, the beams are automatically provided with an extra width at the top which is called a flange. Such beams are known as T-beams. The portion below the flange is called the web.

Flange bending stress is not uniform from beam to beam, being largest over the web and tending to drop off with distance from the web. In part this results from the necessity of transferring all the flange flexural stress by longitudinal shearing stresses on vertical sections parallel to the web. Codes use a reduced effective flange width at an assumed uniform stress distribution laterally in lieu of a wider actual slab with nonuniform stresses. The Code (8.10.2) for symmetrical monolithically cast T-beams limits the usable flange width for calculations to a maximum projection of eight times the slab thickness on each side of the web, or half way to the next beam, or a total width of one-fourth of the span, whichever is smallest.

T-beams are analyzed in much the same way as rectangular beams. The Code intends the same limitation (used in Sec. 3.6) to be placed on tension steel; that is, the tension steel should be limited to 0.75 that of the balanced section for the T-beam. This is usually no problem since the large flange area normally keeps compression on the concrete quite low and economy prevents the engineer from using excessive steel. Hence it is usually safe to assume (and check later) that $N_{tn} = A_s f_y$ and calculate the depth of stress block a which will provide an equal N_{cn}. If a, the resulting depth of stress block, is less than the slab (flange) thickness h_f, as shown in Fig. 3.11a, the entire analysis is identical with that of a very wide rectangular beam of width b, except that the limiting a for ductility will usually not be 0.75 a_b.

If the area within the flange depth does not provide enough compression, the form of the calculation must be modified to take account of the narrower web width b_w below the flange (Fig. 3.11b). The total tension can be equated to the total compression to establish the depth of the stress block.

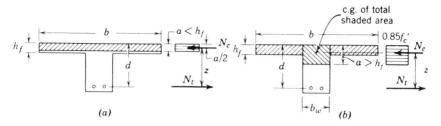

(a)   (b)

FIGURE 3.11 Design of T-beams. (a) Acts as rectangular beam. (b) Direct analysis as T-beam.

$$A_s f_y = 0.85 \, f_c'[ab_w + h_f(b - b_w)]$$

$$a = \frac{A_s f_y - 0.85 \, f_c' h_f(b - b_w)}{0.85 \, f_c' b_w}$$

The resultant compression N_c acts at the centroid of the shaded area in Fig. 3.11b.

Similarly, in checking A_s against the balanced reinforcement, the only change from the rectangular beam procedure is in noting that the area which establishes the balanced N_{cb} is not a single rectangle. The neutral axis, being determined from strains, is located at the same depth c_b as for the rectangular beam. The depth of stress block a_b is 0.85 c_b or less,* since tests indicate that the ratio a/c used for rectangular beams is accurate enough. The total compression N_{cb} is still 0.85 f_c' times the compression area and this N_{cb} can be equated to $N_{tb} = A_s f_y$ to establish the balanced A_s; or, more simply, the usable $N_t = 0.75 \, N_{cb} = 0.75 \, N_{tb}$. The rectangular beam limitation of a to 0.75 a_b is not usable here because of the varying width.

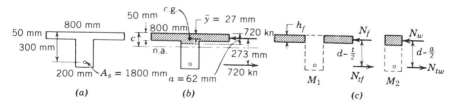

(a)   (b)   (c)

FIGURE 3.12 T-beam of Sec. 3.12.

*Less than 0.85 for $f_c' > 30$ MPa, Code 10.2.7c.

3.12 T-BEAM EXAMPLES

For the T-beam shown in Fig. 3.12a, find the design moment if $A_s = 1800 \text{ mm}^2 = 1.8 \times 10^{-3} \text{ m}^2$, steel is Grade 400, and $f'_c = 20 \text{ MPa}$.

Solution

Since neither span nor beam spacing is given, the only available check on flange width is in terms of flange thickness. First check ρ_{min} (Code 10.5.1).

$$\rho_w = A_s/b_w d = 1800/200 \times 300 = 0.030 > \rho_{min} = 1.4/f_y = 0.0035 \quad \textbf{O.K.}$$
$$\text{Max. flange overhang} = 8h_f = 400 \text{ mm} > \text{actual} \quad \textbf{O.K.}$$
$$\text{Ultimate } N_t = 1.8 \times 10^{-3} \times 400 = 0.720 \text{ MN}$$
$$\text{Max. } N_c \text{ within depth } h_f = 0.85 \times 20 \times 0.8 \times 0.05 = 0.680 \text{ MN}$$

Since $0.720 > 0.680$, the stress block extends below the flange far enough to pick up the remaining compression of $0.720 - 0.680 = 0.040$ MN.

$$0.040 = 0.85 \times 20 \times 0.2(a - 0.05)$$
$$a - 0.050 = 0.012 \text{ m} = 12 \text{ mm}, \quad a = 0.062 \text{ m} = 62 \text{ mm}$$

The resultant N_c acts at the centroid of the compression area, which establishes z and leads to $M_u = N_t z$ or $N_c z$, as indicated in Figs. 3.11b and 3.12b.

$$\bar{y} = \frac{800 \times 50 \times 25 + 200 \times 12 \times 56}{40\,000 + 2400} = \frac{1\,000\,000 + 134\,000}{42\,400} = 27 \text{ mm}$$
$$z = 300 - 27 = 273 \text{ mm}$$
$$M_n = 0.720 \times 0.273 = 0.196\,5 \text{ MNm} = 196.5 \text{ kN m}$$
$$\text{Design } M = \phi M_n = 0.9 \times 196.5 = 177 \text{ kN} \cdot \text{m}$$

Check the ductility requirement $\rho \leqq 0.75 \rho_b$, equivalent to limiting N_c to $0.75 N_{cb}$. Equation 3.1 in Sec. 3.5 is

$$c_b = \frac{600}{600 + 400} \times 300 = 180 \text{ mm} \quad a_b = 0.85 \times 180 = 153 \text{ mm} = 0.153 \text{ m}$$
$$N_{cb} = 0.85 \times 20(0.80 \times 0.05 + 0.20 \times 0.103) = 1.030 \text{ MN}$$
$$\text{Permissible } N_c = 0.75 \times 1.030 = 0.772 \text{ MN} > 0.720 \text{ MN} \quad \textbf{O.K.}$$

As a convenient procedure the resisting moment is calculated as the sum of two couples. The compression N_f* on the outstanding flanges, as shown by the shaded areas in Fig. 3.12c, pairs with a corresponding tension to form couple M_1. The remainder of the tension pairs with the compression N_w in the 200 mm web width to form the second couple M_2.

$$N_f = (0.8 - 0.2) \times 0.05 \times 0.85 \times 20 = 0.51 \text{ MN} = 510 \text{ kN}$$

---

*The complete symbol N_{cf} will be shortened here to N_f because there is no likely confusion. Since in flexure $N_c = N_t$ it is probable that subscripts c and t can often be dropped where a particular emphasis is not demanded.

$$M_1 = 510(0.3 - \tfrac{1}{2} \times 0.05) = 140 \text{ kN} \cdot \text{m}$$
$$N_w = 0.2 \times 0.062 \times 0.85 \times 20 = 0.211 \text{ MN} = 211 \text{ kN}$$
$$M_2 = 211(0.3 - \tfrac{1}{2} \times 0.062) = 57 \text{ kN} \cdot \text{m}$$
$$M_n = M_1 + M_2 = 140 + 57 = 197 \text{ kN} \cdot \text{m}$$
$$\text{Design moment} = \phi M_n = 0.9 \times 197 = 177 \text{ kN} \cdot \text{m}$$

The minimum thickness (overall depth) should be checked against Table 3.2 to determine whether deflection is a probable concern. The old standard of $\rho = A_s/bd$ $\leqslant 0.18 f_c'/f_y$ (now discarded from the Code) also gives a rough idea. The corresponding ρ in terms of the M_2 couple alone (Fig. 3.12c) is often convenient for design tables.

3.13 ANALYSIS OF BEAMS WITH COMPRESSION STEEL

Beams are occasionally restricted in size to such an extent that steel is needed to help carry the compression. Compression reinforcement A_s' is also excellent for reducing the longtime deflection. For this member, the design of A_s and A_s' for a required M_u (Sec. 4.5) is simpler than the analysis problems that follow. For analysis it is convenient to consider two couples M_1 and M_2 similar to the approach used for the T-beam in Fig. 3.12c.

If the member is relatively deep the M_2* steel couple based on A_s' and A_{s2} should be evaluated first. Strictly, the extra compression available because of A_s' must also account for the loss of compression concrete

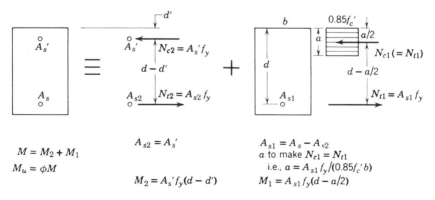

FIGURE 3.13 Analysis of double-reinforced beam if all steel yields.

*The subscript 2 is used because the *usual* problem is in *design* where the M_2 couple is the second operation. It is well not to vary the numbering between the two cases. Both M_1 and M_2 are nominal moments but the omission of the n subscript should cause no confusion here.

displaced by A'_s. Then

$$M_2 = A'_s(f_y - 0.85 f'_c)(d - d') \quad \text{and} \quad N_{t2} = N_{c2}$$
$$A_{s2} = A'_s(f_y - 0.85 f'_c)/f_y$$

If A'_s is small and f'_c is low, it is often close enough to neglect the displaced concrete, as follows (in Fig. 3.13):

$$M_2 = A'_s f_y(d - d') \quad \text{and} \quad A_{s2} = A'_s$$

The remaining tension steel is the basis for the M_1 couple.

$$A_{s1} = A_s - A_{s2} \qquad M_1 = A_{s1}f_y(d - a/2)$$
$$\text{Design } M = \phi M_n = \phi(M_2 + M_1) = 0.9(M_2 + M_1)$$

The depth of the stress block a is calculated from $N_{c1} = N_{t1}$ or $0.85 f'_c ba = A_{s1}f_y$.

If the member is shallow, especially a slab, there is a reasonable chance that the strain at the A'_s level will not develop f_y in compression and this must be verified early in the analysis:

1. Assume $A_{s2} = A'_s$. Then $A_{s1} = A_s - A'_s$.
2. Calculate $a = A_{s1}f_y/(0.85 f'_c b)$.
3. *Calculate c,* using the ratios between c and a at the close of Sec. 3.3.
4. Use the strain triangle of Fig. 3.14 to calculate

$$\epsilon'_s = 0.003(c - d')/c$$

5. If $E_s\epsilon'_s > f_y$, proceed to evaluate M_1 and M_2, using f_y with A'_s.

If $f'_s < f_y$ it becomes more important to evaluate N_{c2} and A_{s2} considering the displaced concrete, which in turn modifies the original assumed A_{s1}—an iteration problem in theory. However, the second example below sets up a possible quadratic equation that can be used.

Deflections are determined by top and bottom strains on the beam and

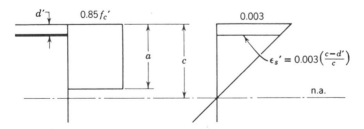

FIGURE 3.14 Check on compression steel stress.

hence can be measured by the M_1 couple developed by A_{s1} and the concrete, ignoring the M_2 couple developed by A'_s and A_{s2}.

Ties (or stirrups) are required around A'_s in beams, as in columns, Code 7.11.1.

3.14 DOUBLE-REINFORCED BEAM EXAMPLES

(a) A rectangular beam has $b = 280$ mm $= 0.28$ m, $d = 510$ mm $= 0.51$ m, $A_s = 5 - \#35 = 5000$ mm$^2 = 5 \times 10^{-3}$ m^2, C35 concrete ($f'_c = 35$ MPa), Grade 400 bars ($f_y = 400$ MPa), $A'_s = 2 - \#35 = 2000$ mm$^2 = 2 \times 10^{-3}$ m^2; with d' (cover to centerline of A'_s) of 50 mm $= 0.05$ m. Calculate the design moment capacity.

Solution

Neglecting concrete displaced by A'_s, consider initially $A_{s2} = A'_s = 2000$ mm$^2 = 2 \times 10^{-3}$ m^2, thus assuming $F'_s = f_y$.

$$M_2 = A'_s f_y (d - d') = 2 \times 10^{-3} \times 400(0.51 - 0.05) = 368 \times 10^{-3} \text{ MN} \cdot \text{m} = 368 \text{ kN} \cdot \text{m}$$

The remainder of the tension steel, $A_{s1} = A_s - A_{s2} = 5000 - 2000 = 3000$ mm$^2 = 3 \times 10^{-3}$ m^2, works with the compression concrete, as in a rectangular beam without A'_s, to develop M_1. Since $N_t = N_c$

$$3 \times 10^{-3} \times 400 = 0.85 \times 35 \times 0.28a$$

$$a = 0.144 \text{ m} = 144 \text{ mm} = 0.28 \, d < 0.36 \, d \text{ (Table 3.1)} \qquad \textbf{O.K.}$$

Since A_{s2} and A'_s do not themselves fail in a brittle manner, the Code (10.3.3) approves their separation into the M_2 couple, assumed ductile, and considers A_{s1} as the only reinforcement subject to the 0.75 ρ_b limitation. The check above shows this criteria more than satisfied. However, the assumption that $f'_s = f_y$ should be verified before going further.

Since $f'_c = 35$ MPa, the depth of stress block a will be less than 0.85 c, as noted near the end of Sec. 3.3. It was assumed there that the 0.85 factor would be decreased 0.08 for each 10 MPa of concrete strength beyond 30 MPa. Hence, $a/c = 0.85 - 0.04 = 0.81$, and with $a = 144$ mm (from above), $c = 144/0.81 = 178$ mm.

$$\epsilon'_s = 0.003(c - d')/c = 0.003(178 - 50)/178 = 0.0022$$
$$\epsilon_y = 400/E_s = 400/200\,000 = 0.0020 < \epsilon'_s$$

Therefore $f'_s = f_y$ and the calculations so far are correct.

$$z_1 = d - a/2 = 510 - \tfrac{1}{2} \times 144 = 438 \text{ mm} = 0.438 \text{ m}$$
$$M_1 = A_{s1} f_y z_1 = 3 \times 10^{-3} \times 400 \times 0.438 = 526 \times 10^{-3} \text{ MN} \cdot \text{m} = 526 \text{ kN} \cdot \text{m}$$
$$M_n = M_1 + M_2 = 526 + 368 = 894 \text{ kN} \cdot \text{m}$$
$$\text{Design moment} = \phi M_n = 0.90 \times 894 = 805 \text{ kN} \cdot \text{m}$$

Check this solution by verifying that tension equals compression, $N_{tn} = N_{ct}$.

$$N_{tn} = A_s f_y = 5 \times 10^{-3} \times 400 \qquad\qquad = 2.00 \text{ MN tension}$$
$$N_{cn} \text{ on } A_s' = 2 \times 10^{-3} \times 400 \qquad\qquad = 0.80 \text{ MN compression}$$
$$N_{cn} \text{ on concrete} = 0.280 \times 0.144 \times 0.85 \times 35 = 1.20 \text{ MN compression}$$
$$\Sigma N = 0 \qquad \textbf{O.K.}$$

(b) A rectangular beam has $b = 280$ mm $= 0.28$ m, $d = 510$ mm $= 0.51$ m, $A_s = 3 - \#35 = 3000$ mm$^2 = 3 \times 10^{-3}$ m^2, C35 concrete ($f_c' = 35$ MPa), Gr 400 bars ($f_y = 400$ MPa), $A_s' = 1 - \#35 = 1000$ mm$^2 = 1 \times 10^{-3}$ m^2, with $d' = 50$ mm $= 0.05$ m. Calculate the design moment capacity.

Solution

Again neglecting concrete displaced by A_s' and assuming $f_s' = f_y$:

$$A_{s2} = A_s' = 1000 \text{ mm}^2, \quad A_{s1} = A_s - A_{s2} = 3000 - 1000 = 2000 \text{ mm}^2 = 2 \times 10^{-3} \text{ m}^2$$
$$a = A_{s1}f_y/(0.85\, f_c' b) = 2 \times 10^{-3} \times 400/0.85 \times 35 \times 0.28$$
$$= 0.096 \text{ m} = 96 \text{ mm}$$

From the close of Sec. 3.3, $a/c = \beta_1 = 0.85 - 0.04 = 0.81$, $c = a/0.81 = 96/0.81 = 119$ mm

$$\epsilon_s' = 0.003(119 - 50)/119 = 0.001\,74, \qquad \epsilon_y = 400/200\,000 = 0.0020, \quad f_s' < f_y$$

Since f_s' is smaller than assumed, it is just as easy now to include recognition of the lost concrete displaced which would carry $0.85 \times 35 = 30$ MPa. An algebraic equation will be set up later, but a trial-and-error technique seems more efficient. Try $f_s' = 350$ MPa (about f_y times the ratio ϵ_s'/ϵ_y)

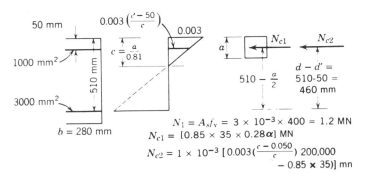

FIGURE 3.15 Basic relations for solution in Sec. 3.14b, using dimensions in meters.

Effective $f_s'' = 350 - 30 = 320$ MPa

$A_s' f_s'' = A_{s2} f_y$, $1000 \times 320 = A_{s2} \times 400$, $A_{s2} = 800$ mm^2, $A_{s1} = A_s - A_{s2} = 2200$ mm^2

Max. ρ_1 from Table 3.1 = 0.0271

Max. $A_{s1} = 0.0271 \times 280 \times 510 = 3870$ mm$^2 \gg 2200$ mm^2 **O.K.**

$a = A_{s1} f_y/(0.85\, f_c'b) = 2200 \times 400/(0.85 \times 35 \times 280) = 106$ mm

For $f_c' = 35$ MPa, $c = a/\beta_1 = a/(0.85 - 0.04) = 106/0.81 = 131$ mm

$\epsilon_s' = 0.003(131 - 50)/131 = 0.001\,85$, $f_s' = \epsilon_s' E_s = 0.001\,85 \times 200\,000 = 370$ MPa

Since the trial started with 350 MPa, it is probably close enough to use as 360 MPa. This leads to effective $f_s' = 360 - 30 = 330$ MPa, $N_{c2} = f_s' A_s' = 330 \times 1 \times 10^{-3}$ MN = 330 kN.

$M_2 = N_{c2}(d - d') = 330(510 - 50) = 152$ kN$\cdot$m

$A_{s2} = A_s' f_s'/f_y = 1000 \times 330/400 = 825$ mm^2 leaving $A_{s1} = 3000 - 825 = 2175$ mm^2
$= 2.175 \times 10^{-3}$ m^2

$a = 2175 \times 400/(0.85 \times 35 \times 280) = 104$ mm = 0.104 m

$M_1 = 2.175 \times 10^{-3} \times 400(0.510 - 0.052) = 3.98 \times 10^{-3}$ MN$\cdot$m = 398 kN$\cdot$m

and $M_y = 0.90\,(M_1 + M_2) = 0.90 \times 550 = 495$ kN$\cdot$m

The solution can also be set up algebraically for dimensions in meters after it is discovered that $\epsilon_s' < \epsilon_y$, as shown in Fig. 3.15.

$N_{c1} + N_{c2} = N_t$, $0.85\, f_c'b(\beta_1 c) + A_s'[E_s \times 0.003(c - d')/c - 0.85\, f_c'] = N_t$.

$0.85 \times 35 \times 0.28 \times 0.81\, c + 1 \times 10^{-3}$
$\times [200\,000 \times 0.003(c - 0.050)/c - 0.85 \times 35] = A_s f_y = 3 \times 10^{-3} \times 400$

The solution of this quadratic equation is

$$6.75\, c - 0.03/c = 0.63$$
$$c^2 - 0.0933 = 0.0044$$
$$c = 0.1280 \text{ m} = 128 \text{ mm.}$$

Thus, this exact solution gives $c = 128$ mm and $a = (0.85 - 0.04)c = 104$ mm, as earlier.

The nominal M_1 and M_2 values can now be found easily.

$M_1 = N_c z = (0.85 \times 35 \times 0.28 \times 0.104)(0.51 - \tfrac{1}{2} \times 0.104) = 397 \times 10^{-3}$ MN$\cdot$m
$= 397$ kN$\cdot$m

$\epsilon_s' = 0.003(128 - 50)/128 = 0.001\,83$

Effective f_s' or $f_s'' = 200\,000 \times 0.001\,83 - 0.85 \times 35 = 336$ MPa

$M_2 = A_s' f_s'' z = 1 \times 10^{-3} \times 336(0.51 - 0.05) = 155 \times 10^{-3}$ MN$\cdot$m = 155 kN$\cdot$m

Design strength = $\phi M_n = 0.90(M_1 + M_2) = 0.90(397 + 155) = 497$ kN$\cdot$m

Although it is not needed in this solution, the above in effect used A_{s2} as 840 mm^2 and A_{s1} as 2160 mm^2.

3.15 ANALYSIS OF SPECIAL BEAM SHAPES

The general analysis presented in Secs. 3.5 and 3.7 and adapted to the T-beam in Secs. 3.11 and 3.12 is adaptable to any shape of cross section. The strain triangles can always be used to establish the balanced section neutral axis.* The stress block depth may be fixed from the a/c ratios of Sec. 3.3, and N_{cb} evaluated even for very irregular member outlines. The usable N_c or N_t should be restricted to 0.75 N_b as an upper limit. For this maximum N_t, or such smaller $A_s f_y$ as may exist, the required depth of stress block a can be established, the resultant compression located, and the moment couple calculated. Since a also establishes c, values of $f_s < f_y$ can be taken into account when specific bar locations will not develop strains as large as ϵ_y. A sample analysis for one shape follows the next paragraph.

Tests show that when a nonrectangular compressive stress area is used, the ϵ_c Code limit of 0.003 could well be larger for a peaked area, as in the example of Fig. 3.16, or could be slightly more limited for an inverted triangular area. The collapse of the extremely narrow compression top is apparently delayed by the bracing effect of the adjacent wider horizontal strips at lower strain levels. However, these variations are not reflected in the Code.

Example

(a) Given the beam of Fig. 3.16a with $f'_c = 30$ MPa, Gr 400 bars. What is the maximum A_s the Code would allow to be counted as effective if no A'_s is used?

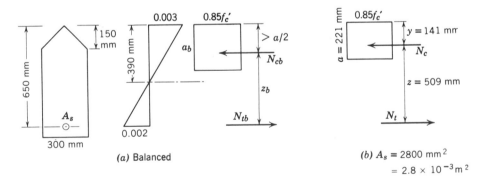

(a) Balanced

(b) $A_s = 2800$ mm^2
$= 2.8 \times 10^{-3}$ m^2

FIGURE 3.16   Example of special beam shape.

*Some uncertainty can arise when there are several layers or depths of tension reinforcing. Rather arbitrarily, the author uses the yield strain on the steel most distant from the compression face to establish his balanced section.

Solution

The maximum A_s would be limited to 0.75 A_{sb} for the balanced section. The balanced strains are shown in Fig. 3.16b, with $\epsilon_y = f_y/E_s = 400/200\,000 = 0.002$

$$c_b = 650 \times 0.003/(0.003 + 0.002) = 390 \text{ mm}$$
$$\text{Depth of stress block} = a_b = 0.85\, c_b = 332 \text{ mm}$$
$$\text{Area under stress block} = A_c = 332 \times 300 - \tfrac{1}{2} \times 2 \times 150 \times 150$$
$$= 77\,000 \text{ mm}^2$$
$$N_{tb} = N_{cb}, \qquad A_{sb} \times 400 = 0.85\, f'_c A_c = 0.85 \times 30 \times 77\,000$$
$$A_{sb} = 4909 \text{ mm}^2$$

With irregular shapes it appears simpler to work with A_{sb} rather than with ρ_b.

$$\text{Max. allowable } A_s = 0.75\, A_{sb} = 3682 \text{ mm}^2$$

(b) If the beam in Fig. 3.16a has A_s of $4 - \#30$ bars ($A_s = 2800 \text{ mm}^2$), what is the permissible design moment?

Solution

$$N_{cn} = 0.85\, f'_{c2} A_c = N_{tn} = A_s f_y, \qquad 0.85 \times 30\, A_c = 2800 \times 400$$

$A_c = 43\,900 \text{ mm}^2$, which shows that the stress block extends below the upper triangle.

$$300a - \tfrac{1}{2} \times 2 \times 150 \times 150 = 43\,900 \quad a = 221 \text{ mm}$$

Since N_{cn} acts more than $a/2$ from the top, locate it at the centroid of A_c. If $y = $ distance to centroid from top,

$$43\,900\, y = \tfrac{1}{2} \times 2 \times 150 \times 150 \times 100 + (221 - 150) \times 300\, (150 + 35)$$
$$= 6\,190\,000 \text{ mm}^2$$
$$y = 141 \text{ mm}$$
$$z = 650 - 141 = 509 \text{ mm} = 0.509 \text{ m}$$
$$M_n = N_{tn} z = 2.8 \times 10^{-3} \times 400 \times 0.509 = 570 \times 10^{-3} \text{ MN} \cdot \text{m} = 570 \text{ kN} \cdot \text{m}$$

Allowable design moment $= \phi M_n = 0.9 \times 570 = 513 \text{ kN} \cdot \text{m}$

3.16 TABLES AND CHARTS

Sample helpful tables and charts are included in Appendix B for rectangular beams and slabs, without compression steel. Tables B.1 and B.2 augment Table 3.1, with Table B.1 giving k_m, c/d, a/d, and z/d for several common materials combinations. These are in terms of $\omega\, (= \rho f_y/f'_c)$ and also in terms of ρ. The use of ω permits the grouping of data into a very compact form, with k_m

tabulations here in MPa, over the entire usable range. Table B.2 shows values of coefficients that can be multiplied by f'_c to give k_n values, or by $0.9 f'_c$ to give k_m values, each quite accurate in terms of ω.

For example, consider the following, similar to the example in Sec. 3.9a. Calculate M_u for a rectangular beam having $b = 275$ mm, $d = 500$ mm, $A_s = 3 - \#25 = 1500$ mm^2, $f'_c = 25$ MPa, $f_y = 300$ MPa.

$$\rho = 1500/275 \times 500 = 0.0109, \quad \omega = \rho f_y / f'_c = 0.0109 \times 300/25 = 0.131$$

Enter Table B.1 with ω and read for given f'_c and f_y, $k_m = 2.70$ (for $\omega = 0.130$) plus 0.1×0.19 (for the last 0.001), a total $k_m = 2.72$. (Note that this table stops at the maximum allowable k_m for $\rho = 0.75 \, \rho_b$, as already tabulated in Table 3.1 in Sec. 3.7.)

$$M_u = k_m b d^2 = 2.72 \times 0.275 \times 0.5^2 = 0.187 \text{ MN} \cdot \text{m} = 187 \text{ kN} \cdot \text{m}$$

Sometimes the values of c, a, and z tabulated at the right of the table are useful.

Alternatively, Table B.2 may be used, but note that it does not indicate the $0.75 \, \rho_b$ limit. For $\omega = 0.131$ as before, read directly a coefficient of 0.1209. Multiply by $0.9 f'_c$ to obtain $k_m = 2.72$, as before. Graphical values of k_m reading less accurately are also given in Figs B.1 and B.2.

The ACI Handbook in the references at the end of Chapter 4, although primarily directed toward design, has many charts and tables that are also useful in analysis but at present these are only in English units.

SELECTED REFERENCES

1. A. H. Mattock and L. B. Kriz, "Ultimate Strength of Nonrectangular Structural Concrete Members," *Jour. ACI*, Jan. 1961; *Proc.*, *57*, p. 737.
2. R. Park and T. Paulay, "*Reinforced Concrete Structures*," John Wiley & Sons, New York, 1975.

PROBLEMS

General Note   Tables 2.1 and 2.2 in Sec. 2.9 list bar sizes, areas, and strengths. Necessary load factors are given in Sec. 2.12e (usually Code Eq. 9.1 unless otherwise specified) and ϕ factors in Sec. 2.12f. Beam constants for many cases are given in Table 3.1 and Tables B.1 and B.2 in Appendix B.

PROB. 3.1.   If $f'_c = 25$ MPa and steel is Grade 400, find ϕM_n, the moment capacity of the beam of Fig. 3.17 if the A_s is:

(a) 2-#25.   (c) 4-#30.
(b) 4-#25.   (d) 4-#35.

FIGURE 3.17   Beam for Prob. 3.1.

PROB. 3.2.   Find moment capacity ϕM_n of the beams of Fig. 3.18, assuming sketch (a) for a and b, sketch (b) for c, d.
(a) Grade 300 steel, $f_y = 300$ MPa
(b) Grade 400 steel, $f_y = 400$ MPa

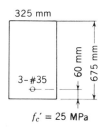

FIGURE 3.18   Beams of Prob. 3.2.

PROB. 3.3.   If $f_c' \times 30$ MPa and steel is Grade 400, calculate the moment capacity ϕM_n of the beam of Fig. 3.19, assuming the steel as follows:

(a) 3-#25.   (c) 4-#35.
(b) 6-#25.   (d) 6-#35.

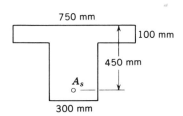

FIGURE 3.19   Beam of Prob. 3.3.

PROB. 3.4.　Calculate ϕM_n on the beams of Fig. 3.20, assuming sketch (a) for a, b, sketch (b) for c, d; A_s as follows:

(a) 4-#35, Grade 300
(b) 3-#25, Grade 300

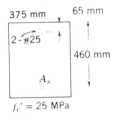

375 mm　　65 mm

2-#25

460 mm

A_s

$f_c' = 25$ MPa

FIGURE 3.20　Beam of Prob. 3.4.

PROB. 3.5.　Check the adequacy of the beam of Fig. 3.21 in flexure assuming $f_c' = 35$ MPa, Grade 300 bars, service live load of 18 kN/m and dead load (including beam weight) of 8 kN/m. (*Suggestion*: Compare calculated ϕM_n with required ϕM_n)

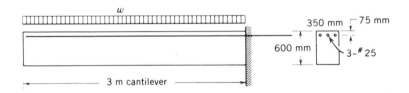

w

350 mm　⌐75 mm

600 mm

3-#25

3 m cantilever

FIGURE 3.21　Beam for Prob. 3.5.

PROB. 3.6.　Find the moment capacity ϕM_n of the joist of Fig. 3.22 if $f_c' = 30$ MPa and A_s is 2-#20 bars of Grade 400.

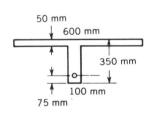

50 mm

600 mm

350 mm

100 mm

75 mm

FIGURE 3.22　Joist of Prob. 3.6.

PROB. 3.7.   Find the allowable moment ϕM_n of a 1-m strip of the slab of Fig. 3.9 if the bars are made #25 at 175 mm on centers and $d = 125$ mm, $f'_c = 25$ MPa, Grade 400 steel.

PROB. 3.8.   Find allowable moment ϕM_n for the beam of Fig. 3.23 if $f'_c = 35$ MPa and Grade 400 steel is used. Ignore the lack of symmetry.
(a) Omitting top bars.
(b) Including top bars.

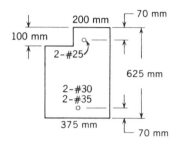

FIGURE 3.23   Irregular beam for Prob. 3.8.

PROB. 3.9.   Prob. 3.8 except consider the beam of Fig. 3.24.
(a) Omitting the 2-#30.
(b) Including the 2-#30.

PROB. 3.10.   If $f'_c = 35$ MPa and steel is Grade 400, what is the moment capacity ϕM_n of the beam of Fig. 3.25 when each leg is reinforced with:
(a) 1-#20.   (b) 1-#30.

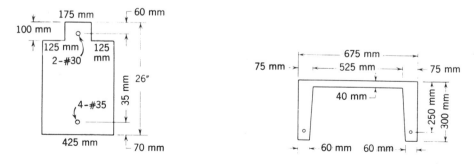

FIGURE 3.24   Irregular beam for Prob. 3.9.

FIGURE 3.25   Beam for Prob. 3.10.

PROB. 3.11. If the precast section of Fig. 3.26 has $f_c' = 35$ MPa, Grade 400 steel, 1-#25 bar in each leg, calculate the moment capacity ϕM_n. Flange $h_f = 40$ mm.

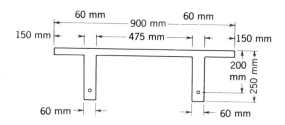

FIGURE 3.26 Precast section for Prob. 3.11.

PROB. 3.12.

(a) Calculate the moment capacity ϕM_n (Fig. 3.27) of a strip of slab 1 m wide if $f_c' = 30$ MPa and the steel is Grade 400.

(b) What would be the service or working live load per square meter if the slab were used on a 3 m simple span? Use the Code load factors. Concrete mass may be taken at 2500 kg/m³ which includes the weight of reinforcement.

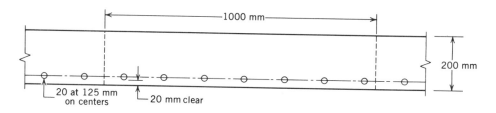

FIGURE 3.27 Slab for Prob. 3.12.

PROB. 3.13. A rectangular beam has $b = 375$ mm, $d = 450$ mm, $f_c' = 25$ MPa, Grade 300 steel, $A_s = 6$-#35.

(a) Calculate the moment capacity ϕM_n.

(b) Calculate the moment capacity ϕM_n if one adds $A_s' = 3$-#35 with cover d' to center of steel = 60 mm.

4
DESIGN FOR FLEXURE

4.1 GENERAL PROCEDURES

Analysis and design are related problems that nevertheless involve fundamental differences and require different techniques. It is essential that the student clearly understands these differences.

In analysis or review for allowable loads or moments the engineer deals with given beams, known both as to dimensions and steel. Each such case has a neutral axis unique to those particular dimensions and not dependent on the engineer's desires.

In design, loads and material strengths are known and some or all the dimensions remain to be fixed. Here designers have some control over the location of the neutral axis. They can shift it where they want it, to the extent that changes in dimensions can alter the balance between tension and compression areas.

After loads are known and the layout of the structure or structural element has been established,* the maximum bending moment can be determined. The member design for moment involves three separate steps:

1. Choice of beam cross section.
2. Choice of reinforcing steel at point or points of maximum moment.
3. Determination of points where bars are no longer needed for moment, that is, points for bending or stopping bars.

---

*The choice of proper design loads and the layout of a structure to support these loads efficiently is a major part of the design. The choice of loads and the general problem of framing layout are not covered in this book. Chapter 10 has a brief discussion of beam-and-girder framing and the proper arrangement of loading for maximum moment on continuous one-way slabs and beams. Various types of slab construction are discussed in Chapters 12, 15, and 16.

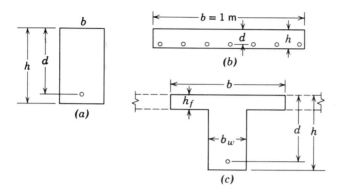

FIGURE 4.1 Typical reinforced concrete beams. (a)
Rectangular. (b) Slab. (c) T-beam.

Often other factors than moment determine the choice of the beam cross
section, such as shear (Chapter 5). Such interrelationships are discussed
later.

In rectangular beams (Fig. 4.1a), both width b and depth d to steel are
to be chosen. In slabs (Fig. 4.1b), the strip width b is fixed and only the
depth d or the over-all thickness h can be selected. In T-beams (Fig. 4.1c),
the flange thickness h_f and flange width b are determined, usually by a
slab already designed, thus leaving the stem width b_w and the depth d to
the steel to be selected. The choice of b_w and d is not usually determined
by the moment alone.

In double-reinforced beams the compression steel is required because b
and d have been chosen smaller than the normal requirements or because
it is desirable to reduce creep deflections. However, when compression steel
in definite amounts is desired for protection against deflection or otherwise, the size of double-reinforced beams can be based on moment
requirements.

For any beam shape the design of steel, once the moment and beam size
are known, is dependent on a proper choice of the internal lever arm z
between resisting tension and compression forces. In rectangular beams
(and slabs) and in double-reinforced beams, the analytical solution for z is
feasible, but the resulting quadratic equation is usually not conducive to
an accurate and rapid solution. Hence the author suggests that the
cut-and-try procedure necessary for T-beams and irregular shapes be used
for all shapes. The value of z is first estimated, the corresponding approximate A_s is calculated, and from this A_s a better value of z is
established; in turn, this better z leads to a better and usually satisfactory
value of A_s. Convergence is rapid.

The bending or stopping of bars is determined by the maximum moments to be resisted at various points along the beam. Bending of bars is discussed separately from choice of beam and steel (Sec. 4.11).

4.2 RECTANGULAR BEAMS—MOMENT DESIGN

The economical dimensions for a rectangular beam are not sharply defined. A shallow beam is expensive because of the weight of steel required; and on long spans it may run into deflection problems. A deep beam is more economical of steel but costs more for side forms and uses up headroom, which means greater story heights. The ratio of over-all depths to span listed in Table 3.2 in Sec. 3.8 (copied from Code Table 9.5a) can serve as a guide.

This chapter deals with beam sizes established in various manners. It is almost wholly concerned with strength, slightly concerned with deflection, and not concerned with over-all economy. However, it will emphasize methods that conserve steel for the particular beam size chosen.

The maximum ρ ratios developed in Sec. 3.7 (Table 3.1) are absolute limits on the steel that may be considered effective because of danger from brittle-type failures.

The minimum reinforcement of $\rho = 1.4/f_y$ (Code 10.5.1) is necessary to insure that the member does not lose some strength when it first cracks. This requirement does not apply if the A_s used is 4/3 that required by analysis, because this implies a lowered probability that the cracking load will ever be reached, in other words, adequate safety. Slabs of uniform thickness are also exempt, but the author questions this exemption in those locations where shears are near the limiting value. A small ρ and a maximum shear make a bad combination.

4.3 RECTANGULAR BEAM EXAMPLES

(a) Design a minimum practical beam for service moments of 70 kN·m for D.L. and 140 kN·m for L.L. using $f'_c = 20$ MPa and Grade 400 steel.

Solution

The load factors of Code Eq. 9.1. (see. Sec. 2.12e) are assumed to govern:

$$M_u = 1.4\,M_D + 1.7\,M_L = 1.4 \times 70 + 1.7 \times 140 = 336 \text{ kN·m}$$
$$M_n \geqslant M_u/\phi = 336/0.9 = 373 \text{ kN·m} = 0.373 \text{ MN·m}$$

From Table 3.1, for these materials, maximum $k_n = 5.25$. The size of the beam can be found by equating $k_n bd^2$ to be external M_n (or $k_m bd^2$ to M_u or ϕM_n).

$$M_n = k_n bd^2 = 5.25\ bd^2 = 0.373$$
$$\text{Reqd. } bd^2 = 71.0 \times 10^{-3}\ \text{m}^3 = 71\,000\,000\ \text{mm}^3$$

This requirement can be met by a wide range of sizes, for example:

If $b = 175$ mm,	$d = 637$ mm	If $b = 275$ mm,	$d = 508$ mm
$b = 200$	$d = 596$	$b = 300$	$d = 486$
$b = 225$	$d = 562$	$b = 325$	$d = 467$
$b = 250$	$d = 533$		

The narrowest beam shown may be too narrow to accommodate the reinforcing steel, it uses up the most headroom, and requires the greatest area of forms; but it uses the smallest volume of concrete and steel. A d/b ratio from 1.0 or 1.5 to 2.5 or 3.0 is usually considered in the desirable range, with the larger ratios for the bigger beams. Since deep beams often require increased story heights, more wall and partition height, and thus other costs than the cost of the beam, there is no simple textbook answer to the *best* depth. There are a number of satisfactory choices possible. For the purpose of this problem, $b = 275$ mm $= 0.275$ m will be chosen although a wider beam might be needed for a lower strength steel, with more bars then necessary.

If not exposed to the weather or the ground such a beam requires 40 mm of clear cover (Code 7.7.1c) over the steel. The steel typically includes some stirrups as reinforcement as shown in Fig. 4.2, and # 10 bars would be a reasonable size for this beam. Then the cross section of the beam would appear as shown in Fig. 4.2a. The minimum over-all depth would be $508 + 15$ (assuming bar diam.* $d_b = 30$ mm) $+ 10$ (stirrup) $+ 40$ (cover) $= 574$. This is an impractical dimension to give a carpenter for forms and the only change permitted is to greater dimensions, since maximum k_n was used originally.

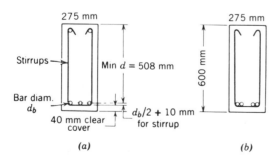

FIGURE 4.2   Dimensions for beam of Sec. 4.3. (*a*) Minimums. (*b*) Final.

*Bar diameters are given in Table 2.2.

USE 275 mm by 600 mm (overall).

$$d = 600 - 40 - 10 - 15 \text{ (assumed } \tfrac{1}{2}d_b) = 535 \text{ mm}$$

With $d = 535$ mm the beam is a little deeper than the theoretical minimum and thus has a greater z than the minimum, which means that N_t and the balancing compression block depth a will be slightly reduced. Note that Table 3.1 shows a maximum $a/d = 0.38$, or $a = 0.38\ z\ 535 = 203$ mm.

Assume $a = 200$ mm, $z = 535 - \tfrac{1}{2} \times 200 = 435$ mm $= 0.435$ m

$$M_n = A_s f_y z, \quad A_s = 373 \times 10^{-3}/400 \times 0.435 = 2.149 \times 10^{-3} \text{ m}^2$$
$$a = A_s f_y/0.85 f'_c b = 2.149 \times 10^{-3} \times 400/0.85 \times 20 \times 0.275 = 184 \text{ mm}$$

This indicates the original guess is adequate, but this can be proved by completing

$$z = 535 - \tfrac{1}{2} \times 184 = 443 \text{ mm} = 0.443 \text{ m}$$
$$A_s = 373 \times 10^{-3}/400 \times 0.443 = 2.10 \text{ m}^2 = 2100 \text{ mm}^2$$

The choice of bars cannot be final without checks on bar development (Chapter 6), but a selection can be made tentatively:

2—#45 = 3000 mm^2 Development apt to be troublesome with large bars, and the area is 45% larger than is necessary.

3—#30 = 2100 mm^2 **O.K.**

5—#25 = 2500 mm^2 **O.K.**

8—#20 = 2400 mm^2 **O.K.**, except that two layers of bars are required, which increases overall depth.

The clear spacing of bars in a layer must not be less than the nominal bar diameter, 25 mm, or $\tfrac{4}{3}$ of the nominal maximum aggregate size (Code 7.6.1 and 3.3.3c). Try 3—#30 for spacing. Clear width inside stirrups is $275 - 2 \times 40$ (cover) $- 2 \times 10$ (stirrups) $= 175$ mm. The three bars require 3×30 (for bars) $+ 2 \times 30$ (for spaces) $= 150$ mm, assuming 20 mm aggregate is used.

USE 3-#30 bars

Deflections must be investigated (Chapter 8).

The student should note that the entire design process could be very compactly presented. This presentation has been greatly expanded by explanations.

(*b*) For architectural reasons the beam in (*a*) is to be made 300 mm $= 0.3$ m wide by 600 mm $= 0.6$ m deep over-all. Find the necessary steel, assuming D.L. M unchanged.

Solution

Available $d = 600 - 40$ cover $- 10$ stirrup $- 15$ for $d_b/2 = 535$ mm $= 0.535$ m

Actual $k_n = M_n/bd^2 = 373 \times 10^{-3}/0.3 \times 0.535^2$

$$= 4.34 \times 10^6 < 5.25 \times 10^6 \text{ from Table 3.1.} \qquad \textbf{O.K.}$$

Therefore the beam can be much underreinforced, but one has no accurate value of a to work from, although it will usually fall in the range from $0.1d$ to $0.3d$ (still more for minimum size beams).

Try $a = 0.20\ d = 0.2 \times 535 = 107$ mm, $z = 535 - \frac{1}{2} \times 107 = 482$ mm $= 0.482$ m

$$A_s = M_n/f_y z = 373 \times 10^{-3}/400 \times 0.482 = 1.93 \times 10^{-3}\ \text{m}^2 = 1\,930\ \text{mm}^3$$

$$a = A_s f_y/0.85 f'_c b = 1\,930 \times 400/0.85 \times 20 \times 300 = 151\ \text{mm}$$

This lowers z, increases A_s and a.

Try $a = 160$ mm, $z = 535 - 80 = 455$ mm $= 0.455$ m

$$A_s = 373 \times 10^{-3}/400 \times 0.455 = 2.050 \times 10^{-3}\ \text{m}^2 = 2\,050\ \text{mm}^2$$

$$a = 151 \times 2\,050/1\,930 = 160\ \text{mm}$$

$$z = 455 \times 2\,050/1\,930 = 483\ \text{mm} \qquad \textbf{O.K.}$$

$$\text{USE } 3\#30 = 2100\ \text{mm}^2$$

Clear inside stirrups $= 300 - 2 \times 40$ cover $- 2 \times 10$ stir. $= 200$ mm

Clear width between bars $= 200 - 3 \times 30 = 110$ mm

$$= 2 \text{ spaces of 55 mm each} \qquad \textbf{O.K.}$$

4.4. BEAMS WITH COMPRESSION STEEL—MOMENT DESIGN

Compression steel is used wherever a given beam requires more tension steel than the engineer thinks desirable in a rectangular beam, either from consideration of approaching brittle failure conditions or from consideration of excessive deflections. In either case the designer knows how much moment he or she must carry or desires to carry without compression steel and this moment is designated as the M_1 couple with steel A_{s1}. Any necessary additional moment M_2 can then be resisted by added tension steel A_{s2} and compression steel A'_s as sketched in Fig. 4.3. Whether A'_s works at yield stress or a lower stress depends upon the ϵ'_s strain, established by the neutral axis required with the M_1 couple.

When compression capacity requires the compression steel, the M_1 couple is based on the k_n value of Table 3.1 and A_{s1} can be calculated from the corresponding ρ value (or from the a value, which establishes z for use in $A_{s1} = M_1/f_y z$). It should be noted that the permissible kbd^2 is normally so large that compression steel is rarely required for strength. When compression steel is used to reduce long time deflection, M_1 and k_1 can be based on the $\rho = 0.18\ f'_c/f_y$ values tabulated in Table 3.1 and Fig. 3.8, or engineers can set up their own criterion following the method of Sec. 4.5*b*.

Whereas *analysis* of beams containing A_s' sometimes requires a cut-and-try procedure (Sec. 3.14*b*), the *design* procedure is always direct and relatively simple.

Code 7.11.1 requires ties (or stirrups) around A_s' in beams, as in columns (Fig. 18.13).

4.5 DOUBLE-REINFORCED BEAM EXAMPLES

(*a*) A double-reinforced concrete beam with $b = 300$ mm $= 0.300$ m, $d = 500$ mm $=$ 0.500 m, $f_c' = 30$ MPa, $f_y = 400$ MPa, must carry a service dead load moment of 200 kN·m and live load moment of 320 kN·m. Calculate the required steel using the Code load factors.

Solution

$$\text{Required } M_u = 200 \times 1.4 + 320 \times 1.7 = 824 \text{ kN·m}$$

$$\text{Reqd. } M_n \geq M_u/0.9 = 824/0.9 = 915 \text{ kN·m} = 0.915 \text{ MN·m}$$

As a rectangular beam without A_s', the maximum steel for A_{s1} is limited to $0.75\,\rho_b$, which is tabulated in Table 3.1 as 2.44%, and which develops

$$M_1 = k_n bd^2 = 7.89 \times 0.300 \times 0.500^2 = 0.592 \text{ MN·m}$$

$$A_{s1} = \rho_1 bd = 0.0244 \times 300 \times 500 = 3660 \text{ mm}^2$$

Or, using $a = 0.38\,d$ (from Table 3.1) $= 0.38 \times 500 = 190$ mm

$$z = d - a/2 = 500 - 190/2 = 405 \text{ mm}$$

$$A_{s1} = M_1/f_y z = 0.592/400 \times 0.405 = 0.003\,65 \text{ m}^2 = 3650 \text{ mm}^2$$

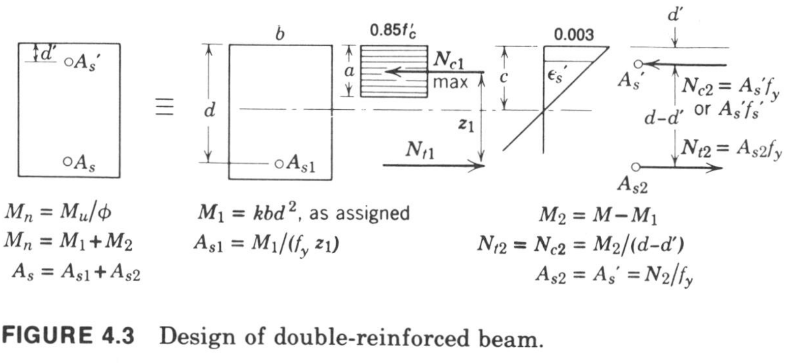

d'
A_s'
A_s
≡
b
d
A_{s1}
0.85f'c
a
N_{c1}
max
c
z_1
N_{t1}
0.003
ϵ_s'
d'
A_s'
$N_{c2} = A_s'f_y$ or $A_s'f_s'$
d−d'
$N_{t2} = A_{s2}f_y$
A_{s2}
$M_n = M_u/\phi$
$M_n = M_1 + M_2$
$A_s = A_{s1} + A_{s2}$
$M_1 = kbd^2$, as assigned
$A_{s1} = M_1/(f_y\, z_1)$
$M_2 = M - M_1$
$N_{t2} = N_{c2} = M_2/(d-d')$
$A_{s2} = A_s' = N_2/f_y$

FIGURE 4.3 Design of double-reinforced beam.

The a value is not quite as accurately stated as the ρ value in the table, but either is as accurate as the loads. The value of a is also needed for f_s' calculations below.

$$M_2 = M_n - M_1 = 0.915 - 0.592 = 0.323 \text{ MN} \cdot \text{m}, \qquad N_{t2} = N_{c2} = M_2/(d - d')$$

Assume the cover d' over A_s', to center of bars, is 65 mm, making $d - d' = 500 - 65 = 435$ mm. Then $N_2 = 0.323/0.435 = 0.743$ MN

$$A_{s2} = N_2/f_y = 0.743/400 = 0.001\,86 \text{ m}^2 = 1860 \text{ mm}^2$$
$$A_s = A_{s1} + A_{s2} = 3660 + 1860 = 5520 \text{ mm}^2$$

Check to see if A_s' develops the yield strain $\epsilon_y = 400/200\,000 = 0.0020$. Use strain triangles based on the above value of $a = 190$ mm, $c = a/\beta_1 = 190/0.85 = 224$ mm, and $\epsilon_s' = 0.003(224 - 65)/224 = 0.0021 > \epsilon_y$.

Neglecting compression concrete, which is displaced by A_s'

$$A_s' = N_2/f_y = A_{s2} = 1860 \text{ mm}^2$$

If the displaced concrete is considered, the effective f_s''—that is, the increased value added by A_s' over the loss of the same area of compression concrete—becomes

$$f_y - 0.85\, f_c' = 400 - 0.85 \times 30 = 374 \text{ MPa}$$
$$A_s' = N_2/f_s'' = 0.743/374 = 0.001\,99 \text{ m}^2 = 1990 \text{ mm}^2$$

Choice of bars involves the same procedure used in Sec. 4.3(a).

USE $A_s = 5520 \text{ mm}^2$, $A_s' = 1990 \text{ mm}^2$, $d' = 65$ mm

The large A_{s1} results from a minimum depth beam for strength and may give a beam with excess deflection. Add cover to the d of 500 mm and check that h against Table 3.2 (Code Table 9.5a), or calculate deflection (Chapter 8).

(b) Recalculate the reinforcing steel for the above case if it is desired to reduce deflection somewhat by limiting A_{s1} to $\rho_1 = 0.18\, f_c'/f_y *$, $d' = 65$ mm.

Solution

The total required M_n is 0.915 MN · m, as calculated before. Table 3.1 and Fig. 3.8 show that this steel limitation gives $M_1 = 0.160\, f_c'bd^2$, $a = 0.21\, d$, and $z_1 = 0.89\, d$. Thus, $k_1 = 0.160 \times 30 = 4.80$ MPa.

$$M_1 = k_1 bd^2 = 4.80 \times 0.300 \times 0.500^2 = 0.360 \text{ MN} \cdot \text{m}$$
$$A_{s1} = (0.18\, f_c'/f_y)bd = (0.18 \times 30/400)0.300 \times 0.500 = 0.002\,04 \text{ m}^2 = 2040 \text{ mm}^2$$
$$M_2 = M_n - M_1 = 0.915 - 0.360 = 0.555 \text{ MN} \cdot \text{m}$$
$$N_{t2} = N_{c2} = M_2/z = 0.555/(500 - 65) = 1.28 \text{ MN}$$
$$A_{s2} = N_2/f_y = 1.28/400 = 0.003\,20 \text{ m}^2 = 3200 \text{ mm}^2 = 0.002\,98 \text{ m}^2 = 2980 \text{ mm}^2$$
$$A_s = A_{s1} + A_{s2} = 2040 + 3200 = 5240 \text{ mm}^2$$

---

*Note that the method of Sec. 3.8 and Fig. 3.8 is available for any desired ρ_1, to give k_1. Likewise any desired ρ' can lead to k_2. The sum $k_1 + k_2 = $ total k resulting.

For compression strain triangles, $a = 0.21 \, d = 0.21 \times 500 = 105$ mm and $c = a/\beta_1 = 105/0.85 = 124$ mm

$$\epsilon'_s = 0.003(124 - 65)/124 = 0.001\,43 < \epsilon_y = 400/200\,000 = 0.0020$$

Using $f'_s = f_y \times 0.001\,43/0.002 = 0.715\,f_y$ and allowing for displaced concrete,

$$A'_s = N_2/(0.715f_y - 0.85\,f'_c) = 1.28/(286 - 0.85 \times 30)$$
$$= 0.004\,92 \text{ m}^2 = 4\,920 \text{ mm}^2$$

USE $A_s = 5\,240$ mm^2, $A'_s = 4\,920$ mm^2

Comparison with the design in (a), which was for maximum M_1 and minimum A'_s, shows that A_s has dropped 5 percent because of the increased z given by a smaller A_{s1} and hence a reduced stress block depth for compression. On the other hand, A'_s is increased 147 percent.

With a larger ρ, or with deeper beams, the required A_{s2} and A'_s are much reduced from those just found. There is no cut-and-try needed in the design of double reinforced beams.

4.6 T-BEAMS—MOMENT DESIGN

The T-beam flange has usually been established by the thickness of the slab design, and the web size (b_w and d) is often determined by shear or requirements other than moment. To some extent b_w, the stem width, is influenced by the number of reinforcing bars used for A_s, but the student probably lacks the experience at this stage to estimate this need in advance. In this section it is assumed that design for moment includes only the choice of the area of steel A_s, a check against brittle failure, and some thought of deflection.

A flange is ordinarily assumed to provide enough compression area to eliminate the possibility of brittle failure in compression. This is frequently confirmed by a neutral axis which falls high in the flange and in effect turns a T-beam design into the design of a wide rectangular beam. Actually, a T-beam must be quite deep relative to the slab and be quite heavily reinforced to make it really act as a T-beam in carrying compression. Even when the balanced condition would lead to a neutral axis deep in the beam (and a rather-large value of $a = \beta_1 c$), the brittleness concept leads to restricting the compression stress block area to 0.75 of that at the balanced condition. Since it is rare that more than 0.25 of the total stress block area lies below the bottom of the flange, the usable stress block tends to be restricted to less than the total depth of the flange. Thus most T-beams carrying positive moment are only rectangular beams in flexural behavior.

The cut-and-try determination of z, as for the rectangular beam, is recommended, as illustrated in Sec. 4.7. The Code provision of transverse steel in the flange (8.10.5) must not be overlooked:

> Where primary flexural reinforcement in a slab that is considered as a T-beam flange (excluding joist construction) is parallel to the beam, reinforcement perpendicular to the beam shall be provided in the top of the slab in accordance with the following:
>
> (a) Transverse reinforcement shall be designed to carry the factored load on the overhanging slab width assumed to act as a cantilever. For isolated beams, the full width of overhanging flange shall be considered.
>
> (b) Transverse reinforcement shall not be spaced farther apart than 5 times the slab thickness, nor 500 mm.

4.7 T-BEAM EXAMPLES

(a) The floor of Fig. 4.4 consists of a 100 mm slab supported by beams of 6.700 m span cast monolithically with the slab at a spacing of 2.440 m center to center. These beams have webs 300 mm wide and a depth of 480 mm to the center of reinforcement plus the necessary cover over the bars. Service dead load moment is 84 kN·m and live load moment is 163 kN·m. Using $f'_c = 30$ MPa and Grade 400 steel, calculate the required area of reinforcement.

Solution

The effective flange width is defined by Code 8.10.2 as the smallest of these three limits (or see Sec. 3.11):

$$\text{Span}/4 = 6700 \text{ mm}/4 = 1670 \text{ mm} \leftarrow \text{Governs } b$$

$$8\, h_f \text{ overhang gives } b = 2 \times 8 \times 100 + 300 = 1900 \text{ mm}$$

$$\text{Beam spacing} = 2.440 \text{ m} = 2440 \text{ mm}$$

$$\text{Required } M_u = 1.4\, D + 1.7\, L = 1.4 \times 84 + 1.7 \times 163 = 395 \text{ kN·m}$$

$$M_n \geqslant M_u/\phi = 395/0.9 = 439 \text{ kN · m}$$

The lever arm z will usually be large and a good trial value will usually be at least

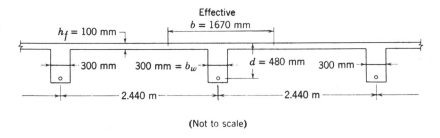

Effective
b = 1670 mm

h_f = 100 mm

300 mm   300 mm = b_w d = 480 mm   300 mm

2.440 m   2.440 m

(Not to scale)

FIGURE 4.4 Floor of Sec. 4.7.

0.9 d or $d - h_f/2$, whichever is the larger.

$$0.9\,d = 0.9 \times 480 = 432 \text{ mm} \qquad d - h_f/2 = 480 - 100/2 = 430 \text{ mm}$$
$$\text{Trial } A_s = M_n/(f_y z) = 439 \times 10^{-3}/400 \times 0.432 = 0.002\,54 \text{ m}^2 = 2540 \text{ mm}^2$$

It must be determined whether the stress block is as deep as the flange. If $a = h_f$, available $N_c = 0.85 \times 30 \times 1.670 \times 0.100 = 4.26$ MN, compared to the necessary $N_c = A_s f_y = 0.002\,54 \times 400 = 1.02$ MN. Thus, the neutral axis is obviously high in the flange and the design becomes simply that for a rectangular beam 1670 mm wide.

$$a = N_c/(0.85\,f'_c b) = 1.02/(0.85 \times 30 \times 1.670) = 0.024 \text{ m} = 24 \text{ mm}$$
$$z = 480 - 0.5 \times 24 = 468 \text{ mm}$$
$$A_s = M_n/(f_y z) = 439 \times 10^{-3}/400 \times 0.468 = 0.002\,34 \text{ m}^2 = 2340 \text{ mm}^2$$

Better $a = 24(2340/2540) = 22$ mm. Close enough.

USE $A_s = 2340$ mm^2

There can be no possibility of brittle compression failure with this small a. Deflection should be checked.

(b) Calculate A_s if the loads in Example (a) are increased to give $M_u = 1.80$ MN $\cdot$ m.

Solution

$$M_n \gtrless M_u/\phi = 1.80/0.9 = 2.00 \text{ MN} \cdot \text{m}$$

If $a = h_f = 100$ mm, available $N_c = 4.26$ MN, as in (a). Compare approximate required

$$N_c \text{ for } z = 0.9\,d = 0.9 \times 480 = 432 \text{ mm}.$$
$$N_c = M_n/z = 2.00/0.432 = 4.63 \text{ MN} > \text{available } 4.26 \text{ MN}$$

Since a is greater than h_f, this acts as a true T-beam. As above, try $z = 0.9\,d$.

$$\text{Approx. } A_s = 2.00/400 \times 0.9 \times 480 = 0.011\,57 \text{ m}^2 = 11\,570 \text{ mm}^2$$
$$\text{Compression block area} = N_t/(0.85\,f'_c) = 0.011\,57 \times 400/0.85 \times 30$$
$$= 0.1814 \text{ m}^2 = 181\,400 \text{ mm}^2$$
$$\text{Area below flange} = 181\,400 - 1670 \times 100 = 14\,400 \text{ mm}^2$$
$$\text{Depth below flange} = 14\,400/300 = 48 \text{ mm, total } a = 148 \text{ mm}.$$

This area is shown in Fig. 4.5 and gives the centroid distance $\bar{y}$ from top as:

$$\bar{y} = (1670 \times 100 \times 50 + 300 \times 48 \times 124)/181\,400 = 56 \text{ mm}$$
$$z = d - \bar{y} = 480 - 56 = 424 \text{ mm},$$
$$A_s = M_n/(f_y z) = 2.00/400 \times 0.424 = 0.011\,80 \text{ m}^2 = 11\,800 \text{ mm}^2$$

Compression block area increases to $(11\,800/11\,570)181\,400 = 185\,000$ mm^2

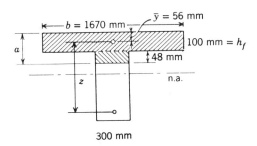

FIGURE 4.5   Trial compression block.

Area below flange $= 185\,000 - 1670 \times 100 = 18\,000$ mm^2, with necessary distance below flange increasing from 44 mm to 60 mm. The change in y and z (about 4 mm more $\bar{y}$, 4 mm less z) is reasonably small and could be neglected here. The design would be satisfied with $A_s = 11\,800$ mm^2 (more exactly about 1 percent more, say 11 900 mm^2) except for the problem covered in the next paragraph.

The neutral axis is quite low and should be checked with regard to $\rho \gtrless 0.75\,\rho_b$, where ρ_b is the balanced condition.*

$$\text{Balanced } c_b = 600\ d/(600 + f_y) = 600 \times 480/(600 + 400) = 288 \text{ mm}$$
$$a_b = 0.85\ c_b = 0.85 \times 288 = 245 \text{ mm}$$

Balanced compression stress block area $= 1670 \times 100 + 300 \times 145 = 210\,500$ mm^2

Usable compression area to match $0.75\,\rho_b = 0.75 \times 210\,500 = 158\,000$ mm^2.

This is substantially less than the 185 000 required above. Under the Code it is definitely not desirable to use this T-beam for the indicated moment. However, if it is necessary, it can be accomplished by using A_{s1} as required to balance the 158 000 mm^2 stress block, in which case this leads to M_1 as for a rectangular beam, since 158 000 mm^2 is less than bh_f. The additional moment $M_2 = M_n - M_1$ should then be provided by A'_s and A_{s2} as for any double reinforced beam.

This T-beam case is far from typical since most are not so heavily loaded with moment. The given moment might well call for a heavier slab design that would automatically increase the moment capacity; otherwise a deeper beam seems appropriate.

(c) Calculate the required A_s for the beam of (b) if beam depth is increased to 700 mm and the required M_u to 2.50 MN · m, while the span is reduced to 6.5 m, which limits b to 1.625 m.†

*Code 10.3.3 limits compression area to 0.75 of the *balanced* compression area as used here. A more consistent value for uniform ductility limit would be to limit c to 0.75 c_b.

†The 100 mm slab still indicated is probably inappropriate for the larger M_u. It is retained to show how to handle the T-shape of compression block on the very unusual cases where this is necessary—or how to handle other irregular shapes.

Solution

$$\text{Required } M_n \geqslant M_u/\phi = 2.50/0.9 = 2.78 \text{ MN} \cdot \text{m}$$

Find the required compression stress block.

Approximate reqd. $N_c = 2.78/(0.700 - \frac{1}{2} \times 0.100) = 4.28$ MN

Reqd. area of stress block $= 4.28/0.85 \times 30 = 0.1678$ m²

Area in flange level $= 1.625 \times 0.100 = 0.1625$ m²

Area reqd. below flange $= 0.1678 - 0.1625 = 0.0053$ m² $= 5300$ mm²

Total stress block depth $a = 100 + 5300/300 = 118$ mm

Check this against the ductility requirement that $\rho \leqslant 0.75 \rho_b$, or, if A_c is area under compression, $A_c \geqslant 0.75 A_{cb}$. Calculate c_b from strain triangles, noting $\epsilon_y = 400/200\,000 = 0.0020$.

$$c_b = 0.003 \, d/(0.003 + 0.0020) = 0.003 \times 700/0.0050 = 420 \text{ mm}$$

$$a_b = 0.85 \times 420 = 357 \text{ mm}$$

Balanced stress block area $= 1625 \times 100 + 300 \times 257 = 239\,600$ mm²

Code limit on $A_c = 0.75 \times 239\,600$

$$= 179\,800 \text{ mm}^2 = 0.1790 \text{ m}^2 > 0.1678 \text{ m}^2 \qquad \textbf{O.K.}$$

Proceed to locate centroid of compression stress block and evaluate z and A_s, using above stress block areas.

From top $\bar{y} = [162\,500 \times \frac{1}{2} \times 100 + 5300(100 + \frac{1}{2} \times 18)]/(162\,500 + 5300) = 52$ mm

$z = 700 - 52 = 648$ versus 650 assumed

Reqd. $A_s = 2.78/400 \times 0.648 = 0.010\,73$ m² $= 1073$ mm²

It is not mandatory for the designer to find the centroid of the compression block, as done above. For any nonrectangular compression area there is the option

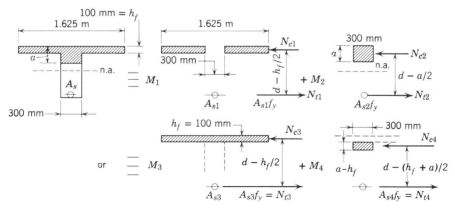

FIGURE 4.6 Breaking down a total moment into couples convenient for calculation.

of considering it as the sum of its parts, but with each part having a common neutral axis. For the above example the designer could use either the N_1 and N_2 couples of Fig. 4.6 or the N_3 and N_4 couples, much as used earlier in Sec. 3.12 and Fig. 3.12.

As an example, verify the above A_s of 10 730 mm^2 by using N_{c1} and N_{c2}.

$$N_{c1} = (1.625 - 0.300) \times 0.100 \times 0.85 \times 30 = 3.38 \text{ MN} = N_{t1}$$
$$A_{s1} = N_{t1}/f_y = 3.38/400 = 0.008\ 45 \text{ m}^2 = 8450 \text{ mm}^2$$

The approximate a of 118 mm from above will be used in calculating A_{s2}.

$$N_{c2} = 0.300 \times 0.118 \times 0.85 \times 30 = 0.903 \text{ MN} = N_{t2}$$
$$A_{s2} = N_t/f_y = 0.903/400 = 0.002\ 26 \text{ m}^2 = 2260 \text{ mm}^2$$

Required $A_s = 8450 + 2260 = 10\ 710 \text{ mm}^2$ versus 10 730 mm^2 earlier, almost a negligible difference. However, for improved accuracy (especially when in new territory), the revised a based on this A_s is readily determined and would lead to a better N_{c2} and A_{s2}.

4.8 BEAMS OF SPECIAL SHAPES

As presented in the preceding sections, the criterion against brittle failure is perfectly general. The nominal compression must be limited to 0.75 N_{cnb}. The only Code exception is that the contribution of A'_s to ductility is recognized by the last sentence in Code 10.3.3 and thus in Section 4.5 the 0.75 factor was used only with the M_1 couple, not with the M_s couple representing A'_s and A_{s2}.

Special shaped beams present small problems for the designer in that the k_m or k_n values, maximum a/d, and the like will not be available in tables unless it happens to be a shape commonly used in precast work or the like. The necessary A_s for a beam such as that of Fig. 3.16 can be found by a cut-and-try process that travels over much of the same ground used in Section 3.15 for finding the allowable design moment to accompany a given A_s. The procedure would be as follows:

1. Assume z, noting that with the narrow top width in Fig. 3.16 N_c acts farther from the top than in a rectangular beam of constant width.
2. Calculate $A_s = M_n/(f_y z)$.
3. Evaluate depth of stress block from $N_{cn} = N_{tn}$, possibly first finding A_c in compression from $A_c = A_s f_y/0.85\ f'_c$.
4. Find centroid of A_c to get location of N_{cn} and a better z.
5. Recycle (if the original estimate of z was poor), to the accuracy desired.

An alternate procedure for step 4 would be to break down A_c into simple pieces, as done for the T-beam in Fig. 4.6, each with its individual A_s and couple. This does not, however, so clearly indicate how much error was in the assumed z of step 1 and what would be the better guess for the next cycle.

4.9 FACE STEEL ON LARGE BEAMS

The primary Code emphasis on deep beams is for relatively short spans, ℓ_n not more than $5d$. (See coverage in Sec. 5.2.1 as primarily a shear problem.) However, Code 10.6.7* relates primarily to flexural cracking in any web more than 900 mm deep, regardless of span length, even in constant moment zones where shear is essentially zero. It calls for longitudinal reinforcement "at least 10 percent of the area of the flexural reinforcement—placed near the side faces of the web and distributed in the zone of flexural tension."

A depth of more than 900 mm is not a common beam size for buildings, but it is common for bridge structures; this specification is written to control flexural cracking on the deep side faces. Unfortunately, it is *not* adequate for this purpose. Actual construction and tests in France and at the University of Texas at Austin show this inaccuracy, although the final report on the latter is not yet issued.

The basic problem is simple, once it is noted. Ordinary beams are designed on the assumption that the strain gradient from flexure is linear, the shear strains being negligible in shallow members. In the deep beam the flexural strains, top and bottom, are about the same as in a shallow beam, which means the flexural strain gradient varies essentially inversely with depth, thus very small in very deep members. The shear strains are no longer negligible in comparison. Instead they can almost totally cancel the flexural strains and relax the longitudinal tension in the concrete just above A_s and widen the flexural crack once it gets well above A_s.

A second pattern, less predictable except from experiments, is that all the cracks forming at the level of the bars do *not* rise to equal heights in the web. With deep beams without horizontal web bars there may be 3 or 4 cracks at the A_s level for each crack continuing farther up in the web. In this way also the longitudinal strains can lead to web crack widths that may be several times as wide as those at the A_s level.

The function of longitudinal face bars in the web is to stretch the concrete (help it carry some tension) and also to cause more web cracks of narrower average width, rather than fewer wide cracks. The author was

*The highway bridge code (AASHTO) discussed in Chapter 17 is essentially the same in this respect.

almost startled by the flexural crack widths in the first 36 in. (=0.914 m) beam he tested, cracks substantially wider at middepth than at A_s level.

Tests in France[3] reported in 1972 on beams 1 m deep and 0.3 m thick seem to show that bars totaling 0.25 percent of the web area and placed between A_s and the neutral axis limited web cracks to the crack width at the level of A_s. Such bars can also be counted as part of A_s to the extent consistent with their lower strains and shorter internal lever arm for flexure.

The author cannot make firm recommendations in this area. He feels that the problem may be real in a 0.75 m deep member when Gr 400 bars are used for A_s. The French tests are the best guide for requirements at this time; but a beam 2 m deep would probably need *more* face steel for the same effectiveness. A thickness of 1 m would also raise questions about whether only face steel (for appearance) would not leave much wider internal cracks (from shear relaxation) which might influence durability under some exposures.

4.10   BEAMS WITH REINFORCING AT SEVERAL LEVELS

When reinforcing is placed in two layers, it is usually satisfactory to consider the f_y stress at the centroid of the steel. However, if the bars are in many layers or are distributed over all faces, as in a column section, some bars will be near the neutral axis and have a stress less than f_y. In such cases strains as well as stresses must be considered. This is not overly complex when strain triangles are used after the fashion already used for compression steel, as illustrated by Fig. 4.3 and the examples of Sec. 4.5 and Fig. 3.15.

4.11   STOPPING OR BENDING OF BARS—FOR MOMENT

(a)   General Considerations

The maximum required A_s for a beam is needed only where the moment is maximum. Insofar as moment is concerned this steel may be reduced at points along the beam where only smaller moments exist. Two other stress considerations are also important in fixing the length of bars. First the individual bars require a specific length in which to develop their yield strength. Development length, the subject of Chapter 7, is as important as moment length; the designer must put the two aspects together. Second, when a bar is cut off where the neighboring bars are still carrying tension, a discontinuity in the tensile resistance occurs, flexural cracks open at earlier loads and become wider and more inclined than normal; shear strength is lowered, often with some loss of the beam ductility. Shear strength loss is discussed in Sec. 5.16, which points out the extra shear reinforcement (stirrups) required by Code 12.11.5 near the end of bars that are cut off. This requirement offsets some of the apparent savings from reduced bar length.

A bent bar anchored in the compression zone causes no loss in shear strength.

In this chapter only the moment aspect of stopping or bending bars is discussed, and this only for statically determinate members.

(b) Maximum Moment Curves

A study of where bars can be stopped, or bent away from the tension zone, must start with a study of the maximum moments which are possible at all points along the beam.

In some simple cases, the maximum moment diagram is simply the moment diagram for full load. In other cases, as for wheel loads (Fig. 4.7a), the determination of the maximum moment diagram requires the calculation of maximum moments at many points; the maximum moment diagram is then the envelope of the maximum moment values from the several loads, as in Fig. 4.7b for a single wheel load W. Each dashed triangle corresponds to a particular position of the load, the maximum moment at any point being given with the wheel at that point. The envelope in this case happens to be a parabola. With several wheels the curve approximates two half parabolas separated by a section of constant moment as in Fig. 4.7c.

Only fixed loadings are illustrated in the remainder of this chapter. Continuous beams are discussed in Chapter 10.

(c) Theoretical Bend Point or Cutoff Point

When a variable depth member is considered, the required A_s curve must be established by calculating the required A_s at representative points from the basic relation

$$A_s = \max M_n \div (f_y z)$$

where z varies somewhat like d. This procedure is illustrated in Sec. 9.6.

With constant depth beams, the denominator becomes nearly a constant, with the usual small variations in the ratio z/d ignored in this connection. Hence required A_s varies directly with the maximum moment, and the shape of the required A_s curve is identical with that of the maximum moment curve. The maximum moment curve may then be used as the required A_s curve, simply by changing the scale. The maximum

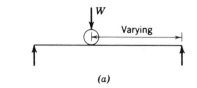

(a)

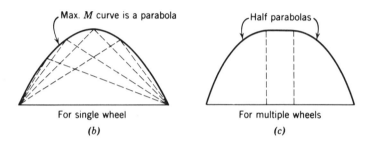

For single wheel   For multiple wheels
(b)   *(c)*

FIGURE 4.7   Maximum moment curves.

ordinate corresponds to the maximum required A_s and this establishes the necessary correlation.

Bars can theoretically be stopped or bent wherever they are no longer needed for moment. The solution can be either graphical (for complex maximum moment diagrams), semigraphical for typical cases, or analytical where desired. The semigraphical process is illustrated in Secs. 4.12 and 4.13.

(d) Shear Complications

Where a bar is cut off, the rather sudden transfer of tension to the continuing bars causes early flexural cracking at the end of the bar. This flexural crack is wider than usual and turns more diagonally to become what is commonly called a shear crack, Fig. 5.2. Shear strength is lowered in a zone around the last portion of the terminated bar (maybe for 0.75 d from the cutoff). See bar chart of 34 test results in Fig. 5.18.

Because of this complication, Code 12.11.5 forbids cutting off bars unless *one* of three possible conditions is satisfied:

1. Shear at cutoff point is not more than 2/3 the allowable.
2. Extra shear reinforcement is added.
3. Continuing bar stress is not more than $f_y/2$ *and* shear is not over 3/4 the allowable.

Research showed that bending the bars from one beam face to the other avoided this shear problem.

(e) Arbitrary Requirements for Extending Bars

The ACI Code (12.11.3) requires that each bar be extended a distance d or 12 bar diameters beyond the point where it is "no longer required to resist flexure." This prohibits the cutting off of a bar at the theoretical minimum point, but can be interpreted as permitting bars to be *bent* at the minimum point. Two sound reasons can be advanced for this arbitrary extension, which the author designates as a, in addition to the shear problem in (d) above.

First, when a bar is simply cut off (not bent away from the main steel), there is a large transfer of stress from this bar to those remaining. This stress concentration will cause a moment crack in the concrete at the end of the bar if the beam is carrying its working load. Bending the bar spreads out this concentration; extending the bar, without bending, removes the concentration from a point of maximum steel stress to a point of lower steel stress. (If a bar is cut off at the theoretical minimum point, the remaining bars necessarily work at maximum stress.)

The author gives greater weight to a second reason, well stated in the long obsolete 1940 Joint Committee Specification:

> To provide for contingencies arising from unanticipated loads, yielding of supports, shifting of points of inflection, or other lack of agreement with assumed conditions governing the design of elastic structures, it is recommended that the reinforcement be extended at supports and at other points between supports as indicated. . . .

Thus interpreted, the arbitrary extension of the bar is the result of an envisioned possible extension of the maximum moment diagram. It would follow that bars must not even be bent until most of this extra length has been provided. Tests show, however, that bending bars is less dangerous than cutting them off in tension zones. The author would be satisfied to see *bends*, not cutoffs, at $d/2$ beyond the theoretical cutoff point. Even this would be a more severe requirement than many engineers observe.

The Code (12.13) requires that one-third of the negative moment reinforcement be extended past the point of inflection by the larger of: beam depth d, 12 bar diameters, or $\ell_n/16$ where ℓ_n is the clear span. (In script and on the typewriter it is desirable to use a script ℓ to distinguish it clearly from the number one.)

The Code (12.12) also requires that at least one-third of the positive moment reinforcement extend into the support in the case of simple spans and one-fourth in the case of continuous beams. Custom usually increases these minimums. However, it must not be overlooked that the Code 12.12.2, for beams which are used as part of the primary lateral load system, requires the one-fourth into the support be anchored for full f_y to give some ductility to the system. Here it is not acceptable to run twice as many bars into the support anchored for half of f_y; this will not give the desired ductility.

(f) Nomenclature

It is common to refer to the first bar cut off or bent, second bar, and so forth. First bar in this context means the one nearest the point of maximum moment, typically closest to the support for top bars and closest to midspan for bottom bars. It is thus possible for the first bar bent up to become the second bar bent down, and the second bent up the first bent down, when only two bars are offset.

Bars are often bent or cut off by pairs instead of singly. The first *pair* can be bent or cut off where the second *bar* is no longer needed, that is, beyond the extensions of (e), which apply to the second bar.

Bars are sometimes bent at points that make them available as web reinforcement to resist shear (diagonal tension). Since present practice uses fewer bent bars, this value is often ignored in the United States, unless the member is a precast one on a production line basis.

(g) Author's Philosophy on Bar Extensions

Beams using the offset bars behave in a more desirable manner than those with bars cut off, but the labor involved in fabrication and erection often make them uneconomical. Where members are designed against some earthquake hazard (which should mean over 90 percent of the United States), the sway motions imposed on a frame may reverse the beam moments at column faces. In such cases, continuous top and bottom bars, with staggered lap splices, appear to make a superior member. If they were to become common, the arbitrary extensions of (e) would have less impact, because fewer bars would be totally cut off.

The Code phrase quoted in the third paragraph of (e) is of long standing with only minor editorial changes. The author believes it has now become desirable to start thinking of the arbitrary extensions more as an *actual* potential shift in the moment diagram. As moment diagrams from frame and computer analyses replace the approximate moment coefficients, these more clearly show the omission of the factors the Joint Committee listed; and these factors do cause real moments. The *idea* of shifted moment diagram becomes essential in some form if we move toward redistribution of moments (Code 8.4) and limit design.

As far as the lengths of bars *for moment* are concerned, the practical effect of the substitution of a shifted moment diagram for an arbitrary extension is nil; but it makes a large difference in the magnitude of the stress considered to exist next to the cutoff point in the bars that do continue farther. It thus substantially influences the development length demands of the continuing bars as discussed in Chapter 7, especially in Sec. 7.11 and 7.12.

4.12 EXAMPLES INVOLVING NO EXCESS STEEL, OR NEGLECT OF THE EXCESS

(*a*) A uniformly loaded beam of 6 m simple span requires 5-#30 bars, with $d = 375$ mm. Considering moment alone, where can the first bar be cut off? The second bar?

Solution

The maximum moment diagram is simply the full load moment diagram and this parabola can be used as the required A_s diagram as in Fig. 4.8. The maximum ordinate becomes five bars. Steel areas could be used as ordinates, but the number of bars provides a more convenient unit when all bars are of the same size.

The first bar can theoretically be cut off at x_1 where only four bars are required. Since offsets to the center tangent vary as the square of the distances from the center,

$$x_1^2/3^2 = 1/5 \qquad x_1 = 3\sqrt{1/5} = 1.342 \text{ m}$$

Since d of 375 mm exceeds $12\, d_b = 12 \times 30 = 360$ mm, use $a = 375$ mm for cutoff (or 188 mm for bend). The minimum distance from the center line of the span to the first bar cutoff is $1.342 + a = 1.342 + 0.375 = 1.717$ m (or 1.530 m to a bend point).

For the second bar stopped,

$$x_2^2/3^2 = 2/5 \qquad x_2 = 3\sqrt{2/5} = 1.897 \text{ m}$$

The minimum distance from the center to the second bar cutoff is $1.897 + 0.375 = 2.272$ m, or 2.085 m to a bend point.

Cutting off bars in a tension zone, as above, requires special shear calculations (Sec. 5.16). Also development lengths must be checked as in Sec. 7.12c and 10.15.

(*b*) A uniformly loaded cantilever beam 2.5 m long requires 7-#20 bars when d is 450 mm. Where can the first pair of bars be bent down for moment?

Solution

The maximum moment diagram is a half parabola, the maximum ordinate corresponding to seven bars of required A_s (Fig. 4.9). The first pair of bars can be bent

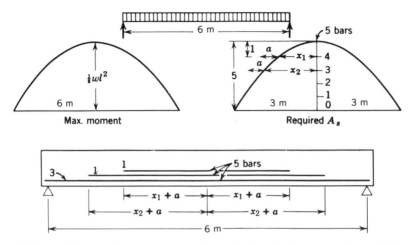

FIGURE 4.8 Cutting off bars in uniformly loaded beam according to the moment diagram.

where only five bars are required.

$$x^2/2.5^2 = 5/7 \qquad x = 2.5\sqrt{5/7} = 2.113 \text{ m}$$

$$d = 0.450 \text{ m} \qquad 12\,d_b = 0.240 \text{ m} \qquad a = 0.450 \text{ m for cutoff, or } 0.225 \text{ m for bend}$$

The minimum distance from the support to the bend point for the first pair is $2.500 - 2.113 + a = 0.387 + 0.225 = 0.612$ m. With the bar bent instead of cut off there is no loss of shear strength.

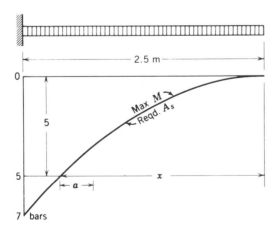

FIGURE 4.9 Bending bars for cantilever beam.

4.13 EXAMPLES PERMITTING REDUCED LENGTHS BECAUSE OF EXCESS STEEL

(*a*) A 5.5 m simple span beam with a fixed concentrated load at midspan (negligible uniform load) and $d = 550$ mm requires 3000 mm² of steel but uses 5-#30 bars = 3500 mm². Where can the first pair of bars be stopped?

Solution

Figure 4.10 shows the maximum moment diagram used as a required A_s curve with the maximum ordinate marked as 3000 mm² = 4.3 bars required. The first pair of bars can be stopped where only three bars are required.

$$x/2.75 = 1.3/4.3 \qquad x = 0.831 \text{ m}$$

$$d = 0.55 \text{ m} \qquad 12\,d_b = 0.36 \text{ m} \qquad a = 0.55 \text{ m for cutoff, or } 0.28 \text{ m for bend}$$

The minimum distance from the mid-span to the stop point for the first pair is $0.831 + a = 0.831 + 0.550 = 1.381$ m

(*b*) A uniformly loaded cantilever beam 2.5 m long with $d = 450$ mm requires 1920 mm² and uses 7-#20 bars. Where can the first pair of bars be bent down?

Solution

The maximum ordinate of the maximum moment diagram is designated as 1920 mm²/300 mm² = 6.4 bars required (Fig. 4.11). The desired point is where five bars are needed.

$$x^2/2.5 = 5/6.4 \qquad x = 2.5\sqrt{5/6.4} = 2.210 \text{ m}$$

$$d = 0.45 \text{ m} \qquad 12d_b = 0.24 \text{ m} \qquad a = 0.45 \text{ m to cutoff, } 0.225 \text{ m to bend}$$

The minimum distance from the support to the bend point for the first pair is $2.500 - 2.210 + a = 0.290 + 0.225 = 0.515$ m.

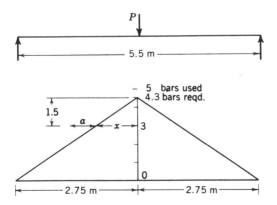

FIGURE 4.10 Stopping bars, excess steel.

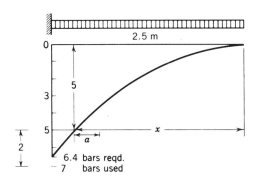

FIGURE 4.11   Bending bars, excess steel,
cantilever beam.

4.14 DESIGN AIDS

Available design aids for rectangular beams have already been discussed
in Sec. 3.16.

Revised design tables are in preparation for a new Ref. 1, which will
probably be available by the time this edition is published. The new tables
will considerably simplify some problems.

The CRSI Handbook[2] has extensive tables for many types of construction and curves that appear very helpful for long columns designed under
the moment magnifier method of Chapter 19.

SELECTED REFERENCES

1. *Design Handbook, Vol. 1*, Special Publication No. 17(73), American
    Concrete Institute, Detroit, 1973.
2. *CRSI Handbook*, 2nd ed., Concrete Reinforcing Steel Institute, Chi-
    cago, 1975.
3. J. Colonna-Ceccaldi and S. Soretz, "Large Reinforced Concrete Beams
    with a Main Reinforcement Consisting of Two Thick Bars," Vol. 46
    from the series *Betonstahl In Entwicklung*, Tor-Isteg Steel Corp.,
    Luxembourg, 1972.

PROBLEMS

PROB. 4.1.   In Prob. 3.5 calculate the exact steel that is needed for the
loads given.

PROB. 4.2. In Prob. 3.5 calculate the exact steel that is needed if Grade 400 steel is used.

PROB. 4.3. Calculate steel for the channel section of Prob. 3.10 and Fig. 3.25 if $M_u = 142$ kN m and the steel is changed to Grade 300.

PROB. 4.4. A simple span rectangular beam 6 m long is to carry a service load of 50 kN/m made up of 24 kN/m dead load (including beam weight) and 16 kN/m live load. Use $f'_c = 30$ MPa, Grade 400 steel, and design a beam, subject to later check on shear and bond, for ρ approximately $0.18\, f'_c/f_y$. Make $b = 325$ mm. Choose bars.

PROB. 4.5. Assuming deflections are not considered serious for the usage contemplated, redesign the beam of Prob. 4.4 with $b = 325$ mm and the minimum effective depth permitted under the Code.

PROB. 4.6. A continuous rectangular beam with $b = 350$ mm, $d = 600$ mm, $f'_c = 35$ MPa, Grade 400 steel, must care for $M_u = -1.5$ MN m. Design the necessary steel, using $d' = 75$ mm if A'_s is necessary.

PROB. 4.7. Redesign the depth in Prob. 4.6 to avoid use of A'_s. Design the steel for moment assumed unchanged.

PROB. 4.8. A rectangular beam 325 mm wide by 500 mm deep to center of steel must carry a total factored moment M_u of 340 kN m with $f'_c = 25$ MPa and Grade 300 steel. Find the required steel.

PROB. 4.9. A slab having $d = 100$ mm must carry a factored moment M_u of 17 kN m per meter width for D.L. and 22 kN m per meter width for L.L. Find the required A_s per meter if $f'_c = 25$ MPa and steel is Grade 400. Specify bar size and spacing.

PROB. 4.10. A slab 200 mm thick with 40 mm cover to center of steel must carry its own weight and a service live load of 5 kPa over a 4.5 m simple span. Find the required A_s per meter if $f'_c = 25$ MPa and steel is Grade 300. Choose bars.

PROB. 4.11. From basic principles establish the value of the Code maximum steel ratio ρ for a rectangular beam with $f'_c = 25$ MPa and $f_y = 400$ MPa, no A'_s.

PROB. 4.12. A rectangular beam 350 mm wide by 625 mm deep overall with 75 mm cover to center of steel must carry a factored moment of 700 kN m. Find the necessary steel if $f'_c = 30$ MPa and steel is Grade 400.

PROB. 4.13. Design a double-reinforced beam 325 mm wide for a total

$M_u = 700$ kN m using $f'_c = 25$ MPa, Grade 300 steel, cover 100 mm to center of A_s, and basing the beam size on an approximate k_n of 5.5. Keep ρ-ρ' not more than $0.18\, f'_c/f_y$, as in Sec. 4.5b.

PROB. 4.14. You wish to design a double reinforced rectangular beam for a known moment (as in Sec. 4.5b) with $f'_c = 30$ MPa and Grade 400 steel. First you want to choose such a beam size that $\rho_1 = 0.18\, f'_c/f_y$ will be consistent with the use of $\rho' = 0.01$ for A'_s. Assuming $d'/d = 0.12$ and that A'_s bars yield, find the numerical k_n to use in establishing the beam size. (*Suggestion:* Base k_n on an equation for $M_{n1} + M_{n2}$.)

PROB. 4.15. If the joist of Fig. 3.22 must care for a factored moment of 34 kN m, calculate the required A_s. Take $f'_c = 35$ MPa and $f_y = 400$ MPa.

PROB. 4.16. A simple span beam with $d = 500$ mm and carrying uniform load over a 6.7 m span requires 1550 mm² of steel and uses 6-#20 bars.

(*a*) Where can the first pair be bent up? The second pair?

(*b*) If the excess steel were neglected, where would these bends be permitted?

PROB. 4.17. A 6.4 m simple span beam carries a large fixed concentrated load at 2.75 m from the left end and requires $A_s = 3000$ mm². The steel used is 5-#30, $d = 500$ mm. If the beam weight is disregarded, where can the first bar be bent up (each side of the load)? The third bar?

PROB 4.18. A 3.65 m cantilever beam carries a large concentrated load at its end, uniform load negligible. Required $A_s = 2500$ mm², 4-#30 used, $d = 475$ mm. Where can one pair of bars be stopped or bent down?

PROB. 4.19. A 2.75 m cantilever beam with $d = 400$ mm carries only uniform load and requires $A_s = 2500$ mm². If 4-#30 are used, where can half of the bars be bent down?

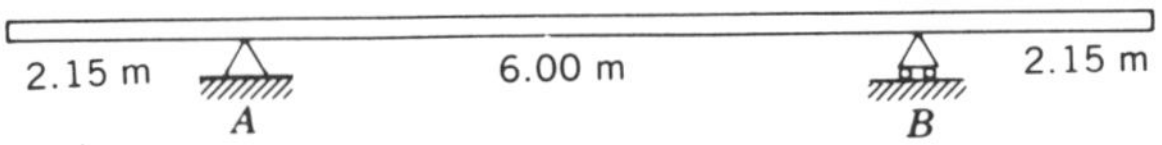

2.15 m   6.00 m   2.15 m
A   *B*

FIGURE 4.12 Simple beam with overhanging ends for Prob. 4.20

PROB. 4.20. The beam of Fig. 4.12 with $d = 450$ mm is subject to a fixed dead load of 15 kN/m and a movable live load of 30 kN/m. The required positive moment steel is 3200 mm² and the required negative moment steel is 1600 mm². If the positive moment steel used is 5-#30 where can the first pair of positive steel bars be stopped? If the negative moment steel is 4-#25 bars, where can this be reduced to 2 bars? (Note that distances on *each* side of the support or reaction are needed. See Sec. 2.12d for load factors.)

PROB. 4.21.   From basic principles, establish the maximum steel ratio ρ that the Code permits for a rectangular beam (without A'_s) when used with $f'_c = 30$ MPa and $f_y = 400$ MPa.

PROB. 4.22.   A rectangular beam 350 mm wide by 650 mm deep overall with 75 mm cover to center of steel must carry a ϕM_n of 700 kN·m. Find the necessary steel if $f'_c = 30$ MPa and $f_y = 400$ MPa.

PROB. 4.23.   A rectangular beam 330 mm wide by 500 mm to steel centroid must carry a total ϕM_n of 350 kN·m with $f'_c = 35$ MPa and $f_y = 300$ MPa. Find the required steel.

PROB. 4.24.   A wide slab with $d = 100$ mm must carry a ϕM_n for dead load of 1800 N·m per 100 mm width of strip and 2300 N·m for live load. Find the required A_s per strip for $f'_c = 35$ MPa and $f_y = 400$ MPa and then specify bar size and spacing.

PROB. 4.25.   Considering moment alone, design a rectangular beam 360 mm wide and of minimum depth (using 10 mm as the minimum step in d) to carry a service load M of 580 kN·m with an average load factor of 1.55, using C25 concrete and Grade 400 bars.

5

SHEAR IN BEAMS
AND ONE-WAY SLABS

5.1 SHEAR STRESS AND DIAGONAL TENSION

Concrete is relatively much weaker in tension than in compression, with real shear strength intermediate between the two. Most failures that would be termed shear failures are diagonal tension failures, occasionally diagonal compression failures. While long usage has established shear stress as standard nomenclature, the phenomena involved will be more clearly understood if diagonal tension is kept in mind.

Shear stresses as normally computed are, except sometimes in prestressed concrete, stress coefficients only nominally related to the actual critical stresses. Almost the only place a real shear stress is computed is in the shear friction theory (Sec. 5.20).

5.2 DIAGONAL TENSION BEFORE CRACKS FORM

In homogeneous beams diagonal stress can be analyzed by well-established relationships. Reinforced concrete beams, prior to the formation of cracks, probably have stresses quite similar to those of a homogeneous beam.* The diagonal tension stresses are emphasized here.

In the beam of Fig. 5.1a a small element at the neutral axis at A would be subject to a shearing stress v, but no bending stress. Figure 5.1b shows that such an element will develop unit diagonal tensile and compressive stresses of magnitude v. An element at B in Fig. 5.1a will have a compressive stress f_c in addition to shear, adding a diagonal compression,

---

*The notation of this section on basic strength of materials does not attempt to follow the new standard ACI notation. C and T are total forces; c, t, v, f_c, and f_t are unit stresses.

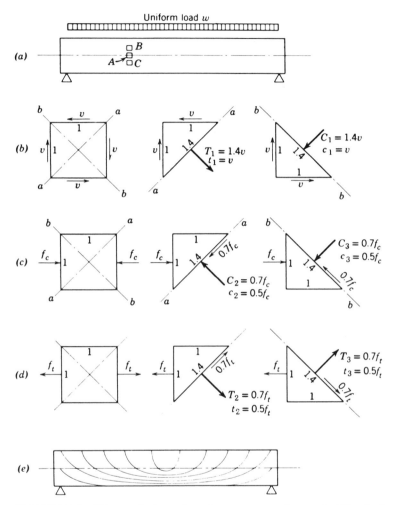

FIGURE 5.1 Diagonal stress in a homogeneous beam. (*a*) Typical beam under uniform load. (*b*) Analysis of stresses at *A*. (*c*) Analysis of added stresses at *B*. (*d*) Analysis of added stresses at *C*. (*e*) Tension stress trajectories.

Fig. 5.1*c*; combined with shear as in Fig. 5.1*b*, the diagonal tension on section *a-a* is reduced and the diagonal compression on section *b-b* is increased. Vertical loading on top of the beam causes a vertical compression that combines similarly, although rotated 90 degrees to that sketched in Fig. 5.1*c*.

Similarly, an element at *C* in Fig. 5.1*a* will carry a tension stress as well as a shear, leading to the added stresses shown in Fig. 5.1*d*. These increase diagonal tension on section *a-a* and reduce diagonal compression on section *b-b*. The combined stresses are maximum on sections other than *a-a* and *b-b*, with tension being maximum on a steeper plane than *a-a* when the (horizontal) direct stress is tension, or on a flatter plane when the (horizontal) direct stress is compression. A load hung from the bottom of the beam adds to the diagonal tension and makes it steeper.

The relation developed in mechanics for maximum diagonal tension, adjusted to the notation used here is:

$$t = f/2 + \sqrt{(f/2)^2 + v^2}$$

In this relation *t* is the unit diagonal tension, *f* is the unit direct stress, taken as positive when it is tension, and *v* is the unit shear. The direction of the maximum diagonal tension is given by the relation:

$$\tan 2\theta = 2\,v/f$$

where θ is the angle *t* makes with the stress *f*, in this case with the horizontal.

Figure 5.1*e* illustrates the approximate trajectory of maximum tensile stresses in a homogeneous rectangular beam under uniform loading. Diagonal tension cracks would be roughly perpendicular to the trajectories shown in Fig. 5.1*e* for a beam uniformly loaded. In a reinforced concrete beam the pattern will be very similar until cracks open, usually vertical cracks due to moment in the lower half of the beam. Diagonal tension cracks usually open at approximately 45° with the axis of the beam, starting typically from the top of a moment crack, but in short shear spans or deep beams starting independently near the neutral axis.

5.3 SHEAR STRENGTH—CODE FORMAT

Much research has been done since the 1950s in the area of shear strength, but there are many different load conditions to be considered. No *overall* theory has yet developed in spite of at least a fourfold increase in our knowledge about shear. Thus shear must still be handled by semiempirical rules. It is not always easy to see why one case is algebraically so different from another. There are, however, research data to back up these rules, but some simpler correlation should eventually be found. Some special cases are discussed separately in Secs. 5.18 to 5.22, and some special behavior problems not yet in the Code in Sec. 5.23.

The 1977 Code expresses all allowable shears in terms of the allowable total shear. (The 1971 Code expressed all allowable shears in terms of unit stresses.) There is almost no change in the allowables except the form of presentation. In the author's opinion calculations in one form are just about the same in length as in the other and some designers have always drawn shear reinforcement curves in terms of unit stresses, some in terms of total shear. In this presentation mor\ examples will remain in terms of stresses, but both formats will be shown and some examples will be in the new format.

Nearly all beams must (since 1971) ha\ e web reinforcement for shear, usually stirrups (Fig. 5.6), occasionally bei t bars from flexural reinforcement.

The basic Code statement on shear strength is that the factored shear force V_u shall be equal to or less than the design shear ϕV_n, that is,

$$V_u \geqslant \phi V_n \quad \text{where} \quad V_n = V_c + V_s$$

V_c = nominal shear strength provided by concrete

V_s = nominal shear strength provided by shear reinforcement (usually stirrups)

$\phi = 0.85$

The bulk of Code Chapter 11 is built around permissible values of V_c and V_s, plus definitions of critical sections.

To understand how stirrups improve strength, it is desirable first to understand how beams fail when they contain no web reinforcement such as stirrups.

5.4 BEAM BEHAVIOR UNDER SHEAR LOADING— WITHOUT STIRRUPS*

(a) Failure Types, by Appearance

A shear failure appears to be least complicated when it occurs away from loads and reaction. It is convenient to classify shear failures in terms of the distance between test load and reaction, a distance designated as shear span or a, Fig. 5.3a.

The simple *diagonal tension failure* of Fig. 5.2 occurs when the shear span is more than 3d or 4d, under which conditions there is ample room for the full crack to develop to failure.

As the shear span is reduced, the load itself begins to influence the diagonal crack more and more, in a manner that increases shear capacity. The failure sketched in Fig. 5.3a is called a *shear-compression failure* and occurs when the shear span is from d to 2.5d, with a rapidly reducing influence in the range of 2.5d to 4d, as indicated in Fig. 5.4a.

When the shear span is less than d, the pattern changes to a *splitting failure* or compression failure at the reaction at a still higher load, and a tendency toward sudden failure, as sketched in Fig. 5.3b.

---

*Stirrups are vertical U-shaped reinforcement that anchor in the compression face of the beam and enclose the flexural steel bars. See Secs. 5.5 and 5.7.

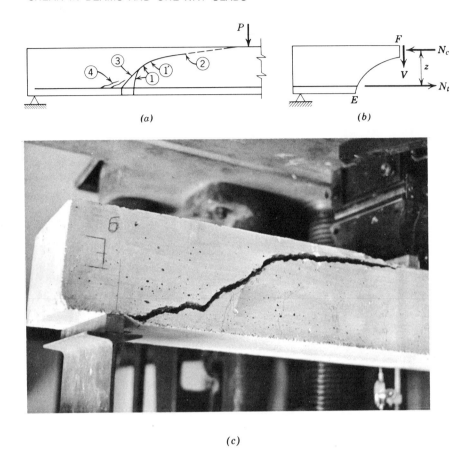

(c)

FIGURE 5.2 Development of a diagonal tension crack when loads and reactions are far apart. (a) Diagram showing sequence in crack formation. (b) Equilbrium sketch for portion of beam. (c) Failure of beam. The failure crack developed from the flexural crack faintly seen about one beam depth from the end. This crack turned gradually into the diagonal crack, as at 1 in the sketch. The final wide crack is comparable to 2–1–3–4 in (a) (The failure picture has been inverted to make this comparison easier).

Since no overall basic theory can be presented, it is essential for general understanding to note details which may some day become parts in a more logical "theory."

(b) Diagonal Tension Failure

Always in the range of a/d above 2, and sometimes at lower a/d values, the diagonal crack starts from the last flexural crack and turns gradually into a crack more and more inclined under the shear loading, as noted in Fig. 5.2a. Such a crack does not proceed immediately to failure, although in some of the longer shear spans this either seems almost to be the case or an entirely new and flatter diagonal crack suddenly causes failure. More typically, the diagonal crack encounters resistance as it moves up into the zone of compression, becomes flatter, and stops at some point such as that marked 1' in Fig. 5.2a. With further load, the tension crack extends gradually at a very flat slope until finally sudden failure occurs, possibly from point 2. Shortly before reaching the critical failure point at 2 the more inclined lower crack 3 will open back, at least to the steel level, and usually cracks marked 4 will develop. Cracks 3 and 4 are further discussed in Sec. 5.23. Figure 5.2c illustrates a failure with the start of the crack nearer the end than the usual $a/2$, with this location resulting because two tension bars were cut off at the crack point (Sec. 4.11d). This crack is called a flexure-shear crack.

(c) Shear-Compression Failure

Very often the development of the diagonal crack described above is stopped by the presence of a nearby load, as indicated in Fig. 5.3a. Then the vertical compressive stresses under the load reduce the possibility of further tension cracking, and the vertical compressive stresses over the reaction likewise limit the bond splitting and diagonal cracking along the steel. Alternatively, a large shear in short shear spans (especially in I-beams) may initiate approximately a 45-degree crack (called a web-shear crack) across the neutral axis before a flexural crack appears. Such a crack crowds the shear resistance into a smaller depth and, by thus increasing the stresses, tends to be self-propagating until stopped by the load or reaction. With either start,* a compression failure finally occurs adjacent to the load.

This type of failure has been designated as a shear-compression failure because the shaded area also carries most of the shear and the failure is caused by the combination. Such a failure can be expected to occur when the shear span a, as indicated in Fig. 5.3a, is less than four times the beam depth, or possibly a little less for lightweight concrete or very high strength concrete. When the shear span is small, the increased shear strength may be significant, with the ultimate shear over twice as much for $a = 1.5\,d$ as for $a = 3.0\,d$ (Fig. 5.4). The width of the critical crack, if there is no crack control steel, becomes large as the load increases, sometimes over 3 mm.

Occasionally with inadequate anchorage of the flexural steel beyond the

---

*The second case is not always classified as shear-compression; the term web-shear crack is often used.

crack, the small diagonal cracks (raveling cracks) will split out the bars before the compression failure occurs—this is called a shear-tension failure. The beam here acts somewhat as a tied arch with a thrust line from the load toward the reaction area, which demands full anchorage of the bars beyond the crack. Stirrups, now required on most beams, almost totally eliminate the raveling.

An overemphasis shows here concerning how flexural compression in the beam chokes off the diagonal tension failure and adds strength in shear (Fig. 5.4, especially at a/d of 1.5 or 2). What happens near a point of inflection (P.I.) where moment is too small to crack the beam? We ran some tests trying to make beams fail in shear near the P.I., but the only shear failure at the P.I. was when we had added shear reinforcement

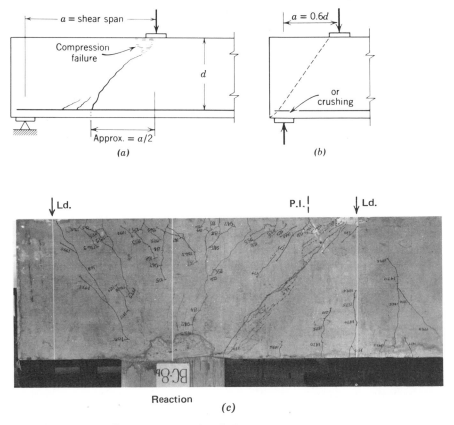

FIGURE 5.3 (*a*) Shear-compression failure when shear span is small. Shear strength is increased. (*b*) Shear span less than *d*. (*c*) Failure in short shear span of semicontinuous beam, from load to reaction.

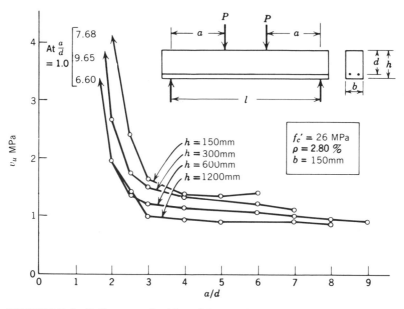

FIGURE 5.4 Influence of a/d and beam depth on shear resistance, without stirrups. (Adapted from References 4 and 5, ACI.)

(stirrups as in Sec. 5.5) everywhere except around the P.I. This failure was at substantially *higher* unit shear stress. It is safe to say that in non-prestressed beams shear weakness develops from the weakness already introduced by a moment crack and this combination makes it necessary to hold shear stresses low or to add stirrups.

(d) Splitting or True Shear Failure

Where the shear span is less than the effective depth d, the shear is carried as an inclined thrust between load and reaction which almost eliminates ordinary diagonal tension concepts. Shear strength is much higher in such cases.[3] The final failure, as in Fig. 5.3b, becomes a splitting failure, almost like the vertical splitting of a compression test cylinder (which occurs when end friction is reduced) or it may fail in compression at the reaction. The analysis of such an end section is closely related to the analysis of a deep beam having a span of $2a$. Reference 3 has shown that it is the clear shear span between bearing and loading plates that is critical.

(e) Code Shear Strength V_c

Because of the objectionable amount of cracking for a member like that in Fig. 5.2, the Code defines the nominal shear strength V_c on the basis of the

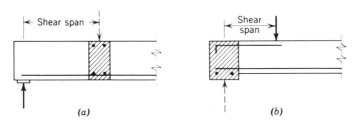

FIGURE 5.5 Cases where small shear span does not significantly increase shear strength. (*a*) Girder loaded by shear from beam. (*b*) Beam supported by girder.

shear when the diagonal crack first occurs. In many cases just when a crack becomes diagonal is a little vague, as in Fig. 5.2; however, it is not at all vague in Fig. 5.3.

(f) Design Considerations Relating to Shear Span

As a result of the shear span effect, the worst position (on the basis of diagonal tension failure) for a concentrated load on the usual simple span test beam is not adjacent to the reaction but at some distance out on the span. The Code uses the distance d from the face of support.

In test beams most of the extra resistance beyond initial diagonal cracking is created by the vertical compressive stress under the load and over the reaction. Loads or reaction applied as shears, as in Fig. 5.5, create very little vertical compression and add very little shear resistance. If loads are applied below mid-depth of a girder, say, from precast beams bearing on a lower flange or bracket attached near the bottom of the girder, diagonal cracking occurs at lower loads and ultimate strength is reduced. Stirrups (Fig. 5.6) as hangers to pick up such loads are essential in such cases.

Tests have also shown that if the loads at a small a/d are applied as shear loads on the side of the beam, as in Fig. 5.5*a*, or if the reaction is

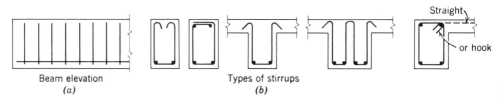

FIGURE 5.6   Vertical stirrups.

picked up in shear (as when a beam frames into the side of a girder, as in Fig. 5.5b), not much increase in shear resistance is obtained as a result of the small shear spans.[15]

5.5 VERTICAL STIRRUPS—INFLUENCE ON BEHAVIOR

To understand how stirrups improve beam behavior and strength, consider the stirrup simply as vertical reinforcement spaced along the length of the beam at not more than 0.5 d on centers, well anchored in the compression zone of the beam and usually bent around the longitudinal tension bars, as shown in Fig. 5.6. In Fig. 5.2 several stirrups, if any had been provided, would have crossed the failure crack.

Prior to concrete cracking, the vertical stirrup carries essentially no stress, possibly even a little compression arising from vertical shrinkage of the concrete. However, the stirrup must go into tension as the diagonal crack crosses it; and this tension controls and limits the progress of the crack, delaying the failure of the beam until higher loads are imposed. The next stirrup (toward the reaction) would also hold the flexural steel in place and not allow the bars to be pushed downward by so-called dowel action creating the "raveling out" mentioned as cracks 3 and 4 in Fig. 5.2a. Until the stirrup crossing the crack yields (under higher load), the beam cannot fail in shear.

The Code defines V_s as the *increase* in total shear capacity above the V_c at first diagonal cracking. The Code equations are given in Sec. 5.6 and practical use in simple situations does not require detailed knowledge of the origin of these equations. Section 5.7 discusses minimum stirrups and maximum stirrup spacings, Sec. 5.8 some Code limits on size of bars used for stirrups, Sec. 5.9 gives some simple examples, and Secs. 5.10 and 5.12 give more detailed background.

5.6 CODE EVALUATION OF V_c AND V_s FOR BEAM CONTAINING VERTICAL STIRRUPS

(a) Normal Weight Concrete

With or without stirrups, the nominal shear strength provided by the concrete is V_c:

$$(V_c = \tfrac{1}{6}\sqrt{f'_c}b_w d) \qquad \text{(Code Eq. 11.3)}$$

or

$$(V_c = \tfrac{1}{6}(\sqrt{f'_c} + 100\,\rho_w V_u d/M_u)bd \leqslant 0.29\sqrt{f'_c}b_w d) \qquad \text{(Code Eq. 11.6)}$$

with $V_u d/M_u$ taken not greater than 1.0;

where b_w = web width

d = effective depth, as for M_n

$\rho_w = A_s/b_w d$ = ratio of reinforcement

V_u = factored shear force at section

M_u = factored moment at section

The designer may use *either* equation* and will soon note that only a few situations give large differences between them.

Attention is called first to the coefficient $\frac{1}{6}$ in Code Eq. 11.3. It represents a judgment decision of code engineers having much detailed knowledge of research in this area. It would probably be lower if the reinforcement ratio A_s/bd were normally as low as 0.005 or if the Code did not require nominal stirrups wherever V_n was greater than half V_c (Sec. 5.7).

The 1971 Code allowed a unit stress $v_c = \frac{1}{6}\sqrt{f'_c}$ or $v_c = \frac{1}{6}(\sqrt{f'_c} + 100\,\rho_w V_u d/M_u)$, again the designers choice. It then stated the total unit stress v_u (taken by concrete and stirrups) = $V_u/(\phi b_w d)$, with $\phi = 0.85$. This differs from the 1977 Code chiefly in form, but also in associating v_u with the factored V_u/ϕ. The 1977 Code, in comparison, eliminates v_u and substitutes the *nominal* shear force $V_n = V_c + V_s$ and requires that $\phi V_n = 0.85\,V_n \geqslant V_u$. In this book the unit stress equivalent will often be used as $v_n = v_c + v_s$. The needed $v_s = v_n - v_c$ then leads to the choice of stirrups from $V_s = v_s b_w d$ and the equation in the next paragraph.

The nominal shear strength added from vertical stirrups is given in Code 11.5.6:

$$V_s = A_v f_y d/s \qquad \text{(Code Eq. 11-17)}$$

where A_v is the shear reinforcement (two bar areas for U-stirrups) and s is the spacing between stirrups in a direction parallel to the axis of the number.

(b) Lightweight Aggregate Concretes

Some lightweight aggregate concretes have a lower shear resistance than ordinary aggregate concrete of the same compressive strength. The Code (11.2) provides alternate procedures: (1) allowable stress multiplied by an arbitrary 0.75 factor for "all-lightweight" and 0.83 for "sand-lightweight" concrete,† or (2) when f_{ct} is suitably specified the use of $1.8\,f_{ct} \gtrless \sqrt{f'_c}$ instead of $\sqrt{f'_c}$ in stating the allowable. The term f_{ct} is the average splitting tensile strength.

---

*The special related problem in Sec. 5.23c should interest the designer or advanced student.
†"Sand-lightweight" is concrete using natural sand for fine aggregate.

5.7 CODE REQUIREMENTS FOR MINIMUM STIRRUPS AND MAXIMUM SPACINGS

Code 11.5.5 requires at least minimum stirrups where the total factored shear force V_u exceeds 1/2 the shear strength provided by the concrete ϕV_c except in:

(a) Slabs and footings.
(b) Concrete joist construction as defined in Code 8.11.
(c) Beams with total depth not more than 250 mm, 2.5 times the flange thickness, or half the web width.

   Where torsion is small (Code 11.5.5.3) the minimum area of shear reinforcement A_v at spacing s is

$$A_v = \tfrac{1}{3} b_w s / f_y \qquad \text{(Code Eq. 11-14)}$$

for b_w and s in millimeters, equivalent to stirrups for $v_s = 0.33\,\text{MPa}^*$. Maximum spacing (Code 11.5.4) of vertical stirrups is $d/2$ in nonprestressed members, but is decreased to $d/4$ where $v_s > \tfrac{1}{3}\sqrt{f_c'}$, that is, where $V_s > \tfrac{1}{3}\sqrt{f_c'} b_w d$. The maximum v_s is limited to $\tfrac{2}{3}\sqrt{f_c'}$.
   The Canadian Code for A_v in mm^2 shows minimum $A_v = 0.35\, b_w s / f_y$.

5.8 CODE LIMITATIONS ON SIZES OF BARS USED FOR STIRRUPS

Although the general treatment of the development lengths necessary to bond the bar and concrete together is deferred to Chapter 7, a limited discussion before other aspects of stirrup capacity are applied to design will be helpful here. The Code requires a development length between point of zero bar stress and a point of f_y stress which increases directly as the product $d_b f_y$† increases. Thus, it requires a development length ℓ_d 15/10 times as long as to develop a #15 bar to a given f_y stress as it does for a #10 bar.

   Code 12.14 deals with development of web reinforcement (stirrups) and only a few subsections are needed here. The stirrup's maximum stress is considered to exist at $\tfrac{1}{2} d$ from the beam compression face. The Code reads:

---

*The author has found it convenient to measure the value of stirrups in terms of their contribution to the unit shear capacity, that is, $v_s = \rho_n f_y$, where $\rho_n = A_v/(b_w s)$. The minimum stirrup must provide $v_s = \rho_n f_y = \tfrac{1}{3}\,\text{MPa}$.

†Code 12.2.2 gives two basic ℓ_d requirements for #10 to #35 bars: (1) $\ell_d = \tfrac{1}{3}\,\text{MPa}\,A_b f_y/\sqrt{f_c'}$, (2) $\ell_d = 0.058\, d_b f_y$. For #10 to #15 bars the second is the larger and governs. There are a number of possible multipliers, for lightweight concrete and so on.

12.14.1—Web reinforcement shall be carried as close to compression and tension surfaces of member as cover requirements and proximity of other reinforcement will permit.

12.14.2—Ends of single leg, simple U-, or multiple U-stirrups shall be anchored by one of the following means:

12.14.2.1—A standard hook plus an embedment of $\frac{1}{2}\ell_d$. The $\frac{1}{2}\ell_d$ embedment of a stirrup leg shall be taken as the distance between middepth of member $\frac{1}{2}d$ and start of hook (point of tangency). (Note: A minimum ℓ_d of 300 mm is not required here, Code 12.2.5.)

12.14.2.2—Embedment $\frac{1}{2}d$ above or below middepth on compression side of the member for a full development length ℓ_d but not less than 24 d_b; or for deformed bars or deformed wire, 300 mm.

12.14.2.3—For #15 bar of D31 wire, or smaller, bending around longitudinal reinforcement at least 135 deg. plus, for stirrups with design stress exceeding 300 MPa, an embedment of $\frac{1}{3}\ell_d$. The $\frac{1}{3}\ell_d$ embedment of a stirrup leg shall be taken as the distance between middepth of member $\frac{1}{2}d$ and start of hook (point of tangency).

12.14.2.4—(omitted here—for welded smooth wire fabric)

12.14.3—Between anchored ends, each bend in the continuous portion of a simple U-stirrup or multiple U-stirrup shall enclose a longitudinal bar.

12.14.4 and 12.14.5—(not reprinted here) cover bent bars acting as web reinforcement and lapping of pairs of U-stirrups to form a closed unit, respectively.

The following ℓ_d values are minimum stirrup values without a hook, in any usual grade of concrete.

Bar	Grade 300	Grade 400	Lightweight
#10	175 mm	235 mm	Multiply by
#15	260 mm	350 mm	factor > 1.0
			See Code 12.2.3c.

If bent at least 135° around a longitudinal bar (close to compression face), the additional development between start of hook and middepth $\frac{1}{2}d$ is:

Bar	Grade 300	Grade 400
#10	0	80 mm
#15	0	120 mm

5.9 SHEAR EXAMPLES IN STATICALLY DETERMINED MEMBERS

(a) The beam of Fig. 5.7 will be checked for shear assuming $f_c' = 25$ MPa and Grade 300 steel.

Solution

Beam mass = $0.325 \text{ m} \times 0.250 \text{ m} \times 2\,500 \text{ kg/m}^3 = 203 \text{ kg/m}$.
Beam weight (multiply by g, acceleration due to gravity) = $203 \times 9.8 = 1.99 \text{ kN/m}$.
Alternatively, beam weight = $0.325 \text{ m} \times 0.250 \text{ m} \times 24 \text{ kN/m}^3 = 1.95 \text{ kN/m}^3$

$$d = 325 - 40 \text{ (cover)} - 10 \text{ (stirrup)} - 10 \left(\tfrac{1}{2}\, d_b\right) = 265 \text{ mm}$$

$$w_u = 1.4\, w_d + 1.7\, w_\ell = 1.4 \times 1.99 + 1.7 \times 15 = 28.29 \text{ kN/m}$$

At d from support, $V_u = 28.29\,(1.8 - 0.265) = 43.42 \text{ kN}$

$$\text{Required } V_n = V_u/\phi = 43.42/0.85 = 51.08 \text{ kN}$$

$$\text{Available } V_c = \tfrac{1}{6}\sqrt{f_c'}\,b_w d = \tfrac{1}{6}\sqrt{25} \times 0.25 \times 0.265 = 0.0552 \text{ MN} = 55.2 \text{ kN}.$$

This V_c should carry this shear without stirrups, but Code 11.5.5 requires minimum stirrups wherever $V_n > \tfrac{1}{2}\,V_c = \tfrac{1}{2} \times 55.2 = 27.6 \text{ kN}$. Find point where this particular shear is reached.

$$\text{Slope of } V_n \text{ curve } 51.08/1.535 = 33.28 \text{ kN/m}$$

x from free end = $27.6/33.28 = 0.829 \text{ m}$, or $1.8 - 0.829 = 0.971 \text{ m}$.

The smallest stirrup bar is #10 ($A_v = 2 \times 100 = 200 \text{ mm}^2$) and maximum spacing is $\tfrac{1}{2}\,d = 133 \text{ mm}$, say 125 mm.

$$\text{Min. } A_v = \tfrac{1}{3}\, b_w s/f_y = \tfrac{1}{3} \times 250 \times 150/300 = 42 \text{ mm}^2 \ll 200 \text{ mm}^2$$

Each stirrup should be at the center of the length s that it serves. Start the first stirrup at 50 mm from the support (roughly $\tfrac{1}{2}\,s$) and use a total of 8 stirrups as shown in Fig. 5.7c. In a design this would usually be stated:

USE 8-#10 U (spaced) at 50 mm, 7 at 125 mm (caring for 997 mm)

The length of stirrups above middepth ($\tfrac{1}{2}\,d$) must be checked for development as

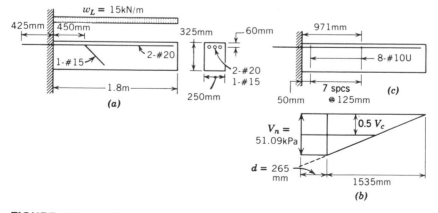

FIGURE 5.7 (a) Beam of Sec. 5.7a. (b) Nominal shear diagram. (c) Stirrups chosen.

discussed briefly in Sec. 5.8. For #15 or smaller stirrups Code 12.14.2.3 says bending around a longitudinal bar may be taken as developing an f_y of 300 MPa (or for a greater f_y an added length $\frac{1}{3} \ell_d$ from middepth to start of bend). The #10 stirrup meets this test. Even if it failed such a test, it could be argued that the design showed only a small portion of the #10 bar area was needed (42/200) and this would justify a decrease in the ℓ_d required in about the same ratio.

If v_c had proved critical and $\frac{1}{6}\sqrt{f_c'}$ a little deficient, the more exact equation for allowable v_c could have been used to a small advantage, as follows:

Permissible $v_c = \frac{1}{6}(\sqrt{f_c'} + 100\ \rho_w\ V_u d/M_u)$ $A_s = 700\ \text{mm}^2$ for 2-#20 + 1-#15 bars

$\rho_w = 700/250 \times 265 = 0.01057$ $M_u = 28.29 \times 10^{-3} \times \frac{1}{2} \times 1.535^2 = 33.33\ \text{kN.m}$

Allowable $v_c = \frac{1}{6}(\sqrt{25} + 100 \times 0.01057 \times 43.43 \times 0.265/33.33) = 0.894\ \text{MPa}$.

(b) For comparison, show the above calculations using the allowable shear stress formulation of the 1971 Code. (The ℓ_d calculation was relaxed in the 1977 Code 12.2.5 and should be used instead of the 1971 requirements.)

Solution

Beam weight = 199 kN/m, as before. $w_u = 1.4 \times 1.99 + 1.7 \times 15 = 28.29\ \text{kN/m}$

At d from support, $V_u = 28.29(1.8 - 0.625) = 33.24\ \text{kN}$

$V_n = V_u/\phi = 33.24/0.85 = 39.10\ \text{kN}$

$v_n = V_n/b_w d = 39.10/0.250 \times 0.265 = 590\ \text{kPa} = 0.77\ \text{MPa} < \frac{1}{6}\sqrt{f_c'} = \frac{1}{6}\sqrt{25} = 0.833\ \text{MPa}$

$v_n <$ allowable v_c but Code requires at least minimum stirrups (able to carry 0.33 MPa as v_s) to a distance x from support, where v_u drops to $\frac{1}{2} \times \frac{1}{6}\sqrt{f_c'} = 0.42\ \text{MPa}$

$x = d + (1.8 - 0.265) \times (0.77 - 0.42)/0.77 = 0.265 + 0.698 = 0.963\ \text{m}$

Try minimum #10 U at $\frac{1}{2} d = \frac{1}{2} \times 265 = 133\ \text{mm}$, say 125 mm

$v_s = A_v f_y/b_w s = 200 \times 300/250 \times 125 = 1.92\ \text{MPa} \geqslant$ min. 0.33 MPa

Say O.K. with #10 U at 125 mm

USE 8 – #10 U (spaced) at 50 mm, 7 at 125 mm

The comments on ℓ_d limitations would be unchanged.

(c) Determine the stirrups for the beam of Fig. 5.8a, using Grade 300 steel and $f_c' = 30\ \text{MPa}$.

Solution (using allowable stress format)

Critical spacing is to be calculated at $d = 0.475\ \text{m}$ from support, 1.325 m from midspan.

Beam weight = $0.225 \times 0.550 \times 2500 \times 9.8 = 3032\ \text{N/m} = 3.032\ \text{kN/m}$

D.L. $V_u = 1.4(30 + 3.032 \times 1.325) = 48\ \text{kN}$

$V_u = 48 + 1.7 \times 65 = 159\ \text{kN}$

$V_n = 159/0.85 = 187\ \text{kN} = 0.187\ \text{MN}$

$v_n = 0.187/0.225 \times 0.475 = 1.75\ \text{MPa} > v_c = \frac{1}{6}\sqrt{30} = 0.91\ \text{MPa}$ ∴ Stirrups

Max. allowable $= 0.91 + \frac{2}{3}\sqrt{30} = 4.56$ MPa   **O.K.** (Code 11.5.6.8)

Just to the left of the load, $V_{u3} = 1.4(30 + 3.032 \times 0.9) + 1.7 \times 65 = 156$ kN

$$V_n = 156/0.85 = 1.84 \text{ kN}, \quad v_3 = 0.184/0.225 \times 0.475 = 1.72 \text{ MPa}$$

The v_n and v_s diagrams are shown superimposed on Fig. 5.8b, with the v_s area shaded. Stirrups must be designed for $v_s = v_n - \frac{1}{6}\sqrt{f_c'}$ over the 0.9 m end length and minimum stirrups must be used between the loads unless $v_n \gtrless \frac{1}{2} v_c$. The variation in v_s is so small that *all* stirrups might well be designed for $v_s = 5.91 - 0.91 = 0.50$ MPa. Code minimum is 0.33 MPa stirrup capacity. Minimum stirrup size is #10. $v_s < \frac{1}{3}\sqrt{f_c'} = 1.83$ MPa (Code 11.5.4.3). Maximum $s = \frac{1}{2} d = 240$ mm.

Try #10 U stirrups, $A_v = 2 \times 100 = 200 \text{ mm}^2 = 0.2 \times 10^{-3} \text{ m}^2$

$$s_0 = 0.2 \times 10^{-3} \times 300/0.5 \times 0.225 = 0.533 \text{ m} > 240 \text{ mm max.}$$

USE #10 U-stirrups.

Sketch (Fig. 5.8c) shows $\frac{1}{2}\ell_d$ (from table in Sec. 5.8) $= \frac{1}{2} \times 175 = 88$ mm required below

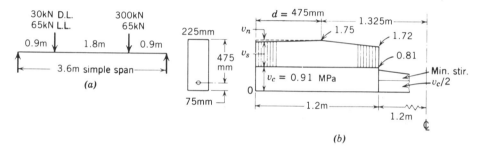

(b)

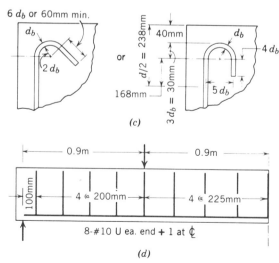

FIGURE 5.8   Beam of Sec. 5.8c.

hook in upper $\frac{1}{2}d$. Stirrup development O.K. Next check beyond the load, the live load to the left being omitted to get maximum shear.

$$V_u = 1.4(30 + 3.032 \times 0.9) + 1.7 \times 65 \times 0.9/3.6 = 73.5 \text{ kN}$$

$V_n = 73.5/0.85 = 86.4 \text{ kN}, \ v_{n3^+} = 0.0864/0.225 \times 0.475 = 0.81 \text{ MPa} < \frac{1}{2} v_c = 0.46 \text{ MPa}$

Because $v_n > v_c/2$, use minimum stirrups across the middle length. By inspection #10 at 2.25 mm will care for more than the minimum 0.33 MPa required by Code 11.5.5.3, Code Eq. 11-4.

A formal stirrup spacing curve could be prepared as in Fig. 5.17, but it is totally unnecessary for this short length and this simple pattern of v_s values. The s_0 value calculated at the critical section must be used back to the support and the small change in v_s to $x = 0.9$ m makes further calculations of s totally unnecessary.

USE $8 - $ #10 U at 100 mm, 4 at 200 mm, 4 at 225 mm at each end with the last one at midspan, total of 17 stirrups. This arrangement is sketched in Fig. 5.8d.

Figure 5.8c could have been plotted with allowable V_c (unshaded) and V_s (shaded) just about as easily as with v_c and v_s. The author likes to use allowable v_c, which is a familiar number, rather than V_c which changes as the beam size changes; however, this is not very important.

5.10 SPACING RELATIONS AND COMMENTS ON USE OF VERTICAL STIRRUPS

The special case of vertical stirrups is presented first because they are almost universally used in the U.S.A. The beam of Fig. 5.9a shows a stirrup intercepting a typical 45° diagonal tension crack. The stirrup is

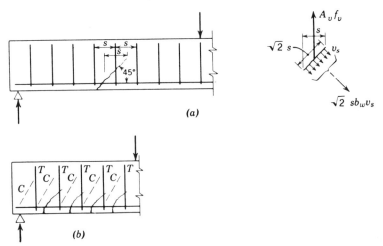

FIGURE 5.9   Basis for calculation of vertical stirrups. (a) From stresses. (b) Truss analogy.

assumed to carry the vertical component of the diagonal tensile stresses originally acting across the crack over the horizontal length s. The horizontal component was automatically included in the design of the longitudinal A_s. The unit tension on the crack is taken as v_s and the area contributing stress to the stirrup is $\sqrt{2}\, sb_w$, where b_w is the web thickness. The vertical component of concrete stress is $(1/\sqrt{2})(\sqrt{2}\, sb_wv_s) = sb_wv_s$, which is equal to the vertical components of stirrup stress $= A_vf_y$.

$$sb_wv_s = A_vf_y \quad s = \frac{A_vf_y}{b_wv_s} \tag{5-1}$$

The area of the stirrup A_v includes two bar areas for a U-type stirrup. As already mentioned in Sec. 5.5, v_s is the *increase* in total shear resistance, since tests have shown some shear is still carried by the concrete, partially by concrete above the crack, partially by aggregate interlock along the crack, partially by dowel action of longitudinal bars across the crack into the adjacent concrete.

A common analogy considers the stirrups as tension truss members, with a longitudinal compression chord on one face of the beam and a tension chord formed by the longitudinal tension reinforcement. Figure 5.9*b* shows dotted diagonal compression members between cracks, marked C and forming a statically determinate truss; or C could be flatter (from the top of one stirrup to the tension steel level two spaces away). This latter concept would give two diagonal members in every space and thus a statically indeterminate truss. For vertical stirrups the author prefers Eq. 5.1 above, although the truss analogy does give an excellent overall picture of the member behavior.

European practice is moving toward fewer stirrups as a result of low measured stirrup stresses in test beams.[12] They reason that (1) the compression members (concrete) act at flatter angles than 45° and thus deliver a smaller component to the stirrup verticals, and (2) the upper chord of the "truss" forms as an inclined thrust line that also leads to lower stress in the verticals. The second reason is similar to ACI's reasoning but is more specific as to *how* the concrete carries part of the shear.

No stirrups are required even in major beams until v_n exceeds $0.5\,v_c$, as noted in Sec. 5.7, although some consider minimum stirrups through all beams good practice. In major beams minimum stirrups are required wherever v_n lies between $\frac{1}{2}v_c$ and v_c. In any member stirrups are also necessary for v_s wherever there is a $v_s = v_n - v_c$.

The 1977 Code stirrup spacing formula is written in terms of V_s where $V_s = v_sb_wd$. The substitution of V_s/b_wd for v_s in Eq. 5.1 for s gives

$$s = A_vf_yd/V_s \quad \text{or} \quad V_s = A_vf_yd/s \quad \text{(Code Eq. 11-17)}$$

Stirrup spacing for stress varies inversely with v_s or V_s, with the stirrup

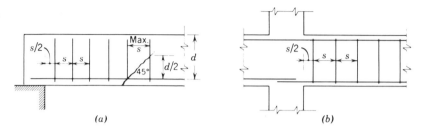

FIGURE 5.10   Arrangement of vertical stirrups.

located in the middle of the length it serves. The first stirrup should thus be placed at $s/2$ from the support as in Fig. 5.10.

The maximum spacings listed in Sec. 5.7 insure that every 45° line representing a potential crack will be crossed in the tension side of the beam by at least one stirrup, as shown in Fig. 5.10a, by two stirrups where v_s is high. In many beams #10 U-stirrups spaced at $\frac{1}{2}d$ will automatically provide more than V_s calls for. Where deformed wire is available this may prove economical in some cases.

One vital function of stirrups for which vertical stirrups are ideal is that of picking up any shear carried by dowel action of the longitudinal steel across a diagonal tension crack. They prevent the raveling out just above the steel at the end of a diagonal crack by transferring the dowel action load into a stirrup stress instead of into a vertical tension stress on the concrete.

Vertical stirrups are adaptable to shear forces acting either upward or downward since, in terms of the truss analogy, only the diagonal crack and the compression diagonal must take a different direction.

5.11 SPACING RELATIONS AND COMMENTS ON USE OF· INCLINED STIRRUPS

(a) Inclined Stirrups at 45°

For the 45° bent bars most frequently used, the spacing can be derived much as for vertical stirrups, that is, for the vertical component of the bent bar stress to resist the vertical component of the diagonal tensile stresses which existed where the crack forms. The forces are shown in Fig. 5.11a.

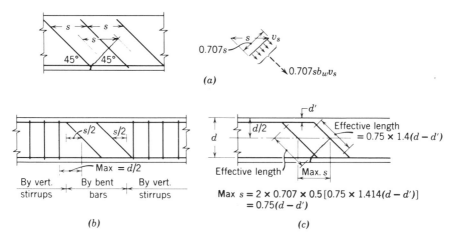

FIGURE 5.11 Inclined stirrups at 45°. (*a*) Basis for calculation. (*b*) Arrangement in combination with vertical stirrups. (*c*) Maximum spacing to intercept cracks.

$$\text{Vertical component of concrete stress} = \tfrac{1}{\sqrt{2}}(v_s b_w s \times \tfrac{1}{\sqrt{2}}) = \tfrac{1}{2} v_s b_w s$$

$$\text{Vertical component of bar stress} = \tfrac{1}{\sqrt{2}} A_v f_y$$

$$\tfrac{1}{2} v_s b_w s = \tfrac{1}{\sqrt{2}} A_v f_y$$

$$s = \frac{1.414\, A_v f_y}{v_s b_w} = \frac{1.414\, A_v f_y d}{V_s}$$

Diagonal or inclined stirrups (Fig. 5.12*a*) are aligned more nearly with the principal tension stresses in the beam. They share in carrying this tension at all stages of loading and slightly delay the formation of diagonal tension cracks. Except where the shear force may reverse direction, inclined stirrups would be the preferred type where preassembled cages of steel are used, as in some present construction; but actually they are little used in this country. Where shears reverse, as in earthquake construction, the inclined stirrup would be parallel to the diagonal crack that forms for one case.

The truss analogy is applicable with inclined stirrups acting as tension diagonals that alternate with concrete compression diagonals, as in the Warren type of truss (Fig. 5.12*b*).

Longitudinal bars bent up where they are no longer needed for moment act also as inclined stirrups. Many designs use them, but few designers calculate their value as stirrups. One reason is that usually only a few bars are bent and these may not be convenient for use as web reinforcement. Another reason is that when only a single bar is bent up or when all bent bars are bent at the same point, the value for diagonal tension is

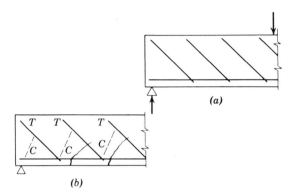

(a)

(b)

FIGURE 5.12 Diagonal stirrups.

discounted to Code Eq. 11.19:

$$V_s = A_v f_y \sin \alpha d / s \lesssim \tfrac{1}{4}\sqrt{f'_c} b_w d$$

This is equivalent to saying that the vertical component of the bar stress carries the shear V_s. Such a bar is considered effective over the center three-fourths of its inclined length (Fig. 5.11c), making maximum $s = 0.75(d - d')$ or half this value where V_s exceeds $\tfrac{1}{3}\sqrt{f'_c} bd$.

(b) Inclined Stirrups at More or Less Than 45 Degrees

The general spacing formula is derived on a similar basis for the more general geometry of Fig. 5.13. The tension resisted acts over the distance y.

$$y = \frac{s \sin \alpha}{\sin (135 - \alpha)} = \frac{s \sin \alpha}{\sin (45° + \alpha)} = \frac{s \sin \alpha}{\sin 45 \cos \alpha + \cos 45 \sin \alpha}$$

$$= \frac{s \sin \alpha}{0.707(\cos \alpha + \sin \alpha)}$$

Vertical component of stress $= 0.707\, v_s b_w y = \dfrac{v_s b_w s \sin \alpha}{\cos \alpha + \sin \alpha}$

Vertical component of bar stress $= A_v f_y \sin \alpha$

$$\frac{v_s b_w s \sin \alpha}{\cos \alpha + \sin \alpha} = A_v f_y \sin \alpha$$

$$s = \frac{A_v f_y(\cos \alpha + \sin \alpha)}{v_s b_w} = \frac{A_v f_y d(\cos \alpha + \sin \alpha)}{V_s}$$

This formula yields the special relations above for vertical and 45° stirrups

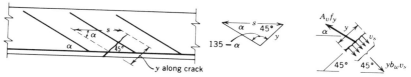

FIGURE 5.13 General case of inclined stirrups.

if $\alpha = 90°$ and $\alpha = 45°$, respectively. The Code limits the use of this relation to a series of bars bent up at different points.

Bent-up bars are normally bent at a 45° angle, but in some members, usually light joists, a flatter angle is used to secure some effect as diagonal tension reinforcement over a greater length.

5.12 DESIGN OF VERTICAL STIRRUPS

The practical problem of designing stirrups requires:

1. Determination of maximum shears and the length over which stirrups are needed.
2. Choice of desirable size of stirrup bar.
3. Selection of a series of practical spacings.

The maximum shear diagram may involve partial as well as full span loads.

Stirrup bar sizes are almost never mixed in a given beam. Hence, with uniform loads, close spacings are required at points of maximum shear while at points of lesser shear the spacing may be limited by the maximum spacing, that is, the interception of all potential cracks. The stirrup size must be large enough to give a minimum spacing adequate to pass the aggregate readily and, for practical reasons, rarely as small as 50 mm, usually 75 mm or more. Larger spacings are more economical if they do not involve too many spaces fixed by crack interception instead of by stress capacity. Also, larger spacings may require stirrup bar sizes that cannot meet the development requirements of Code 12.14, as quoted in Sec. 5.8.

For a member with a constant shear the stirrup spacing would be constant and preferably near the maximum permissible. When the shear varies, as is more usual, a sound general method is to calculate the required stirrup spacing at enough points to establish a stirrup spacing curve, including thereon the specification limits on maximum spacing. From this curve the practical spacings can be worked out. A detailed design of stirrups for a continuous T-beam span is given in Secs. 5.13 and 10.17.

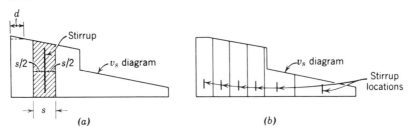

FIGURE 5.14   Stirrup spacing related to area of v_s diagram.

The theoretical number of stirrups required for stress is often useful. For any v_s diagram (not total v_u diagram), such as Fig. 5.14a, the area under the diagram is a direct measure of the number of stirrups theoretically required. Since $s = A_v f_y \div b_w v_s$

$$A_v f_y = s b_w v_s = b_w \text{ (area of } v_s \text{ diagram for length } s)$$

Each stirrup thus cares for an equal area under the v_s diagram and the total number of stirrups needed is

$$n = \frac{b_w(\text{total area of } v_s \text{ diagram})}{A_v f_y}$$

This is a theoretical number. Spacings used will be stated in practical units, usually full inches (except for small dimensions, say, under 125 mm), and will thus average less than the theoretical spacings. There also will usually be some spaces kept smaller than the theoretical in order to intercept potential cracks. Extra stirrups will always be required beyond the theoretical in the length where $v_c > v_n > 0.5\, v_c$. The practical number of stirrups in a carefully designed section will usually be from three to six more than the theoretical number.

5.13 STIRRUP DESIGN EXAMPLE—CONTINUOUS T-BEAM

A continuous T-beam with $b_w = 250$ mm, $d = 415$ mm, $f'_c = 20$ MPa, Grade 400 steel must provide for the unit shears shown in Fig. 5.15. (Similar shears are calculated as in Sec. 10.17.) Design and space the necessary stirrups.

Solution

The critical v_s is at a distance d from the face of support where v_s is $3.00 - 0.75 = 2.25$ MPa. Stirrups are needed for a distance ℓ_v, out to the point where $v_c = \frac{1}{6}\sqrt{f'_c} = 0.75$ MPa, the nominal allowable without stirrups. However, minimum stirrups are

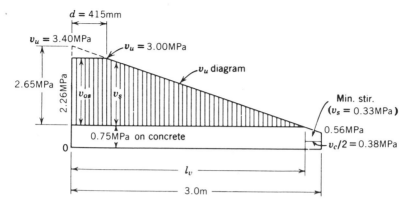

FIGURE 5.15 v_s diagram for beam similar to Sec. 10.17.

also required where $v_u > \frac{1}{2} v_c$, which means to midspan in this case.

$$\frac{\ell_v}{3.0} = \frac{3.40 - 0.75}{3.40 - 0.56}, \quad \ell_v = 2.80 \text{ m}$$

Stirrups must be designed for $v_s = v_u - 0.75$, as indicated by the shaded area in Fig. 5.15 and then be extended at maximum spacing to midspan.

For vertical stirrups, the maximum spacing permitted by Code 11.5.4.1 is $\frac{1}{2} d = 208$ mm for zones where v_s is $\frac{1}{3}\sqrt{f_c'} = 1.50$ MPa or less, which is true only at 1.2 m or more from the support. For this 1.2 m Code 11.5.4.3 limits spacing to $\frac{1}{4} d = 104$ mm.

For #10 U-stirrups,

$$s_o = 2 \times 100 \times 400/2.26 \times 250 = 142 \text{ mm}.$$

The #10 stirrups must be investigated for available development length in the upper half depth of $\frac{1}{2} \times 415 = 208$ mm. The required $\frac{1}{2} \ell_d$ between middepth and start of hook (Code 12.14.2.1) is $\frac{1}{2} \times 0.058 \, d_b f_y = \frac{1}{2} \times 0.058 \times 10 \times 400 = 116$ mm,* A sketch similar to Fig. 5.8c shows available distance is $\frac{1}{2} d - 40$ mm cover $- 3 \, d_b$ for hook $= 208 - 40 - 3 \times 10 = 138$ mm > 116 mm. This #10 stirrup is O.K.

The #10 stirrups will next be investigated at the practical maximum spacing of 200 mm with respect to whether they meet the 0.33 MPa v_s requirement for capacity specified in Code 11.5.5.3.

$$v_s = A_s f_y / b_w s = 200 \times 400/250 \times 200 = 1.6 \text{ MPa} > 0.33 \text{ MPa} \qquad \textbf{O.K.}$$

The detailed stirrup spacing curve will be developed for establishing the final spacings graphically. Theoretical spacings will be calculated at several points by noting that $s \propto 1/v_s$, this being simpler than numerical calculation of v_s values.

---

*Note also the similar requirement of $\frac{1}{3} \ell_d$ in Code 12.14.2.3 for small bars hooked around another bar.

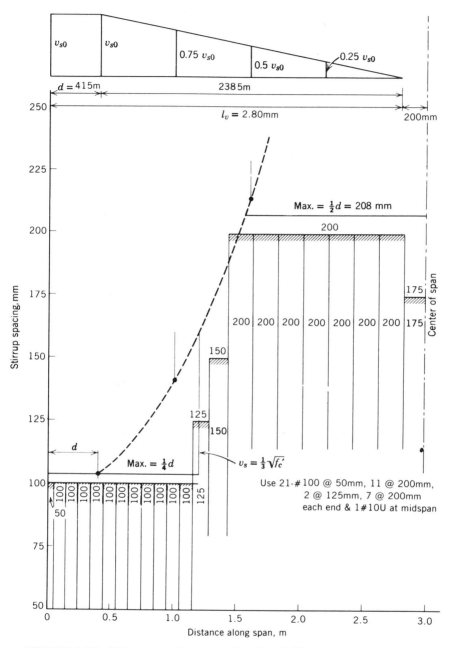

FIGURE 5.16 Stirrup spacing curve for Sec. 5.13.

118

The critical spacing also applies for the 415 mm nearest the support. The remaining v_s triangle is $2\,800 - 415 = 2\,385$ mm long, which can be subdivided into four equal lengths of 596 mm each. At these subdivision lines the shear v_s starting at the right, is successively $\frac{1}{4} v_{so}$, $\frac{1}{2} v_{so}$, and $\frac{3}{4} v_{so}$ leading to stirrup spacings:

v_s	$\times$ from midspan	Spacing
$\frac{1}{4} v_{so}$	$200 + 596 = \quad\ 796$ mm	$4\,s_o = 4 \times 142 = 568$ mm $>$ Maximum 208 mm
$\frac{1}{2} v_{so}$	$796 + 596 = \ 1\,392$ mm	$2\,s_o = 2 \times 142 = 284$ mm > 208 mm
$\frac{3}{4} v_{so}$	$1\,392 + 596 = 1\,988$ mm	$1.33\,s_o = 1.33 \times 142 = 189$ mm
$1\,v_{so}$	$1\,988 + 597 = 2\,585$ mm	

These values are plotted in Fig. 5.16 and a theoretical curve sketched through the points. The maximum spacings drawn in at 208 mm and 104 mm show that the spacing curve governs only for a very short distance. Use the spacings marked at the bottom of the figure and indicated by the shaded practical spacing curve. It is obvious that these data have scarcely justified the formality of drawing the spacing curve.

Practically, the spacings shown in Fig. 5.16 are not the best. It does not seem possible to stretch the 125 mm and 150 mm spacings enough to reduce the 175 mm at midspan to half 200 mm spacing, thereby eliminating the center stirrup. The next best move is to simplify the spacings by changing 150 mm to 125 mm and the 175 mm to 200 mm, thus calling for USE 21-#10 U at 50 mm, 11 at 100 mm, 2 at 125 mm, 7 at 200 mm each end plus 1-#10 at midspan (200 mm spacing maintained).

5.14 SHEAR IN ONE-WAY SLABS

Since a one-way slab is really a wide shallow beam, no shear behavior different from that in beams should occur, but three comments are appropriate.

1. Shear stresses are usually low in one-way slabs, except where heavily loaded as in one-way footings under heavy walls.
2. Stirrups are rarely needed or feasible to use in thin members like most slabs.
3. Concentrated loads require transverse reinforcement and develop shear demands similar to those discussed in Chapter 6 for two-way slabs.

Two-way slabs supported on beams are even less apt to have shear problems, but two-way slabs supported on columns (without beams) are totally different and have major shear problems, treated in Chapter 6.

5.15 SHEAR IN JOISTS

Joists are defined in Code 8.11.1 as consisting "of a monolithic combination of regularly spaced ribs and a top slab arranged to span in one direction or two orthogonal directions." Clear spacing is limited to 800 mm and any wider spacing disqualifies for the Code title of joists. Two design concessions are made for joists.

Because of their close spacing, joists are better able to distribute local overloads to adjacent joists than are typical beams. Accordingly, joists may be designed for 10 percent higher shear stresses than are permitted for v_c in beams.

Joists are also exempted from the requirement that stirrups be used wherever the shear exceeds 0.5 v_c.

5.16 SHEAR LOSS FROM CUTTING OFF BARS IN A TENSION ZONE

As already pointed out in Sec. 4.11d, when flexural tension bars are cut off in the tension zone of a beam a sharp discontinuity in the steel is created; the sharpness is dependent on the percent of bars cut off. The terminated bar opens an early flexural crack at the cutoff point and then appears to act as an eccentric pull on the surrounding concrete that in some way changes the flexural crack into a diagonal crack, again prematurely. The mechanism is not quite clear, but the end result, a reduced shear strength, is well documented,[7] as in Fig. 5.17. The beams represented were designed to reach f_y and the Code shear strength simultaneously; hence the steel stress ratio (ordinate) is also the shear ratio.

This study of 64 beams[7] shows that the usual extension of bars by $12d_b$

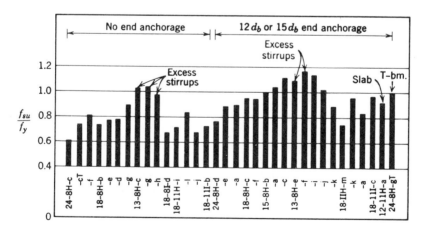

FIGURE 5.17 Summary of beams with bars cut off and without remedial measures. (The five beams marked with arrows had excess $\rho_n f_y$ of 0.5 MPa to satisfy minimum spacing rule.)

or $15d_b$ beyond the moment cutoff point kept the ill effects from being more severe. Without remedial steps only 2 out of 33 beams developed the design ultimate strength, with losses generally in the order of 15% to 25%, a few higher. In contrast, bars bent up caused no losses. Extra stirrups could erase losses, but seemingly at only 50 percent efficiency.

The extra stirrups mentioned in Sec. 4.11d are particularly needed over the tail of the bars cut off because Code 12.11.5 specifys excess stirrups having A_v of at least $0.4\, b_w s/f_y$ over the last 0.75 d of bars cut off, with s limited to not more than $d/8\, \beta_d$, where β_d is the ratio of bar area cut off to total A_s near the cutoff point.

5.17 MEMBERS OF VARYING DEPTH

The relations for shear, bond, and even moment resistance must be modified for members in which the depth is varying, that is, members in which the bottom and top surfaces are not parallel. The moment effect is not large unless the angle between the faces is at least 10° or 15°, but shear for diagonal tension and bond may be modified as much as 30% by 10° slopes. The 1940 Joint Committee Specification gave the following formula for the effective total shear V_1 to be used for V in the usual relations for v and u:

$$V_1 = V \pm \frac{M}{d}\,(\tan c + \tan t)$$

where V and M are the external shear and moment to be resisted, d is the depth to tension steel, and t and c are the slope angles of the top and bottom of the beam as shown in Fig. 5.18. The plus sign before the parenthesis is used when the beam depth decreases as the moment increases, as in Fig. 5.18a, and the minus sign is used for the more usual case of increasing depth with increasing moment, as in Fig. 5.18b. When the sum of the angles becomes as much as 30°, such a formula becomes very inexact.

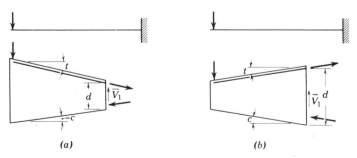

(a)   (b)

FIGURE 5.18 Shear in beams of varying depth.

5.18 ALLOWABLE SHEAR WITH AXIAL STRESS SUPERIMPOSED

Axial compression increases the diagonal tension resistance and axial tension lowers it. Overlooked axial tension from shrinkage or temperature variation often plays a prominent part in beam failures that occur in structures, in spite of the relief creep may give.

A quick picture of the influence of axial compression or tension is given in Fig. 5.19, adapted from the ACI Commentary on the 318-71 Code. The zero axial load is shown at the bottom with $v_c/\sqrt{f'_c}$ plotted on the vertical dividing line at $0.17 = \frac{1}{6}$, corresponding to $V_c = \frac{1}{6}\sqrt{f'_c}b_w d$.

1. *Axial tension.* Where axial tension N_u is large, as in the tension chord of a Vierendeel truss (no diagonals), the concrete may be cracked through in tension and unable to help the stirrups carry the tension aspects of shear. Stirrups must then carry the full shear. The permissible V_c can be evaluated from:

$$V_c = \frac{1}{6}(1 + 0.3N_u/A_g)\sqrt{f'_c}b_w d \qquad \text{(Code Eq. 11.9)}$$

where N_u is *negative* for tension, with units of MPa for N_u/A_g. For 3.5 MPa axial tension this v_c drops to zero. This shows as the sloping dashed line on the right side of Fig. 5.19.

2. *Axial compression.* The axial compression forces available in a

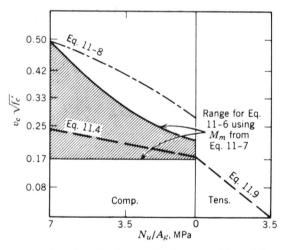

FIGURE 5.19 Design unit shears with axial load.

compression strut or column or compression chord of a Vierendeel truss greatly improve its capacity to resist shear. The better analysis is given by the more exact equation for allowable V_c modified by using the M_m of Code Eq. 11.7 for M_u and at the same time not limiting $V_u d/M_u$ to 1.0:

$$V_c = \tfrac{1}{6}(\sqrt{f'_c} + 100\, \rho_w V_u d/M_u)b_w d \qquad \text{(Code Eq. 11.6)}$$

$$M_c = M_u - N_u(4h - d)/8 \qquad \text{(Code Eq. 11.7)}$$

N_u is positive for axial compression. The second term in the M_m equation is approximately the moment of N_u about the center of the resisting compression. Where M_u is negative it is a signal to discard it and go directly to:

$$\text{Max. } V_c = 0.3\sqrt{f'_c}b_w d \sqrt{1 + 0.3\, N_u/A_g} \qquad \text{(Code Eq. 11.8)}$$

In Fig. 5.20 this maximum V_c shows as the uppermost dot-dash curve and the shaded area shows approximately the range of Code Eq. 11.6 using M_m of Code Eq. 11.7 in place of M_u.

Because Code Eq. 11.6 is difficult to apply, the simpler Code Eq. 11.4 (the dashed straight line sloping upward to the left in Fig. 5.20) is permitted and usually gives lower values:

$$V_c = \tfrac{1}{6}(1 + 0.07\, N_u/A_g)\sqrt{f'_c}b_w d \qquad \text{(Code Eq. 11.4)}$$

For 3.5 MPa compression this gives $0.20\sqrt{f'_c} = \tfrac{1}{5}\sqrt{f'_c}$ compared to $0.17\sqrt{f'_c} = \tfrac{1}{6}\sqrt{f'_c}$ without compression. This equation is not required if the other is used, it is an alternate.

Where the compression is from prestress, Code 11.4 controls. Comment on the equations there is left to Chapter 21 on prestressed concrete members.

5.19 BRACKETS AND SHORT CANTILEVERS

The Code covers brackets and corbels (limited to a/d of unity or less) and the author includes other short cantilevers as an extension of the same general behavior. Their normal resistance near ultimate consists of a tension tie across the top with an inclined compression strut forming a triangle, with normal bending making only slight variations. The inclination of this strut would determine the tension in the tie if it were simply a truss; the flexural calculation at the face of the column gives essentially the same tension, but the triangular truss idea emphasizes the anchorage problem; and this anchorage problem is the most critical one unless the bar extensions of Fig. 5.20b are possible.

The PCA bracket tests[8] took account of the high shrinkage and expan-

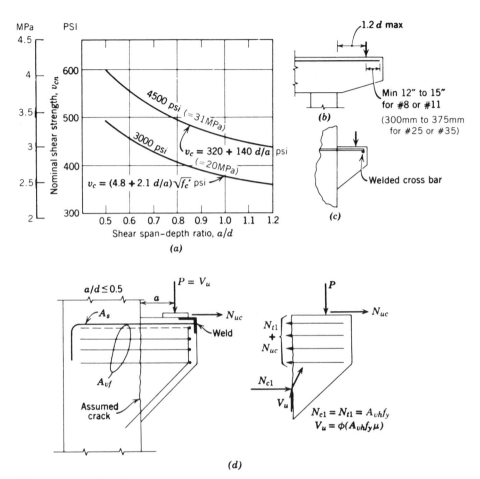

Figure 5.20 Brackets and short cantilevers. (*a*) Author's[9] unit shear strength, v_c, (*b*) Minimum end anchorage for values in (*a*). (*c*) Preferred end anchorage in short bracket. (*d*) Shear-friction applied to bracket design.

sion stresses frequently occurring from beams supported on brackets. The Code, based on these tests, requires a horizontal force equal to $0.2\,V_u$ and provides a complex equation for the allowable shear, at the same time limiting the ρ to be used in this equation. The Commentary also suggests that the bracket plate be welded to the top bars where such tension is possible. Where provision is made to avoid the horizontal loading, the allowable shear may be simplified to

$$V_n = 0.54(1 - 0.5\,a/d)(1 + 64\,\rho_v)\sqrt{f'_c}b_w d \qquad \text{(Code Eq. 11.32)}$$

where ρ_v is based on the sum of the flexural steel plus lower horizontal ties of area A_h (equal at least to $0.5 A_s$ and distributed over the upper two-thirds of the effective depth) and is limited in this equation to $0.20 f_y/f'_c$. The Code limits the effective depth d used at the column face to twice the depth at the outside edge of the bearing and the Commentary suggests that this bearing not start closer than 50 mm from the bracket face. The extension of the tension bars as close to the face as possible and an anchor provided by a welded cross bar of the same diameter is also recommended (Fig. 5.20c).

The problem area in a long bracket is the anchorage of bars because the variable depth (inclined strut idea) makes the bar tension nearly as critical at the load as at the column face.

If the anchorage can be extended as in Fig. 5.20b, the shear recommendations shown in Fig. 5.20a are still valid[9] in the author's opinion; they are simpler and a little less restrictive. For the longer brackets the author would normally calculate the required A_s at the face of column, keep the bars of a diameter that can be anchored into the column (or try for a deeper bracket if necessary to make them smaller), and if a horizontal tension pull were expected add enough bars to resist it. Finally the shear should be checked and should rarely control. For a short bracket the shear may be the chief control on the bracket depth. For a heavy bracket, say more than 450 mm deep, additional horizontal U-bars or ties in the upper half will avoid wide cracks, add to the bracket strength, and even help with the moment. Vertical stirrups are essentially useless. The Code requirement is horizontal steel A_h equal to half the top A_s.

5.20 SHEAR-FRICTION

The shear-friction theory (Code 11.7) is particularly useful for precast assemblies and composite construction.[10, 11, 16] The Code suggests this approach may be used for brackets cast-in-place (with the same reinforcement limitations of Code 11.7) provided $a/d \lesssim 0.5$. This method is especially appropriate where the resistance needed is against pure shear or sliding tendency, rather than primarily against a diagonal tension.

The shear-friction approach is to assume a shear plane already cracked as in Fig. 5.20d, with a coefficient of friction μ given in Code 11.7.5 (1.4 if monolithic, 1.0 if placed against hardened concrete (Code 11.7.9), or 0.7 against as-rolled structural steel. The depth must then be such that the shear stress on the area bd will not exceed $0.2 f'_c$ or 5.5 MPa. The necessary A_{vf} must be large enough to develop the normal force necessary to support the shear by friction:

$$A_{vf} f_y \mu = V_n$$

Closed stirrups or ties are appropriate for A_{vf} and these should be distributed uniformly within two-thirds of d adjacent to A_s.

The moment of P about N_{c1} must be resisted by A_s plus A_{vf} and, if N_{u2} acts, this increases this moment. (Values of P and N_{u2} should be factored loads divided by the strength reduction value ϕ.)

5.21 DEEP BEAMS

Deep beams are now covered in Code 11.8. Although the Code deals only with shear, a few general ideas first might be helpful. A deep beam is simply a member short enough to make shear deformations important in comparison to pure flexure; the Code considers a clear span ℓ_n of up to $5d$. Plane sections in these beams do not remain essentially plane under loading; but this is also true at the end of every simple span beam. If one has a clear picture of how any beam fails in shear at $a/d = 1$, one can get a rough idea of how a deep beam of $\ell_n = 2\,d$ would fail in shear by imagining two such end sections of beam joined together with the load at the junction. At $a/d = 1$ in a long beam, flexure will not be of much interest; in the deep beam of $\ell_n = 2\,d$, flexure at the load is the critical flexure. If the load is applied to the top of the deep beam the lever arm z of the internal couple will be smaller than usual, say $0.8\,d$; if the load is applied to the bottom of the beam in tension, z may drop as low as $0.4\,d$ or $0.5\,d$. The Code provisions are limited to loading on top of the beam. Anchorage of tension steel becomes critical. Leonhardt[12, 13] suggests *horizontal* hooks on tension bars (vertical compression helping to avoid splitting) or U-shaped bars lap spliced at midspan.

If such a deep beam is provided with stirrups that are able to deliver the bottom load to the upper part of the beam, the beam will behave nearly like a top loaded beam. It is always desirable in any kind of beam to provide hangers (stirrups) to pick up bottom loads in this fashion. Experimental evidence indicates that stirrups that must act as hangers *and* as web reinforcement need *not* be designed for the sum of the two requirements, but simply for the larger of the two.

Higher shear strengths are available when a/d is small, as already pointed out in Secs. 5.4c,d. For the deep beams a multiplier is given for application to the usual allowable v_c:

$$v_c = [3.5 - 2.5\,M_u/(V_u d)] \times [\text{usual formula values for } v_c]$$

where $[3.5 - 2.5\,M_u/(V_u d)]$ has here been designated as the magnifier, to be limited to a maximum of 2.5 and the magnified v_c also to be limited to $\frac{1}{2}\sqrt{f_c'}$. The values of this multiplier are plotted in Fig. 5.21a and should be applied to the usual allowable of $\frac{1}{6}\sqrt{f_c'}$ or to $\frac{1}{6}(\sqrt{f_c'} + 100\,\rho_w V_u d/M_u)$.

The critical beam section for shear is to be taken at $0.5\,a$ for concentrated loads and 0.15 ℓ_n for uniform load. For simple spans these appear quite reasonable values, closely akin to what is done on ordinary beams. The M_u and V_u in the formula above are to be taken at this critical section. The shear reinforcement required at the critical section is to be used throughout the span (Code 11.8.10).

The Code also limits v_n to $\frac{2}{3}\sqrt{f_c'}$ when $\ell_n/d \lesssim 2$, and for larger ℓ_n/d to

$$v_n = 0.06(10 + \ell_n/d)\sqrt{f_c'}$$

which is plotted in Fig. 5.21b. So long as v_n on ordinary beams may be used up to $\frac{2}{3}\sqrt{f_c'} + v_c$ at an a/d of 1, the author regards this deep beam limit as probably too strict.

Vertical stirrups are less effective in these beams because the shear cracks are steeper;[14] $horizontal$ bars become more effective as a/d or ℓ_n becomes smaller. The Code (11.8.7) accordingly expresses the shear resistance as the weighted sum of the two types of reinforcement in a modification of the basic equation:

$$\frac{A_v}{s}\left(\frac{1 + \ell_n/d}{12}\right) + \frac{A_{vh}}{s_2}\left(\frac{11 - \ell_n/d}{12}\right)*$$
$$= [(A_v/s)(1 + \ell_n/d) + (A_{vh}/s_2)(11 - \ell_n/d)]/12 = v_s b_w/f_y$$

The horizontal reinforcement A_{sh} is weighted much the heavier, as shown in Fig. 5.21c. The Commentary suggests the precise weighting is not critical but that the weighted sum should be maintained. In addition the spacing may not be more than 450 mm nor $d/5$ for s or $d/3$ for s_2, and the minimum A_v must be at least $0.0015\,bs$, the minimum A_{vh} at least $0.0025\,bs_2$. The designer should also note the requirement of Code 10.6.7

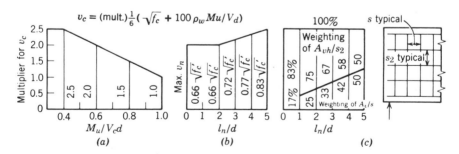

FIGURE 5.21 Deep beam shear. (a) Multiplier for usual v_c. (b) Maximum total v_n. (c) Weighting of vertical and horizontal shear reinforcement; spacings.

*The 1977 Code writes this as $V_s = f_y d$ times the form shown here, and in the Commentary shows that the form here equals $V_s/f_y d$.

calling for extra side face reinforcement in a longitudinal direction, already discussed in Sec. 4.9, for *any* web over 900 mm deep, *not* limited to ℓ_n of 5 d or less.

5.22 SHEAR WALLS

Shear walls are considered as cantilevers of deep beam type but carrying axial load, with horizontal wall length ℓ_w as beam depth and the wall height h_w as the beam length. Shear is most important in walls having small ratios of h_w/ℓ_w because flexure usually controls for high ratios.

The critical design section for shear is at a height of $\ell_w/2$ or $h_w/2$, whichever is smaller, and the lower wall is reinforced as this section requires. Shear strength is limited to a maximum of $0.8\sqrt{f_c'}hd$, where h is overall thickness and d is not more than $0.8\,\ell_w$, unless a strain compatibility check is made. Minimum shear reinforcement is required wherever V_n exceeds 0.5 V_c. Unless a more detailed calculation is made, unit shear on concrete is limited to $\frac{1}{6}\sqrt{f_c'}$ where a compression N_u exists or to $\frac{1}{6}(1+0.3\,N_u/A_g)\sqrt{f_c'}$ where N_u is negative for tension, as already given in Sec. 5.18.1.

Wherever $V_s = V_n - V_c$ exists, horizontal reinforcement is required in the amount $A_v f_y = V_s s_2/d$, with s_2 the vertical spacing. Two allowable V_c equations which give more exact limits are given in Code 11.10.6 and Code 11.10.9 gives detailed design rules for shear reinforcement, both horizontal and vertical.

Typical wall reinforcement is considered adequate only if v_n is less than $0.5\,v_c$.

5.23 BEHAVIOR PATTERNS NOT YET INCORPORATED IN CODE

(a) How Shear is Transferred in Concrete Beams

The internal stress transfer in an uncracked beam (Fig. 5.1) is greatly complicated by the flexural cracking; and the flexural cracking itself is complicated by the presence of shear. In a constant moment region, flexural cracks are essentially perpendicular to the tensile reinforcement. In a shear zone, say at a/d of 4, the flexural cracks form first under the load and then, as moment increases, form progressively farther from the load. The last-formed flexural cracks do not initially run as high into the web as the earlier ones and they tend to slope some toward the load, this slope moving more away from the vertical when shear is important; then the critical crack is like that shown in Fig. 5.2a. Kani has spoken of these cracks as forming small cantilevers downward from the uncracked compression part of the beam, like prongs or teeth on a comb.

In the absence of stirrups, vertical shear can be carried by (1) the uncracked compression concrete of the beam, (2) dowel action of the bars carrying shear across the flexural cracks, and (3) shear between the cantilever teeth across the curved crack interface. Paulay and Fenwick in New Zealand established[4] that the aggregate interlock across these tension cracks accounted for substantial shear transfer, most effectively when cracks were small, but always significantly since the crack surface is far from smooth. They found at ultimate from 33% to 60% of the vertical shear was transferred by this interlock. Dowel action can be large but near ultimate appears to account for not more than 10% to 15% of the external shear.

Beyond the last flexural (and shear) crack the shear is carried as an inclined thrust, downward toward the reaction, actually a mixture of thrust and some flexural action and shear. As a/d becomes smaller, the shear failure begins to relate more and more to the strength of an inclined compression member carrying some bending and shear. The limiting case is the splitting or pure shear failure when a is less than d.

(b) Internal Stress Complications Which Stirrups Reduce

Especially in the absence of stirrups, a problem exists at the bottom of the crack in Fig. 5.2a (point E in Fig. 5.2b) that involves the increased bar stress, increased bond stress, and local vertical tension stress. If the dowel action of the bar and the aggregate interlock along the crack were considered minor (and omitted), moments about N_c at F show that this moment M_F would measure N_t, which is the tension at E. Thus the steel at E suddenly picks up all the tension that to the left was carried by the concrete (a normal situation at the first flexural crack) and *also* all the normal tension increment that should develop from E to F. This creates a severe local bond stress (between steel and concrete) just to the left of E.

This large bond stress leads to large localized horizontal shear stress at the same spot and the localized diagonal cracking (or raveling) indicated at points 3 and 4 in Fig. 5.2a. Dowel action of the bars and the vertical load transferred across the diagonal crack by aggregate interlock (as shown by Pauley and Fenwick) add to this local cracking. This local shear failure, always observed, indicates approaching general failure.

Such localized raveling can extend and create the effect of a bond splitting failure. Such raveling seems to have been prominent in some diagonal tension failures that, in the absence of stirrups, occurred in the negative moment reinforcement near the point of inflection of continuous beams in actual structures.

The code requirements that at least minimum stirrups be used in most beams wherever v_n exceeds 0.5 v_c tends to limit this raveling phenomenon; the stirrups pick up in tension any dowel or aggregate interlock shear action.

Those stirrups crossing the diagonal crack itself in Fig. 5.2c would act as downward vertical loads that reduce the moment about F and thus the tension N_t at E. This in turn also reduces the general problem.

(c) Vd/M Term for Shear in Continuous Beams

The allowable shear stress equation of the 1971 Code, $v_c = \frac{1}{6}(\sqrt{f'_c} + 100\, \rho_w V_u d/M_u)$, was first proposed[2] in 1962 as an evaluation of the shear cracking stress, with $V_u d/M_u$ representing the influence of the shear span. The 1977 Code uses this form as the basis for its allowable V_c (Code Eq. 11.6, shown here in Sec. 5.6). It was primarily based on simple span tests.

In a continuous span Vd/M and d/a are quite different and only d/a lends itself to a plausible physical interpretation. For the simple span with concentrated loads one can visualize a reducing shear span as making feasible an inclined strut to carry much of the shear; the smaller a/d (larger d/a) means a steeper strut and one leaving less of the shear to cause diagonal tension problems. However, applied to a continuous beam, the equation seems to imply that the point of inflection acts as a reaction for the positive moment length and as a load point for the negative moment length; but the P.I. (point of inflection) furnishes neither the reaction below nor the load above to produce the compression forces that make the simple span concept valid. Instead, as in Fig. 5.3c, the failure line goes from real load to real reaction and the P.I. seems to have shown no influence.

To think in terms of the slope of the end strut for a continuous beam, the equation must be in terms of d/a and not Vd/M. If there is a P.I. midway between load and reaction, the strut can be no steeper than if it were a simple span, as indicated in Fig. 5.22, but the use of Vd/M would

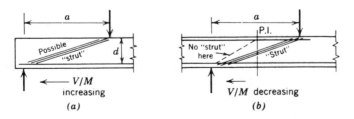

FIGURE 5.22   Use of shear equation. (a) Simple span has $Vd/M = d/a$. (b) Continuous span $Vd/M = d/($distance to P.I.$) \neq d/a$.

mean the influence of this term was doubled by the change from simple span to continuous. This appears on the *unsafe* side, even for the diagonal cracking load.

Accordingly, for *continuous beams* the author now *recommends*:

1. Further study and tests to look again at the Joint Committee report equation.
2. That, in the interim, d/a be used in the equation instead of Vd/M.
3. That, for uniform load, d/a be used as 0.25 or the overall $\frac{1}{6}\sqrt{f_c'}$ be used, the particular d/a being a judgment decision.

The ASCE-ACI Task Committee 426 seems to be interested, at least in some measure, in the use of d/a as suggested in item 2 above (Sec. 3.43a of Ref. 1).

(d) Lower Shear Strength Where ρ_w is Less than 0.01

Tests have shown[6] that the Code overvalues shear resistance of the concrete without stirrups where flexural steel ρ is less than 0.01 and the a/d ratio is as large as 2.75. These lower strengths are probably the result of higher steel stresses and wider cracks that accompany low steel percentages, with resultant loss in the aggregate interlock. On the basis of Fig. 5.23 the author recommends that for all beams (except cantilevers uniformly loaded) the usual allowable stress value of $\frac{1}{6}\sqrt{f_c'}$ be reduced where $\rho < 0.012$ to

$$v_c = (\tfrac{1}{15} + 8\,\rho_w)\sqrt{f_c'}$$

Cantilevers such as in wall footings have been excepted because they have lower flexural steel stresses at their critical shear section than is the case in the usual test specimen; test verification would be desirable. The ASCE-ACI Task Committee 426 has endorsed the allowable v_c equation above except for using a coefficient of 10 instead of 8 as the ρ_w multiplier.

This matter now is less critical in that stirrups able to carry 0.33 MPa shear are required wherever v_n exceeds the allowable $v_c/2$. However, in the exempted cases (Sec. 5.7 or Code 11.5.5) it is as significant as ever.

(e) Interaction of Bond, Shear, and Moment

Bond, shear, and moment resistance must not be regarded as independent responses to given loads. It is convenient for calculation purposes to treat each as a separate calculation. It is possible to design sections that in the laboratory will fail in any one of these three designated manners, that is, specimens that appear to be free from any significant influence from the

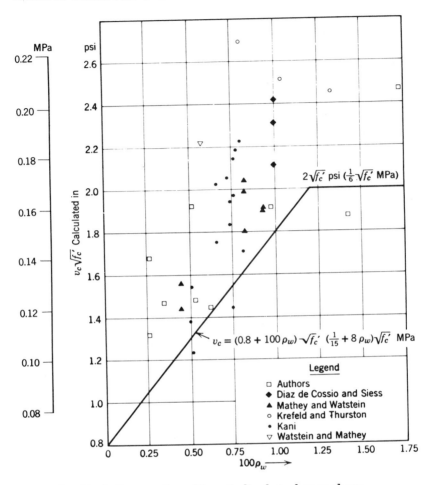

FIGURE 5.23   Influence of ρ_w of longitudinal steel upon shear resistance, for $a/d = 2.5$.

other two responses. However, when other combinations of dimensions and loading are used, unexpected interactions between these responses show up. The following three cases have been noted.

1. In Sec. 5.4*b* it was noted that flexural cracking always precedes diagonal tension weakness, except for small a/d values. It must follow that the diagonal tension capacity is influenced by the height of these flexural cracks and the beam length that does crack. The smaller steel areas appropriate with the use of high strength steel result in deeper cracks and cracks over more of the length. On the other hand, the

uncracked zone at a point of inflection is somewhat protected and stronger, in spite of its reduced compression stresses.

2.   A study of the bond failure shown in Fig. 7.3c and the special stresses resulting from a diagonal crack (Fig. 5.2b) indicates that the diagonal crack has almost surely affected the bond resistance adversely.

3.   The severe local bond conditions indicated in Fig. 5.3c contribute to the local diagonal cracking (raveling out) along the top bars which lowers both bond resistance and shear resistance. The failure can be definitely bond, definitely shear, or a mixture. Even minimum stirrups improve resistance here.

Such interrelations, some actual and some possible, point up the need for further research. The change of bar stress must develop a bond stress over the total bar perimeter that transfers into the beam as horizontal shear over the total beam width. Because of this geometric relationship between bond and shear stresses it is probable that neither can be well understood until such interactions are thoroughly explored.

SELECTED REFERENCES

1.   ASCE-ACI Committee 426, "The Shear Strength of Reinforced Concrete Members," *Proc. ASCE, Jour. Struct. Div.*, June 1973, ST6, pp. 1091–1187.

2.   ACI-ASCE Committee 326, "Shear and Diagonal Tension," *Jour. ACI, Proc.*, 59, Jan. 1962, p. 1; Feb. 1962, p. 277; Mar. 1962, p. 532.

3.   H. A. R. dePaiva and C. P. Siess, "Strength and Behavior of Deep Beams in Shear," *Proc. ACSE*, Vol. 91, No. ST5, Part 1, Oct. 1965, p. 19.

4.   R. C. Fenwick and T. Pauley, "Mechanisms of Shear Resistance of Concrete Beams," *Proc. ASCE*, Vol. 94, No. ST10, Oct. 1969, p. 2325.

5.   G. N. J. Kani, "How Safe Are Our Large Reinforced Concrete Beams," *Jour. ACI, Proc.*, Vol. 64, No. 3, Mar. 1967, p. 128.

6.   K. S. Rajagopalan and Phil M. Ferguson, "Exploratory Shear Tests Emphasizing Percentage of Longitudinal Steel," *Jour. ACI, Proc.*, Vol. 65, No. 8, Aug. 1968, p. 634.

7.   Phil M. Ferguson and Syed I. Husain, "Strength Effect of Cutting Off Tension Bars in Concrete Beams," Research Report 80-1F, Center for Highway Research, University of Texas at Austin, June 1967, 37pp.

8.   L. B. Kriz and C. H. Raths, "Connections in Precast Concrete Structures—Strength of Corbels," *Jour. Prestressed Concrete Institute*, Vol. 10, No. 1, Feb. 1965, p. 16.

9. Phil M. Ferguson, "Design Criteria for Overhanging Ends of Bent Caps—Bond and Shear," *Research Report No. 52-1F*, Center for Highway Research, The University of Texas at Austin, Aug. 1964.

10. J. A. Hofbeck, I. O. Ibrahim, and Alan H. Mattock, "Shear Transfer in Reinforced Concrete," *Jour. ACI, Proc.*, Vol. 66, No. 2, Feb. 1969, p. 119.

11. R. F. Mast, "Auxiliary Reinforcement in Precast Concrete Construction," *Proc. ASCE*, Vol. 94, No. ST6, June 1968, p. 1485.

12. F. Leonhardt, "Reducing the Shear Reinforcement in Reinforced Concrete Beams and Slabs," *Magazine of Concrete Research*, Vol. 17, No. 53, Dec. 1965, p. 187.

13. F. Leonhardt and R. Walther, "Deep Beams," *Bulletin* 178, Deutscher Ausschuss fur Stahlbeton, Berlin, 1966 (in German).

14. Fung-Kew Kong, Peter J. Robins, and David F. Cole, "Web Reinforcement Effects on Deep Beams," *Jour. ACI, Proc.*, Vol. 67, No. 12, Dec. 1970, p. 1010.

15. S. M. Fereig and K. N. Smith, "Indirect Loading on Beams with Short Shear Spans," *Jour. ACI, Proc.*, 74, No. 5, May 1977, p. 220.

16. Alan H. Mattock, "Design Proposals for Reinforced Concrete Corbels," *Jour. Prestressed Concrete Institute*, Vol. 21, No. 3, May–June 1976.

17. National Standard of Canada, "Code for Design of Concrete Structures for Buildings," Canadian Standards Association, Rexdale, Ontario, Canada, Dec. 1977.

PROBLEMS

Note   The problems in this group relate only to shear; flexural considerations are not included, unless specifically stated. Whenever actual stresses are called for, these should be compared with the specified allowables of the Code.

PROB. 5.1.

(*a*) For shear only and $v_c = \frac{1}{6}\sqrt{f'_c}$, what is the permissible uniform live load on the beam of Fig. 5.24 if minimum stirrups by Code Eq. 11–14 are used?

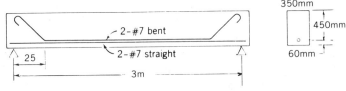

FIGURE 5.24   Simple span beam for Prob. 5.1.

Assume $f'_c = 30$ MPa, Grade 400 steel, and $w_d = 20$ kN/m (including beam weight).

(*b*) Where could the stirrups in (*a*) then be discontinued by Code 11.5.5.1? Ignore partial span loads.

(*c*) Repeat (*a*) if #10 U at 225 mm are used (minimum practical bar size and maximum spacing by Code 11.5.4.1).

(*d*) Repeat (*a*) with Code Eq. 11.6 (quoted in Sec. 5.6) for allowable v_c.

PROB. 5.2. In the beam of Fig. 5.25, if the 130 kN is all live load, $f'_c = 30$ MPa, and Grade 400 bars are used, design the stirrups and on an elevation (as in Fig. 5.25) show their arrangement and spacing.

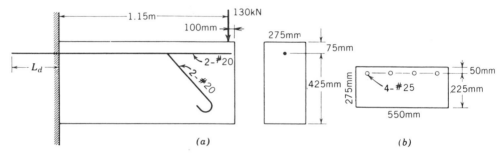

FIGURE 5.25 Cantilever beam. (*a*) For Probs. 5.2 and 5.3 (*b*) For Prob. 5.4.

PROB. 5.3. If the load on the beam of Fig. 5.25 and Prob. 5.2 is changed to $w_d = 30$ kN/m (including beam weight) and $w_\ell = 60$ kN/m, and ℓ overall becomes 2 m (with more A_s), design the stirrups and sketch their arrangement.

PROB. 5.4. If the cross section of Fig. 5.25*b* is used on a 1.3 m cantilever, $f'_c = 30$ MPa, Grade 400 steel, under a dead load $\omega_d = 30$ kN/m (including beam weight) and $w_\ell = 60$ kN/m, check for shear and, if needed, choose and position stirrups.

PROB. 5.5. Evaluate V_u for the beam of Fig. 3.19, using $f'_c = 25$ MPa $f_y = 400$ MPa.

(*a*) For minimum stirrups; design these stirrups for region near support.

(*b*) For maximum stirrups; design these stirrups for region near support.

PROB. 5.6. Evaluate maximum V_u for the joist of Fig. 3.22 if $f'_c = 35$ MPa.

PROB. 5.7. Design and space vertical stirrups for a 6 m simple span rectangular beam, $b = 375$ mm, $d = 675$ mm, $f'_c = 25$ MPa, Grade 300 steel, w_d (including beam weight) $= 30$ kN/m, and $w_\ell = 120$ kN/m. Use stirrup spacing curve.

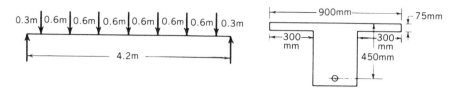

FIGURE 5.26   Simple span beam for Prob. 5.8.

PROB. 5.8.   The loads shown in Fig. 5.26 are at fixed points, not moving loads. Each load consists of 13 kN of dead load and a possible 20 kN of live load. Establish the maximum shear curve and design and space the necessary stirrups for $f'_c = 25$ MPa and Grade 300 steel. Consider uniform load negligible.

PROB. 5.9.   Given the essentially simple span beam of Fig. 5.27 carrying the column loads shown, $f'_c = 30$ MPa, Grade 400 steel. Design shear reinforcement required and sketch its placement on elevation of beam.

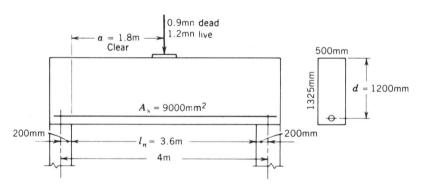

FIGURE 5.27   Beam for Prob. 5.9.

6
SHEAR IN TWO-WAY FOOTINGS AND SLABS—TORSION COMBINED WITH SHEAR

6.1 TWO-WAY FOOTINGS AND SLABS—SHEAR STRENGTH

When a two-way slab is heavily loaded with a concentrated load or where a column rests on a two-way footing, diagonal tension cracks form that encircle the load or column.[4] These are not visible on the surface, except as flexural cracks. Such cracks extend into the compression area of the slab and encounter resistance near the load similar to the shear-compression condition shown for the beam in Fig. 5.3a. The slab or footing continues to take load and finally fails around and against the load or column, punching out a pyramid of concrete as indicated in Fig. 6.1. Diagonal cracks do not form further out from the load or column because of the rapid increase in the failure perimeter, which under a square load always totals eight times the distance from the center of load. The initial diagonal cracks thus proceed to failure in shear compression (or punching shear) directly around the load.

In compromising between the initial cracking location and the final punching shear condition at failure for different ratios between column (or load) diameter and footing (or slab) thickness, the Joint Committee[1] recommended a single-strength calculation at a pseudo-critical distance $d/2$ from the column face or edge of the load as shown in Fig. 6.2. Because there is no danger aside from the shear compression type of failure (with the diagonal crack not visible from outside the concrete), these slab shear

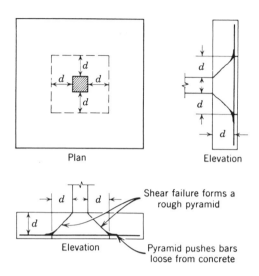

Plan

Elevation

Shear failure forms a
rough pyramid

Elevation

Pyramid pushes bars
loose from concrete

FIGURE 6.1   A square column tends to shear out a pyramid from a footing.

strengths are based on ultimate strengths rather than cracking loads:

$$\text{Allowable } V_c = \tfrac{1}{6}(1 + 2/\beta_c) \sqrt{f_c'}b_o d$$
$$\text{Max. } V_c = \tfrac{1}{3} \sqrt{f_c'}b_o d, \text{ for } \beta_c \text{ of 2 to 1 or smaller}$$

where

β_c = ratio of long side to short side of the loaded or reaction area
b_o = perimeter of pseudocritical area, as in Fig. 6.2.

This increased shear strength is available only where two-way bending occurs at the load considered. The 1977 Code introduces the β_c term to warn that a 400 mm × 1500 mm wall-like column has a zone on the long side that forces the slab to act as a one-way slab locally. It would at first appear that a similar behavior would result around a 1500 mm square column; and it probably would with a thin slab. Slab thickness is obviously an important variable here and more data are needed to cover all cases. Reference 9 is an excellent discussion.

For an L-shaped interior column the present Code would be interpreted rather arbitrarily as shown in Fig. 6.2, using the maximum diagonal and the perpendicular width to fix β_c.

Wall and corner columns involve not only large moment transfers (treated for interior columns in Sec. 6.4) but also torsional problems in the slab. The latter must be covered separately.

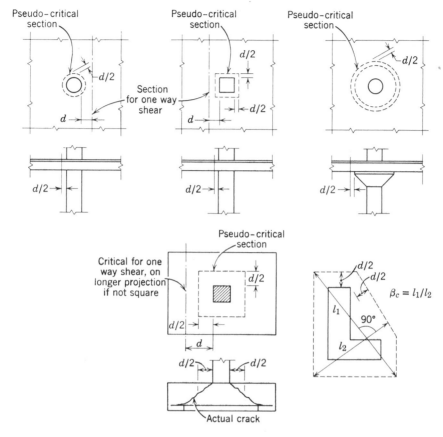

FIGURE 6.2   Pseudocritical diagonal tension sections in slabs supported directly on columns and in footings.

In special cases footings or slabs may fail from (one-way) beam action at a distance d from the face of the load or support, on a straight section all across the member. Such failure is most apt to happen in long narrow footings or long span slabs of narrow width. The limiting shear here, as given in Sec. 5.6, is usually $V_c = \frac{1}{6}\sqrt{f'_c}bd$.

Examples involving shears covered in this section are included in Chapter 15 on flat plates and flat slabs and in Chapter 20 on footings.

6.2 SHEAR REINFORCEMENT IN SLABS

Slabs can be reinforced at interior columns for shear in at least two different fashions. The more common is the use of shearheads. These are small structural steel channels or I-shapes that can be totally embedded in the slab. They are welded into a crossing pattern (at the same level) with cantileverlike projections into the slab as though they were the start of beams to adjacent columns. These pick up both moment and shear near the column and greatly increase the periphery on which shear on the concrete will be critical, moving it out to a periphery at 0.75 of the spearhead projection. The maximum shear there is limited to $\frac{1}{3}\sqrt{f'_c}b_od$, but it may be as much as $0.6\sqrt{f'_c}b_o$ for V_n at the section near the column.

The other reinforcement pattern might be called the formation of shallow concrete "beams" of slab depth that have both top and bottom bars around which closed stirrups can be anchored. Here V_n is limited to $\frac{1}{2}\sqrt{f'_c}b_od$ and V_c to $\frac{1}{6}\sqrt{f'_c}b_od$. Adequate anchorage of the shear reinforcement is a most important element.

The Code Commentary has good discussions of both shearhead and the shallow beam type and these will not be repeated here.

Substantial research has greatly increased the knowledge about shearheads at exterior columns and to some extent corner columns. These are substantially different problems from those at interior columns under vertical loads. The 1977 Code has not been extended to cover exterior columns and its present shearhead requirements should not be extrapolated to such cases.

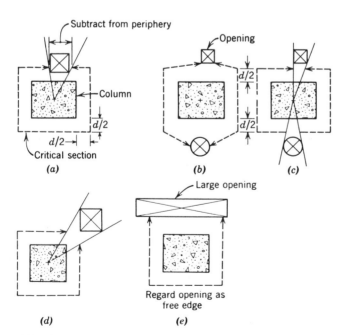

FIGURE 6.3 Effect of openings on shear perimeter. (From Reference 1, ACI.)

6.3 OPENINGS IN SLABS THAT INFLUENCE SHEAR STRENGTH

The Code suggests a simple way of handling openings near columns (Code 11.11.5). Radial lines are drawn from the edges of the openings to the center of the column as shown in Fig. 6.3. These radial lines intersect the pseudo-critical sections around the column, and all the resisting perimeter caught between such radial lines is considered as ineffective in resisting shears. If too much perimeter is lost, the designer must be sure adequate two-way bending is really present; otherwise the lower shears permitted in beams become the limiting values for the slab.

Openings more than 10 times the slab thickness from the concentrated load or reaction (unless within the column strips of flat slabs) need not be deducted.

For slabs with shearheads only half the above deductions are required.

The 1962 Joint Committee Report[2] and Sec. 5.4.3 of the 1974 report[1] both show more detailed background.

6.4 TRANSFER OF MOMENT BETWEEN COLUMN AND SLAB

Another condition that adds locally to the shear around the column in flat plate construction arises where a moment is carried to or from a column. The Code requires an analysis of the resulting shears. It recognizes that much of the moment is transferred directly by flexure in the slab, but allows for the extra punching type shear by specifying the following portion γ_v of the moment M_u (transferred) be used in the shear calculation:

$$\gamma_v = \left(1 - \frac{1}{1 + \frac{2}{3}\sqrt{\frac{c_1 + d}{c_2 + d}}}\right) M_u$$

where c_1 is the column width parallel to span of the slab strip that will resist the moment by flexure and c_2 is in the perpendicular direction, as in Fig. 6.4. For a square column this moment becomes $0.4M_u$ applied as a torsion on the slab.

The ACI-ASCE Committee 326 in 1962 recommended[2] essentially that the slab shear be analyzed as follows.*

1. Consider the usual critical shear perimeter at $d/2$ from the column to resist this shear as a torsion.

---

*The proportion of the column moment assigned to torsion has been increased since then.

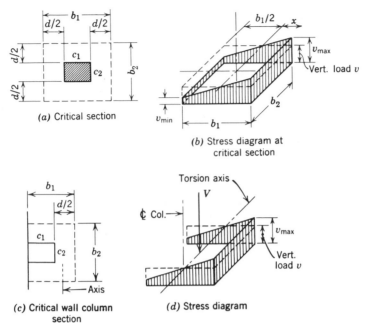

(a) Critical section

(b) Stress diagram at
critical section

(c) Critical wall column
section

(d) Stress diagram

FIGURE 6.4 Shears around a column which must introduce moment (and torsion) into a slab or accept moment from the slab.

2.   The resisting area becomes the four vertical sides of the slab cut by the perimeter. The axis of torsion is a horizontal line, through the column center for the usual interior column.

3.   Calculate the polar moment of inertia of the resisting slab areas. From Fig. 6.4a:

$$J = 2(b_1 d^3/12 + db_1^3/12) + 2\ b_2 d(b_1/2)^2$$

4.   Obtain the unit shear from Tx/J where T is the portion of M_u computed above and x is the distance from the torsion axis, with the maximum at $x = b_1/2$.

5.   This torsional shear is to be added to the usual vertical shear as in Fig. 6.4b, quite similar to the P/A and Mc/I combination in elastic theory.

If one considers an edge column, the following adjustments are to be made.

1.   The torsion axis must be the centroid of the three resisting slab faces.

2.   The polar moment of inertia and shear area are correspondingly reduced.

TORSION

6.5 PURE TORSION

(a) General

Torsion is important in some spandrel beams, in flat plates at exterior columns, in curved stair slabs or curved beams, and wherever large loads must be carried off the axis of the member.[7] In box girder bridges and in any isolated beam that must resist unbalanced loading, torsion is an important problem. When torque is small, producing no larger torsional shear stress than $v_t = 0.12\sqrt{f'_c}$, the Code suggests that it be ignored. Where it is possible to devise framing that can take the load without torsion, this is a preferred solution (by the author). Because torque design is a new area in reinforced concrete, it is given more space here than the seriousness of the problem warrants.

(b) Elastic Theory of Pure Torsion

Elastic analysis of a round shaft under torsion T is simple and results in tangential shearing stresses all around the member as shown on one diameter in Fig. 6.5a. The equation is $v_{tn} = Tr/(\int r^2 dA)$. Any other shape is increasingly complex because such a cross section warps and the member

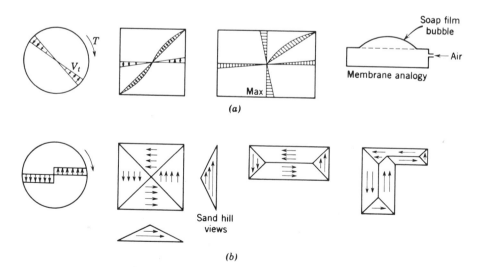

FIGURE 6.5 Elastic and plastic analysis for pure torsion shears. (a) Elastic analysis. (b) Plastic analysis, v_t equal over entire cross section.

lengthens as it twists. For analysis and visualization the soap film (membrane) analogy is useful. A soap film is stretched across a hole cut in the top of a box to the shape of the member cross section. A small inside air pressure then raises a bubble over the hole. The volume between the bubble and the original plane, by analogy of the governing differential equations, is proportional to the total torque resistance; the slope of the membrane (which is under uniform tension) measures the unit shear stress, a steep slope representing a high stress.

(c) Plastic Analysis for Pure Torsion

A plastic analysis assumes uniform shear intensity all around the surface and all over the cross section, as in Fig. 6.5b. The analysis can be envisioned in terms of the sand heap analogy. This gives uniform side slopes with the stress at 90 degrees to the slope, and again a volume proportional to the torque. For the slope equal to the maximum in the elastic analysis, volume and torque indicated is considerably larger. The equation can be written $v_{tn} = T/J_p$ where J_p is twice the volume of the sand heap divided by the slope.

The final failure is a diagonal crack following around the surface to form a helix with about a 45° slope.

(d) Code Analysis

Neither of the above analyses is realistic, partially because of the properties of concrete but chiefly because concrete at design loading will be cracked. The best—that is, the most consistent—calculation method has not yet been fully established. The author in early investigations found these two methods about equally consistent, but prefers the lower stress numbers the plastic method gives. The analysis is not fully adequate because it shows small, but consistent, differences in the ultimate stress for the T-beam, L-beam, and the rectangular beam, in a systematic decreasing manner that indicates the analysis does not reflect the shape properly. The value is also a little higher for uniform distributed torque load than for a concentrated torque load.

The Code uses an approximate relationship for the torque T_n that lies between the elastic and plastic analysis:

$$v_t = T_n/[\Sigma(x^2 y)/3] = 3T_n/(\Sigma x^2 y) \quad \text{or} \quad T_n = v_t \Sigma x^2 y/3$$

where x is the short dimension of the rectangle and y the long one and Σ indicates the sum of all the component rectangles making up the beam cross section, Fig. 6.6. The beam shape may be subdivided to give the largest total.

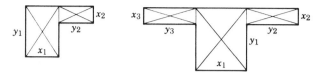

FIGURE 6.6   Rectangles for Code v_{tu} calculation.

TORSION COMBINED WITH SHEAR AND MOMENT

6.6 MEMBER BEHAVIOR UNDER TEST

Because the crack patterns of torque, shear, and moment are quite different, it is necessary to test under all three in combination. To the author test behavior seems best described physically and mathematically from the torque and shear interaction. On one web face the torque shear adds (and on the other subtracts) relative to the vertical shear stresses. A little torque causes a diagonal crack to open on one side of a beam earlier than on the other, but the shear failure pattern is not much altered, possibly a little steeper diagonal crack on the opposite face. The torque tends to make diagonal cracks out of the flexural cracks on the tension face. As the ratio of torque to shear is increased the failure pattern is more sensitive to the shape of the beam, and the tension face cracks become full diagonals.

With the rectangular beam without stirrups the load difference between initial diagonal crack and ultimate may not be large and the diagonal crack on the opposite face may be replaced by an inclined compression crack (evidence of a skew bending effect); or a compression crack on a skew plane may finally develop on the compression face of the beam and a steeper diagonal crack (nearly vertical or of reversed slope) may show on the opposite web face.

With a T-beam the failure mechanism is slower in developing (more load after first diagonal cracking) and is not yet well understood. The diagonal crack usually eventually opens on the far side of the web, always steeper, sometimes of reversed slope, dependent upon the ratio of torque to shear.

To the author torsion appears primarily as a special shear problem, one that is troublesome because many variables influence the available resistance, including beam shape, the stirrups present, and even the longitudinal beam steel. Progress has been made in defining ultimate strength in terms of skew bending.

When closed stirrups are added, which in practice must be everywhere that torsion is important, the spread between first diagonal crack load and ultimate load increases. The few diagonal cracks present in the test members without stirrups (Fig. 6.7a) become amazingly numerous with stirrups (Fig. 6.7b), member stiffness against rotation drops to from 3 to 7 percent of that before diagonal cracking,* and longitudinal steel stresses then increase rapidly, in proportion to the angle. Most stirruped beams rotate so much before reaching ultimate strength as to appear worthless as structural members well before they fail.

6.7 STRENGTH INTERACTION OF TORSION WITH SHEAR

(a) Without Stirrups

When torsion is calculated by plastic analysis of the gross (uncracked) concrete section, a pure torsion, without flexural shear, leads to a calculated v_{tc} of from $0.4\sqrt{f_c'}$ to $0.5\sqrt{f_c'}$, compared to about $0.15\sqrt{f_c'}$ for flexural shear cracking strength v_c without any torsion.† The combinations of torque and shear lead to the ellipse of Fig. 6.8a which could be replaced with a circle (Fig. 6.8b) in terms of v_{tn}/v_{tc} and v_n/v_c. The circle is accepted generally in the United States and Canada, but not so widely in Europe. The 1977 Code uses *total* shear and torque for its equations, but the unit stress format seems to give the clearer picture here. (The author also finds the ellipse appropriate for strengths with stirrups, both v_{tn} and v_u being then increased to v_{to} and v_o for a specific quantity of stirrups, as in Fig. 6.8c. This view is not generally accepted as yet.)

(b) With Stirrups, Code Design

When *pure* torsion tests are run in which closed *stirrups and longitudinal reinforcement are kept equal in volume*, which is the ideal ratio for pure torsion, the strength curve of Fig. 6.9a is found. The plotted points trace a straight line until an overreinforced condition is reached. When these strengths are projected back to zero stirrups, the straight line portion intersects at about v_{tc} of $0.2\sqrt{f_c'}$. The Torsion Committee concluded†† that not

---

*Measured as the change in rotation angle ϕ relative to the torque increment applied, that is, $d\phi/dT$, similar to the tangent modulus idea.

†That the ultimate v_{tc} might be three times the ultimate v_c long concerned the author, since each led to diagonal tension cracks. Finally he realized that neither calculation had a rational approach and that both numbers were simply crude stress coefficients, not stresses at all. Now he calls them "imaginary stresses" whenever someone raises the question. Both normally relate to flexurally cracked members, but both use properties of the uncracked section. A calculation based on imaginary areas (with sloppy theory as well) necessarily leads to imaginary stresses.

††Based on work by Dr. Hsu at the PCA Skokie Laboratory.

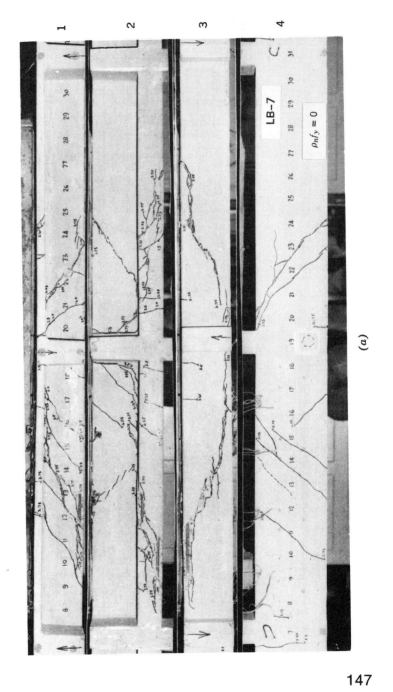

(a)

147

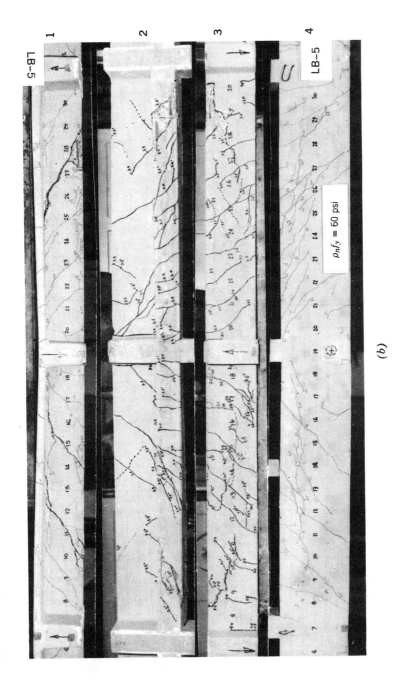

FIGURE 6.7   L-beams with $b_w = 75$ mm loaded as semicontinuous beams with load eccentricity of 125 mm. (*a*) No stirrups. (*b*) Light stirrups. Corresponding torque-rotation curves are shown for LB-7 and LB-5 in Fig. 6.10. The four views of each beam are: 1, web on flange side; 2, bottom of beam and flange; 3, side of web opposite flange; 4, top.

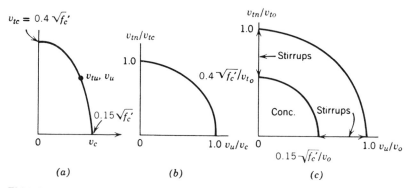

FIGURE 6.8 Shear-torsion interaction diagram.

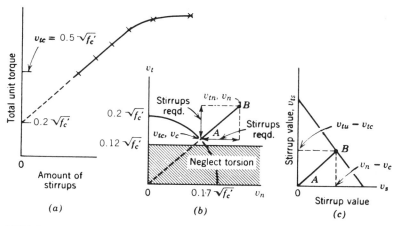

FIGURE 6.9 Stirrup strength and design.

all of $v_{tc} = 0.5\sqrt{f_c'}$ could be used with stirrups on the straight-line basis and the Code rules are based on this finding. The concrete stress usable for resisting torsion is limited to $0.2\sqrt{f_c'}$ and for general design the ellipse of Fig. 6.9b is used.

6.8 Stirrup Design for Torsion

Although the Code allows $0.12\sqrt{f_c'}$ in torsion without any special provisions, above this value any combination of v_{tn} and v_n falling outside the ellipse requires the design of stirrups both for shear and for torsion. These limiting values without stirrups are given by these equations, with the same 1977 Code equations in terms of total torque and shear to the right:

$$v_{tc} = \frac{\frac{1}{5}\sqrt{f_c'}}{\sqrt{1 + (1.2\, v_n/v_{tn})^2}} \qquad T_c = \frac{\frac{1}{15}\sqrt{f_c'}\,\Sigma x^2 y}{\sqrt{1 + \left(\frac{0.4\, V_u}{C_t T_u}\right)^2}}$$

$$v_c = \frac{\frac{1}{6}\sqrt{f_c'}}{\sqrt{1 + (v_{tn}/1.2\, v_n)^2}} \qquad V_c = \frac{\frac{1}{6}\sqrt{f_c'}\, b_w d}{\sqrt{1 + \left(2.5\, C_t \frac{T_u}{V_u}\right)^2}}$$

where $C_t = b_w d/\Sigma x^2 y$

The critical section for torsion, as for shear, is a distance d from the face of support. If the design combination of stresses falls at point B in Fig. 6.9b, stirrups for torsion are to be based on the vertical projection of the overrun AB and for shear on the horizontal projection of AB. This stirrup requirement is equivalent to using the triangular interaction of Fig. 6.9c for the stirrup portion. The line AB from Fig. 6.9b can be superimposed on the triangular arrangement if the scales are each in terms of v_{tn} and v_n.

The minimum area of stirrups for torsion outside the ellipse in Fig. 6.9b must satisfy the equation

$$A_v + 2\, A_t = \tfrac{1}{3}\, b_w s/f_y$$

where A_v is the (two-leg) area of stirrups for shear and A_t is the area of one leg of the stirrups for torsion.

For a given $v_{ts} = v_{tn} - v_{tc}$ the required area for A_t is

$$A_t = \frac{v_{ts}s\,\Sigma x^2 y}{3\,\alpha_t x_1 y_1 f_y} = \frac{T_s s}{\alpha_t x_1 y_1 f_y}$$

since $v_{ts} = 3\, T_s/\Sigma x^2 y$,

where

x_1 and $y_1 =$ the narrow and long sides, respectively, of the closed stirrups
$s =$ the stirrup spacing
$f_y =$ the stirrup yield point
$\alpha_t = 0.66 + 0.33\, y_1/x_1 \lessgtr 1.50$

Torsion reinforcement must extend at least a distance $d + b$ beyond the point theoretically required.

6.9 Longitudinal Steel for Torsion

When beams fail in torsion their length increases and extra stress moves into the longitudinal bars. Torsion stirrups cannot act effectively unless a tension truss "chord" is available in each corner of a rectangular beam, since the tension in pure torsion may be needed in any (or all) corners as the diagonal crack wraps around the member.

The Code sets two requirements for extra longitudinal steel for torsion; the first is a volume of longitudinal steel for torsion equal to the volume of stirrups provided for torsion:

$$A_\ell = 2\,A_t(x_1 + y_1)/s$$

If the stirrup volume is very low, the following requirement defines a larger required A_ℓ:

$$A_\ell = \left[\frac{2.75\,xs}{f_y}\left(\frac{v_{tn}}{v_{tn} + v_n}\right) - 2\,A_t\right]\frac{x_1 + y_1}{s}$$

where "$2\,A_t$ ... need not be taken less than $\frac{1}{3}\,b_w s/f_y$."

The second equation is better understood in terms of its development. The Torsion Committee originally recommended that the torsion stirrups used not be less than

$$A_t = \frac{1.35\,b_w s}{f_y}\cdot\frac{v_{tn}}{v_{tn} + v_n}$$

and that longitudinal reinforcement A_ℓ for torsion be used with a volume equal to that of the stirrups used for torsion. The Committee later agreed to the Code version, which uses $\frac{1}{3}$ instead of 1.35 $v_{tn}/(v_{tn} + v_n)$ on condition that A_ℓ be increased enough to make up the deficiency in $2\,A_\ell$ brought about by the change. This requires something *more* for A_ℓ than would have been required from the simpler equation. The result is a rather awkward equation. Its use is covered in the design of Sec. 6.11.

The 1977 Code notation has been inserted in the 1971 equations above, using nominal values of torsion and shear stress instead of ultimate values; the ratio is the same either way. The 1977 Code also replaces this ratio with the ratio $T_u/(T_u + V_u/3C_t)$, which is an identity whether written this way or in terms of nominal torque and shear. (C_t represents $b_w d/\Sigma x^2 y$.)

6.10 DESIGN EXAMPLE—TORSION AND SHEAR

A 450 mm wide by 675 mm deep rectangular cantilever beam 2.75 m long has been designed to pick up at its end a 180 kN hoisting load, which includes a proper allowance for impact. It is found operating conditions may place the load 250 mm off the axis of the beam to either side. Design the torsion and shear reinforcement, assuming Grade 400 reinforcement and f_c' of 20 MPa.

Solution

The critical section for both torsion and shear is at a distance d from the support.

$d = 675 - (\text{say})\ 70 = 605$ mm

Beam weight $= 0.450 \times 0.675 \times 2\ 500 \times 9.8 = 7\ 442$ N/m $= 7.44$ kN/m

$V_u = 1.4 \times 7.44(2.750 - 0.605) + 1.7 \times 180 = 328.3$ kN

$V_n = 328.3/0.85 = 386.2$ kN $= 0.3862$ MN

$v_n = 0.3862/0.450 \times 0.675 = 1.271$ MPa

$T_u = 1.7 \times 180 \times 0.250 = 76.5$ kN $\cdot$ m

$T_n = 76.5/0.85 = 90.0$ kN $\cdot$ m 0.090 MN $\cdot$ m

$v_{tn} = T/\frac{1}{3}\Sigma x^2 y$ See sec. 6.5.c.

$\frac{1}{3}\Sigma x^2 y = \frac{1}{3} \times 0.450^2 \times 0.675 = 0.0456$ m^2

$v_{tn} = 0.090/0.0456 = 1.974$ MPa

From ACI Code 11.6.9.4 the maximum value

$T_s = 4\ T_c$

From Code Eq. 11.2.1

$T_n = T_s + 4\ T_c = 5\ T_c$

From Code Eq. 11.2.2

$$T_n = 5\ T_c = \tfrac{1}{3}\sqrt{f_c'}\ \frac{\Sigma x^2 y}{\sqrt{1 + \left(\dfrac{0.4\ V_n}{C_t T_n}\right)^2}}$$

From Section 6.8, the maximum allowable

$v_{tn} = 5\ v_{tc} = \sqrt{f_c'}/\sqrt{1 + (1.2\ v_n/v_{tn})^2}$
$\quad\quad = \sqrt{20}/\sqrt{1 + (1.2 \times 1.271/1.974)^2} = 4.472/1.264$
$\quad\quad = 3.538$ MPa > 1.974 MPa   **O.K.**

Allowable $v_{tc} = v_{tn}/5 = 3.538/5 = 0.708$ MPa

$v_{ts} = v_{tn} - v_{tc} = 1.974 - 0.708 = 1.266$ MPa

$T_s = v_{ts} \times \Sigma \frac{1}{3} x^2 y = 1.266 \times 0.0456 = 0.057\ 73$ MN $\cdot$ m

Assume #15 stirrups with 40 mm clear cover

$\alpha_t = 0.66 + 0.33\ x_1/y_1$ from Code 11.23

$\quad = 0.66 + 0.33\ (675 - 2 \times 40)/(450 - 2 \times 40) = 1.191$

$A_t = T_s s/\alpha_t x_1 y_1 f_y$ from Code Eq. 11.23

$2\ A_t = 2 \times 0.057\ 73\ s/1.191 \times 370 \times 0.595 \times 400$

$\quad\quad = 1.101 \times 10^{-3}\ s$ in m units $= 1.101\ s$ in mm units

This stirrup area must be combined with the stirrup area for shear:

Allowable $v_c = \frac{1}{6}\sqrt{f_c'}/\sqrt{1+(v_{tn}/1.2\,v_n)^2}$ from Code Eq. 11.5 divided by $b_w d$

$$= \frac{1}{6}\sqrt{20}/\sqrt{1+(1.974/1.2 \times 1.271)^2} = 0.745/1.636 = 0.456 \text{ MPa}$$

$$v_s = 1.271 - 0.456 = 0.815 \text{ MPa} < \frac{1}{3}\sqrt{f_c'} = 1.490 \text{ MPa}$$

Maximum $s = \frac{1}{2} d = 303$ mm

$$A_v = (v_n - v_c)b_w s/f_y = (1.271 - 0.456) \times 0.450\, s/400$$

$$= 0.917 \times 10^{-3} \text{ in m units} = 0.917\, s \text{ in mm units}$$

For torsion plus shear the sum of $(1.101 + 0.917)\, s = 2.018\, s$ mm^2 will be provided by one size of stirrups at a regular spacing.

For #15 stirrups:

$$2 \times 200 = 2.018\, s$$

$$s = 198 \text{ mm}$$

The Code requires a torsion spacing not greater than $\frac{1}{4}(x_1 + y_1) = \frac{1}{4}(370 + 5.95) = 421$ mm, nor greater than 300 mm.
Development of closed transverse torsion reinforcement is not specifically covered in the Code; the same provisions applied to stirrups for shear will be used (Sec. 5.8 or Code 12.14).

$$\tfrac{1}{2} d = 303 \text{ mm} \qquad \tfrac{1}{2}\ell_d = 174 \text{ mm}$$

From Code 12.2.2 $\ell_d = 0.058\, d_b f_y = 348$ mm for #15

Deduct cover and 3 d_b from $\frac{1}{2} d = 303 - 40 - 45 = 218 > 174$ mm. This available depth plus the anchorage around the corner is O.K.

USE #15 closed stirrups at 175 mm

Calculate extra longitudinal reinforcement for torque.

$$A_\ell = 2\, A_t(x_1 + y_1)/s = 1.101\, s(370 + 595)/s = 1\,062 \text{ mm}^2$$

$$A_\ell = \left[\frac{2.75\, x\, s}{f_y}\left(\frac{v_{tn}}{v_{tn} + v_n}\right) - 2\, A_t\right]\frac{x_1 + y_1}{s} \text{ from Code Eq. 11.24*}$$

$$= \left[\frac{2.75 \times 450\, s}{400}\left(\frac{1.974}{1.974 + 1.271}\right) - 1.101\, s\right]\frac{965}{s} = 753 \text{ mm}^2$$

The $1\,062$ mm^2 governs. The Code requires at least #10 longitudinal bars, spaced not farther apart than 300 mm, and at least one bar in each corner of the stirrups. Assume approximately $1\,062/3 = 354$ mm^2 at top, at bottom, and at midheight on the sides.

For A_ℓ USE 1 – #15 = 200 mm^2 at midheight *each* side

ADD $\frac{1}{2}(1\,062 - 200) = 431$ mm^2 *each* to the top and to bottom bars. Because of beam width, there must be at least three bars on top and bottom face, that is, one at each stirrup corner and one midway between, since the spacing is limited to 300 mm. The parts of A_ℓ can be added to the longitudinal steel required for moment and the bars selected for the totals.

*With parentheses converted from total forces to unit stresses.

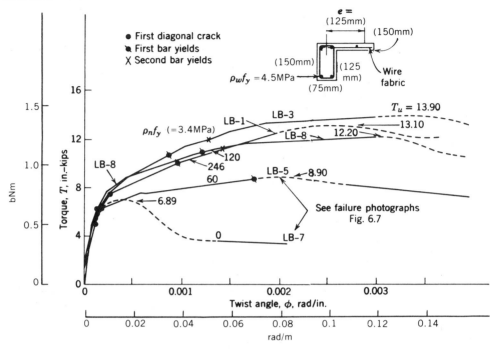

FIGURE 6.10 Effect of amount of stirrups on T-ϕ curves. (f_c' from 4060 to 4 400 psi = 28.0 to 30.7 MPa) $\rho_n f_y = A_v f_y/(b_w s)$. Beams are semi-continuous, also carrying M and V.

6.11 CALCULATION OF TORQUE DEVELOPED IN STRUCTURES

A meaningful calculation of the torque to be used in design is complex unless the torque is almost statically determined, like an awning slab cantilevered from the side face of a beam. Any statically determined case should be conservatively designed by the Code provisions already discussed.

The more general case is the spandrel beam, loaded in torque by the twisting it receives from the deflecting slab. The resulting torque is a matter of relative stiffness, which varies sharply as the members crack, much more sharply in torque than in flexural stiffness. For example, Fig. 6.10 shows torque-twist curves[8] for semicontinuous beams loaded in M, V, and T. For member LB-5 with minimum stirrups ($\rho_w f_y = 0.4$ MPa) the rotation curve shows that to increase torque resistance by 35 percent above the cracking point required an 1100 percent increase in the unit rotation. With stirrups having $\rho_w f_y = 120$ psi(0.8 MPa)(member LB-8) the torque increased only 75 percent for a 1000 percent increase in rotation. Turning this into

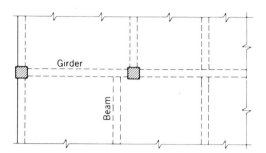

FIGURE 6.11   This beam framing produces severe torsional stresses in girder between beam and adjacent column.

analysis terms, one can say that, while flexural stiffnese decreases maybe 50 percent from cracking, torsional stiffness drops down to 5 or 10 percent its uncracked value. After diagonal cracking, stirrups will absorb relatively little additional torque from an indeterminate structure and might then serve chiefly as crack control steel. The next section presents an alternate approach the Code now permits for cases similar to the spandrel beam.

A special caution should be given for any case similar to that in Fig. 6.11 where one beam rotates the girder while the nearby column joint is either fixed or rotating in the opposite direction. Such a girder will be sharply twisted in the length between beam and column. Diagonal cracks will occur and stirrups as crack control should be liberally supplied.

6.12 ALTERNATE CODE DESIGN WITH REDISTRIBUTION OF TORQUE

Where the member torque is not required to maintain equilibrium, Code 11.6.3 relaxes the design demands by permitting design for T_n of $\frac{1}{3}\sqrt{f_c'}x^2y/3$, the equivalent of v_{tn} of $\frac{1}{3}\sqrt{f_c'}$ which is close to the cracking torsion. This torque, resisted partially by concrete (Sec. 6.8) and the remainder by stirrups, may be considered uniform for the first distance d from the support, then if the torque comes to the beam from a slab, decreasing linearly toward zero at midspan. The minimum stirrups specified in Code 11.5.5.5 for the distance $(d + b)$ beyond the theoretical point where v_{tc} would be adequate are still required.

With stirrups as specified, the cracking torque is not increased much by additional twist and the stirrups then act as crack control steel. Any theoretical end moment on the slab in excess of the nominal torque resistance assumed must be distributed back to the slab, thus increasing the slab positive moment demands.

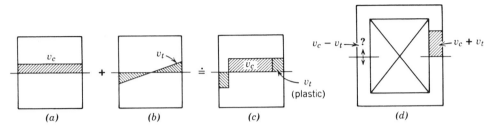

FIGURE 6.12 Shear (*a*) plus torque (*b*) on solid section results probably as in (*c*). (*d*) Equivalent on a box section.

6.13 BOX SECTIONS

No detailed design of box sections resisting torsion will be shown. Instead, related comments on member behavior under combined torsion, shear, and flexure are made because it is such a *combination* that is critical in every practical torque case the author can envision. Consider first a solid cross section near the ultimate failure zone with a shear resistance uniformly distributed across the member, in the absence of torque, as in Fig. 6.12*a*. Then, on the same section, consider the added shear resisting the torque, at the elastic stage, as in Fig. 6.12*b*. Failure will *not* occur when v_c plus v_t approaches the ordinary shear limit, because more of v_c can shift to the center of the member, leaving the side faces chiefly for torque (probably with a plastic distribution somewhat as in Fig. 6.12*c*). Such a distribution of stresses explains why the elliptical or circular interaction (Figs. 6.8 and 6.9) in the ACI Code is rational, as well as experimentally justified.

Compare the response of an open box section under the same combination of loading. If, in Fig. 6.12*d*, one superimposes the vertical shear and the torque shear, there is no nearby area to relieve the local overstress; the top and bottom slabs are too far away. A local wall failure will probably occur from the *sum* of the two stresses, equivalent to a straight line interaction diagram. This is the basis used for many European codes. The author rates this as the better design concept for the box; he does object to applying the same concept to solid sections. For solid sections the ACI interaction approach is preferred.

ACI Code 11.6.1.2 assumes an open box section (with interior corner fillets) with wall thickness at least 0.25 *b* may be considered the same as a solid section. Tests have shown the lost area not very helpful in resisting pure torsion. However, the missing area would be by no means useless in carrying shear with torsion, and its absence must weaken the section for the practical case. Fortunately, Sec. 6.14*b* indicates the Code solid section design as overly safe. Code 11.6.1.2 gives a reduced torque capacity for thinner side walls.

6.14 AUTHOR'S COMMENTS ON STATE OF THE ART AND CODE PROVISIONS

(a) Narrow Deep Members

It is possible that, as the ratio of h/b for a rectangle increases to where it represents more of a wall-like section, the circular (Code) interaction may prove less tenable; an L-shape with h/b_w of 3 for the web, for example, shows at the diagonal *cracking load* (but not at ultimate) less of the circular interaction than does one with h/b_w of 2. It is also imaginable that the redistribution beyond initial diagonal cracking might well be much smaller for a big box than for a smaller solid section.

(b) Need for $A_\ell = A_t$

None of the author's tests with stirruped L-beams has given results similar to those of Fig. 6.9a, probably because he has never tested with the steel limitations there imposed, that is, volume of longitudinal steel equal to volume of stirrup steel. When adequate longitudinal reinforcing for moment is used, the total longitudinal reinforcing is usually much more than equal in volume to the transverse reinforcement.* A number of the author's tests have shown that stirrups valued at $\rho_n f_y = 60$ psi (0.4 MPa) (such as beam LB-5 in Fig. 6.10) did proportionately increase strength under combined torsion, shear, and flexure loading beyond the nominal v_{tc} value of $0.42\sqrt{f'_c}$.† He considers that the conditions that led to the straight line from $0.20\sqrt{f'_c}$ were the result of test beam designs not practical for real design. The design of longitudinal steel primarily for the moment seems automatically to eliminate any need for the $0.20\sqrt{f'_c}$ limitation. A limit for v_{tc} of approximately $0.42\sqrt{f'_c}$ would be more appropriate and could also reduce the stirrup requirements. The use of an elliptical interaction for the stirrups would give a further saving, although the author would not recommend this for box beams.

(c) Influence of Distributed Loads

Continuous L-beams with stirrups[4] and continuous T-beams without stirrups[5] both seem to show better under distributed torque and shear loading than under constant torque and shear. Tested under T, V, and M, with

---

*Note that the author compares *total* $A_s + A'_s$ with total stirrups, *not at all* the same as the Code A_ℓ compared with A_t steel. The author's only limitation is to keep $A_s + A'_s \geqslant 1.25 A_s$, which appears to be a safe rule.

†The increase in Fig. 6.10 was from T_u of 6.89 in.-k (0.78 kN · m) for LB-7 to 8.90 in.-k (1.01 kN · m) for LB-5, or 29 percent. These two beams are shown at failure in Fig. 6.7 where numerous fine cracks show with the light stirrups, which is desirable.

critical T and V stresses assumed to be at a distance d from the support, the strengths showed at least 25% better than with the constant T and V. This increased strength probably results from the fact that critical stress combinations then exist over much shortened lengths of the beam, thus possibly damping the spread of cracks after they initially form.

(d) The Spandrel Beam as an L-beam

Divided opinion exists as to whether a spandrel beam should be designed for torsion as a rectangular or an L-beam. One investigator[6] noted large yield cracks in the slab at the face of spandrel web and no diagonal cracks in the slab, which caused him to consider the beam as a rectangular beam for torque resistance. However, he found on that basis a substantial *excess* torque strength for his "rectangular" beam.

The author notes that a slab has two functions in this case, one to deliver load to the beam and the other to help carry the torque that results. The slab probably works imperfectly to resist the torque at midspan where the torque is small, but near the column it should be more effective. Whether that is true or not, the slab forces the spandrel to twist about an axis at the level of the slab and not about the natural center of rotation as a rectangular beam. This alone increases the torque resistance through transverse bending. Also, after diagonal cracking, when the spandrel starts to lengthen under torsion, the monolithic slab tends to resist this lengthening and adds a counteracting axial force. In summary, the slab in this case probably works differently from the flange in a freestanding L-beam, but the flange does act significantly and effectively in the matter of strength. In fact, the Code limitation on flange overhang to $3\,h_f$ is about half of what might safely be used in the author's opinion.

(e) The Large Rotation Necessary to Develop Stirrups

The author doubts that the full torque value of stirrups can actually be developed in a structure, because of the excessive rotations required to bring them fully into effect (Fig. 6.10). Possibly one needs to discount the last 40% of rotation capacity and the gain of resistance associated with it. This might mean discounting, say, the last 20% of stirrup strength, although what percentages are proper certainly needs to be further explored.

(f) Relation of Stress on A_ℓ to Rotation Angle

Longitudinal steel stress from torsion appears to be wholly a function of rotation angle (under combined loading) rather than a linear function of the torque carried. Before diagonal cracking this longitudinal stress is negligible, in the order of 15 or 20 MPa. It seems to be roughly the same for a given angle of twist regardless of whether the minimum or the maximum stirrups are used. Any necessary limitation the structure imposes on ultimate rotation should thus relieve also the present demands for longitudinal steel added in proportion to the stirrups used for torsion.

SELECTED REFERENCES

1. ASCE-ACI Committee 426, "The Shear Strength of Reinforced Concrete Members—Slabs," *Proc. ASCE, Jour. Struct. Div.*, Aug. 74, ST8, pp. 1543–1591.
2. ACI-ASCE Committee 326, "Shear and Diagonal Tension," *Jour. ACI, Proc.*, 59, Jan. 1962, p. 1; Feb. 1962, p. 277; Mar. 1962, p. 352.
3. Thomas T. C. Hsu and E. L. Kemp, "Background and Practical Application of Tentative Design Criteria for Torsion," *Jour. ACI, Proc.*, 66, Jan. 1969, p. 12.
4. K. S. Rajagopalan and Phil M. Ferguson, "Distributed Loads Creating Combined Torsion, Bending, and Shear on L-Beams with Stirrups," *Jour. ACI, Proc.*, Vol. 69, No. 1, Jan. 1972, p. 46.
5. David J. Victor and Phil M. Ferguson, "Beams Under Distributed Load Creating Moment, Shear, and Torsion," *Jour, ACI, Proc.*, Vol. 65, No. 4, April 1968, p. 295.
6. John Minor and James O. Jirsa, "A Study of Bent Bar Anchorages," *Structural Research at Rice*, No. 9, Department of Civil Engineering, Rice University, Mar. 1971.
7. *Torsion of Structural Concrete*, SP-18, American Concrete Institute, Detroit, 1968, 505 pp. (A collection of 19 papers.)
8. Umakanta Behera, K. S. Rajagopalan, and Phil M. Ferguson, "Reinforcement for Torque in Spandrel L-Beams," *Proc. ASCE, Jour. Struct. Div.*, Feb. 1970, ST2, p. 371.
9. Frederick P. Wiesinger's discussion of 318-77 Code, *Jour. ACI, Proc.*, V. 74, No. 7, July 1977, p. 305.

PROBLEMS

PROB. 6.1 A 150 mm thick flat plate floor (Fig. 15.1), with $d = 120$ mm (average) brings the load from a 6×6 m area to a 450 mm square column at the center of the area, $w_d = 4$ kPa, $f'_c = 25$ MPa. (a) For a $w_\ell = 5$ kPa, check whether the slab is safe for shear. (b) At its ultimate in shear what w_ℓ would be permissible on the slab?

PROB. 6.2   A rectangular beam freestanding except for being fixed against any rotation at each end, must carry a midspan live load of 150 kN which can be as much as 300 mm off the axis of the beam. Given $b = 300$ mm, $d = 500$ mm, $h = 575$ mm, $\ell_n = 6$ m, $f'_c = 30$ MPa, and Grade 400 steel. Design the necessary shear and torsion reinforcement.

PROB. 6.3   A spandrel beam 300 mm wide by 500 mm deep ($d = 460$ mm) with a slab 100 mm thick available (on one side) to act as a flange must carry ultimate loads producing a shear V_u of 22 kN and a torque T_u of 24 kN · m. Using $f'_c = 30$ MPa and Grade 400 steel, design stirrups and longitudinal steel to be added (to the flexural demands).

7

DEVELOPMENT AND SPLICING OF REINFORCEMENT

DEVELOPMENT LENGTH

7.1 DEVELOPMENT LENGTH ℓ_d AND ITS RELATION TO BOND

Bar development length ℓ_d is the embedment necessary, under specific surrounding conditions, to assure that a bar can be stressed to its yield point, with some reserve to insure member toughness. The necessary length is a function of variables beyond those the Code now uses, that is, bar size d_b, f_y, and f'_c; certainly needed are bar spacing and cover[1, 2], and the influence of stirrups.

The state of the art is such that the 1977 Code differentiates between spacings only by a reduced ℓ_d where spacing is at least 150 mm, and varying cover is not mentioned. ACI Committee 408 is currently reviewing this area, including a further study[3] that finds meaningful ratios of $u/\sqrt{f'_c}$ (clear cover)$/d_b$, and d_b/ℓ_d in a regression analysis, with a further term suggested to reflect the strengthening influence of transverse reinforcement. At this writing it appears that clear cover (or one-half the clear spacing) and the influence of transverse reinforcement or stirrups will be in the Code before too long. The author's suggestions in Ref. 2 appear practical but are not expected to reflect closely or accurately Code changes finally adopted.

Bond stress as such is not in the 1971 or 1977 Code, although the specified ℓ_d lengths are directly based on 1963 allowable bond stresses. Bond stress is the local longitudinal shear stress, per unit of bar surface, transferred from the concrete to the bar to change the bar stress from point to point. Discussion in Secs. 7.4 and 7.6 shows bond stress very varied along the bar. The ℓ_d concept is based on the experimental evidence that the *average* bond stress* is critical, not the higher peak bond stress usually adjacent to cracks.

Whether the ℓ_d concept (Sec. 7.9) or the bond concept is considered, approaching failure in a flexural member is usually indicated by longitudinal cracks that start from flexure cracks and extend toward the lower stressed end of the bar. The typical final failure (Fig. 7.3) is a continuous longitudinal splitting crack; these cracks develop from the ring tension in the concrete (Sec. 7.2) around the bar. It is hoped that in the future, calculation methods for average splitting stresses or values derived from them may be available. Such an analysis should reveal that the present ℓ_d values based on the splitting failure really discriminate against bars embedded in mass concrete (perpendicular to the surface).

In such a changing and developing area it is important to understand what is known (and what is unknown) about bond stress, even though the Code simplifies this in such a way that bond stress itself is not usually calculated.

For development length the presentation follows this sequence:

Sec. 7.2   Internal behavior and cracking from bond acting on tension bars.

Sec. 7.3   Relation between development length and bond stress.

Sec. 7.4   Bond stress concentrations under normal conditions.

Sec. 7.5   The problem of testing for bond stress or development length.

Sec. 7.6   Bond pullout tests.

Sec. 7.7   Bond beam tests

Sec. 7.8   Semibeam tests.

Sec. 7.9   The shifting emphasis in bar development field.

Sec. 7.10–Sec. 7.16   Design considerations.

Those interested only in design can skip Secs. 7.3 through 7.9.

---

*The designer would probably profit at times by knowing the old standard equation for bond stress u:

$$u = V/(\Sigma o \cdot z)$$

where Σo is the total perimeter of the bars at the section considered and z is the arm of the internal resisting moment couple (on the cracked section).

7.2 INTERNAL BEHAVIOR AND CRACKING FROM BOND ACTING ON TENSION BARS

The internal stress distribution for many years could only be rationalized on the basis of the longitudinal failure crack. In 1971 Goto published[3] photographs of internal cracking patterns that resulted from pulling both ends of a tension bar embedded in a concrete prism encasement, as shown in Fig. 7.1. Transverse cracks from axial tension developed at intervals, much like the appearance of flexural cracks in a beam, and in effect created a number of small length specimens. Adjacent to each side of each crack and adjacent to each end these specimens developed the same internal crack pattern.

The systematic internal crack pattern, with cracks starting just behind each lug and progressing diagonally a limited distance toward the nearest transverse crack, is impressive. Since cracking in concrete results from a principal tensile stress, there must be compressive stresses parallel to these cracks, forces directed outwardly around the bar to form a hollow truncated cone of pressure. This pressure is directed inwardly against the lug and outwardly must be resisted by ring tension in the concrete.

The pictured angle between crack and bar axis is surprisingly large, considerably more than 45 degrees. This slope points to radial splitting

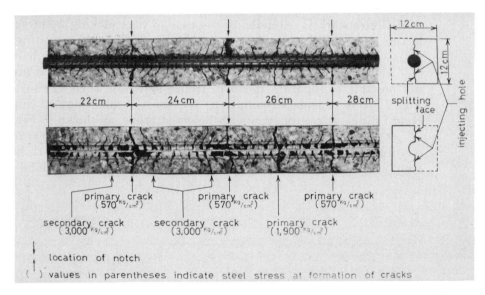

FIGURE 7.1 Internal cracks created by tension pull on both ends of the bar. Cracks were dyed with an injection of red ink which, in the bar imprint below, also spread over part of the bar surface where slip occurred. (Courtesy ACI, Reference 4.)

forces much exceeding the longitudinal bond component. It seems rational to hypothesize that these crack directions are actually functions of the thickness of cover, most steeply angled to the bar when the cover is large enough to produce a large resisting ring tension and at flatter slopes when cover is small. In any event, the radial pressures are large, and an idealized sketch is shown in Fig. 7.2. These outward forces, like water pressure in a pipe, lead to splitting on weak planes along the bar unless the bar cover is unusually large or the aggregate very soft.

The behavior at lower load levels had been rationalized much earlier from pullout specimens (Fig. 7.6). The supported face of the cylinder or prism around the bar was assumed to be somewhat like the face of a tension crack (ignoring the compression from the support). The end of the bar when loaded first pulled out enough to break the adhesion of the concrete close by and shift that resistance to a frictional resistance to sliding and some bearing on the nearest bar lug. Further pulling out then resulted from crushing against the lug or from shearing off the concrete to form a cylinder having a diameter matching that of the bar out-to-out of lugs, the latter especially with small bars or with lightweight aggregate. For large bars the cylinder or prism usually split lengthwise, in spite of lateral confinement from friction at the bearing plate. For many years this specimen splitting was not considered related to behavior in the beam. Now the splitting is nearly always the key to the member behavior.

If Goto's work is extended to the case of bars picking up stress, as in a beam, and to splices, it may open the possibility of computing *splitting* stresses and a more rational design procedure for developing bars.

Although splitting should be considered apart from bond, at present it is not possible to separate the two. Hence splitting, such as that shown in Fig. 7.3a, is considered simply as the most visable sign of approaching bond failure. It occurs even with heavy stirrups, as shown in Fig. 7.3d.

Likewise, Fig. 7.3c indicates that diagonal tension (shear) cracking and possibly resulting dowel action can be closely associated with bond splitting; stirrups would help. Traditionally bond and shear have been treated

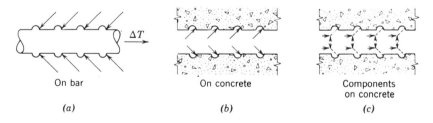

On bar	On concrete	Components on concrete
(a)	(b)	(c)

FIGURE 7.2 The forces between a deformed bar and concrete that may cause splitting, as in Fig. 7.3.

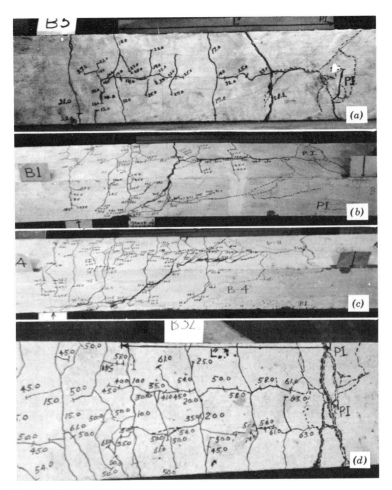

FIGURE 7.3   Bond splitting photographs of beams tested as shown in Fig. 7.9. (*a*) Splitting is directly over bar and runs lengthwise. The cross cracks are flexural cracks. (*b*) Failure chiefly from bond splitting. (*c*) Diagonal tension (inclined side crack) combined with longitudinal splitting to produce failure. (*d*) Splitting over two bars in beam with heavy stirrups. (The L'' marked is the development length ℓ_d.)

as separate phenomena. This procedure has been followed here for the most part. However, Sec. 5.23*e* points out that bond and shear are interrelated topics and that in many ways neither can be understood fully without knowledge of the other.

7.3 RELATION BETWEEN DEVELOPMENT LENGTH AND BOND STRESS

A bar with enough embedment in concrete cannot be pulled out. If, after slip at the loaded end has progressed far enough to develop bond over a considerable length, this bar reaches its yield strength, it will fail in tension; it is then described as fully anchored in concrete. The algebra relating ℓ_d to allowable bond stress is hidden in the constants in Sec. 7.10 (Code 12.2), but some related discussion should improve the designer's understanding in this area.

The development of a bar in a structure where the stress varies from zero at bar end to a maximum, usually considered f_y, at some critical cross section, is a similar concept. The required development and anchorage lengths would be the same if the two lengths related to the same materials and surroundings. To be specific, a top bar in an overhanging (cantilever) end of a beam must be developed in the overhang. If it is anchored back into the adjacent beam span, the bar size, yield strength, and concrete strength are automatically the same. If the bar spacing, bar cover, and possibly the shear forces* are the same, the minimum anchorage and development lengths are considered the same.

The basic concept of anchorage length considers a bar embedded in a mass of concrete, as in Fig. 7.4. The actual bond stress will be distributed similarly to that of the pullout test (Sec. 7.6), quite large near the surface and nearly zero at the embedded end until fairly close to failure. If the average bond stress u were limited to a permissible value determined from comparable pullout tests, safe results should be obtained. Based on this logic, at ultimate

$$A_b f_y = u \ell_d \pi d_b \text{ (for one bar)}$$

Substituting for bars of diameter d_b, $A_b = \pi d_b^2/4$, leads to

$$\ell_d = \frac{f_y}{4u} d_b \tag{7.1}$$

This ℓ_d is the minimum permissible anchorage length. However, even the deformed bar anchored in mass concrete exhibits several flaws in this logic, some on the safe side, some unsafe insofar as design is concerned.

On the safe side, the mass concrete does not fail like the cylinder or prism cast around a pullout bar, which always splits open without actually failing the bar (unless the bar is *very* small or the aggregate is weak against local crushing). The same anchorage length in ordinary mass concrete should thus be stronger. One must note, however, that even in

---

*Interaction of shear and anchorage is mentioned in Sec. 5.23e.

$$A_b f_y = u\, \ell_d\, \Sigma o$$

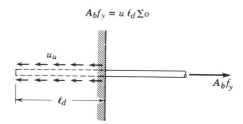

FIGURE 7.4   Anchorage of a bar.

mass concrete a bar parallel to the surface and with only nominal cover is as free to split off this cover as if it were in the top of a thin slab.

On the unsafe side, doubling the pullout embedment length will not double the pull it will resist. The longer the specimen the less efficient it becomes and linear extrapolation of data is normally on the unsafe side. Also, groups of bars are probably less effectively anchored than individual bars.

Development lengths computed on the above basis (Eq. 7.1) can be no better than the values of bond stress used in deriving them. Since the development may occur in very diverse surroundings, as to spacing, cover, edge distance, and enclosing stirrups, no single value of u or ℓ_d is really appropriate over a large range.

This discussion may appear to be an argument for calculating the bond stress itself and limiting it to an acceptable value. Codes as late as 1963 included computation of the flexural bond stress from the formula in the footnote to Sec. 7.1. But design control by bond stress is no more reliable than by development length.

Basically, strength is not sharply influenced by *local* bond stresses or local splitting. Although concrete appears to be brittle where failing in tension, it has a surprising ability to carry local overstress and readjust to such a change that *average* stress appears to be the sounder failure criterion. The average attainable, however, is subject to variations for length, cover, spacing, stirrups, and other factors. The only reliable way to determine an allowable average bond stress is to test a similar situation. Any such test establishes a maximum change in bar tension over a given development length; the allowable bond stress can then be evaluated from Eq. 7.1 rearranged to

$$u = f_s d_b / 4 \ell_d$$

Because a varying shear is not usually practical in a test situation, the test moment variation is usually linear; one then obtains the same answer as from the bond stress equation, $u = V/(\Sigma o \times z)$.

If one uses the flexural bond equation, the allowable stress must thus come from a development length equation. There is no advantage in going from test development length to allowable bond and then back to required development length. Since no improved accuracy results from the allowable bond concept, the 1977 Code no longer uses it. The development of the specific ℓ_d values used in the 1977 Code is covered in Sec. 7.9.

7.4 BOND STRESS CONCENTRATIONS UNDER NORMAL CONDITIONS

For a better understanding of member behavior, the very uneven distribution of bond stress at service loads and probable readjustment prior to failure are important. Consider a simple beam as shown in Fig. 7.5. Although at B in the constant moment length the nominal steel stress is constant, the bond stress is large (not zero) adjacent to each flexural crack.

At the crack the steel carries most of the tension; but between cracks, tension in the concrete shares significantly. Thus the bond stress condition on each side of a crack is almost identical with that at the loaded end of a pullout specimen, with a near ultimate value of bond stress existing for a short distance adjacent to the crack,[4] of opposite sign on the two sides of the crack. The author calls these the "in-and-out" bond stresses. At a moment crack in a section carrying a shear, as at C, the bond stresses needed to build up the steel stress are superimposed on those caused by the crack itself, but some readjustment is necessary on the side of the lesser bar stress since the total bond stress at any point cannot exceed the ultimate already existing next to the crack. With #55 bars of grade 400 steel, even service load may lead to some splitting over the bars on each side of the cracks where moment is constant, and on the low moment side of the cracks elsewhere. On closely spaced bars the splitting would show on the side of the beam near the level of the steel. With #35 bars the service load will probably produce no such cracks, but some will show in a pure flexure zone before the yield stress of Grade 75* bars is reached. The probability of the splitting cracks decreases with decreasing bar size and lower steel stress.

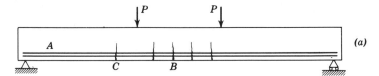

FIGURE 7.5   Crack pattern in a simple span.

*In p.s.i. units; corresponding to $f_y = 517$ MPa.

At point A in Fig. 7.5 the concrete carries most of the tension and both steel stress and bond stress are low. At C, the first flexural crack from the reaction, the steel stress increases to carry almost all of the tension and the short length to the left of C must marshal bond stress to offset all the deficiency between C and the reaction. The smaller the percent of steel in the beam the longer the necessary section of this high bond stress and the greater the possibility of some local splitting.*

The special cases of brackets, of variable depth members, deep beams, and bar cutoff points, where special design precautions are necessary, have already been discussed in connection with shear, Secs. 5.16 to 5.20.

It appears probable that bond stresses become less variable as slip takes place and loads approach the ultimate. Goto's crack pictures seem to show an equalizing process. Near ultimate, splitting will have started and will have lowered some of the peak bond stresses. The detail of what occurs has not been mapped for development length, but for splices some conclusions are developing (Sec. 7.18).

Tests show that some of the in-and-out bond stresses between cracks relax substantially under repeated loading, but they reappear when a higher load is applied; the relaxation is local and develops only at the load level which is often repeated. With reversal of forces this decay in bond resistance is more severe. Study is continuing with reversing stresses.

7.5 THE PROBLEM OF TESTING FOR BOND STRESS OR DEVELOPMENT LENGTH

Except in relation to splice tests, a good bond test has not yet been devised. The problems are several:

a. Inability to scale down specimens with confidence. (Note discussion below.)
b. Some interaction between shear and bond stress.
c. Nonlinear response, either with bar size, steel stress, or development length (which in design are related quantities).
d. A wide range of failure modes that can range from crushing against the lugs or longitudinal shearing of a surface just outside the lugs to the more usual splitting of concrete, either over individual bars or through an entire layer of bars.

(a) Possibly because much higher bond stresses can be developed around small bars than large ones, or probably because the spacing of flexural

---

*To the extent smaller ρ means smaller bars, this is an offsetting factor.

cracks does not scale down properly as specimen size is reduced, scaled specimens react differently from full size ones.

(b) As already pointed out in Sec. 5.23e, particular combinations of shear and bond stress can lead to lower shear strengths. The mixture of shear and bond problems in tests is evident in Fig. 7.3b,c.

(c) The longer the test length of the bar in the specimen the lower the average bond stress attained at failure. A Grade 300 bar can be developed in less than two-thirds the length required for the same size of a Grade 400 bar. (In some splice tests of Grade 300 bars 83 percent of the two-thirds yield stress ratio was entirely adequate.)

(d) The tensile splitting stress and the influence of cover thickness on ultimate strength,* neither of which can yet be evaluated with confidence, make generalizations difficult.

The only sure test at present is a full size specimen combining the given mix of variables, which is a very expensive procedure and one difficult to generalize to a different mix of variables. The next several sections discuss various specimens that have been used and their weaknesses.

7.6 BOND PULLOUT TESTS

Permissible bond stresses were formerly established largely from pullout tests with some beam tests as confirmation. A bar was embedded in a cylinder or rectangular block of concrete and the force required to pull it out or make it slip excessively was measured. Figure 7.6 shows such a test schematically, omitting such details as hemispherical bearing plates. Slip of the bar relative to the concrete is measured at the bottom (loaded end) and top (free end). Even a very small load causes some slip and develops a high bond stress near the loaded end, but leaves the upper part of the bar totally unstressed, as shown in Fig. 7.6. As more load is applied, the slip at the loaded end increases, and both the high bond stress and slip extend deeper into the specimen. The maximum bond is somewhat idealized in these sketches; its distribution depends on the type of bar and probably varies along the bar more than shown.

When the slip first reaches the unloaded end, the maximum resistance has nearly been reached. Failure will usually occur (1) by longitudinal splitting of the concrete in the case of deformed bars, or (2) by pulling the bar through the concrete in the case of a very small bar or very light-weight aggregate, or (3) by breaking the bar, if the embedment is long enough.

The average bond resistance u is always calculated just as though it were uniform over the bar embedment length. In this test simulation of

---

*But see Sec. 7.18 for cover effect on splices.

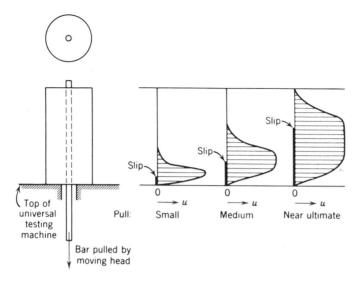

Figure 7.6 Bond pullout test, with bond stress distributions.

usage conditions is not attempted, no tension cracks cross the bar, and adjacent concrete in compression increases loaded end slip. Friction on the base restrains splitting of the specimen; many tests have even included spirals to avoid splitting collapse. The test appears useful chiefly where relative rather than real bond resistance is acceptable, as in comparing the slip resistance of various lug sizes and patterns. The principal problem of splitting is not realistically handled.

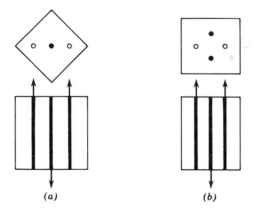

FIGURE   7.7 Tension   pullout   tests, schematic.

Modifications of this test, called the tensile pullout specimen, have also been used (Fig. 7.7) eliminating the compression on the concrete. The pull at one end is on one bar in (a) or two bars in (b) while the other end is held by pulls on the remaining bar. Although better than the Fig. 7.6 test, each introduces some of the special problems of spaced splices and any crack pattern is influenced by this interaction.

7.7 BOND BEAM TESTS

Beam tests are now considered more reliable, since these reflect the influence of flexural tension cracks. Two types of beams have been used: that of Fig. 7.8 at the Bureau of Standards and the one of Fig. 7.9 at The University of Texas at Austin. A major consideration in each was to remove reaction restraints that might confine the concrete over the bar and thereby increase splitting resistance.

The Bureau of Standards beams were heavily reinforced with stirrups and failed as sketched in Fig. 7.8b; the diagonal crack created an increased steel stress much closer to the end of the bar and concentrated bond stresses nearer the end.

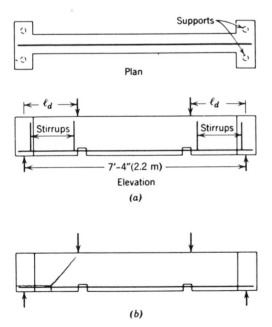

FIGURE 7.8   National Bureau of Standards bond test beam. (a) Load and support system. (b) Failure pattern.

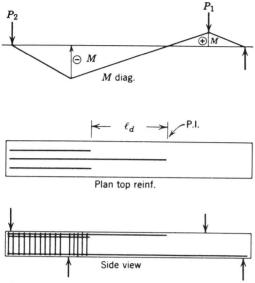

FIGURE 7.9   The University of Texas beams.
(Failure shown in Fig. 7.3.)

The University of Texas beams (Fig. 7.9) placed the bar in a negative moment region where there was no external restraint against splitting except by the use of a rather wide concrete beam.[6] Some beams were with stirrups, some without. Figure 7.3b and d shows beams that failed by splitting out to the end of the bar, whereas the beam of Fig. 7.3c first developed a diagonal crack that increased the bar stress as in the Bureau of Standards tests. When two bars were placed in the same beam width, with stirrups to carry the expected shear, the influence of bar spacing nearly offset the advantage of the stirrups. More of these tests with narrower members and stirrups appropriate to the shear would be helpful.

While the National Bureau of Standards tests incorporated an excessive amount of stirrups to avoid diagonal tension failures, the University of Texas beams were generally excessively wide and represented bar spacings too wide to be fully practical. Both specimens were expensive to make and heavy to handle, since both used full-size members with the larger sizes of bars, which create the most serious bond problems.

The results from these two widely different test specimens were reasonably in agreement and had much influence on the 1963 Code bond provisions, especially on fixing bond stresses varying as $\sqrt{f'_c}$ and decreasing with increasing length ℓ_d. In a following section it is pointed out that in tension members the permissible bond varies as $1/d_b$, where d_b' is the bar diameter up to #35 bar size. Thus ℓ_d lengths will actually vary as d_b^2 instead of the apparent d_b shown in Sec. 7.3. In other words, these beams acted as though the splitting of the concrete limited the pull that could be put into a bar in a given length to the same total number of pounds whether the bar was large or small.

For #35 bars and above beam tests were quite few in number, but seemed to indicate that a further decrease in bond stress for these bar diameters was not necessary.

7.8 SEMIBEAM SPECIMENS

To reduce specimen size and expense, partial beam specimens have been popular, such as the one sketched in Fig. 7.10. Although various details are used, those in this figure convey the general idea. The bottom reaction may bear against the end of the bar, may be spread such as not to be near the bar or, as sketched, the bar may be shielded and isolated by a soft covering. The test pull is put directly into the bar.

The overall length and the part of it used for the test length can each be varied, anything from the full length of the specimen to the reduced length. One or several bars, with or without stirrups may be used.

The advantage of this specimen lies in its simplicity. The disadvantages lie in the confining pressure against the bar (if the shield is not used), increases in the length over which splitting resistance tends to be mobilized, and in the unrealistically low ratio of shear to bond. If stirrups are used, their influence on the bond resistance should be noted. In the specimen in the sketch, there is a greater length of concrete subject to splitting for a given bar test-length than will exist in the actual member, making the apparent strength too large.

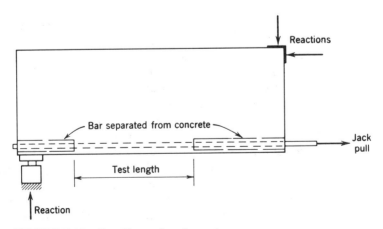

FIGURE 7.10 Cantilever bond specimen.

7.9 THE SHIFTING EMPHASIS IN BAR DEVELOPMENT FIELD

Development length or development bond was a part of the 1963 Code, but more emphasis was then placed on flexural bond. Section 7.1 (and its footnote) calls attention to omission of bond stress as such from the Code since that issue. In effect the 1977 Code (and that of 1971) specifies development lengths ℓ_d of about 1.2 times those of the 1963 bond stress would demand to assure built-in moment ductility when the steel strain exceeds f_y; a failure over the development length tends to be a brittle failure. For example, for #10 to #35 (other than top bar), with an allowable $u = 0.8\sqrt{f_c'}/d_b$:

$$\text{Reqd. } \ell_d = 1.2\, f_y d_b/4u = 1.2\, f_y d_b/(4 \times 0.8\sqrt{f_c'}/d_b)$$
$$= 0.375\, f_y d_b^2/\sqrt{f_c'} = 0.375\, f_y (4\, A_b/\pi)/\sqrt{f_c'}$$
$$= 0.477\, f_y A_b/\sqrt{f_c'} = \text{say } 0.5\, f_y A_b/\sqrt{f_c'}$$

For spacing at least 150 mm on center, a length 0.8 times this value is permitted. Since $1.2 \times 0.8 = 0.96$, this returns the development length for the wider spacings essentially to the 1963 status.

The above basis sounded logical, but in the early 1970s it began to be clear that no single development length (or allowable bond stress) for a given bar could accurately represent bar covers from 20 mm to 75 mm nor clear bar spacings from 25 mm to 150 mm, nor the absence or presence of stirrups (in a wide range of spacings). Furthermore, none of the test specimens described above had accumulated adequate data covering any one of these variables much less all three*.

Orangun et al.[3] in 1977 summarized a different approach to this problem, already mentioned in the opening paragraph of this chapter, using a regression analysis of data from many research reports. One variable was eliminated by assuming that the term C could be taken as the smaller of clear cover or clear "half-space" between bars and the larger value could be ignored. He initiated his analysis from lap splice data and found the equation below. Then he tried this approach again for development length and found the *same* equation, needing only a multiplier for very large ratios of bar spacing to cover. The usual equation for closer spacings, *the same for both splices and development length*†, was

$$u_{\text{calc}}/\sqrt{f_c'} = 0.1 + 0.25\, C/d_b + 4.2\, d_b/\ell_{\text{splice}}$$

---

*Hence the previous edition of this book could only discuss some early splice tests by the author where spacing and cover were found to be significant and then to include preliminary warning about small values of either.

†This finding surprised the author, but when he remembers how unrealistically wide the University of Texas beams were he can see the result could well be an unconservative bond stress. He would like to see additional confirming tests with narrower beams and stirrups just adequate to meet the Code demands for the ultimate unit shear.

with an added term for the influence of transverse reinforcement (stirrups, etc.). The resulting design development lengths are longer than in the 1977 Code and the design splice lengths are shorter, and the transverse steel usually required in beams lowers each. The ℓ_d equation is not given here because ACI Committee 408 on Bond and Anchorage seems now to be finding a simpler presentation about as accurate. Whatever stirrups are appropriate will be counted as transverse steel increasing u and lowering ℓ_d.

An amendment to the 1977 Code will almost surely result, but acceptance into a standard takes two years at least *after* it passes the Code Committee. In the meantime the 1977 development lengths are presented here in the next section.

7.10 DEVELOPMENT LENGTHS FOR TENSION BARS

The 1977 Code specifies four basic development lengths for tension bars, derived from allowable bond stresses listed in the 1963 Code for bars other than top bars. The version below is slightly condensed.

Code 12.2 Development length of deformed bars and deformed wire in tension
12.2.2 Basic development length shall be:
(a) For #35 or smaller bars $0.019\,A_b f_y/\sqrt{f'_c}$
     but not less than $0.058\,d_b f_y$
(b) For #45 bars $26\,f_y/\sqrt{f'_c}$
(c) For #55 bars $34\,f_y/\sqrt{f'_c}$
(d) Deformed wire $0.36\,d_b f_y/\sqrt{f'_c}$
12.2.1 Development length ℓ_d ... shall be computed as the product of ... Sec. 12.2.2 and the applicable ... factors of Secs. 12.2.3 and 12.2.4, but ℓ_d shall not be less than specified in Sec. 12.2.5.
12.2.3 Basic development length shall be multiplied by the applicable factor or factors for:
(a) Top reinforcement*   1.4
(b) Reinforcement with f_y greater than 400 MPa $(2-400/f_y)$
(c) Lightweight aggregate concrete (when f_{ct} is not specified†):
     "all-lightweight" concrete   1.33
     "sand-lightweight" concrete   1.18
     (interpolating when part sand replacement is used)
12.2.4 ... (multiply by applicable factors for ...)
(a) Reinforcement being developed in length under consideration and spaced laterally at least 140 mm on center with at least 70 mm clear from face of member to edge bar ... in direction of spacing 0.8

---

*See following discussion.
†Code also includes case where f_{ct} is specified.

   (b)   Reinforcement in excese of required (A_s reqd./A_s provided)
   (c)   Reinforcement enclosed within spiral ... not less than 6 mm diam. and
         not more than 100 mm pitch ... 0.75
   12.2.5 Development ... shall be not less than 300 mm, except in ... lap splices
         and in .. web reinforcement by Code 12.14.

   In Sec. 12.2.2a the second limitation of 0.058 $d_b f_y$ governs only for very
small bars, such as #15, or very high strength concrete.
   Because of the nonlinear response (Sec. 7.5) these equations are a bit
severe for Grade 300 bars, but not severe enough for high-strength steel.
Thus Code 12.2.3b specifies that for $f_y > 400$ MPa the lengths are also to be
increased by the factor 2–400/f_y applied to a basic ℓ_d value, which itself
includes the larger f_y, as indicated in Fig. 7.11.
   For top bars, that is, horizontal reinforcement having more than 300 mm
of concrete cast below the bars, 1.4 times the basic ℓ_d must be used. An
accumulation of air and water that rises beneath such bars becomes
entrapped on their undersides. Such bars are not as tightly held and slip at
lower loads.
   As already noted in Sec. 7.9, an 0.8 factor may be used for bars spaced
at least 150 mm on center (and not less than half that clear distance from an
edge). Reduced development length is also specified when a spiral is
available to control splitting of the concrete or where excess steel is
available and it is not necessary to develop the full yield strength.

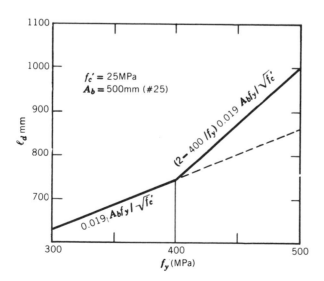

FIGURE 7.11   Development length variation with
f_y for No. 25 bar.

Particular attention is called to the increased development lengths needed with lightweight concrete, values of 1.33 for "all lightweight" (including lightweight for the fine aggregate) and 1.18 for "sand-lightweight" concrete. When the mix is well-controlled and split cylinder data are available this rule is relaxed (Code 12.2.3c). More data on the development length requirements in lightweight concrete are badly needed.

Table B.3 in Appendix B summarizes many of the ℓ_d values for tension bars; Table B.4 covers compression bar ℓ_d values (Code 12.3).

In the formulation of required ℓ_d values, allowable bond stresses, and lap splice lengths, some protection against a brittle type of failure must always be considered. To design a beam that is somewhat ductile in flexure but allow it to fail suddenly in bond splitting at the yield stress is inconsistent. Hence reserve strength in ℓ_d above f_y is essential to maintain member ductility. The Code values aim at making this extra strength about 25% of f_y. This should maintain the local bond strength and cause the failure section to occur at the moment section that has no corresponding large strength reserve above that calculated. However, especially in splices and possibly also in simple development lengths, the steel strain that can be accommodated is not unlimited. If the steel has a long yield plateau on its stress-strain curve the 25% extra strength may not be attainable, but the ability to accept a high strain would serve the same function. Strain limit information is skimpy and deserves further study.

7.11 CRITICAL STRESSES AND CRITICAL SECTIONS FOR DEVELOPMENT

The proper use of development lengths requires a clear picture of how bar tensile or compressive stresses change along a beam. Bar stresses will be a maximum where moment is a maximum; obviously no maximum stress point should be closer than ℓ_d to the end of a bar, in either direction; ℓ_d must exist to put stress into a bar or to take it out.

Equally important, and easier to overlook, is the peak stress that develops in bars wherever a neighboring tension bar is cut off or bent. At that point the total tension must crowd into the (reduced number of) continuing bars. Since the cutoff for moment is usually made where the remaining bars have just enough capacity to handle the total tension, the full yield stress in the continuing bars determines their ℓ_d requirement. The design stress just beyond the cutoff depends on what has controlled that location.

First, if the cutoff has been determined by the development length of the bar that is cut off, the cutoff is farther than needed for moment and the f_s in the remaining bars (less than f_y) determines ℓ_d. It is safe (sometimes also wasteful) to ignore this extra extension and measure the next ℓ_d as though the bar had been cut at the point determined for moment.

The second control is involved with design philosophy, already introduced at the close of Sec. 4.11g. If the arbitrary extension of the bars is considered entirely arbitrary, not related to stress, then stresses permit a cutoff at the theoretical moment diagram used; the development length of the continuing bars can then be considered to start there and this satisfies the Code. The author recommends instead that the arbitrary extension be considered as defining a new (shifted) moment diagram; this concept leads to ℓ_d measured from this shifted diagram rather than the basic one. The examples show the use of both concepts—the legal Code requirement and the author's recommendation.

7.12 EXAMPLES SHOWING USE OF DEVELOPMENT LENGTHS

Examples illustrate more specifically some of the above general ideas.* The basic discussion is continued within the example format.

(a) A 300 mm concrete wall heavily loaded requires a footing 1.7 m wide as sketched in Fig. 7.12. The footing is designed for $f'_c = 20\,\text{MPa}$ and Grade 400 reinforcement. What maximum size bars will conform to the development length requirements?

Solution

With a concrete wall, the critical footing section for flexure is at the face of wall (Sec. 20.7). The bars are there stressed to maximum and project 625 mm, leaving 50 mm end cover. Based on the values of Sec. 7.10, a bar must be found that has an ℓ_d of 625 mm or

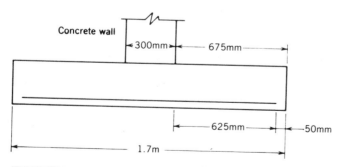

FIGURE 7.12   Concrete wall footing.

*See also Fig. 15.9 (taken from the Code) showing good practice for slabs supported directly on columns rather than beams.

less. In the absence of a table of ℓ_d values of various bars under various conditions (which a design office will surely develop), the required ℓ_d is:

$$\ell_d = 0.019\ A_b f_y / \sqrt{f_c'} \text{ but not less than}$$
$$\ell_d = 0.058\ d_b f_y.$$

The second value governs only with small bars and will not be checked initially.

$$0.019\ A_b\ 400 / \sqrt{20} = 625 \text{ mm}, \quad A_b = 368 \text{ mm}^2$$

A #20 bar ($A_b = 300$ mm^2) is the largest permissible without a hook. Checking the other limit with d_b for the #20 bar,

$$0.058\ d_b f_y = 0.058 \times 20 \times 400 = 464 \text{ mm} < 625 \text{ mm} \qquad \textbf{O.K.}$$

It is noted that a hook at the end of bar would reduce the requirement for straight bar development length and permit larger bars (Sec. 7.14).

(*b*) Assume the footing is made of sand-lightweight concrete but the example is otherwise the same.

Solution

Required development length is increased by a 1.18 factor when lightweight concrete with a sand fine aggregate is used. If the spacing is under 150 mm,

$$1.18\ (0.019\ A_b \times 400 / \sqrt{20}) = 625 \text{ mm}$$
$$\text{Max. } A_b = 311 \text{ mm}^2$$

A #20 bar remains the largest usable bar (300 mm^2) at this spacing.

(*c*) The 4-#30 bars in Fig. 7.13 are designed for the moment diagram shown. The 0.44 m extension of one-third the bars beyond the point of inflection (P.I.) is required by the $\ell/16$ rule* (Code 12.13.3). May two bars be cut off as indicated on the moment diagram? Assume that the shear requirements of Code 12.11.5 have been satisfied.

Solution

If the bars are cut as shown, the spacing of the extended bars in a 450 mm wide beam width will be near 225 mm. Also, since these would then be developed at or near B, the two short bars would be contributing all the important splitting forces from B to the support and each of these could also mobilize splitting resistance over a 225 mm width. Hence, for the bars cut off the effective spacing would also be 225 mm, justifying the 0.8 coefficient on ℓ_d for $s \geqslant 150$ mm. As top bars they require a 1.4 coefficient.

$$\text{Reqd. } \ell_d = 0.8 \times 1.4 (0.019\ A_b f_y / \sqrt{f_c'})$$
$$= 0.8 \times 1.4 \times 0.019 \times 700 \times 400 / \sqrt{20}$$
$$= 1332 \text{ mm} = 1.332 \text{ m}$$

*Or d or 12 d_b, if larger.

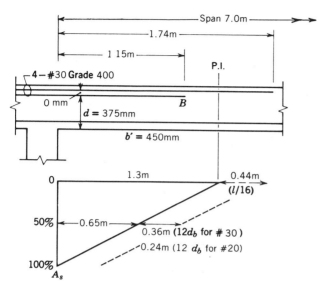

FIGURE 7.13   Data for example (*c*).

The "short" bars must extend 1.35 m which is beyond the P.I., making a moment *cutoff not feasible.* The 1.35 m cutoff also satisfies the following consideration.

The author has recommended that the dashed diagram be considered as the potential or shifted moment diagram that might result from the conditions discussed in Sec. 4.11e, something beyond the legal requirement of the Code. The 1.35 m dimension would then be so near the P.I. that one would not need to worry about the capacity of the longer bars to pick up the small necessary moment at this location. Regardless of whether a designer chooses to accept the shifted moment diagram idea or not, a stagger of cutoff points is desirable, of the order of 150 mm when closely spaced or the bar spacing where wider than 150 mm. If external forces produce any significant tension at the section, the discontinuity from stopping all bars at one length will produce an early crack. Even if flexural tension never occurs, shrinkage tends to open the crack at such a cutoff.

(*d*) If in Fig. 7.13 the moment requires only 4-#25 bars, may the given dimensions be used with the #25 bars?

Solution

ℓ_d in this case involves the same 0.8 and 1.4 coefficients as in (*c*).

$$\ell_d = 0.8 \times 1.4 \times 0.019 \times 500 \times 400/\sqrt{20} = 952 \text{ mm}$$

The 1.15 m dimension is larger and satisfactory for the shorter bars.

The development of the longer bars must also be considered. The author recommends that these bars be considered fully stressed at *B*, because of the shifted (dashed) moment diagram. The longer bars would then need ℓ_d of 952 mm beyond the *required* distance to B, or a total from the column of $(0.65 + 0.375)* + 0.952 = 1.977$ m, say 2.0 m. The longer bars should then be extended 1.85 m instead of 1.74 m.

If one totally disregards the shifted moment diagram concept, the long bars could be considered fully developed at $0.65 + 0.952 = 1.60$ m and the 1.74 m dimension could be accepted under the Code as giving 140 mm leeway. The extra 140 mm would make the author feel more comfortable than no extra. His concern exists because settlements do frequently occur and the usual moment diagrams, even those from moment distributions, are approximate. They neglect the cracking of the beams, the axial shortening of columns, and the differences between a joint as a point and as a substantial portion of the member lengths.

Since cutting off some bars frequently requires others to be run farther, each bar cutoff point becomes a critical section to be checked as regards continuing bars.

The implications of a situation often encountered (which might be overlooked in the case just discussed) is emphasized in the next example.

7.13 POSITIVE MOMENT BARS

(a) Continuous Beams†

When uniform loads are considered in simple spans or within the positive moment length of continuous spans, there is a problem in applying the basic development length concept in a meaningful way. The question finally emerging is what is the maximum bar size where moment is nearly zero. The answer is essentially the equivalent of the old flexural bond concept, which is presented below in terms of development length and some Code liberalization.

Consider the case shown in Fig. 7.14a and a simple case where ℓ_d is exactly ℓ_0 to midspan. Such a bar would develop its yield strength at midspan but would be unsafe at $\ell_0/2$ from the P.I. There the bar would be developed for only half its f_y but would need to care for $0.75\,M_n$, which requires $0.75\,f_y$. Nor is it safe to use a bar developing this $0.75\,f_y$ in exactly $\ell_0/2$, because then a somewhat similar overstress exists at points closer to the P.I. The simple answer developed below is identical with the old flexural bond demands, but encourages realistic adjustments to relax these demands.

The slope of the moment diagram at the P.I. is the shear V_n at that point. If the bars at the P.I. develop toward f_y at the rate M is initially developing, they can develop their full value f_y in the length M_n/V_n, as indicated in Fig. 7.14a. If the bars extend beyond the P.I., as some must

---

*The smaller #25 bars cause the extension to be governed by d rather than 12 d_b.

†The detailing of continuous beam reinforcement, including detailed examples, is covered in Secs. 10.13 to 10.16.

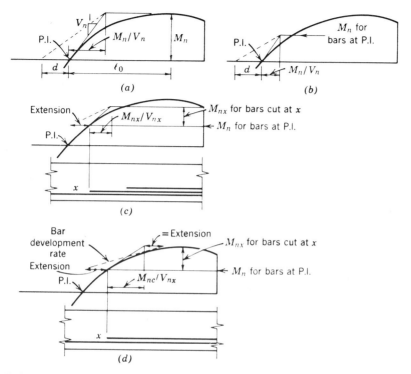

FIGURE 7.14 Available development lengths for positive moment bars at a point of inflection (P.I.). (*a*) No bars cut. (*b*) Bars at P.I. less than total A_s. (*c*) General case. (*d*) For first bars cut off (nearest point of maximum positive moment).

according to the Code, the development length concept implies that local "overstress" in bond at the P.I. is not a problem because it no longer occurs at the end of the bar. The extra length beyond the P.I. which may be counted is a judgment decision, but the Code limits it to ℓ_a equal to the effective beam depth d. The criterion becomes:

$$\text{At a P.I. choose bars with } \ell_d \gtrless M_n/V_n + d$$

The dotted development rate line indicates this would be safe even with a slightly shifted moment diagram. The shifted moment diagram is less critical in a positive moment region, but still valid in the author's opinion.

If not all bars extend to the P.I., Fig. 7.14*b* indicates the only change needed is to include in M_n only the bars actually reaching the P.I.

Usually, if bar size at the P.I. is satisfied and other positive moment bars are not too widely different in size, the required bar lengths for flexure will control over development lengths. However, if larger bars are used away from the P.I. they could be selected by the same procedure used at the P.I. with one modification indicated in Fig. 7.14c. These bars must care for all or part of the moment in excess of M_n developed by the other bars at the P.I. If their moment capacity is designated M_{nx}, the local demands* indicate ℓ_d equal to or less than M_{nx}/V_{nx}, where V_{nx} is the shear at x. Any further extension of the bar obviously adds to the available ℓ_d to give the criterion for bars not extending to the P.I. as:

For the added bars $\ell_d \gtrless M_{nx}/V_{nx}$ + extension used

If such a criterion is used for the first bars cut off, the moment diagram will be rather flat. Any added arbitrary bar extension not only moves the critical bond stress away from the nominal cutoff point but also permits a flatter development line as sketched in Fig. 7.14d. It appears safe, *for a moment diagram that can shift very little*, to use the criterion:

For the added bars *first* cut off $\ell_d \gtrless M_{nx}/V_{nx}$ + twice the extension

It is noted again that the need for this approach occurs only when the moment diagram has this general shape and when larger bars are used here than at the P.I. If the moment diagram were a straight line, as for a concentrated midspan load, the procedures used for negative moment would be adequate.

(b) Simple Spans

At the support of a simple span that is hung from above by an embedded bar hanger, the condition is exactly that at the P.I. of the continuous span, except that usually the bar ends with an extension ℓ_a less than d (Code 12.12.3). However, more commonly the reaction is from below and provides a compressive force that restricts possible splitting of concrete caused by bond stresses. The Code recognizes this by specifying that the bars may be chosen from the criterion:

For bars confined by a reaction use $\ell_d \gtrless 1.3\, M_n/V_u + \ell_a$.

The 1.3 factor leaves this less strict than at a P.I. *Simple* beams can rarely be supported by framing into a girder which picks up the reaction in shear, but where such a case exists the 1.3 factor must not be used.

7.14 HOOKS AND END ANCHORAGE

Where bars must be anchored (or developed) in tension within a limited distance, a hook can add some capacity. A hook in compression is useless.

---

*The old flexural bond requirement would not even indicate this problem unless Σo were used as only the bars being cut off, not the total number at x.

Although in tension a hook is not as good as an equal length of straight bar, a tension hook can be used to advantage where no room exists for the equal length of straight bar. For example, the footing bars of the example in Sec. 7.12a can be made larger if hooks are added at the bar ends.

A standard hook is defined (Code 7.1) to have one of the forms shown in Fig. 7.15a. A general evaluation of a hook is difficult. In mass concrete or other confined conditions a hooked bar slips more than the same length of straight bar, although the ultimate strength may be reached. In a structural member where the bar is close to an exposed face the hook pressure on the concrete tends to split off the concrete face and a lowered strength results. In such a place the value is limited to the splitting resistance of the concrete, a value not well established for this condition but definitely lower than the hook value in mass concrete.

The Code (12.5) assumes either standard hook of Fig. 7.15a develops a tensile resistance $f_h = \xi\sqrt{f_c'}$, where the ξ values in Table 7.1 are those tabulated in the Code for #10 to #35 bars. A tie or stirrup hook (Fig. 7.15b) is valued in Code 12.14.2.1 as the equal of $0.5\ \ell_d$.

Analysis of the top and bottom beam bars in Fig. 7.15c emphasizes several differences that are important. End anchorage of such bars is not as simple as at interior columns and may be quite troublesome.

For the top bars the designer has the option of developing fully (anchoring) smaller diameter bars or of using larger bar diameters at less than 100% efficiency. The upper #25 bars must be classified as *top* bars and Table 7.1 shows ξ of 30 for either Grade 300 or Grade 400 bars. The value of the hooks, assuming normal weight concrete and $f_c' = 20$ MPa is

$$f_h = \xi\sqrt{f_c'} = 30\sqrt{20} = 134.2\ \text{MPa}$$

The remaining length of straight bar is 250 mm, which can be used in the ℓ_d relationship to evaluate the f_s obtainable.

$$\ell_d = 1.4(0.019\ A_b f_s/\sqrt{f_c'})$$
$$250 = 1.4 \times 0.019 \times 500\ f_s/\sqrt{20},\ f_s = 84.1\ \text{MPa}$$

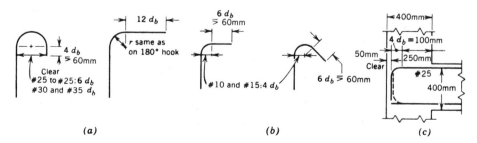

(a)   (b)   (c)

FIGURE 7.15 Standard hooks. (*a*) General case. (*b*) Ties and stirrups only. (*c*) Use of hooks. (Radius and diameter are each to inside of hook.)

TABLE 7.1 ξ Values

Bar Size	$f_y = 400$ MPa		$f_y = 300$ MPa
	Top Bars	Other Bars	All Bars
#10 and #15	43	43	33
#20	37	43	31
#25	30	43	30
#30	30	39	30 .
#35	30	34	30

The total value for the #25 bar is thus $134.2 + 84.1 = 218.3$ MPa, about 73% of the normal value of Grade 300 bars or 55% of the value of Grade 400 bars. The design moment requirements for A_s must then be based on this lowered efficiency if #25 bars are used.

If the column has a dependable compression load, this might improve the value of the straight bar portion of the top bar, as discussed for bottom bars in simple spans in Sec. 7.13. For such a case the value of the 250 mm length might be taken as $1.3 \times 84.1 = 109.3$ MPa. The hook portion is probably not improved by the vertical compression. Note that if the column ever loses its compression, as in an overturning or earthquake situation, the extra strength is *not* available.

The bottom bars might need evaluation on either of two bases. If the frame has shear wall bracing, the value of a straight #25 bar in compression* would be established from the available 350 mm embedment:

$$\ell_d = 0.24\, f_y d_b / \sqrt{f_c'}$$
$$350 = 0.24\, f_s' \times 25 / \sqrt{30}, \qquad f_s' = 319.5 \text{ MPa}$$

This would be the limiting value of such a bar used as compression steel, if f_y is assumed to be greater. A hook would *not* improve its compression strength; even if a hook were used to meet the requirements of the next paragraph, the author would use the straight 250 mm portion then available to compute the usable f_s'.

The second basis of evaluation would apply if the beam were a part of a primary lateral load resisting system, that is, if external shears were not taken by a shear wall. The 1977 Code then considers the possibility of a reversing moment on the column and beam and requires (for the first time) a built-in energy absorbing detail (12.12.2). At least 25% of the bottom steel required for positive moment resistance must not only be carried *into*

*See Sec. 7.16 or Code 12.3.1.

the column, but must also be anchored there *for its full* f_y. This requirement will fix the maximum bar size usable at the end because the substitution of excess bars less well anchored is not acceptable; the energy absorption of the yielding bars is the end objective. In this case a hook is helpful. Except in massive construction such bars must usually be smaller than the #25 bar which in Table 7.1 leads essentially to $f_s = 0.5\,f_y$ for either Grade 300 or Grade 400 bars. Precisely, for Grade 400,

$$f_h = 43\sqrt{20} = 192.3 \text{ MPa}$$

The straight bar section, say 250 mm, must develop the remaining part of f_y, equal for Grade 400 to $400 - 192.3 = 207.7$ MPa

$$\ell_d = 0.019\,A_b f_y/\sqrt{f_c'}$$
$$250 = 0.019\,A_b \times 207.7/\sqrt{20} \qquad \text{Max. } A_b = 283 \text{ mm}^2$$

The bar size is thus limited to #15.

7.15 MIXED BAR SIZES AND BUNDLED BARS

The variation in bond stress that elastic analysis has indicated as existing at the same cross section on bars of different diameter (a larger bond stress on the larger bars) is not believed to be significant at the ultimate stage. These inequalities can be ignored just as are various other bond stress concentrations (Sec. 7.4) on individual bars. The development length of each size bar must be maintained as separately calculated.

The Code permits up to four bars to be bundled or clustered together. In the absence of test data, it is assumed that the bars must be developed individually but that the lengths must be increased 20% for a three-bar bundle and 33% for a four-bar bundle. The increase is to cover any uncertainty as to how well mortar will penetrate into the core of the bundle. All bundled bars must be within ties or stirrups.

Bundles are limited to bar sizes #35 of smaller and a maximum of four bars. Within a span of a beam the cutoff point of individual bars must be staggered at least 40 diameters, and splices of individual bars must not overlap each other. In applying bar spacing rules, the group is treated as having a single diameter which would give the actual total area of the group.

7.16 DEVELOPMENT LENGTH IN COMPRESSION

Little recent exploration has been done on the development of compression bars. The absence of tension cracks eliminates a major weakness and permits shorter development lengths in compression than in tension. The 1963 Code allowable bond stress has been directly transformed into a development length requirement with no modification.

$$\ell_d = 0.24\, f_y d_b / \sqrt{f'_c} \geq 0.044\, f_y d_b$$

The $0.044\, f_y d_b$ controls wherever $f'_c > 30.6$ MPa. An absolute minimum of 200 mm is specified (Code 12.3.1). Inside of spirals the equation length may be reduced 25%. Reductions may be made when excess compression steel is used.

Attention is called to Code 12.18.3 which says definitely that not all column reinforcing may be considered as compression steel for the purpose of splicing (and presumably for development).

At horizontal cross sections of columns where splices are located, a minimum tensile strength at each face equal to one-fourth the area of vertical reinforcement in that face multiplied by f_y shall be provided.

SPLICES

7.17 SPLICES IN TENSION

It is generally necessary to splice tension bars, partly because of the limited (usually 18 m) length of commercial bars, but more because of the awkwardness of interweaving long bars on the job. Splicing may be by welding, by Cadweld (fusion type) mechanical connections, or most frequently, for bars #35 and smaller, by lapping bars as in Fig. 7.16a. The

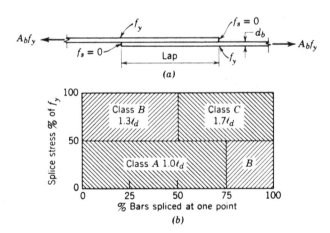

FIGURE 7.16 Tension lap splices. (a) Diagram of stress transfer. (b) Requirements in terms of stress and percent of area spliced at one section.

lapped bars are commonly tied in contact with each other, but may be spaced up to 150 mm apart, with an upper limit of one-fifth the lap length.

With development lengths already rather carefully specified, the splice lap in tension has been specified (Code 12.16) as a multiple of ℓ_d, where ℓ_d in this case must be based on the *full* f_y.* Unless this splice can be made away from a point where the bar stress is high, this multiple must be from 1.3 to 1.7, as diagramed in Fig. 7.16*b* and as outlined below. Two factors lead to the necessary increase above the 1977 ℓ_d values. Obviously there is a severe stress condition at *both* ends of a lap splice. Second, the concrete at a splice must take out of the two bars a total stress of $2\,A_b f_y$ and this may increase the splitting forces. Research[8,9] confirming the need for even greater lengths where cover is small is summarized in Sec. 7.18.

The Code has broken down splices into three† classes in increasing order of severity of conditions, as follows:

For stress always less than $f_y/2$:

> Class A if not over 75% of bars are spliced within one lap length. Use 1.0 ℓ_d.
>
> Class B if more than 75% are spliced within one lap length. Use 1.3 ℓ_d.

For stress exceeding $f_y/2$:

> Class B if not over half the bars are spliced within one lap length. Use 1.3 ℓ_d.
>
> Class C for more than half the bars spliced within one lap length. Use 1.7 ℓ_d.

Note that 300 mm is the minimum for the splice lap, but is not used here as the minimum for the ℓ_d used in calculating the splice.

Attention is directed to the saving in splice length if splices are staggered (Fig. 7.16), a reduction from 1.7 ℓ_d to 1.3 ℓ_d for a net saving of 0.4 ℓ_d. The wider spacing of splices should then legally qualify for the 0.8 factor on ℓ_d in many cases, but Sec. 7.18c indicates this is a problem area and the author recommends otherwise. The behavior of a member with closely spaced splices is much improved if only half the bars are spliced, an improvement that automatically results from staggering them and nearly doubling the splitting width per splice. However, if the failure when not staggered was already initiated through the cover (Sec. 7.9), added clear lateral spacing would still leave this weak link and less improvement would result. The failure mechanism is further discussed in the following section.

---

*See Sec. 7.19 for author's additional recommendations.

†The 1971 Class D splice for tension tie member has been voted out of the Code.

7.18 SPLITTING FAILURE GOVERNS SPLICE STRENGTH— A RESEARCH REPORT

Research under the author[8,9] and others for the Texas Highway Department has somewhat clarified the picture of splice behavior and ultimate strength. This section reports some of the conclusions, and Sec. 7.19 gives some related design recommendations.

(a) Failure Modes

A basic finding is that all tension lap splices fail from splitting of the surrounding concrete in one of the basic patterns of Fig. 7.17. At very close spacings the side split failure will occur and additional cover changes it little (Fig. 7.18a); lessened cover tends to change it to the face-and-side split type. In the latter the face splitting over part of the length leaves weak corner sections and finally reaches enough of the length to cause the side split failure at a lowered efficiency (Fig. 7.17b). Extra cover in this case increases the resistance offered to the face splitting; lesser cover tends to move the failure toward the V-type (Fig. 7.17c) where a splice is not seriously influenced by further spacing changes.

At failure, the concrete splitting stress over the full splice length is far from uniform. The steel stress is sharply changing in the spliced bars and just before failure becomes remarkably linear in both bars (Fig. 7.19a),

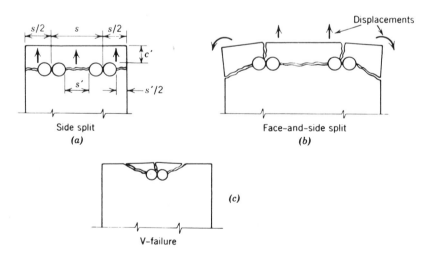

Side split
(a)

Face-and-side split
(b)

V–failure
(c)

FIGURE 7.17   Splitting around splices.

although not totally so. Slip of one bar relative to the other must occur except near the middle of the lap where the two bar stresses are nearly equal. The slip must be large, near $0.5 f_y/E_s$ times half the splice length, as shown in Fig. 7.19b. This slip implies loss of the "tight" shear connection in the slip plane directly between the two bars over, at least, portions of

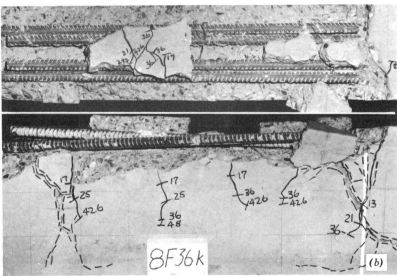

FIGURE 7.18 Splice failure in test beams. (a) Spaced splice with side split failure. Arrows mark ends of bars where they have separated from the concrete as they slipped at failure. (b) A longer contact splice, also in a constant moment length; a face-and-side split failure.

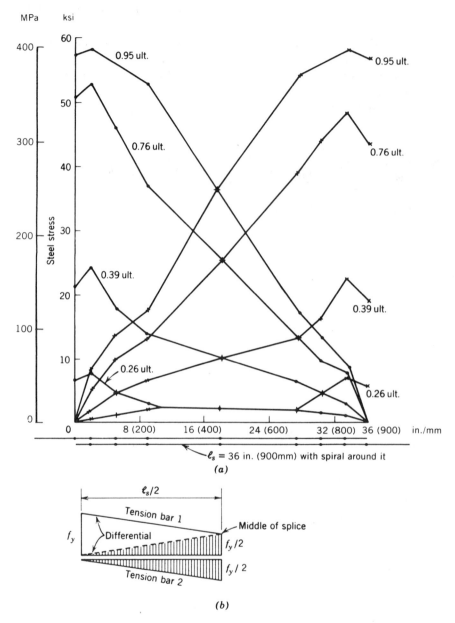

FIGURE 7.19 (*a*) Typical stress distribution along No 14 (45 mm) splice at various load levels. Constant moment zone; face-and-side-split failure. (*b*) Stress differential between two bars (roughly).

192

the length. At least the end section of long splices must have failed in the slip plane earlier than the time of the general failure.

An adequate splice of a large bar goes through several stages in addition to flexural cracking:

1. Splitting starts at the ends of the splice, beginning from the end flexural cracks. The splitting may be on the tension face or on the sides of the beam.
2. The splitting progresses toward the middle of the splice, although not usually completely to the next crack before splitting starts there also, etc. Near failure, splitting may extend over 30 to 80% of the splice length.
3. Final failure is sudden and complete unless confining steel is present.

The concrete tensile splitting stresses existing just prior to failure must vary considerably over the splice length, probably being most effective near midlength where cracking is less, and least effective where splitting has started. A bar is not released by a single face split over it, but it would seem to be held less effectively.

(b) Semiempirical Analysis

An oversimplified approach to analysis is helpful, as follows.

1. Assume a uniform bond stress over each bar representing in total a pull of $A_b f_y$ as indicated in Fig. 7.20a.
2. Assume also a radial component equal to this bond stress, the concept of the bar pressing on the concrete just like water pressure held in a pipe (Fig. 7.20b).
3. On a horizontal plane through the splices calculate the total splitting per unit length as in Fig. 7.20c.
4. Assume a *uniform* resisting tension in the concrete across the section and calculate this tensile stress.
5. Calculate the ratio of this stress to the concrete tensile splitting strength.

There are too many approximations for this calculated stress to be realistic:

1. Bond stress is not uniform (Fig. 7.19a)—it is different near the splice ends.
2. The Goto picture (Fig. 7.1) would indicate radial forces greater than the bond stresses for thick cover (maybe smaller for thin cover?).

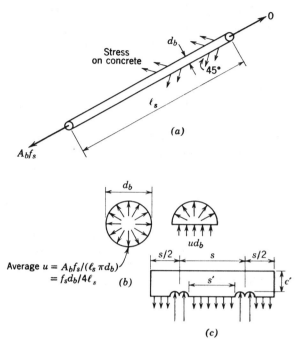

FIGURE 7.20 Analysis of splice. (*a*) Forces on concrete from bond. (*b*) Equivalent loading on concrete. (*c*) Equilibrium on short length of splice.

3. Tensile resistance over the member width is not uniform where the clear spacing s' is large, nor over the length (less where splitting has occurred).

The calculated stress ratios begin to take meaningful form when plotted against s'/c', where s' is the clear spacing between splices and c' is the clear cover over the splice, as shown in Fig. 7.21. It was noted that the side split failures were all at s'/c' of 1.7 or less, the face-and-side split failures from 1.5 to 8, and the V-type only beyond 7.5. Also, a reasonable lower bound curve can be drawn for f_y for 400 MPa, the small cluster of points excluded being cases where yield stresses have led to large strains. Short splices leading to low steel stresses plot higher, and the points are loosely stratified in terms of stress or splice length; but the lower bound furnishes a reasonable base for design. When relevant data are replotted in terms of the inverse of the stress ratio above, an upper bound curve can be drawn, as shown by the center diagonal line in Fig. 7.22, which forms a good design base. To allow for ductility needs, this base was raised, as noted in the next subsection, and was used to set up design equations.

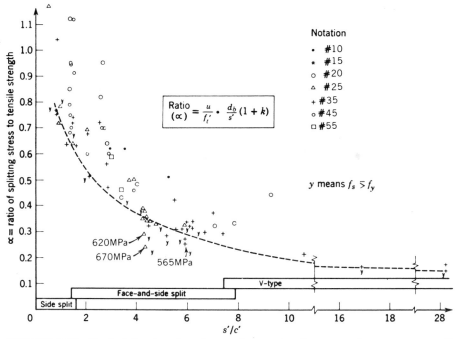

FIGURE 7.21 Test data showing decreasing ratio of computed stress to tensile strength of concrete as s'/c' increases. See Sec. 7.18d for k terms in the box.

(c) Development of Design Equations

On the basis just discussed, for $f_y = 400\,\text{MPa}$ and $f'_c = 20\,\text{MPa}$, design equations can be set up.

Splitting force per linear inch of lap $= u\,2\,d_b = (f_s d_b/4\,\ell_s)2\,d_b = f_s d_b^2/2\,\ell_s$

Resisting tensile stress $= (f_s d_b^2/2\,\ell_s)/s' = f_s d_b^2/(2\,\ell_s s')$

Tensile strength $= 0.53\sqrt{f'_c} = 2.37\,\text{MPa}$ for $f'_c = 20\,\text{MPa}$

For $f_y = 400\,\text{MPa}$, the basic ratio $(1/\alpha)$ from tests (Fig. 7.22) is 0.9 $(1 + 0.5\,s'/c')$, which ought to be increased for ductility in the ratio of $500/400 = 1.25$ plus a small further factor for dropping efficiency at higher stresses, to give a ratio, say,

$$1.30 \times 0.90(1 + 0.5\,s'/c') = (\text{tensile strength})/(\text{splitting stress})$$

$$= \frac{2.37}{400\,d_b^2/(2\,\ell_s s')} = \frac{2.37 \times 2\,\ell_s s'}{400\,d_b^2}$$

$$\ell_{s400} = \frac{1.30 \times 0.90(1 + 0.5\,s'/c')400\,d_b^2}{2.37 \times 2s'} = 98.7\,d_b^2\left(\frac{1}{s'} + \frac{1}{2c'}\right)$$

$$= 100\,d_b^2\left(\frac{1}{s'} + \frac{1}{2\,c'}\right)\ \text{approx.}$$

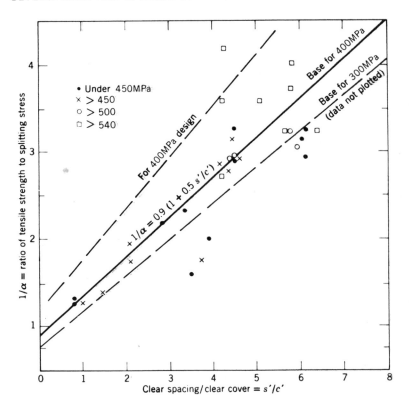

FIGURE 7.22 For particular f_y values, design limits on ratio of tensile strength of concrete to calculated splitting stress, the inverse of the ratio used in Fig. 7.21.

A similar basic ratio for a limiting stress of 300 MPa leads to

$$\ell_{s300} = 57\, d_b^2 \left(\frac{1}{s'} + \frac{1}{2\,c'} \right)$$

Note that an *upper limit* for the parenthesis in each required ℓ_s is $1.5/s'$.

The author does not claim precision for these equations but they do emphasize that clear spacing and clear cover are both important, and the 1977 Code could not give adequate guidance here. Unfortunately, the most scatter in the data falls in the region that most needs modification of practice—the small lateral spacing or small cover. Orangun's[3] regression analysis in effect separates s' and c' and uses only the smaller of c' or $s'/2$. Thus, his shift from one variable to the other occurs at $s'/c' = 2$. To the left of $s'/c' = 2$ in Figs. 7.21 and 7.22 he would be using only s' as the control, to the right only c'. A weakness in the equations presented here probably

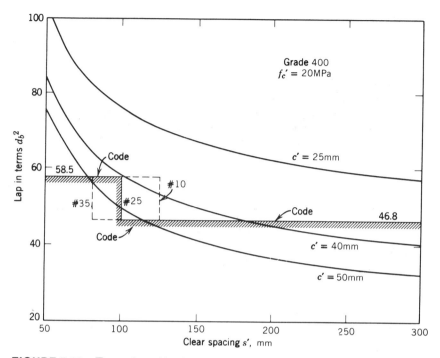

FIGURE 7.23 Formula splice lap compared to ACI Code, Grade 400 bars. $\ell_s = 100\, d_b^2(1/s' + 1/2\, c')$ with $(\cdots) \gtrless 1.5/s'$.

lies in the *summation* in the parenthesis $(1/s' + 1/2\, c')$. If cover is thick and bar spacing small $(s'/c' < 2)$, further increase in cover should not greatly influence strength, because the splitting crack would lie in the plane of the bars, not through the cover.

For other f_c' values the ℓ_s equations should be multiplied by $\sqrt{20/f_c'}$. For lightweight concrete the multipliers of Code 12.2.3c are needed, 1.33 for "all lightweight" and 1.18 for "sand-lightweight" concrete. Top bar splices need to be lengthened by the multiplier 1.4.

Although separation of the two empiricals, $1/s'$ and $1/2\,c'$, may decrease confidence level, Figs. 7.23 and 7.24 have been prepared by using s' for abscissa and separate c' values for several curves. The ordinate is the constant that must be multiplied by bar diameter squared (d_b^2) to get needed splice length; nothing has been included to cover assistance from stirrups. The 1977 Code requirements are shown by the shaded line, with a transition between the two controlling ℓ_d equations at $s = 150$ mm. Note that with Grade 400 bars, the Code can be unsafe with clear cover of 25 mm almost anywhere. Such stirrups (or spirals) as are used can be helpful, although quantitatively this is not too firmly established and the Code is not yet

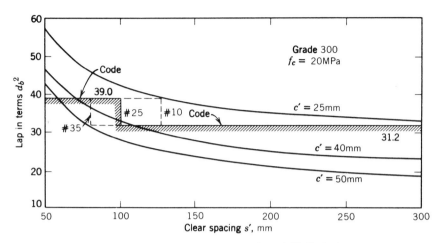

FIGURE 7.24 Formula splice lap compared to ACI Code, Grade 300 bars.
$\ell_s = 57\, d_b^2(1/s' + 1/2\, c')$ with $(\cdot \cdot \cdot) \gtrless 1.5/s'$.

helpful. With Grade 300 bars the unsafe margin is reduced, but at *clear* spacings less than 75 mm the problem appears real.

(d) Different Steel Stress at Two Ends of Splice

In a retaining wall stem, or a similar situation, the moment varies significantly over the splice length, and this leads to a stress at the upper end of $kf_y < f_y$ where k is a ratio <1. The derivations above still apply except that, instead of two bars at f_y giving a total stress of $2\,f_yA_b$, there are two bars with a total stress of $(1 + k)f_yA_b$. The ℓ_s equations can be multiplied by $0.5(1 + k)$ for either the required ℓ_{s400} or ℓ_{s300}. The result is a shorter lap required.

7.19 SPECIAL PROBLEMS IN TENSION SPLICES

(a) Splices for Walls Carrying Moment

At a construction joint in a retaining wall it appears that a uniform vertical bar spacing results in the splice of the last bar being something in the order of 15 percent understrength because it tends to have a face-and-side split failure (Fig. 7.17b). This does not lower the overall wall strength seriously, but the loss can be avoided by staggering the end bar splice to give the arrangement of Fig. 7.25 or by lengthening the end bar splice to increase its strength by 15 or 20 percent.

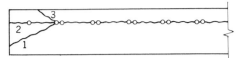

FIGURE 7.25 Strengthening wall splices by elimination of end splice at weak section.

(b) Transverse Reinforcement Around Splice To Reduce Length

Splice length can be reduced by providing confinement reinforcement in the form of ties or spirals over the lap. Spirals are at least a little more effective, but they are more awkward in the field. Tentative design recommendations included in the Orangun, Jirsa, and Breen paper[3] should be helpful; space limitations do not permit adequate coverage here. They indicate a potential reduction of splice lap lengths "by about 20 percent for #20 bars and about 50 percent for #35 bars" when the maximum recommended transverse reinforcement is used. With stirrups mandatory now in most beams, this could be a significant improvement. It is not yet clear whether stirrups needed for shear can also function for reducing splice length or whether the sum of the two requirements must be used.

(c) Staggered Splices

No change in the basic equation is needed for staggered splices, but s' then becomes $2s - 3d_b$, where s is the basic bar spacing and $2s$ the splice spacing center-to-center.

7.20 RECOMMENDATIONS FOR TENSION SPLICE DESIGN

Splices with greater cover are shown in Figs. 7.23 and 7.24 as requiring less length than those with reduced cover. The Code requirements fit best the needs with a 50 mm cover.

In view of the above research, the author recommends that the following further restrictions be incorporated into the use of the Code requirements for tension lap splices:

1. That for Grade 400 bars the Code rules be considered not applicable for clear covers of less than 40 mm; that the 0.8 factor for wide spacing (Code 12.2.4) not be used unless checked against Fig. 7.23 or modified in accordance therewith.

2. That for Grade 300 bars the Code rules be considered not applicable for clear covers less than 30 mm; that the 0.8 factor for wide spacing (Code 12.2.4) not be used for Class B splices (1.3 ℓ_d) unless checked against Fig. 7.24 or modified in accordance therewith.

3. That, with the *minimum* covers of 1 and 2 above, the *clear* spacing between splices not be less than 100 mm for Class C (1.7 ℓ_d) splices or 200 mm for Class B (1.3 ℓ_d).

The curves in Figs. 7.23 and 7.24 reflect the best available information for splices in 20 MPa concrete. Where the Code is not mandatory the curves are recommended, with the usual factors for other strengths of concrete, lightweight concrete, and top bars to be imposed.

When the Code revision appears reflecting the regression analysis mentioned in Secs. 7.1, 7.9, and 7.19b, this should certainly supercede the special recommendations above.

7.21 SPLICING OF STIRRUPS

In some congested spots the use of two U-stirrups (without hooks) turned together to form closed stirrups is feasible. Tested in either shear or torsion, members so reinforced show reasonable ductility, but final failure tends to be more sudden and complete when it occurs.

Code 12.14.5 covers the requirements:

Pairs of U-stirrups or ties so placed as to form a closed unit shall be considered properly spliced when the length of laps are 1.7 ℓ_d. In members at least 450 mm deep, such splices with $A_b f_y$ not more than 40 kN per leg may be considered adequate if the stirrup legs extend the full available depth of member.

The last sentence definitely accepts stirrup laps of #10 bars of either Grade 300 or 400 and #15 bars of Grade 300. The #10 stirrup lap for Grade 400 would even be some 10% less than 1.7 times the theoretical ℓ_d. This is possible because the lap in such a beam cannot be subject to much error in placement and it is known that the small bars are slightly penalized in fitting them into a general specification that also covers large bars.

7.22 COMPRESSION SPLICES

Where bars are required only for compression the bar ends may be cut square within 1.5°, then butted together, and positively held in place to transmit the stress in end bearing. The restrictions are detailed in Code 12.17.5.2, but Code 12.18.3 is probably more important. This requires some

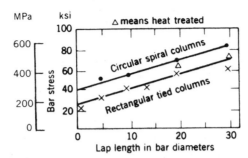

FIGURE 7.26   Compression splice test strengths in columns. (Modified from Reference 5).

tensile steel on each face at column sections where splices are located, as already mentioned in Sec. 7.16.

Compression lap splices transmit a substantial portion of their load in end bearing on the concrete, as indicated by the PCA test results shown in Fig. 7.26 with the intercept high on the load axis. This bearing is evidently dependent on how well the concrete is confined, spirals showing better than ties.

The Code (12.7.1) in effect specifies for f'_c of 20 MPa or better 20 d_b lap for Grade 300, and 30 d_b lap for Grade 400, but in no case a lesser lap than 300 mm. Within ties or spirals of specified makeup, these laps may be reduced to 0.83 or 0.75, respectively, of these amounts, but still must be not less than 300 mm.

SELECTED REFERENCES

1. Raymond E. Untrauer and Geo. E. Warren, "Stress Development of Tension Steel in Beams," *ACI Jour.*, 74, No. 8, Aug. 1977, p. 368.
2. Phil M. Ferguson, "Small Bar Spacing or Cover—A Bond Problem for the Designer," *ACI Jour.* 74, No. 9, Sept. 1977, p. 435.
3. C. O. Orangun, J. O. Jirsa, and J. E. Breen, "Reevaluation of Test Data on Development Length and Splices," *ACI Jour.*, 74, No. 3, Mar. 1977, p. 114.
4. Yakimasa Goto, "Cracks formed in Concrete Around Deformed Tension Bars," *ACI Jour.*, 68, No. 4, April 1971, p. 244.
5. ACI Comm. 408, "Bond Stress—The State of the Art," *ACI Jour.*, 63, No. 11, Nov. 1966, p. 1161.
6. John A. Hribar and Raymond C. Vasko, "End Anchorage of High Strength Steel Reinforcing Bars," *Jour. ACI*, 66, No. 11, Nov. 1969, p. 875.

7. Phil M. Ferguson and J. Neils Thompson, "Development Length of Large High Strength Reinforcing Bars," *Jour. ACI*, 62, No. 1, Jan. 1965, p. 71.
8. C. N. Krishnaswamy, "Tensile Lap Splices in Reinforced Concrete," Ph.D. dissertation, The University of Texas at Austin, Dec. 1970.
9. Phil M. Ferguson and C. N. Krishnaswamy, "Tensile Lap Splices, Part 2: Design Recommendations for Retaining Wall Splices and Large Bar Splices," *Research Report 113-3*, Center for Highway Research, The University of Texas at Austin, April, 1971, 60 pp.
10. Ralejs Tepfers, "A Theory of Bond Applied to Overlapped Tensile Reinforcement Splices for Deformed Bars," Publication 73.2, Div. of Concrete Structures, Chalmers University of Technology, Göteborg, Sweden, 1973, 328 pp.

PROBLEMS

PROB. 7.1. In Fig. 5.25 with $f'_c = 30$ MPa, Grade 400 steel, $w_d = 20$ kN/m, $w_\ell = 65$ kN/m, do the #25 bars at the end satisfy development length requirements? Assume that bars extend 150 mm beyond the reaction.

PROB 7.2. In Fig. 5.26a, for $f'_c = 30$ MPa, Grade 400 steel fully stressed, a triangle (assumed) for the moment diagram, find the anchorage ℓ_d into the support. Check development length and moment length to find whether the arrangement shown can be adequate. If not, modify the arrangement of the #20 bars to make it work properly, (other than by using straight bars full length).

PROB. 7.3. Same as Prob. 7.2 but assume the beam loading changed to full length uniform load that gives the same maximum moment.

PROB. 7.4 How large a Grade 400 stirrup can be developed in a beam having a 600 mm overall depth? $f'_c = 30$ MPa (Fig. 7.15 and Code 12.14).

PROB. 7.5. Some frames must have the bottom beam bars anchored into the column for their full f_y. (Code 12.12.2)

(a) In a 400 mm-square exterior column, how large a Grade 400 bar can be fully anchored without exceeding Code evaluations? $f'_c = 30$ MPa.

(b) If the vertical load from the column is assumed to increase bar resistance by 30%* as allowed at the support of a simple span, what maximum size bar is acceptable?

---

*No such increase should apply on a hook turned vertically, but on a hook turned horizontally it would be as logical as on the straight bar.

PROB. 7.6.

(*a*) If you wish to anchor Grade 400 top bars from a beam into an exterior girder 300 mm wide, what is the maximum size bar which can be used at full value? $f'_c = 30$ MPa.

(*b*) If Grade 300 top bars are used?

PROB. 7.7 A tied column 500 mm square contains 8-#30 bars of Grade 400 steel, equally on all four faces, $f'_c = 30$ MPa. The column design indicates no computed tension on the bars. Sketch and detail the arrangement of splices. (Sec. 7.22)

PROB. 7.8. The beam in Fig. 7.27 is heavily loaded with balanced overhanging end loads such that for this problem the negative moment is essentially constant across the center span and requires 8-#35 bars of Grade 400 steel arranged in two layers of 4 bars each. $f'_c = 30$ MPa. Because the maximum length of available bars is 18 m, tension splices of all bars are necessary.

(*a*) Design and sketch splices, assuming splices are staggered into two groups.

(*b*) As in (*a*) with three groups of 3, 3, and 2 splices.

(*c*) Compare (*a*) and (*b*) with the case of all splices made at a single section. Is the latter detail permissible under the Code?

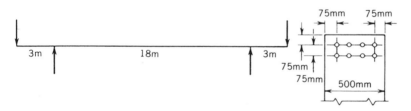

FIGURE 7.27   Beam of Prob. 7.8.

8
SERVICEABILITY OF SLABS AND BEAMS

8.1 INCREASING IMPORTANCE OF SERVICEABILITY

As design procedures become more accurate and more efficient, practice tends to move into the realm of smaller members; greater deflections and more need for attention to serviceability result.

Strength design methods were significantly introduced into practice in the United States in the middle 1950's. These improved strength evaluations made smaller columns possible. The increased column slenderness led to emphasis on column deflections and increased secondary moments in the 1971 Code.

The changes in slabs and beams were less pronounced because (1) steel costs increased substantially as member thickness was reduced, and (2) long-term deflections had already been realized as a problem, at least, in slabs. But the possibility of reduced depths without loss of strength, coupled with the possibility of offsetting increased steel costs with smaller areas of high strength steel, made desirable a more definitive statement of what constitutes poor behavior. For serviceability is almost as much a problem of proper standards as of proper calculations. What constitutes poor serviceability?

Aside from corrosion and poor weathering or wearing properties, which are normally avoided by proper controls in mixing and placing, poor serviceability usually relates to excess deflections, occasionally to extensive cracking or to excessive crack width.

8.2 THE DEFLECTION PROBLEM

Excessive deflections can lead to crushing of partitions, bulging and buckling of glass enclosure walls or metal partitions, sagging of floors, and unsightly drooping of overhanging canopies. ACI Committee 435 on deflections lists[1] four categories of deflection problems for which it discusses proper limits.

1. Sensory problems that include vibrations (a function of member stiffness) and appearance of droopy members.
2. Serviceability problems such as roofs that do not drain and floors that are not plane enough for their intended use, exemplified by the requirements for bowling alleys or for sensitive equipment installations.
3. Effects on nonstructural elements such as masonry and plaster, movable partitions. This category must include beam deflections caused by lateral building deflection and vertical movements of columns from temperature differentials.
4. Effects on structural behavior, instability, etc., which is not a serviceability but a strength problem.

The report suggests proper limits for each category.

Several factors unique to reinforced concrete make exact deflection prediction very difficult. (a) Unequal top and bottom reinforcing automatically leads to shrinkage deflections as a normal phenomenon. (b) Creep of concrete under stress leads to a gradual increase in deflection of members left under loads. (c) Even the somewhat uncertain stage at which reinforced concrete begins to crack significantly adds uncertainty as to the actual deflections to be expected at service loads.

Many slabs or beams are designed for use where deflections may be less critical. The ACI Code (Sec. 9.5.2.1) provides a table listing minimum thickness for slabs or beams "not supporting or attached to partitions or other construction likely to be damaged by large deflections." This table is reproduced here as Table 3.2 in Sec. 3.8, along with a few comments as to its use. Its simple limits often take the place of the lengthy deflection calculations presented in the following sections.

8.3 THE CODE SOLUTION FOR DEFLECTION

The Code takes an overall approach in terms of the immediate deflection plus the expected overall percentage increase with shrinkage and time effects.

The immediate deflection that is the starting point is quite sensitive to whether the member is uncracked or cracked, and if cracked how severely cracked. This severity of cracking necessarily varies along the span as the moment changes. The load-deflection curve shown in Fig. 8.1a is typical for a moderately reinforced slab that is simply supported and loaded for

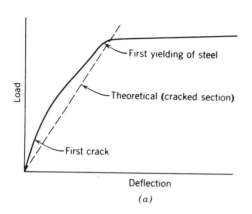
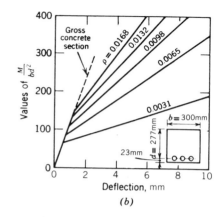

FIGURE 8.1 (*a*) Simple span slab deflection under increasing load (*b*) Schematic of influence of ρ on deflection. (From Reference 4, Portland Cement Assn.)

the first time. Diagrammatically it could be simplified about as shown in Fig. 8.1*b*, where the importance of the percentage of tensile steel is shown. The deflections of interest are usually around a steel stress of 50 to 60% of the yield stress or less.

The Code uses an *average* as the effective moment of inertia and then applies the usual methods of elastic analysis to compute the immediate deflection. The Code then requires that the deflection from the sustained portion of the load be multiplied by a factor that reflects the ratio of A_s'/A_s to determine the added long-term effects from creep and shrinkage.

The average moment of inertia used in Sec. 8.5 involves the transformed area concept which must first be explained in general and then applied for the beam. The application to deflections is picked up again in Sec. 8.5.

8.4 TRANSFORMED AREA CONCEPT—ELASTIC ANALYSIS

(a) General

Where concrete and steel are deformed to the same unit strain ϵ, their initial stresses vary as their modulus of elasticity.

$$f_c = \epsilon E_c \qquad \text{and} \qquad f_s = \epsilon E_s = f_c(E_s/E_c) = nf_c$$

where the modular ratio n is defined as the ratio E_s/E_c. The total immediate stress on a bar of area A_s then can be written as

$$f_s A_s = n f_c A_s = f_c(nA_s)$$

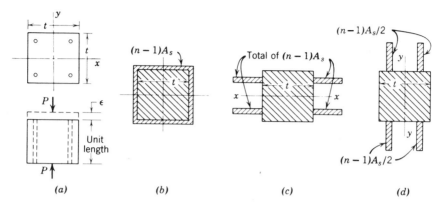

FIGURE 8.2 Transformed area of a column, elastic theory.

The last form associates n with the steel area instead of with the steel stress. It shows that the steel area A_s acts initially* as would a concrete area nA_s. This concrete area is called the *transformed area* of the steel. When the steel is thus replaced in a member by its transformed area, the result is a total area of homogeneous material, in this case concrete, that is relatively easy to analyze for stress or deformation. To be exact, the effective added area is $(n - 1)A_s$ because A_s displaces an equal volume of concrete, but the difference between $n - 1$ and n is beyond the accuracy otherwise attainable in deflection calculations.

In the symmetrical column of Fig. 8.2a the transformed area of the column steel may be sketched as shown in Fig. 8.2b, c, or d, that is, at any point around the cross section provided that symmetry is maintained and only an axial load is considered. This is true only because the unit deformation ϵ is everywhere the same. Normally, eccentricity or moment must be considered.

With bending added, f_s again equals nf_c provided both stresses correspond to the same strain ϵ. The nA_s area must then be spread *parallel* to the neutral axis to hold to the same level of strain. In Fig. 8.2a for moment or eccentricity about axis $x - x$ the transformed area would have to be sketched parallel to the axis of bending as shown in Fig. 8.2c and as shown in Fig. 8.2d for moment or eccentricity about axis $y - y$. Bars in different planes would then carry different stresses in accordance with their locations.

---

*The time effects of concrete creep and shrinkage modify this relationship.

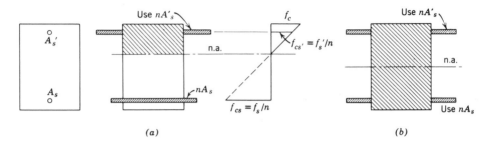

FIGURE 8.3 Transformed area concept for effect of short-time loads on a beam. (a) Usual cracked section. (b) Before cracking.

(b) Applied to Beams

With bending, either in beams or columns, any tension concrete is normally assumed as cracked and omitted from the calculations, but the tension nA_s area is assumed active in tension to any level. The transformed area of a double-reinforced beam would be as shown in Fig. 8.3a with A'_s replaced by an effective transformed area of $(n-1)A'_s$ and A_s replaced by nA_s, there being no useful concrete displaced on the tension side. Prior to cracking with the concrete still carrying tension, the area would be that of Fig. 8.3b.

(c) Neutral Axis of Beam—Mechanics of Elastic Analysis or Review

The transformed area concept makes it possible to replace any reinforced concrete member (under moderate loads) with an equivalent member of homogeneous elastic material to which basic strength of materials relationships apply. For flexure without axial loading, the neutral axis must lie at the centroid of the effective cross section to make the total tension equal to the total compression and thus provide a resultant couple. The moment and stress are related by the equation

$$Mc = fI$$

where M = bending moment, for dimensional homogeny in MN · m
     c = distance to extreme fiber, m
     f = bending stress, MPA, at extreme fiber (at distance c from neutral axis). This term is used as s or σ in engineering mechanics notation, but f is standard notation in reinforced concrete in the United States.
     I = area moment of inertia, $\int y^2 dA$, about the neutral axis, m⁴.

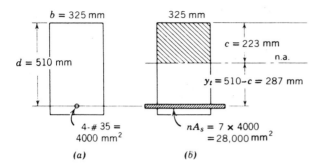

FIGURE 8.4   Analysis of a rectangular beam.

(d) Review Example with Transformed Area

Find f_c and f_s for the rectangular beam shown in Fig. 8.4a for $M = 130$ kN · m, C30 concrete, Grade 400 reinforcement, $E_s = 200\,000$ MPa.

Solution

From Sec. 2.6, $E_c = 5000\sqrt{f'_c} = 5000\sqrt{30} = 27\,400$ MPa

$$n = E_s/E_c = 200\,000/27\,400 = 7.29, \text{ say } 7$$

This transformed area is first sketched (Fig. 8.4b) and the neutral axis indicated at the unknown depth c. The neutral axis lies at the centroid of the transformed area. The moment of the areas about the neutral axis* gives a quadratic equation.

$$325\,c \cdot c/2 = 28\,000\,(510 - c)$$
$$c^2 + 172\,c = 87\,900$$
$$c + 86 = \pm\sqrt{86^2 + 87\,900}, \qquad c = 223 \text{ mm (or negative)}$$

From Fig. 8.4b the moment of inertia is:

$$\begin{aligned}
\tfrac{1}{12} \times 325 \times 223^3 \quad &= \quad 300 \times 10^6 \text{ mm}^4 \\
325 \times 223 \times 112^2 \quad &= \quad 909 \times 10^6 \\
28\,000(510 - 223)^2 \quad &= 2306 \times 10^6 \\
\text{Total} \quad &= \overline{3515 \times 10^6 \text{ mm}^4} = 3.52 \times 10^{-3} \text{ m}^4
\end{aligned}$$

The transformed area of the steel is considered negligibly thin, so that its moment of inertia about its centroidal axis can be neglected.

$$f_c = Mc/I = 130 \times 10^{-3} \times 0.223/3.52 \times 10^{-3} = 8.24 \text{ MPa}_{\text{l}}$$
$$f_s/n = My_t/I = 130 \times 10^{-3} \times 0.287/3.52 \times 10^{-3} = 10.60 \text{ MPa}$$
$$f_s = 7 \times 10.60 = 74.20 \text{ MPa}$$

---

*Although moments about an axis not yet located may seem awkward, it should be noted that this gives a simpler equation. In general, $\bar{y} \int dA = \int y\,dA$. If y is measured from the centroid, $\bar{y} = 0$ and the equation becomes simply $\int y\,dA = 0$.

8.5 DEFLECTION COMPUTATIONS FOLLOWING THE CODE

(a) Algebraic Recommendations

The effective moment of inertia is specified as an *average* value to be used all across a simple span and is a weighted value dependent on the extent of probable cracking under moment (Code Eq. 9.7):

$$I_e = (M_{cr}/M_a)^3 I_g + [1 - (M_{cr}/M_a)^3] I_{cr} \gtrless I_g$$

in which $M_{cr} = f_r I_g/y_t = 0.7\sqrt{f'_c} I_g/y_t$ except that f_r must be modified (Code 9.5.2.3a or b) when lightweight concrete is used.

M_a = max. M in member at loading stage considered in MN m
I_{cr} = I based on transformed area of cracked section in m^4
I_g = I based on gross area of concrete section (without steel), in m^4
y_t = distance from centroid to extreme fiber, in m

The ACI Committee 435 report noted that I_g in this relation would be more accurate if it included the transformed area of the reinforcement, especially where this was heavy. The example in (c) uses this transformed area.
The above equation can be more simply written as

$$I_e = I_{cr} + (M_{cr}/M_a)^3 (I_g - I_{cr})$$

In this form notice that M_{cr}/M_a is most important when M_a is not too far different from M_{cr}.
The use of I_e gives the immediate deflection when used in the usual formulas for elastic deflections. Both shrinkage and creep will result in substantial increases unless the load is primarily a transient one. The Code requires the following multiplier to establish the *increase* in the deflection (added to the basic deflection) caused by the usual load carried:

$$(2 - 1.2 A'_s/A_s) \geqslant 0.6$$

For continuous beams I_e may be computed at mid-span and at supports and the average used in calculating the immediate deflection. Branson's Code discussion[2] indicates that a weighting of the two I_e values in terms of the proportion of negative to positive moments is also possible.
A later study by Committee 435[11] reports that "for controlled laboratory conditions, there is a 90% chance that the deflections of a particular beam will be within the range of 20% less to 30% more than the calculated value" as given by the 1971 Code and retained in the 1977 Code.

(b) Practical Complications

The Code procedure covers the simplest case, an immediate unfactored sustained load and its time effects, plus a later live load regarded as

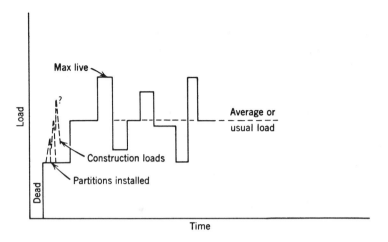

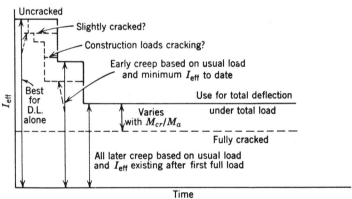

FIGURE 8.5 Relation between loading and I_{eff} in diagrammatic form.

transient. The designer will recognize practical complications. How much will the immediate deflection and time effects be increased by other cracking induced by normal construction loading? Should not much of the time effect be based on cracking that goes with the normal full live load, even if such loading is itself transient? When live load is heavy, does this not imply manufacturing or storage usage where much of this live load is probably applied *after* the early initial period, which means this portion of the creep starts on an older concrete.

For the heavy live load case, the graphs of Fig. 8.5 show the complications to be expected. In the long run the dead load deflection becomes that based on the maximum cracking condition plus any accumulated time effects. Hence, early smaller calculated deflections are useful only for (1) evaluating the maximum increase in deflections that partitions must accept and (2) for evaluating time effects. As a nonexpert in this area the author wonders whether the sustained load deflection base with I_e based on M_a for that sustained load is actually the best for use in determining time effects. Early construction loads or transient live loads will often induce cracking that lowers I_e even where it is not a part of the sustained load. It appears that the Code procedure might give an overly precise value of the early I_e, one that may well be too high in view of the actual physical complications.

(c) Example Based on Continuous T-Beam

Given the cross sections of Fig. 8.6, C25 concrete, $E_c = 4730\sqrt{f_c'}* = 4730\sqrt{25} = 23\ 600$ MPa, $n = 8.5$, $f_r = 0.623\sqrt{f_c'}*$ (for f_c', $\sqrt{f_c'}$, and f_r each in MPa), $w_d = 17$ kN/m, $w_\ell = 38$ kN/m, total service load $w = 55$ kN/m, $\ell_n = 6$ m, design moments of $+0.072\ w\ell_n^2$ and $-0.091\ w\ell_n^2$. Does the beam deflection satisfy the Code requirement for a beam attached to nonstructural members such as partitions? Code Table 9.5b limits such deflection to $\ell_n/480$.[†]

Solution

The final short time deflection, involving the total unfactored load, probably with cracked cross sections, is the most straightforward starting step. To this must be added the shrinkage and creep deflection from the sustained load, which is slightly more involved. From this total, insofar as allowables for partitions are concerned, any deflection existing when they are installed is deductible. The short-time full load deflection requires values of I_{cr} and I_g both at midspan and at support for averaging. For convenience I and c will be computed in mm and the results then converted to meters for I (m⁴) and c.

Midspan Section

$nA_s = 8.5 \times 1600 = 13\ 600$ mm² $\qquad (n-1)A_s' = 7.5 \times 400 = 3000$ mm²

Cracked section I_{cr}:

Centroid of transformed area of Fig. 8.6d, by moments about n.a. (c below top).

$$1500\ c^2/2 + 3000(c - 65) - 13\ 600(455 - c) = 0$$
$$c^2 + 22.13\ c + 11.06^2 = 8329 + 11.06^2 = 8451$$
$$c = \sqrt{8451} - 11.1 = 91.9 - 11.1 = 81\ \text{mm}$$

---

*This example was written by Profeseor Ferguson in metric units, using an earlier draft of the ACI Code with slightly different constants for E_c and f_r.

[†]Code limits are from $\ell/180$ to $\ell/360$ on *immediate deflection caused by live load* for members not attached or supporting "nonstructural elements likely to be damaged by large deflections." For members attached to or supporting such nonstructural elements the *deflection after attachment* is limited to from $\ell/240$ to $\ell/480$.

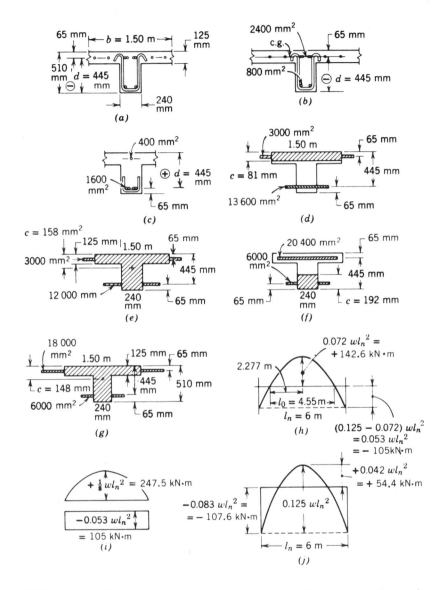

FIGURE 8.6 Details considered in deflection of T-beam. (*a*) General layout, negative moment section. (*b*) Reinforcement for negative moment. (*c*) Reinforcement for positive moment. (*d*) Positive moment transformed area, cracked section. (*e*) Positive moment gross area. (*f*) Negative moment, cracked area. (*g*) Negative moment gross area. (*h*) Maximum positive moment diagram for service load. (*i*) Equivalent moment areas for easy use. (*j*) Moment diagram for dead plus half live load on all spans.

213

$$I_{cr}: \quad 1500 \times 81^3/3 \qquad = 266 \times 10^6$$
$$3000(81-65)^2 \quad = \text{neglect}$$
$$13\,600(445-81)^2 = \underline{1802 \times 10^6}$$
$$I_{cr} = 2068 \times 10^6 \text{ mm}^4 = \text{say, } 2.070 \times 10^{-3} \text{ m}^4$$

Gross section I_g* (Fig. 8.6e). Find c from top.

$$(1500-240)125(c-62.5) + 240 \times 510(c-255) + 3000(c-65) - 12\,000(445-c) = 0$$
$$295 \times 10^3 \ c - 46\,600 \times 10^3 = 0, \ c = 158 \text{ mm}$$

$$I_g: \ (1/12)(1500-240)125^3 \qquad\qquad = 205 \times 10^6 \text{ mm}^4$$
$$+ (1500-240)125(158-62.5)^2 = 1436 \times 10^6$$
$$(1/12)240 \times 510^3 \qquad\qquad = 2653 \times 10^6$$
$$+ 240 \times 510(255-158)^2 \qquad = 1151 \times 10^6$$
$$300(158-65)^2 \qquad\qquad\qquad = 26 \times 10^6$$
$$12\,000(445-158)^2 \qquad\qquad = \underline{988 \times 10^6}$$
$$I_g = 6459 \times 10^6, \text{ say, } 6460 \times 10^6 \text{ mm}^4$$
$$= 6.460 \times 10^{-3} \text{ m}^4$$

$$f_r = 0.623\sqrt{f_c'} = 0.623 \times 5 = 3.12 \text{ MPa}$$
$$M_{cr} = f_r I_g/y_t = 3.12 \times 6.460 \times 10^{-3}/(0.510-0.158) = 0.0572 \text{ MN} \cdot \text{m} = 57.2 \text{ kN} \cdot \text{m}$$

Support Section

$nA_s = 8.5 \times 2400 = 20\,400 \text{ mm}^2 \qquad (n-1)A_s' = 7.5 \times 800 = 6000 \text{ mm}^2$
Cracked section I_{cr} (Fig. 8.6f). Find c above bottom.

$$240\,c^2/2 + 6000(c-65) + 20\,400(c-445) = 0$$
$$c^2 + 220\,c + 110^2 = 78\,900 + 110^2 = 91\,000$$
$$c = \sqrt{91\,000} - 110 = 192 \text{ mm}$$

$$I_{cr}: \ 240 \times 192^3/3 \qquad = 566 \times 10^6 \text{ mm}^4$$
$$6000(192-65)^2 \qquad = 97 \times 10^6$$
$$20\,400(445-192)^2 = \underline{1306 \times 10^6}$$
$$I_{cr} = 1969 \times 10^6 \text{ mm}^4 = \text{say, } 1.970 \times 10^{-3} \text{ m}^4$$

Gross Section I_g at Support (Fig. 8.6g).

Although slightly debatable, the tension flange in which part of A_s is placed will be counted. The use of I_g calculated for midspan (above) would not be a serious error. The c below is used from the top, to locate the centroid.

$$(1500-240)125(c-62.5) + 240 \times 510(c-255) + 18\,000(c-65) - 6000(445-c) = 0$$
$$304\,000\,c - 44\,900\,000 = 0, \ c = 147.7, \text{ say, } 148 \text{ mm}$$

*See comment closing first paragraph of Sec. 8.5a.

$$I_g: (1/12)(1500 - 240)125^3 \quad = 205 \times 10^6 \text{ mm}^4$$
$$+ 1260 \times 125(148 - 62.5)^2 = 1151 \times 10^6$$
$$(1/12)240 \times 510^3 \quad = 2653 \times 10^6$$
$$+ 240 \times 510(255 - 148)^2 \quad = 1401 \times 10^6$$
$$18\,000(148 - 65)^2 \quad = 124 \times 10^6$$
$$6000(445 - 148)^2 \quad = \underline{529 \times 10^6}$$
$$I_g = 6063 \times 10^6 \text{ mm}^4 = \text{say, } 6.060 \times 10^{-3} \text{ m}^4$$

$$f_r = 0.623\sqrt{25} = 3.12 \text{ MPa}$$
$$-M_{cr} = f_r I_g/y_t = 3.12 \times 6.060 \times 10^{-3}/0.148 = 0.127 \text{ MN} \cdot \text{m} = 127 \text{ kN} \cdot \text{m}$$

Full Load Short Time Deflection

The loading that gives maximum positive moment also gives maximum deflection and the most cracking at midspan.

$$\text{Service load} + M = 0.072\,w\ell^2 = 0.072(55 \times 10^{-3})6^2 = 142.6 \times 10^{-3} \text{ MN} \cdot \text{m}$$
$$+ M_{cr}/M_a = 57.2/142.6 = 0.40$$
$$+ I_e = I_{cr} + (M_{cr}/M_a)^3(I_g - I_{cr}) = 2.070 \times 10^{-3} + (0.40)^3(6.460 - 2.070)10^{-3}$$
$$= 2.070 \times 10^{-3} + 0.280 \times 10^{-3} = \text{say, } 2.35 \times 10^{-3} \text{ m}^4$$

At the support, the simultaneous negative moment (Fig. 8.6h) is

$$-M_a = -(0.053/0.072)142.6 = -105 \text{ kN} \cdot \text{m}$$

From the gross section I_g calculation at support (above), $M_{cr} = -127 \text{ kN} \cdot \text{m} > M_a$. This negative moment will not crack the section. However, in an actual structure (not in a laboratory test) this maximum deflection does not preclude an earlier cracking from an earlier maximum negative moment. In this example the author assumes the support region will have cracked. Then this I_e for maximum negative moment will be the one averaged with I_e for maximum positive moment.

$$\text{Max.} -M_a = -(0.091/0.072)142.6 = 180 \text{ kN} \cdot \text{m} \qquad M_{cr}/M_a = 127/180 = 0.71$$
$$-I_e = I_{cr} + (M_{cr}/M_a)^3(I_g - I_{cr}) = 1.970 \times 10^{-3} + (0.71)^3(6.060 - 1.970)10^{-3}$$
$$= 1.970 \times 10^{-3} + 1.460 \times 10^{-3} = 3.430 \times 10^{-3} \text{ m}^4$$

Average of center and end $I_e = (2.35 + 2.43)10^{-3}/2 = 2.89 \times 10^{-3} \text{ m}^4$

The total short time deflection from maximum $+M_a$ (Fig. 8.6h) is calculated as the sum of two simple values* as shown in Fig. 8.6i:

$$y_1 = (5/384)w\ell^4/EI = (5/384)(55 \times 10^{-3})6^4/EI = 928 \times 10^{-3}/EI$$
$$y_2 = -M\ell^2/8EI = 142.6 \times 10^{-3} \times 6^2/8EI \qquad = -642 \times 10^{-3}/EI$$
$$\text{Short time } y = \overline{286 \times 10^{-3}/EI}$$

With given E_c, $EI = (23\,600 \times 2.89 \times 10^{-3} = 68 \text{ MN} \cdot \text{m}^2$
Without creep or shrinkage, the pattern loading maximum is:

$$y_{d+\ell} = 286 \times 10^{-3}/68 = 0.0042 \text{ m, or } \ell_n/1420 \text{ for } \ell_n = 6 \text{ m}$$
$$y_d = (17/55)0.0042 = 0.0013 \text{ m}$$

---

*With unequal end moments, the average (as here) is usually adequate, except in end spans.

Long Term Deflection

The creep and shrinkage deflection is to be taken as the multiplier $(2 - 1.2\, A'_s/A_s)$ times the deflection under the usual or sustained load. For this beam the sustained load is here assumed to be the dead load plus half the live load. The 38 kN/m live load implies something like storage or light manufacturing. It seems reasonable, although only a guess, that all or part of this load will be present enough of the time to average half of the load all the time, that is, $17 + 38/2 = 36$ kN/m.

There is no reason to assume this usual load will be in a pattern producing maximum moment; it will be assumed over all spans and this, for uniform spans, gives the fixed end moments of Fig. 8.6j.

$$\text{Usual negative moment} = -(1/12)\text{w}\ell^2 = (1/12)36 \times 6^2 = 108 \text{ kN} \cdot \text{m}$$

Although the effective I_e calculation above is lengthy, the Code cannot reflect all the possibilities or probabilities. At the very start, creep accruing from dead load and any construction loading may reflect partial cracking. After building occupancy, I_c should reflect cracking from full live load. Theoretically and greatly oversimplified, one needs the sum of two creep and shrinkage calculations. First, the dead load deflection alone for the limited time prior to occupancy (or loading) would be computed on the limited stage of cracking with the multiplier term estimated from the curves of Fig. 8.7; the product then gives the increased deflection before live loads come on. Then the 36 kN/m (dead plus half live) load deflection would be computed on the *final* cracked stage, with $EI = 68$ MN $\cdot$ m^2 as for the pattern loading, since much of this creep would occur after full live load had been experienced.

Such a procedure gets into much detail and many assumptions. Instead,

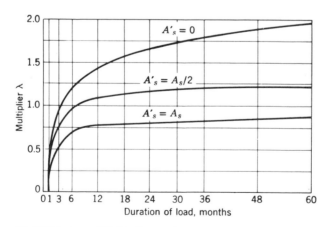

FIGURE 8.7 Multipliers for long-term deflection. (From Code Commentary)

the author will first calculate the second of these basic deflections and use it as a basis for a judgment decision about the first one. The immediate deflection for the sustained loading is as follows, using data from the previous calculations and Fig. 8.6j, with dead plus half live load on all spans:

$$y_1 = (5/384)(36 \times 10^{-3})6^4/EI \qquad\qquad = 608 \times 10^{-3}/EI$$
$$-M = 0.083(36 \times 10^{-3})6^2 = 107.6 \times 10^{-3} \text{ MN} \cdot \text{m}$$
$$y_2 = 107.6 \times 10^{-3} \times 6^2/8EI \qquad\qquad = -489 \times 10^{-3}/EI$$
$$\text{Sustained load} \quad = 119 \times 10^{-3}/EI \text{ initially}$$

For $EI = 68$ MN $\cdot$ m^2, this initial $y = 1.75 \times 10^{-3}$ m $= 1.8$ mm

Creep for the worst possible case, assuming that the cracking here fits the I_e calculations made earlier, will be used as a trial. (Less cracking would indicate a lower concrete stress and lower resultant creep.) Although some of this y occurs before partitions are built in, this will also be temporarily ignored, in an attempt to get the general picture as to what precision is needed.

The Code multiplier for added time effects. unless better calculations are made, is $2 - 1.2(A_s'/A_s) \geqslant 0.6$. From Fig. 8.6$b,c$:

$$A_s'/A_s \text{ for } -M = 800/2400 = 0.33$$
$$A_s'/A_s \text{ for } +M = 400/1600 = 0.25$$
$$\text{Average } A_s'/A_s = 0.29$$

Time thus adds $y = (2 - 1.2 \times 0.29)(119 \times 10^{-3}/EI) = 196 \times 10^{-3}/68$
$$= 0.0029 \text{ m}$$

This adds to the initial y calculated above to give $y = 0.0043 + 0.0029 = 0.0072$ m $= 7.2$ mm $= \ell_n/833$ for $\ell_n = 6$ m. This is much less than the permissible $\ell_n/480$ in Code Table 9.5b and makes further refinement unnecessary.

If this final y were more then $\ell_n/480$, a reduction would be proper to recognize the initial dead load deflection, including creep occurring before partitions were built in. Also, a more exact calculation might replace the Code multiplier, since creep from loads applied several months after the beam becomes self supporting is considerably reduced, as indicated in Fig. 8.8.

It has already been noted at the close of Sec. 8.5a that deflections cannot be as accurately calculated as the above numbers might suggest. The designer must make many estimates in computing a beam deflection, and the end result will only by chance be as good as indicated in Ref. 11. Some questions have already been noted concerning the loading sequence and the age at loading, neither of which is under the designer's control. It is also the *average* concrete strength that determines the E rather than

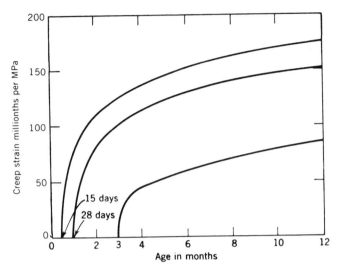

FIGURE 8.8   Variation in creep with age of concrete at loading.

the design f'_c. To the extent that curing differences do not completely obscure the picture, the average cylinder strength is the most usable value for deflection calculations. Much more data and study are needed in this area.

8.6 FURTHER DATA ON COMBINED CREEP AND SHRINKAGE

In 1960 Yu and Winter[5] presented Table 8.1 to show their evaluation of the proper multipler to be used in terms of the duration of loading and the amount of compressive reinforcement. As in the Code, these multipliers

TABLE 8.1   Multipliers for Additional Deflection Due to Shrinkage-plus-Creep

Duration of Loading	$A'_s = 0$	$A'_s = 0.5\,A_s$	$A'_s = A_s$
1 month	0.53	0.42	0.27
3 months	0.95	0.77	0.55
6 months	1.17	0.95	0.69
1 year	1.42	1.08	0.78
3 years	1.78	1.18	0.81
5 years	1.95	1.21	0.82

TABLE 8.2 Multiplier for Additional Long-Time Deflections Due to Shrinkage and Creep

Concrete Strength f'_c at 28 Days	Average Relative Humidity, Age When Loaded								
	100%			70%			50%		
	$\leq 7d$	$14d$	$\geq 28d$	$\leq 7d$	$14d$	$\geq 28d$	$\leq 7d$	$14d$	$\geq 28d$
20 to 30 MPa	2.0	1.5	1.0	3.0	2.0	1.5	4.0	3.0	2.0
>30 MPa	1.5	1.0	0.7	2.5	1.8	1.2	3.5	2.5	1.5

For the period:	1 month or less	3 months	1 year	5 years or more
Use:	25%	50%	75%	100% of table values

The 50% humidity values may normally be used for lower relative humitities, as in a heated building, for example.

lead to the *additional* (not the total) deflection when applied to an initial deflection calculated from the cracked section I_{cr}. The 1963 ACI Code used essentially the values indicated for the five-year period.

ACI Committee 435 on Deflections is reported[3] as suggesting the Table 8.2 multipliers to give the additional deflection for either normal or lightweight concrete with original deflections based on I_{eff}, but adding: "For unusual cases (particularly for very shallow members such as canopies) or when very early application of load is necessary, it is suggested that shrinkage and creep deflections be considered separately and that the choice of appropriate shrinkage and creep coefficients be made by the designer (preferably from local test data)."*

8.7 SEPARATE CALCULATION OF CREEP DEFLECTION

The above discussions have combined creep and shrinkage deflections because the numerical values of the variables are not easily determined in advance, and also because exact methods are even more involved to use. The age at which loading starts is very important as Table 8.2 suggests. The practical simplification is to say that the creep deflection from two cumulative loadings is the sum of what would be the separately computed deflections. But one must beware of any case where f_c becomes much in excess of $0.5 f'_c$, because above $0.5 f'_c$ to $0.6 f'_c$ the influence of microcrack-

*See p. 732 of Reference 3.

ing begins to build up and creep increases more rapidly than the stress increases. Fortunately, most sustained loads fall below this level.

The designer in the past has frequently used a reduced modulus, sometimes called the sustained or effective modulus as a creep approximation. In effect, this is equivalent to assuming E_c effectively reduced to $E_c/2$ or $E_c/2.5$, which is a cruder approximation than those already discussed.

Branson[3] notes that, since creep in the compression concrete shifts the neutral axis toward the tensile steel (this shift in turn lowering f_c), the creep deflection is not as large with respect to percentage as the creep on a concrete prism under constant stress. He suggests that this reduction factor k_r be taken as

$$k_r = 0.85 - 0.45\,A_s'/A_s \gtrsim 0.4$$

for creep alone. The corresponding factor for creep plus shrinkage is

$$k_r = 1 - 0.6\,A_s'/A_s \gtrsim 0.4$$

The latter, with a basic factor of 2 comparable to data from Table 8.2, is the basis for the code multiplier $2 - 1.2\,A_s'/A_s$, although the Code uses a limiting value of 0.6 instead of 2×0.4.

8.8 SHRINKAGE DEFLECTION PHILOSOPHY

Shrinkage deflection is discussed at some length here. For cantilever slabs where deflection can be particularly serious, the empirical method of Sec. 8.9 has the best correlation according to ACI Committee 435 on Deflections. It is recommended for this particular type of slab. The semielastic method of Sec. 8.10*, although not so accurate in predicting the results from available test data[6] on 14 beams, is still reasonable, considering the accuracy of the shrinkage coefficient that must be assumed in most cases. The error in predictions may be of interest, compared in terms of the distribution in each group:

Percent Error	0–9%	10–13%	14–20%	21–25%	Over 25%	Max	Mean
Empirical	11	0	1	1	1	29% low	0.989
Semielastic	2	4	4	2	1	44% high	1.064

*The method of Sec. 8.10 uses a transformed area for moment of inertia with $2\,n$ or $2.5\,n$ while the Committee formulated the comparison on the basis of an I_g using gross area of concrete alone with $E_c/2$ for the modulus. The author has not made similar comparison for the method of Sec. 8.10, but does consider it a more viable approach for varied use.

The unequal intervals are selected to emphasize the different groupings of the results in the two cases.

The semielastic method is flexible in use and able to handle any shape, although sections cracked to an unknown depth become a trial-and-error solution after a first estimate of the extent of the cracking. The method can handle any assumed distribution of shrinkage, although the basic estimate of shrinkage will normally be a crude one at best. The basic process is equally usable for uniform or nonuniform temperature changes.

8.9 BRANSON'S EMPIRICAL METHOD FOR SHRINKAGE OR UNCRACKED SECTIONS

This method is quite suitable for slabs with or without compression steel. The cover over the bars is not one of the variables included, and the equations probably fit unusual cover conditions less accurately. However, the design value of the assumed shrinkage (Sec. 2.8) is the largest probable source of error for any method.

For ρ expressed as a percent of steel rather than a ratio, the curvature is expressed for a singly reinforced member in terms of the shrinkage strain ϵ_{cs} as

$$\theta_{cs} = 18 \, \frac{\epsilon_{cs}}{h} \, \sqrt[3]{\rho} = \text{(coefficient)} \, \epsilon_{cs}/h$$

or

ρ	Coefficient
0.5%	14
1	18
1.5	20
2	22

With θ_{cs} known and uniform over the length, this leads to the very simple deflection calculation of $\theta_{cs}\ell^2/8$ for a simple span and $\theta_{cs}\ell^2/2$ for a cantilever.

With compression reinforcement the curvature is

$$\theta_{cs} = 18 \, \frac{\epsilon_{cs}}{h} \, \sqrt[3]{\rho - \rho'} \, \sqrt{(\rho - \rho')/\rho} \quad \text{for} \quad \rho - \rho' \lessgtr 3\%$$

$$\theta_{cs} = 25 \, \epsilon_{cs}/h \quad \text{for} \quad \rho - \rho' > 3\%$$

The method gives no information as to stresses that, because of creep, are rather uncertain.

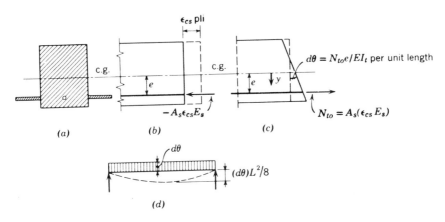

FIGURE 8.9 Shrinkage deflection calculation. (*a*) Transformed area. (*b*) Free shrinkage of concrete, forced shortening of steel. (*c*) Elimination of forcing force. (*d*) Deflection from uniform shrinkage.

8.10 SEMIELASTIC METHOD FOR SHRINKAGE OR TEMPERATURE

(a) General Theory

The shrinkage deflection can be approximated for a given uniform shrinkage by using a reduced modulus of elasticity in lieu of creep effects, as suggested in Sec. 2.7. The usual transformed area under elastic stresses can be used in several ways under this artificial assumption of elastic behavior of the beam. A method of superposition is used here.

If a member is symmetrically reinforced, it shortens as the concrete shrinks but does not become curved or deflect laterally. Consider then a member with unsymmetrical reinforcing as shown in Fig. 8.9*a*. Assume the shrinkage ϵ_{cs} (mm per mm) shortens the concrete uniformly, without restraint from the steel,* as in Fig. 8.9*b*; this can be accomplished by applying an external compressive force $A_s(\epsilon_{cs}E_s)$ on the steel to shorten it an equal amount. This gives $f_{c1} = 0$ and $f_{s1} = -\epsilon_{cs}E_s$, with the minus sign representing compression. Then release the restraint of this unwanted external force by applying an equal and opposite (tensile) force N_{to} to the entire transformed area (A_t, I_t) as shown in Fig. 8.9*c*. The stresses for this loading at any distance y from the centroid are given, as for any ec-

---

*The basic idea is to retain or develop a plane section, and an unrotated plane simplifies the later steps. In the given case the elongation of the concrete to match the unstressed steel would also be an easy starting step.

centrically loaded member, by

$$f_{cy} = \frac{N_{t0}}{A_t} + \frac{(N_{t0}e)y}{I_t}$$

for tension positive and y positive toward the steel. The total concrete stresses are the algebraic sum of the corresponding stresses from the two cases and the steel stress can be obtained similarly, of course, with $f_{s2} = nf_c$ at the level of the steel.

For curvature, the moment $N_{t0}e$ is the only term of significance, leading to $d\theta = M\,ds/EI_t = N_{t0}e\,ds/EI_t$ or $N_{t0}e/EI_t$ per unit length. For constant reinforcing the member shape is a circular curve which, for a simply supported member as in Fig. 8.9d leads to maximum deflection $y = (d\theta)\ell^2/8$ or for a cantilever $y = (d\theta)\ell^2/2$. No significant modification of the method is needed for steel at more than one level. The uncertainties of the method come from (1) the assumption of a uniform ϵ_{cs} and its magnitude, (2) the assumption of the value of n to include the effect of creep, and (3) the necessity of assuming the member either cracked or uncracked in establishing A_t and I_t for the calculation of $d\theta$.

For beams or slabs with varying A_s over the length, deflections can be calculated by area-moment methods by noting that a $d\theta$ curve simply takes the place of an M/EI curve. The sign of $d\theta$ is positive where A_s is in the bottom of the beam, negative with top tension steel.

It should be noted that compression steel reduces shrinkage deflection, eliminating it completely for uniform shrinkage in slabs and rectangular bams when $A_s' = A_s$. Good practice calls for A_s' in *all* cantilever slabs to avoid shrinkage deflection and to reduce creep deflection from dead load.

(b) Example of Shrinkage Deflection, Uncracked Section

A canopy 100 mm thick (Fig. 8.10) extends 3 m as a cantilever. If $E_s = 200\,000$ MPa and $E_c = 23\,000$ MPa, find the deflection at the outer end from uniform shrinkage $\epsilon_{cs} = 0.0004$, using 2.5 n, to approximate the creep effects and assuming an uncracked slab section and a horizontal tangent at the support.

Solution

$$n = 200/23 = 8.7 \qquad 2.5\,n = 21.7$$

Transform the area as in Fig. 8.10a and locate the centroid:

$$\bar{y} = \frac{21.7 \times 1000 \times 25}{21.7 \times 1000 + 1000 \times 100} = 4.46 \text{ mm above middle}$$

$$N_{t0} = 1000 \times 10^{-6} \times 0.0004 \times 200\,000 = 0.08 \text{ MN/m at } e = 75 - 54.46 = 20.54 \text{ mm}$$

$$M = N_{t0}\,e = -0.08 \times 20.54 \times 10^{-3} = -1.64 \times 10^{-3} \text{ MN} \cdot \text{m/m}$$

$$I_t = \tfrac{1}{12} \times 1000 \times 100^3 + 1000 \times 100 \times 4.46^2 + 21\,700 \times 20.54^2$$

$$= (83.33 + 1.99 + 9.16) \times 10^6 = 94.48 \times 10^6 \text{ mm}^4/\text{m} = 94.48 \times 10^{-6} \text{ m}^4/\text{m}$$

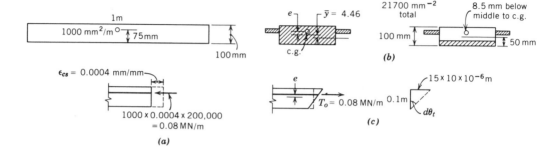

FIGURE 8.10 Shrinkage deflection calculation. (*a*) With cracked section. (*b*) Transformed area of cracked section. (*c*) Temperature deformation pli.

Effective $E_c = 23\,000/2.5 = 9\,200$ MPa

$d\theta = M/E_c I_t = -1.64 \times 10^{-3}/9\,200 \times 94.48 \times 10^{-6} = -1.887 \times 10^{-3}$ radians/meter

$y = \frac{1}{2}\,d\theta\,\ell^2 = -\frac{1}{2} \times 1.887 \times 10^{-3} \times 3^2 = -8.5 \times 10^{-3}$ m $= 8.5$ mm

More significant digits are not justified.

(c) Stresses from Shrinkage

Shrinkage stresses in structural members are rarely large unless member deformations are restrained, but shrinkage stresses add to flexural tension to induce member cracking earlier than would otherwise be expected. In investigating shrinkage stresses, it is the tensile stresses that are the more important. Hence it would be proper to consider all the transformed area shown in the middle upper sketch of Fig. 8.10. In this slab the shrinkage stresses can be found as the sum of those indicated at the bottom of Fig. 8.10*a*. In the first sketch,

$f_{c1} = 0, \quad f_{s1} = N/A_s = -0.08/1\,000 \times 10^{-6} = -80$ MPa
(minus representing compression).

In the second sketch, for axial load:

$$f_{c2} = N/A = +0.08/(21\,700) + 1\,000 \times 100) \times 10^{-6} = +0.66 \text{ MPa}$$

$$f_{s2} = nf_{c2} = +21.7 \times 0.66 = +14.3 \text{ MPa}$$

and for flexure, with y positive toward the compression face:

$$f_{c3\text{top}} = My/I = -1.64 \times 10^{-3}(4.46 - 50) \times 10^{-3}/94.48 \times 10^{-6} = +0.79 \text{ MPa}$$

$$f_{c3\text{bott}} = -1.64 \times 54.46/94.48 = -0.95 \text{ MPa}$$

$$f_{s3} = -21.7 \times 1.64 \times (-20.54)/94.48 = +7.74 \text{ MPa}$$

$$\text{Total } f_{c,\text{top}} = 0 + 0.66 + 0.79 = 1.45 \text{ MPa (tension)}$$
$$f_{c,\text{bott}} = 0 + 0.66 - 0.95 = -0.29 \text{ MPa (compression)}$$
$$f_s = -80.0 + 14.3 + 7.74 = -58 \text{ MPa (compression)}$$

However, if other loading has cracked the concrete, the transformed area in the middle sketch is not valid; all resulting tension areas already cracked must be omitted.

(d) Example of Shrinkage Deflection, Cracked Section

The above canopy may be considered as being cracked to a depth of 75 mm from earlier live loading. Calculate shrinkage deflection.

Solution

For the transformed area of Fig. 8.10b, with all the concrete below the crack as effective.

$$A = 1\,000 \times 25 + 21.7 \times 1\,000 = 46\,700 \text{ mm}^2/\text{m}$$
$$\bar{y} \text{ from middepth} = (25\,000 \times 37.5 - 21\,700 \times 25)/46\,700 = 8.5 \text{ mm}$$
$$I = \tfrac{1}{12} \times 1\,000 \times 25^3 + 1\,000 \times 25(50 - 8.5 - 12.5)^2 + 21\,700(25 + 8.5)^2$$
$$= 46.6 \times 10^6 \text{ mm}^4$$
$$\ell = 25 + 8.5 = 33.5 \text{ mm}$$
$$M = N_{to}e = -0.08 \times 33.5 \times 10^{-3} = 2.68 \times 10^{-3} \text{ MN} \cdot \text{m/m}$$

Stress at bottom of original crack $= N/A + My/I$
$$= +0.08/46\,700 \times 10^{-6} - 2.68 \times 10^{-3} \times 16.5 \times 10^{-3}/46.6 \times 10^{-6} = 0.76 \text{ MPa (tension)}$$

This stress is too low to crack the concrete (usually taken as $0.7\sqrt{f_c'} = 3.5$ MPa for a C25 concrete), and no tension can exist at the crack itself. Assumed section is O.K. The deflection calculation could be next, but the other stresses will be calculated as a matter of possible interest.

$$f_{c,\text{bott}} = 1.71 - 2.68 \times 41.5/46.6 = -0.68 \text{ MPa (compression)}$$
$$f_s = -0.08/1\,000 \times 10^{-6} + 21.7 \times 1.71 + 21.7 \times 2.68 \times 33.5/46.6$$
$$= -80 + 37.1 + 41.8 = -1.1 \text{ MPa (compression)}$$

Notice that with a cracked section the shrinkage stresses tend to dissipate when no A_s' is present, but this very fact means an increase in curvature.

$$\theta_{cs} = M/EI = -2.68 \times 10^{-3}/9\,200 \times 46.6 \times 10^{-6} = 6.25 \times 10^{-3} \text{ radians/m}$$
$$y = \tfrac{1}{2}\theta_{cs}\,\ell^2 = -\tfrac{1}{2} \times 6.25 \times 10^{-3} \times 3^2 = 28 \times 10^{-3} \text{ m} = 28 \text{ mm}$$

The probable deflection lies between -28 mm and the original -8.5 mm, because the entire length of the beam will probably not be cracked from loading (probably nearer the larger value because the most influencial part of the beam is the one which load will crack).

Deflection will be further increased by a differential temperature from the sun on the top surface. For example, assume that the direct sun raises the temperature of the top to 15°C above the underside with a linear temperature gradient between top and bottom. The coefficient of expansion of steel is 12×10^{-6} per degree (C) and of concrete around 10×10^{-6}, making this difference of little importance in most cases. As shown in Fig. 8.10c,

$$d\theta_t = 10 \times 10^{-6} \times 15/0.1 = 1.5 \times 10^{-3} \text{ radians/m}$$
$$\text{Temperature } y = -\tfrac{1}{2} d\theta_t \, \ell^2 = \tfrac{1}{2} \times 1.5 \times 10^{-3} \times 3^2$$
$$= 6.75 \times 10^{-3} \text{ m} = 7 \text{ mm}$$

The total addition to deflection from dead or live load and creep is thus $-28 - 7 = -35$ mm.

(e) Temperature Curvature and Stresses

As a sample for discussion, consider a roof frame where the upper part of the beam is exposed above the roof for architectural reasons while the lower part is within the enclosure. With seasonal temperature changes outside and air conditioning inside, the concrete temperature shifts around. Assume an analysis is needed when the internal concrete temperature is that of Fig. 8.11. Any distribution, straight line or curving, uniform or nonuniform, is possible under the method used above.

As part of a frame, both changes in length and curvature are important; both should be relative to the basic analysis conditions. For simplicity here, but not essential to the method, consider that creep has adjusted the frame to an average 20°C temperature. Step 1 would add on the transformed area the forces required to change the beam and slab section to a plane one, as if all were at 20°C, that is, $f_{c1} = \epsilon_{ct}$ (50–20)E_c at the top. E_c is the most troublesome term, probably the elastic value for daily variations, a reduced value for seasonal variations. (In an extreme case parts of the temperature could be assigned to each.) Step 2 is to apply reversed forces (or their resultant) to the transformed area, cracked or uncracked as is

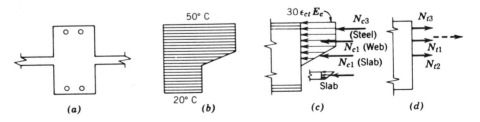

FIGURE 8.11 Temperature stress example. (a) Beam. (b) Temperature in concrete. (c) Step 1 forces. (d) Step 2 forces.

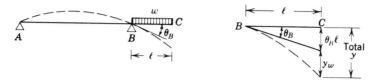

FIGURE 8.12   Extra cantilever deflection caused by rotation of supports.

realistic. For a statically determinate system, elongation, angle change, and stresses of interest are determined as above from these reversed forces. For a continuous frame one must introduce these elongations and angle changes into the system as part of the applied load. Always, for deformations, deflections, or stresses it is the sum of step 1 and step 2 values that must be considered, plus any statical redundants induced by them.

8.11 EXTRA DEFLECTION OF CANTILEVERS AND OVERHANGING ENDS

In contrast to the reduction in deflection from continuity, cantilevers and overchanging ends often have their normal deflections y_w increased by rotations of their supported ends, as indicated in Fig. 8.12.

8.12 EFFECT OF INELASTIC ACTION ON DEFLECTION

The calculation of deflections is more complex when inelastic action occurs. If the relation between moment and angle change is known, as in Fig. 11.1, it is possible to calculate deflections by a summation process in which the real $d\theta$ (or ϕ) for each given moment is obtained from the curve. Each of these $d\theta$ values contributes to the member deflection just as the $M\,ds/EI$ increments do in ordinary area-moment procedures. The area-moment calculation simply changes from $\Sigma x(M\,ds/EI)$ to $\Sigma x\,d\theta$, as indicated in Fig. 8.13. Of course, in indeterminate structures the process is more involved because the moment itself is influenced (shifted) by inelastic deformations.

When the reinforcing steel reaches the yield point, ϵ_s and $d\theta$ increase rapidly with small changes in load. The increasing steel deformation crowds the neutral axis toward the compression face at this section and brings about increasing concrete deformations and finally a secondary compression failure. For an ordinary simple span slab the difference between the load at steel yield point and the failure load will be only a few percent, say, 5% or possibly 10%. The deflection before failure will be many times that at the first yielding of the steel. The total relative

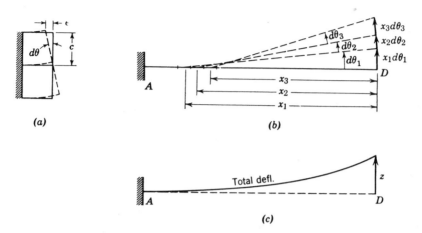

(a)   (b)

(c)

FIGURE 8.13 Deflection from deformation of beam elements, positive moment.

deflection will depend on the percentage of steel, the span, and the type of loading, but ten times the yield-point deflection would be quite usual. This added deflection will be almost entirely due to the large angle change θ_2 at the yielding section, the slab tending to fold as in Fig. 8.14 about this point as though it constituted a stiff hinge. This is the general basis for limit design (Chapter 11) and the yield-line method (Chapter 13).

Except for the general uncertainties as to partially elastic action prior to yielding, the total deflection is easy to calculate for any given θ_2 value. If the moment-ϕ diagram became truly horizontal, θ_2 would increase indefinitely to failure. If, however, a definite known relation exists between M and the angle change $d\theta$ or ϕ per unit length, as in Fig. 11.1 the simple beam moment diagram permits $d\theta$ to be established for each section and particularly for those critical sections near the point of maximum moment, that is, at the so-called plastic hinge section.

The area-moment theorems are restated in Reference 12 in a manner consistent with inelastic and plastic hinge action.

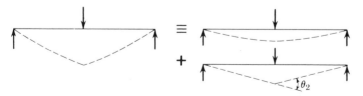

FIGURE 8.14 Beam deflection after a plastic hinge forms equals elastic deflection plus hinge deflection.

8.13 CRACK WIDTH

(a) Code Requirement Related to Crack Width

Both for appearance and for resistance to corrosion, the Code discourages large crack widths without dealing directly in terms of this width.

In 1968 Gergely and Lutz developed an equation[7] for expected maximum crack width at the tension face of a flexural member, as follows:

$$w = 0.076\, \beta f_s\, \sqrt[3]{d_c A}$$

where w = expected maximum crack width in 0.001 in. (0.025 mm) units

β = ratio of distances to the neutral axis from the extreme tension fiber and from the centroid of A_s

f_s = steel stress in ksi

d_c = cover of outermost bar of A_s measured to the center of bar, in in.

A = tension area per bar in in.² measured as in Fig. 8.15, that is, equal to 1/5 of the shaded area for the 5 bars used, centered on c.g. of bar areas, in mm².

The Code (10.6.4) uses this equation with β taken as 1.2 as the basis for computing limiting values of

$$z = f_s \sqrt[3]{d_c A}$$

which are set at 30 MN/m for interior exposure and 25 MN/m for exterior

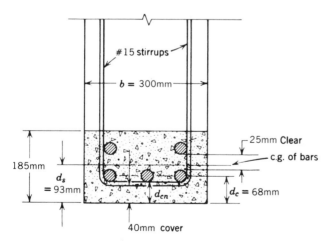

FIGURE 8.15 For computing the area A, the shaded area is divided by the number of bars which is five. (Modified from ACI Commentary.)

exposure. These correspond to w values of 0.4 mm and 0.3 mm, respectively. Since crack widths scatter over ±50%, the object of specifying z is to push toward more and smaller bars rather than toward fewer large ones. In increasing from Grade 300 to Grade 400 bars it has been easy to keep bars as large as before and to use fewer of them. Better crack widths result if smaller bars are used, and this is the remedy where the requirement for z is not met. Both positive and negative moment bars must be checked. For mixed bar sizes use the number of bars as A_s/(area of largest bar).

This limitation on z was never intended to control crack width between the A_s level and the neutral axis. For d greater than 900 mm, see the face steel requirements discussed in Sec. 4.9.

The Code permits the use of $0.6 f_y$ in lieu of a calculated f_s if the designer wishes, and it does not require this check when Grade 300 bars are used. Appendix Fig. B.9 uses the $0.6 f_y$ and gives curves that fix the maximum value of the third variable when any two are known.[14]

The value of β in the Gergely-Lutz equation may logically be used in establishing the limiting z in lieu of the 1.2 value used in fixing the Code limits. The Commentary suggests for one-way slabs that 1.35 would be better than 1.2, which would mean that the permissible z given would be multiplied by the ratio 1.2/1.35, to hold the same nominal crack width. Something similarly done for a beam, using a more "exact" β, might be regarded as simply using a more refined method than that in the Code.

Research by Nawy[8,9] has indicated that cracking in two-way slabs follows a pattern somewhat different, a pattern much influenced by the spacing of the reinforcement grid in the slab. This agrees with the author's observation from research testing that, with transverse steel somewhat widely spaced, flexural cracks nearly always develop first over the stirrups or ties rather than between them. Transverse steel exerts a considerable influence on crack spacing.

(b) Crack Research at University of Texas

The internal shape of cracks were investigated[10] by holding a beam under known loads while flexural cracks in a constant moment region were filled with a colored epoxy. The beams were then sawed such as to cut across the cracks directly over the negative moment bars and permit measurements with a microscope. Cracks varied greatly in thickness and contours, and the mean values were used for plotting. Such mean widths for six beams are superimposed in two groups (200 MPa and 140 MPa stresses) in Fig. 8.16. The surface crack width varies almost linearly with the stress and also almost linearly with the cover over the bar. The interior crack width *at the surface of the bar* is very narrow in comparison. Twenty cycles of loading made little difference in any of these findings.

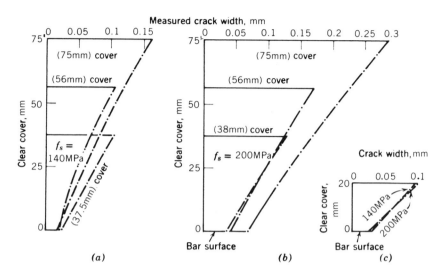

FIGURE 8.16   Measured widths of cracks from bar surface to tension face of member. The shape of crack varied more than these comparisons suggest. (*a*) Beams at 140 MPa. (*b*) Beams at 200 MPa. (*c*) Thick slabs.

Corrosion studies related to daily salt spraying of stressed beams (outdoors), a very serious exposure, are reported in Reference 15. A water-cement ratio of 0.65 by weight and 50 mm clear cover showed some corrosion in two years, only 10 or 15% more in stressed sections than in unloaded and uncracked sections. It appeared that a water/cement ratio not exceeding 0.57 by weight is needed for protection and that the ratio of clear cover to bar diameter is also critical. The indication, based on observing #20, #25, and #35 bars, was that cover of 2.5 to 3.0 bar diameters might be necessary, although this was from extrapolated data, especially with respect to time, with tests running only two years. The results are necessarily limited to chlorine (saltwater) corrosion and probably do not apply to carbonation problems.

SELECTED REFERENCES

1. ACI Committee 435, "Allowable Deflections," *Jour ACI, 65*, No. 6, June 1968, p. 433.
2. Dan E. Branson, Discussion of "Proposed Revision of the ACI 318-63: Building Code Requirements for Reinforced Concrete," by ACI Committee 318, *Jour. ACI, 67*, No. 9, Sept. 1970, p. 692.

3. Dan E. Branson, "Design Procedures for Computing Deflections," *Jour. ACI*, *65*, No. 9, Sept. 1968, p. 730.

4. *Deflection of Reinforced Concrete Members*, Bulletin ST-70, Portland Cement Assn., 1947.

5. Wei-Wen Yu and George Winter, "Instantaneous and Long-Time Deflections of Reinforced Concrete Beams Under Working Loads," *Jour ACI*, *57*, No. 1, July 1960, p. 29.

6. ACI Committee 435, "Deflections of Reinforced Concrete Flexural Members," *Jour. ACI*, *63*, No. 6, Jan. 1966, p. 637.

7. P. Gergely and L. A. Lutz, "Maximum Crack Width in Reinforced Concrete Flexural Members," *Causes, Mechanism, and Control of Cracking in Concrete*, SP-20, American Concrete Institute, Detroit, 1968, p. 87.

8. Edward G. Nawy and G. S. Orenstein, "Crack Width Control in Reinforced Concrete Two-Way Slabs," *Proc. ASCE*, *96*, ST3, Mar. 1970, p. 701.

9. Edward G. Nawy and Kenneth W. Blair, "Further Studies in Flexural Crack Control in Structural Slab Systems," *Cracking, Deflection, and Ultimate Load of Concrete Slab Systems*, SP-30, American Concrete Institute, Detroit, 1971, p. 1.

10. Syed I. Husain and Phil M. Ferguson, "Flexural Crack Width at Bars in Reinforced Concrete Beams," Research *Report 104-1F*, Center for Highway Research, The University of Texas at Austin, June 1968, 35 pp.

11. ACI Committee 435, "Variability of Deflections of Simply Supported Reinforced Concrete Beams," *Jour. ACI*, *69*, No. 1, Jan. 1972, p. 29.

12. George C. Ernst, "Ultimate Slopes and Deflections—A Brief for Limit Design," *ASCE Trans.*, *121*, 1956, p. 605.

13. G. D. Base, J. B. Read, A. W. Beeby, and H. J. P. Taylor, "An Investigation of the Crack Control Characteristics of Various Types of Bars in Reinforced Concrete Beams," Research Report No. 18, Cement and Concrete Assn., London, 1966.

14. Peter Gergely, "Distribution of Reinforcement for Crack Control," *Jour. ACI*, *69*, No. 5, May 1972, p. 275.

15. Ergin Atimtay and Phil M. Ferguson, "Early Chloride Corrosion of Reinforced Concrete—A Test Report," *Jour. ACI*, *70*, No. 9, Sept. 1973, p. 606.

9
CANTILEVER RETAINING WALL DESIGN

9.1 TYPES OF RETAINING WALLS

Retaining walls provide soil stability at a change in ground elevation. Dead weight in such a wall is a major requirement, both to resist over-turning from the lateral earth pressures and to resist horizontal sliding from the same forces. (The curved-plane sliding of soil on soil well below the retaining wall constitutes the most common kind of sliding failure,* but this is strictly a matter of soil mechanics, not of reinforced concrete. A wall can also slide *over* the soil.)

A *gravity* retaining wall (Fig. 9.1a) depends entirely on its own weight to provide the necessary stability. Plain concrete or even stone masonry constitutes an adequate material. Design is then concerned chiefly with keeping the thrust line within the middle third of the cross section.

The *cantilever* retaining wall (Figs. 9.1c and 9.2) is a reinforced concrete wall that utilizes the weight of the soil itself to provide the desired weight. Stem, toe, and heel are each designed as cantilever slabs, as indicated in Fig. 9.8.

The *semigravity* type of wall uses very light reinforcement and is intermediate between the cantilever and gravity types (Fig. 9.1b).

The *counterfort* retaining wall looks something like a cantilever wall and likewise uses the weight of the soil for stability. The wall and base are tied together at intervals by counterforts or bracing walls (Figs. 9.1d and 9.3). These act as tension ties and totally change the supports for stem and

---

*See Fig. 24–2 of Ref. 1.

233

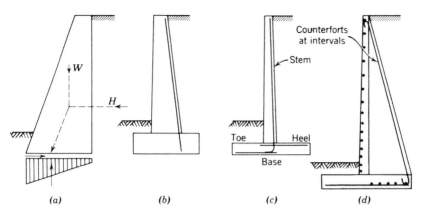

FIGURE 9.1 Common types of retaining walls. (a) Gravity. (b) Semigravity. (c) Cantilever. (d) Counterfort.

FIGURE 9.2 Cantilever retaining wall. (Courtesy Texas Highway Department.)

(a)   (b)

FIGURE 9.3 Counterfort retaining wall. (Courtesy Texas Highway Department.) (a) Under construction. (b) Before backfilling.

heel slabs. The stem becomes a slab spanning horizontally between counterforts and the heel becomes a slab supported on three sides. This type of wall becomes more economical than the cantilever type somewhere in the 6 to 7.5 m height range.

A buttressed wall is similar to the counterfort wall except that the bracing members are on the opposite side of the wall and act in compression.

Crib-type retaining walls may be made of precast concrete, timber, or metal. The face pieces are supported by anchor pieces extending back into the soil for anchorage.

Anchored retaining walls have been prominent in the literature in recent years. The face wall is anchored back into rock or even into a large mass of earth by rods or wire strands that are prestressed to an anchorage in the rock or to an anchor embedded deep in the soil.

9.2 ACTIVE SOIL PRESSURE

The reader who has a background in soil mechanics may proceed at once to Sec. 9.5 for the cantilever wall design example, which also includes some discussion of alternative procedures.

For the purpose of this text only the case of cohesionless soil will be considered, that is, essentially a dry sand. Cohesion in the soil theoretically reduces the demands on a retaining wall, but cohesion generally goes with other adverse factors, such as reduced friction and expansive-type soils, which increase the total effect. An expansive type of soil, such as

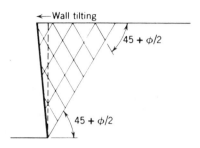

FIGURE 9.4 Rankine's plastic equilibrium for active soil pressure.

many clays, is totally unsuited for a backfill behind a wall, often introducing problems beyond economic solution.

Dry sand left unbraced will not stand steeper than a certain slope which depends on its internal friction. When confined behind a wall, such a sand tends to slide. Closely enough, the sliding can be considered as taking place on plane surfaces, with the slope to be established.

Rankine in 1857 analyzed active soil pressure on a smooth wall as one of plastic equilibrium in the soil. Sliding occurs on two sets of planes in a wedge behind the wall, as sketched in Fig. 9.4. For a smooth (frictionless) wall these planes make an angle of $45 + \phi/2$ with the horizontal, where ϕ is the friction angle for soil on soil, and produce the maximum pressure that can follow through against a wall. (For a wall that does not deflect at all, larger pressures can exist.) The plastic equilibrium idea of sliding planes is a fundamental of soil mechanics. However, Coulomb's theory presented in 1773 gives a better result than that of Rankine's for all conditions except those of a vertical frictionless wall and a horizontal fill. Since for this special case the two theories give the same result, only Coulomb's will be developed.

Coulomb considered possible sliding planes at different slopes and found the one that demanded the greatest holding force on the part of the wall. When friction on the wall is neglected, the holding force is perpendicular to the wall, that is, horizontal for a vertical wall face. Neglect of wall friction is on the safe side for active pressure (but not for the passive pressures of Sec. 9.4).

Consider a wall at AB in Fig. 9.5a. The sand behind will tend to slide on some plane such as BC_1. The wedge of sand ABC_1 is taken as a free body in equilibrium under three forces: (1) W_1, the weight of the soil; (2) R_1, the reaction from the soil below BC_1, which may be considered as a normal reaction N_1 plus a friction force F_1 resisting sliding; (3) H_1, the holding

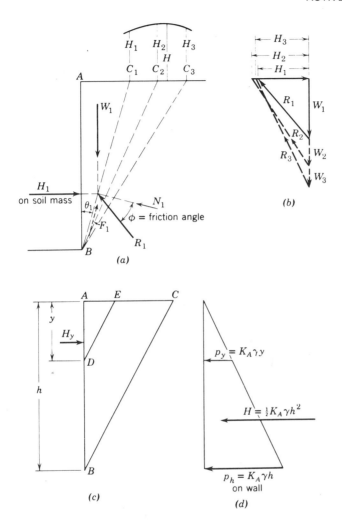

FIGURE 9.5 Maximum active earth pressure.

force from the wall. For any given slope angle, θ_1, the force H_1 (and R_1 if desired) can be found from the force triangle as shown in Fig. 9.5b. Similar force triangles constructed for other potential sliding planes such as BC_2 or BC_3 lead to other holding forces H_2, H_3, and so forth. A number of trials will lead to the determination in Fig. 9.5b of the maximum possible value of H. A helpful visualization is a plot of ordinate H_1 over C_1, H_2 over C_2, and so on, with a curve sketched through the points so located. The peak value H also locates the sliding plane, say BC in Fig. 9.5c.

A similar analysis of the soil above point D will lead to a critical sliding plane through D parallel to BC. The holding force for the depth AD will be proportional to the weight of sand in the wedge ADE, that is, proportional to y^2 and the unit weight of soil γ. If the constant of proportionality is indicated as $K_A/2$,

$$H_y = K_A \gamma y^2/2$$

The coefficient K_A is called the coefficient of active earth pressure and is usually in the order of 0.27 to 0.34, depending on the sliding friction angle.

A total pressure increasing with the square of the depth corresponds to a unit pressure increasing directly with the depth, that is,

$$p_y = K_A \gamma y$$

and the pressure on the wall is that shown in Fig. 9.5d.

With the pressure on the wall like that from a fluid (one weighing somewhat less than water), many designers have used the term *equivalent fluid weight* or *equivalent fluid* for the term $K_A\gamma$. Call this fluid weight w_f. Then

$$p_y = w_f y \qquad H_y = w_f y^2/2$$

Equivalent fluid weight is often assumed without adequate knowledge of the factors that enter into a calculated value of K_A. As presented here for strength design, no great advantage from using the w_f concept seems to accrue.

These pressures are called active soil pressures because they continue to act on a wall after it deflects or slides. Pressures in a confined soil may be higher, because the active earth pressure has been calculated on the favorable basis of a considerable holding force developed by friction. Some small movement on plane BC in Fig. 9.5c is necessary to develop this friction; without it H will be larger. For the triangular pressure distribution to be possible, there must be some sliding on all parallel planes, such as DE, above BC. Hence, the wall must deflect more at the top than at the base, by approximately 0.001 times its height, this necessary theoretical deflection corresponding to a rotation of the wall about the base at B.

It should be noted that many practical constructions fail to satisfy this deflection requirement, for example, basement walls when supported at or near the ground level by the first floor framing. Another case is the usual braced trench construction, where excavation starts with a brace placed near the top of the trench. In both these cases, for the ideal cohesionless soil, the total pressure is roughly the same, say 10% more than discussed above, or in extreme cases on an individual strut in loose sand possibly as much as 45%. The resultant acts nearer mid-depth than at the lower third point. The distribution may vary considerably, but it may be thought of as somewhat parabolic as shown in Fig. 9.6.

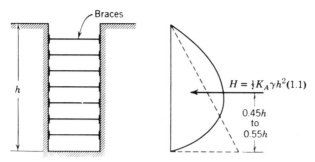

FIGURE 9.6 Earth pressure in a braced trench.

9.3 SURCHARGE

Loads on the surface of the ground over a possible sliding plane, as in Fig. 9.7, increase the horizontal pressure by adding to the ordinary soil weight W in Fig. 9.5a. Uniform surcharge over the entire area adds the same effect as an additional height of soil weighing a like amount. The concept is convenient, however, in visualizing the effect of the surcharge, which is simply that of an added height of earth weighing the same amount. Such a surcharge adds a uniform pressure to the triangular soil pressure already discussed, as shown in Fig. 9.7a. The use of unequal load factors for surcharge and soil weight adds a complication.

Surcharge far enough removed from the wall causes no pressure on the wall. For example, a surcharge far to the right of C in Fig. 9.5c cannot influence the sliding plane BC or the pressure H. A load just to the right of C would influence the sliding on a slightly flatter plane and might make such a plane critical.

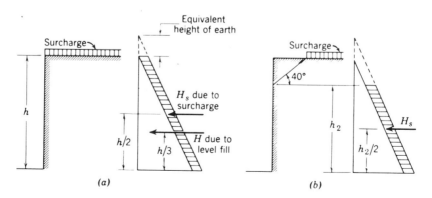

FIGURE 9.7 Effect of surcharge on earth pressure.

Engineers commonly assume that a surcharge cannot influence the pressure above the point where a line sloping downward from the load intersects the wall, as shown in Fig. 9.7b. A slope of 40° or 45° has been used in the past, although now somewhat better methods are available in soil mechanics which recognize the extent of the surcharge. The actual pressure does not change as abruptly as shown in Fig. 9.7b, but this assumption is within reason and indicates a greatly reduced overturning effect compared to Fig. 9.7a.

9.4 PASSIVE EARTH PRESSURE

If the wall is pushed against the soil, the resistance is very much higher than the active pressure because the soil friction then resists the wall movement. In this case the sliding plane is much flatter $(45 - \phi/2$ according to Rankine) and an analysis similar to that of Fig. 9.5 involves a greatly increased soil weight W. Wall friction is more important in this case and true failure is on a curved surface. Passive resistance is several times as large as active pressure.

9.5 DESIGN OF CANTILEVER RETAINING WALL

(a) Data

Overall height $= 5.5$ m.

Level fill, 20 kPa surcharge, soil weight of 16 kN/m³.

Horizontal pressure: $K_A = 0.31$.

Sliding friction of concrete on soil, minimum coefficient assumed as 0.5; of soil on soil, minimum coefficient assumed as 0.62.

Permissible soil pressure under toe = 165 kPa for service load.

Weight of concrete = 24 kN/m³.

$f'_c = 20$ MPa, Grade 400 steel.

Design wall for general provisions of ACI Building Code, using about 1.0% steel ($\rho = 0.01$).

(b) Design Sequence

The design of a cantilever wall involves the choice of heel and toe lengths and the separate design of stem, heel, and toe slabs. Each of these three slabs acts as a cantilever, as shown in Fig. 9.8. The design steps can follow this sequence:

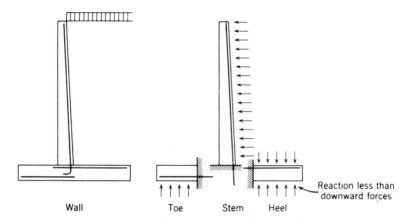

FIGURE 9.8 Design parts of cantilever retaining wall.

1. Foundation conditions[8] determine the elevation of the bottom of the base and thus the overall height.
2. The thickness of the base must then be estimated to establish the height for the stem.
3. A tentative stem thickness can then be calculated.
4. Length of heel and toe for stability against sliding and overturning can next be established. Do not count on uncertain back pressure from soil in front of toe.
5. Heel can then be completely designed.
6. Toe can be completely designed.
7. Stem design can be completed on the basis of the actual base thickness used in 5 and 6.

The design of this wall follows, with discussion interspersed, especially on:

The special problem of safety as reflected in the use of load factors for retaining wall equilibrium, in (d).

Allowable soil pressure for use with involved ultimate loads, in (e).

Conventional soil pressure distribution under eccentric reactions, in (g).

(c) Stem Design

The base thickness will be roughly 7 to 10% of the over-all height with a minimum of about 300 mm. Assume a 400 mm base, giving a stem height of $5.5 - 0.4 = 5.1\,\text{m}$ as shown in Fig. 9.9. The 20 kPa surcharge will be

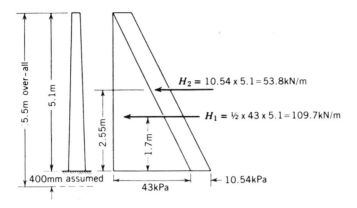

FIGURE 9.9 Earth pressure on a 1 m width of stem, including load factors.

treated as live load with a load factor of 1.7 and the soil weight as dead load with a load factor of 1.4. The Code (9.2.4) provides a separate load factor of 1.7 for the lateral earth pressure H which is to be applied to the service load values of lateral pressure.*

Because of the surcharge there is an added lateral soil pressure of $1.7 \times 0.31 \times 20 = 10.54$ kPa from top to bottom of the wall. This adds to the basic triangular soil pressure from the weight of the soil, which is a pressure increasing from zero at the top to $1.7 \times 0.31 \times 5.1 \times 16 = 43$ kPa at the base of the stem. The resultant lateral pressures per m length of wall are calculated in Fig. 9.9.

Maximum moment and maximum shear occur at the bottom of the stem and the initial design will be at this section for a 1 m length of wall. Stem dead load creates no moment in the stem, and the small direct compression it causes, about 150 kPa, would be generally ignored. Strictly under Code 9.3.2d the ϕ factor of 0.9 for flexure would be slightly decreased toward the column value of 0.70 by using $P_n/A_g = 0.15$ MPa in the equation:

$$\phi = 0.90 - 0.20 \ \phi P_n/(0.10 \ f'_c A_g)$$
$$\phi = 0.90/[1 + (2 \ P_n/f'_c A_g)]$$
$$= 0.90/[1 + 2 \times 0.15/20] = 0.886$$

The reduction from 0.90 to 0.886 is *not* significant, especially if the axial

*The last clause of Code 9.2.4 that is related to reducing load factors for D and L in some cases is discussed in detail in subsection (d).

load is ignored, as here. From Fig. 9.9:

$$M_u = 53.8 \times 2.55 + 109.7 \times 1.7 = 323.7 \text{ kN} \cdot \text{m/m}$$

$$M_n \geqslant M_u/\phi = 323.7/0.886 = 365.3 \text{ kN} \cdot \text{m/m} = 0.3653 \text{ MN} \cdot \text{m/m}$$

For $\rho = 0.01^*$, $N_t = 0.01 \, bd \times 400 = 4 \, bd$

$$a = 4 \, bd/(0.85 \times 20 \times b) = 0.235 \, d$$

$$z = d - 0.118 \, d = 0.882 \, d$$

$$M_n = 0.882 \, d \times 4 \, bd = 3.53 \, bd^2, \quad k_n = 3.53 \text{ MPa}$$

With $b = 1$ m, $bd^2 = d^2 = M_n/3.53$

$d = \sqrt{0.3653/3.53} = 0.322$ m

$h = 322$ mm $+ 50$ mm cover (Code 7.7.1b) $+ 0.5 \, d_b$ (bar diameter, say 25 mm)

$= 385$ mm*

USE $h = 400$ mm, $d = 337$ mm (for estimated #25 bars)

The design would normally proceed to the choice of base length at this stage, but, at the risk of necessary revision later, stem design will be carried further here in order to show it as a complete unit of design.

Since d is slightly increased above the required, a is slightly reduced and z slightly increased above 0.882 d, say, to 0.89 d = 300 mm

$$A_s = \frac{M_n}{f_y z} = \frac{0.3653}{400 \times 0.3} = 3044 \times 10^{-6} \text{ m}^2/\text{m} = 3044 \text{ mm}^2/\text{m}$$

$$a = 1.42 \times 3044 \times 10^{-6} \times 400/0.85 \times 20 \times 1 = 0.072 \text{ m}$$

$$z = 337 - \tfrac{1}{2} + 72 = 301 \text{ mm} \qquad \text{O.K. without revision}$$

Bar spacing for #20 at 75 mm = 4000 mm²/m

#25 at 150 mm = 3333

#30 at 200 mm = 3500

#35 at 300 mm = 3333

If the designer can foresee the development requirement for dowels into the base at this stage, it would be helpful. (This is one of several practical reasons for holding the choice of bars until the base design is complete.) Subject to a later check on bar development requirements, it is noted that the #25 fits nicely, although the 150 mm spacing means many bars to handle. Some excess A_s is thus required for all sizes. For simplicity of spacing

USE #25 at 150 mm

---

*Since the 1.0% of steel is strictly a judgment decision, not a Code limit, h could be made either 350 mm with slightly more steel or 400 mm with slightly less steel.

In slab design, shear rarely governs thickness unless the span is short or loads are unusually heavy. Hence shear can usually be delayed until a late check. It is critical at a distance d from the support, but the calculation of critical shear there may be avoided if the larger (but more easily calculated) shear *at* support is below the allowable. At the support

$$V_u = H_1 + H_2 = 53.8 + 109.7 = 163.5 \text{ kN/m}$$
$$V_n = V_u/\phi = 163.5/0.85 = 192.4 \text{ kN/m}$$
$$v = 0.1924/1 \times 0.337 = 0.57 \text{ MPa} < \tfrac{1}{6}\sqrt{20} = 0.75 \text{ MPa}$$

Development length into the base and splice lengths are discussed in subsection (j).

(d) Heel and Toe Length—Design Philosophy

Heel and toe length are interrelated but can be established in sequence. The heel has often been made just long enough to cause the resultant of the service load forces on the wall to strike the ground under the stem (Fig. 9.10a). In addition to necessary further provision against overturning, this can lead to complications from insufficient sliding resistance which require the use of a key into the soil to take a major part of the horizontal force. Such a key is rather uncertain to calculate and it functions chiefly to change the critical friction surface from concrete resting on soil to soil resting on soil, which usually increases the available friction. Alternatively, the heel length can be lengthened to increase vertical load and thus frictional resistance. Such a long heel increases wall costs and sometimes excavation costs.

The method proposed in the following subsection compromises between

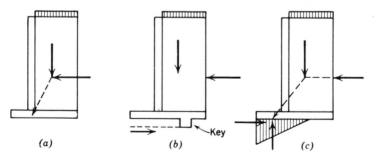

(a)   (b)   (c)

FIGURE 9.10 Design criteria. (a) Heel length to put resultant under wall, service loads. (b) Heel length to limit sliding under factored loads. (c) Toe length to give satisfactory soil pressure under factored loads.

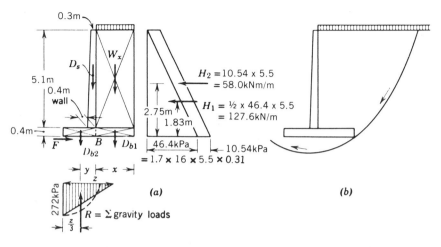

FIGURE 9.11 Retaining wall stability. (*a*) Horizontal sliding and overturning. (*b*) Sliding failure in soil.

these two approaches, basing the heel length on friction of soil-on-soil, with a key added simply to avoid sliding of the concrete at the soil-concrete interface (Fig. 9.10*b*).

The toe length must then be based on maintaining soil pressures within the allowable ultimate while avoiding overturning from overloads, Fig. 9.11. In poor soils, the minimizing of settlement problems [subsection (g)] may control the design length. The length of toe is intimately tied in with the length of heel, since a long heel is a counterbalance to overturning.

Both these equilibrium conditions, for heel and for toe, involve some loads offsetting the effects of other loads. Horizontal loads tend to cause sliding, while vertical loads mobilize more frictional resistance. Horizontal loads cause overturning moments while vertical loads develop counterbalancing moments. As a result, two serious considerations exist outside the basic areas of member design for strength and the allowable soil pressure at service load. These considerations are not, in the author's opinion, yet solved to the point of defining good practice in precise terms:

1. The open question of how best to apply a theory of safety, discussed in the next paragraphs.
2. The correlation of service load pressure with factored loads is here not as clear as in the case of column footings. Ultimate soil pressure is discussed with the choice of toe length, in subsection (f).

With regard to ϕ the Code (9.2.4) appears quite specific:

If lateral earth pressure H is included in design, required strength U shall be at least equal to

$$U = 1.4\,D + 1.7\,L + 1.7\,H \qquad \text{(Code Eq. 9.4)}$$

Except that where D or L reduce the effect of H, $0.9\,D$ shall be substituted for $1.4\,D$ and zero value of L shall be used to determine the greatest required strength U. For any combination of D, L and H, required strength U shall not be less than $1.4\,D + 1.7\,L$.

This Code section, written with the design of building members in mind, seems not adequate for the equilibrium problem of the wall. Two examples will be outlined to define the problem.

The author considers surcharge as a live load, since it may be present or absent as a loading. When present, it both (1) creates a lateral pressure on the wall, similar to H_2 of Fig. 9.9, and (2) adds to the frictional resistance below the base. Item 2 definitely reduces the sliding potential of item (1) but the Code seems to prescribe a zero load factor for this live load. It is illogical here to take this load factor on surcharge as zero and yet include the H it produces; the wall is acted on by both or by neither. (The Code is well designed for a different case, where, say, earth pressure increases a wall moment but a floor load introduces a counteracting moment.)

Likewise, in the case of the soil weight on the heel, the load tends to counteract the sliding tendency created by the horizontal pressure. Although here the Code gives 0.9 for the load factor (not zero) this concept is not really valid in evaluating the wall equilibrium. If D were a floor weight corresponding to the live load in the paragraph above, the 0.9 factor would be a sound value for use. Here, however, $1.7\,H$ will never exist, that is, have a measurable chance of occurring, if only $0.9\,D$ (as a weight of soil) exists. The $1.7\,H$ is intended to cover not only a possible increase in K_A but also a possible increase in the soil weight in $1.4\,D$. One might analyze the 1.7 as the result of the sum* of 1.4 times the nominal soil weight added to an 0.3 factor for variation in K_A.

The author recommends, therefore, that the last half of Code 9.2.4 be considered not applicable here except that 0.9 should be applied to any weight of concrete that tends to balance the effect of H. He further recommends *for the equilibrium equations* (not for the design of cross sections) the use of $1.7\,H$ for all lateral pressures with 1.4 for surcharge weight and 1.4 for soil weight.

---

*Probability theory does not directly add two factors, but for the present discussion the basic concept is correct even if 0.3 is not quantitatively correct.

The less satisfactory alternate discussed in the opening paragraph deals with service loads, requires investigation of overturning when horizontal loading is increased (usually considered as doubled), and may still leave sliding problems. Many do use this approach.

(e) Choice of Heel Length

The heel length will be chosen such that adequate friction of soil-on-soil is provided. A key will be added below the base to prevent failure from the lesser frictional resistance of concrete on soil. The method can be easily adjusted to count on either more or less help from the key. The coefficient of friction given is assumed to be a reasonable lower bound value, not a value greatly reduced to provide a large factor of safety.

In Fig. 9.11a the summation of horizontal forces gives the simple equation:

$$H_1 + H_2 - F = 0$$

where F is given by the coefficient of friction μ times the total weight $\Sigma(W + D)$. The given 0.62 coefficient for soil on soil will be used.

$$
\begin{aligned}
W_x &= 1.4(20 + 16 \times 5.1)x & &= 142.2x \\
D_s &= 0.9 \times 0.5(0.3 + 0.4) \times 5.1 \times 24 & &= 38.6 \\
D_{b1} &= 0.9 \times 0.4 \times 24x & &= 8.6x \\
D_{b2} &= 0.9 \times 0.4 \times 24 \times 1.2 \text{ (assumed)} & &= 10.4 \\
\end{aligned}
$$

$$H_1 + H_2 - F = 127.6 + 58.0 - 0.62\overline{(150.8x + 39.0)} = 0$$

$$150.8x + 39.0 = 185.6/0.62 = 299.4$$

$$x = 260.4/150.8 = 1.730 \text{ m}$$

Use heel projecting 1.73 m beyond B

While the distance x protects against horizontal sliding of the wall on the soil, a soil mechanics check is also needed against sliding on a curved plane such as that indicated in Fig. 9.11b. A complete design requires that both kinds of sliding be investigated.

(f) Choice of Toe Length

The *total* horizontal forces of Fig. 9.11a (somewhat larger than those on the stem) tend to overturn the wall and the gravity loads tend to prevent such overturning. The reaction under the base must equal the total downward load, and the necessary location of the resultant reaction can be calculated by moments about any convenient center, say B.

	Load		Arm	Moments about B	
W_s	$1.4(20 + 5.1 \times 16)1.73$	246.0	$+0.87$ m	$+214.1$	
D_s	$0.9 \times 0.3 \times 24 \times 5.1$	33.0	-0.15		-5.0
	$+0.9 \times 0.5 \times 0.1 \times 24 \times 5.1$	5.5	-0.33		-1.8
D_{b1}	$0.9 \times 0.4 \times 24 \times 1.73$	14.9	$+0.87$	$+13.0$	
D_{b2}	$0.9 \times 0.4 \times 24 \times 1.2$ (estimated)	10.4	-0.60		-6.3
	$(W + D) =$	309.8			
	$H_2 = $	58.0	-2.75		-159.5
	$H_1 = $	127.6	-1.83		-233.5
		185.6		$+227.1$	-406.1
					-179.0

$$R \text{(from gravity loads)} - 309.8 \text{ up}$$
$$M_B = -179.0 - 309.8\, y$$
$$y = -0.578 \text{ m (to left)}$$

Thus the resultant reaction acts 578 mm to the left of B.

The soil cannot support a concentrated reaction. It must be distributed in some fashion over some distance z, as in Fig. 9.11. Assume that the allowable ultimate soil resistance is developed at the toe before the soil fails. This allowable at ultimate is not readily apparent and must be discussed before calculating z.

In the case of a concentrically loaded footing one senses automatically that the allowable service load pressure should increase at the same rate as the factored loads increase, that is:

$$\left(\frac{\text{Allow. } q \text{ at ultimate loads}}{\text{Equiv. } q \text{ at service loads}}\right) = \frac{1.4\,D + 1.7\,L}{D + L} \quad \text{or} \quad \frac{1.4\,D + 1.7\,L + 1.7\,W}{D + L + W}$$

This relation is used in footing design (Chapter 20) to obtain the equivalent soil pressure at ultimate. For the retaining wall stability, the relationship is more difficult because the effect of the H load enters into the solution as a moment, and the use of the above relation in terms of moment seems totally impractical. The permitted increase in soil pressure should probably be somewhere between the 1.4 factor used for L and soil weight W_x (in the moment equation) and the 1.7 factor used for H, since the concrete weight (0.9 factor) is relatively small. The best weighting is obscured by the fact that the moments as used are of opposite signs. Here a 1.65 factor will be used,* this being a pure judgment decision. One does not escape this problem by using service loads and allowable soil pressure in the equilibrium equation. One then faces the same decision when picking

---

*A better solution would be to obtain from the soil mechanics specialist a *safe* ultimate soil pressure, something comparable to his (or her) suggested value of ϕ multiplied by the ultimate that could be expected 95% of the time. The correlation factors otherwise are simply mathematical equivalents without real physical significance, an equivalent in this particular case only roughly approximated.

the appropriate soil pressure for the check against overturning under overloads.

USE allowable soil q at ultimate $= 1.65 \times 165 = 272$ kPa

Although the real reaction distribution is uncertain, designers usually assume a straight-line distribution, which in this case means a triangle of pressure as shown in Fig. 9.11a.

$$0.5 \times 272\, z = \Sigma \text{ gravity loads above} = 309.8$$
$$\text{Min } z = 2.278 \text{ m back from the toe}$$

The resultant vertical reaction R of 309.8 kN acts at $z/3$ from the toe, or 759 mm. Thus the toe extends beyond B by $y + z/3 = 578 + 759 = 1.337$ m, or beyond the front of wall by $1.34 - 0.4 = 0.94$ m. The effect of the weight D_{b2} is so small that the weight for the assumed 1.2 m length need not be revised for these calculations.

USE toe projection $= 1$ m
USE total base length $= 1.00 + 0.40 + 1.73 = 3.13$ m

(g) Service Reaction Pressure Conditions

With the permissible soil pressures as large as 165 kPa (at service loads) settlements are probably not a serious matter. Hence, for this particular wall this subsection can be skipped. However, where settlement is a serious problem, the soil reaction under service loads can govern the toe length. This omits the load factors on all loads. The first step is to locate the resultant reaction, that is, the term y as calculated in Sec. 9.5f but here based on service loads. The reaction pressure is usually assumed to vary linearly although the actual pressure will probably follow some curved line. The resulting pressure distribution will necessarily vary with the location of the resultant R. If the resultant is at the third point of the base or nearer the toe, a nominal triangular reaction will be assumed as shown in Fig. 9.12a or b. If the resultant falls within the middle third of the base as in Fig. 9.12c, a trapezoidal pressure results; if at the center as in Fig. 9.12d, a uniform pressure results.

The more serious the settlement problem the more attention must be given to the possibility of tilting; obviously, only the last case of uniform reaction pressure eliminates this problem. The position of the resultant is little changed by a change in toe length. Hence, lengthening the toe will leave the resultant nearer the middle of the base. The designer can thus increase toe length as much as service load conditions seem to require. A sample of such a reaction calculation follows.

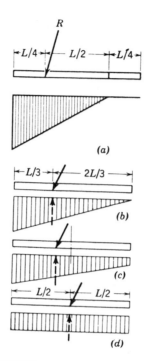

FIGURE 9.12 Reaction pressure distributions for various locations of resultant of load.

(h) Design of Heel

The load on the heel is predominantly a downward load. The supporting upward reaction comes from the tension in the *stem* steel. Hence, the effective cantilever length is 1.73 m plus the stem cover to center of steel = 1.73 + 0.06 = 1.79 m as shown in Fig. 9.13.

The downward load of earth and surcharge totals 148.2 kPa on the heel cantilever strip. The reaction was calculated in Sec. 9.5f as extending back 2.278 m from the toe or 0.938 m into this cantilever length with a triangle of reaction pressure is shown in Fig. 9.13a. In the following design this upward pressure has been omitted because (1) there is no assurance that the straight-line pressure really exists* and (2) it reduces the calculated

*Note that the 0.938 m of upward load might largely disappear if the reaction pressure were more parabolic as sketched in Fig. 9.14.

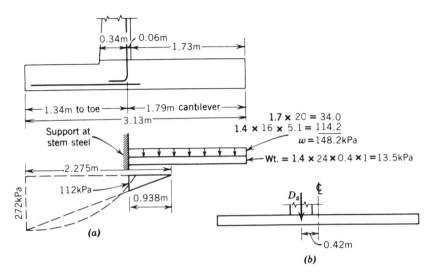

FIGURE 9.13 Heel cantilever and loading. (*a*) Maximum. (*b*) Before backfilling.

M_u and V_u on the heel, but by such a small percentage that its omission is not really wasteful.

$$M_u = (148.2 + 13.5) \times 1.79^2/2 = 259.1 \text{ kN} \cdot \text{m/m}$$

$$M_n \geqslant M_u/\phi = 259.1/0.9 = 287.8 \text{ kN} \cdot \text{m/m} = 0.2878 \text{ MN} \cdot \text{m/m} = k_n b d^2$$

$$3.53 \, d^2 = 0.2878 = 57\,900 \times 12, \quad d_m = \sqrt{0.08\,153} = 0.285 \text{ m} = 285 \text{ mm}$$

Since shear might control under such heavy loads on a short span, the unit shear equation $v_c = V_n/bd$ will be solved for $d_v = V_n/v_c b$, with the v subscript on d_v added here for identification and the m subscript added to the moment depth for the same reason. Although the critical design section for shear is specified at a distance d from the support, this is not safe when the reaction is a tension "hanger" as it is here. Hence, shear will be calculated essentially at the bar support (recognizing that 0.06 m of the length is protected from vertical load by the stem)

$$V_u = (148.2 + 13.5) \times 1.79 = 289 \text{ kN/m};$$

$$V_n \geqslant 289/0.85 = 340 \text{ kN/m} = 0.34 \text{ MN/m}$$

Permissible $v_c = \frac{1}{6}\sqrt{f_c'} = \sqrt{20}/6 = 0.75 \text{ MPa without stirrups}$

$$d_v = V_n/v_c b = 0.34/0.75 \times 1 = 0.453 \text{ m}$$

Since shear does govern, check the more exact permissible v_c and ρ_w

assumed as 0.01 (0.285/0.453) = 0.0063 where the parenthesis is the ratio of d_m/d_v.

$$v_c = \tfrac{1}{6}[\sqrt{f'_c} + 100(\rho_w V_u d/M_u)]$$

$$v_c = \tfrac{1}{6}\left[\sqrt{20} + 100 \times \frac{0.0063 \times 289.0 \times 0.453}{259.1}\right]$$

$$= 0.745 + 0.053 = 0.80 \text{ MPa}$$

(For usual cases, unless ρ is large the more exact value differs so little that it is probably more practical just to ignore it.)

$$d_v = 0.34/0.80 = 0.425 \text{ mm}$$
$$h = 425 \text{ mm} + 50 \text{ mm cover} + d_b/2 = 488 \text{ mm for } d_b = 25 \text{ mm}$$
$$\text{or } 485 \text{ mm for } d_b = 20 \text{ mm}$$
$$\text{USE } h = 525 \text{ mm}, \; d = 462 \text{ mm}$$

This increases the assumed heel weight by $1.4 \times 24 \times 0.1 \times 1 = 3.4$ kPa, but since it also reduces the earth cover by 100 mm or $1.4 \times 16 \times 0.100 \times 1 = 2.2$ kPa, it is obvious that no revision in M_n or V_n is really needed.

Since the d selected is considerably more than the required d_m, z is increased. Assume $z = 0.94 \, d = 0.94 \times 0.462 = 0.434$ m

$$\text{Reqd. } A_s = \frac{M_n}{f_y z} = \frac{0.2878}{400 \times 0.434} = 1658 \times 10^{-6} \text{ m}^2/\text{m} = 1658 \text{ mm}^2/\text{m}$$

$$a = A_s f_y/0.85 \, f'_c b = 1658 \times 400/0.85 \times 20 \times 1000 = 39 \text{ mm}$$

$$A_s = 0.2878/400(0.462 - 0.020) = 1628 \times 10^{-6} \text{ m}^2/\text{m} = 1628 \text{ mm}^2/\text{m}$$

Since, when shear is large, it is possible for bar development length to limit the choice of bars, this requirement will be investigated before choosing bars. Since these are top bars (more than 300 mm of concrete below the bars) the required development length is 1.4 times the basic value of $0.019 \, A_b f_y/\sqrt{f'_c}$, but an 0.8 factor is also probable, since spacing (hopefully) will be 150 mm or more. The available length for development is 1.79 m less 50 mm of end cover.

$$\text{Available } \ell_d = 1790 - 50 = 1740 \text{ mm}$$
$$= 0.019 \, A_b \times 400/\sqrt{20} \times 1.4 \times 0.8$$
$$\text{Max. } A_b = 1147 \text{ mm}^2$$

This A_b permits bars up to #35 (1000 mm² each)

$$\#35 \text{ at } 600 \text{ mm} \qquad \#25 \text{ at } 250 \text{ mm}$$
$$\#30 \text{ at } 400 \text{ mm} \qquad \#20 \text{ at } 175 \text{ mm}$$

The 600 mm spacing for #35 is too large, greater than the heel d. The #20 at 175 mm is reasonable, but to reduce the number of bars to handle

USE #25 at 250 mm for heel (1667 mm²/m)

With *high* walls the possible need for bottom steel in the heel before backfill is placed should be investigated. The factored weight of stem is

$$1.4 \times \tfrac{1}{2} \times (0.3 + 0.4) \times 24 \times 5.1 = 60.0 \text{ kN/m}$$

acting $1.73 + \tfrac{1}{2} \times 0.525 = 1.99$ mm from heel, that is, 0.42 m off center as indicated in Fig. 9.13*b*. Using the equation at the close of Sec. 9.5*g*.

$$
\begin{aligned}
q &= (P/A)(1 \pm 6 \; e/h) \\
&= (60.0/3.13 \times 1) \times (1 \pm 6 \times 0.42/3.13) = 19.17 (1 \pm 0.805) \\
&= 34.60 \text{ kPa at the toe and } 3.73 \text{ kPa at the heel}
\end{aligned}
$$

q at heel side of stem $= 3.73 + \tfrac{1}{2}(34.60 + 3.73) \times 1.73/3.13 = 14.32$ kPa
Positive moment in heel at face of wall:

$$M_{\text{heel}} = \tfrac{1}{2} \times 3.73 \times 1.73^2 + \tfrac{1}{6}(14.32 - 3.73) \times 1.73^2 = 10.86 \text{ kN} \cdot \text{m/m}$$

$M \geqslant 10.86/0.65 = 16.70$ kN·m/m (plain concrete $\phi = 0.65$, Code 9.3.2f)

The weight of the base adds reaction pressure but no moment. As an unreinforced beam the effective thickness should neglect the 25 mm next to ground, leaving $525 - 25 = 500$ mm, the section modulus $= bt^2/6 = 1 \times 0.5^2/6 = 41.7 \times 10^{-3} \text{ m}^3$

$$f = 16.70 \times 10^{-3}/41.7 \times 10^{-3} = 0.40 \text{ MPa}$$

Code 15.11.2 permits a flexural stress of $0.4\phi\sqrt{f_c'}$ with $\phi = 0.65$, to be used with M_u (not M_n). Here M_n has been calculated to be consistent with reinforced concrete sections. With M_n the allowable flexural stress becomes $0.4\sqrt{f_c'} = 1.79 \text{ MPa} \gg 0.40 \text{ MPa}$. Hence, footing is O.K. for this temporary loading.

(i) Design of Toe

Toe design is similar to that of heel design, except that diagonal tension is critical at a distance d from the face of the wall. Earth fill over the toe has been neglected thus far since it may not always be present. This fill adds also to the toe reaction, but this reaction and fill weight cause little change in M and V on the toe.

The maximum moment on the toe occurs where the shear changes sign. Since the large compression from the stem causes this shear reversal to occur just inside the face of the stem, the case is usually simplified by considering the cantilever support *at* the face of stem (Fig. 9.14).

The toe weight is assumed to be 13.5 kPa with the 1.4 load

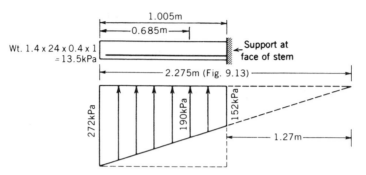

FIGURE 9.14 Toe cantilever and loading.

factor. The load factor and the footing weight is already included in the reaction pressure. Shear stress is almost sure to control the toe depth. Check d_v at distance d from support, say $d = 400 - 75 - 10 = 315$ mm leaving 685 mm to toe.

From the geometry of Fig. 9.14, the shear force due to the soil pressure at a section 0.315 m from the support is

$$V_{0.315} = 190 \text{ kN/m}$$

Subtract the shear force due to the weight of the concrete slab:

$$V_u = 190 - 1.35 \times 0.685 = 189 \text{ kN/m}$$

$$V_n = 189/0.85 = 222 \text{ kN} = 0.222 \text{ MN/m}$$

$$d_v = V_n/v_c b = 0.222/0.75 \times 1 = 296 \text{ mm} \qquad \textbf{O.K.}$$

Check next for moment, using the same 13.5 kPa term for weight. Both weight and reaction change in almost equal amounts, but the calculations are based on net pressures are ignoring the minor change in gross pressures.

$$M_u = \tfrac{1}{2} \times 152 \times 1.005^2 + \tfrac{1}{3}(272 - 152) \times 1.005^2 - \tfrac{1}{2} \times 1.35 \times 1.005^2$$

$$= 116.4 \text{ kN} \cdot \text{m/m}$$

$$M_n = 0.9\, M_u = 129.4 \text{ kN} \cdot \text{m/m} = kbd^2$$

$$3.53\, d^2 = 0.1294 \qquad d_m = 0.191 \text{ m} = 191 \text{ mm} < d_v$$

Most designers, including the author, normally prefer to use the same toe and heel thickness, but this is not essential. A little thicker toe simplifies the shear key between wall and base; a little thinner toe complicates it slightly, as discussed in the next subsection. In this design, to introduce this complication, while not maintaining the extreme thickness difference that is theoretically possible,

USE $h = 400$ mm, $d = 315$ mm (based on #20 bars)

With d_m increased more than 50%, try $z = 0.95 \times 315 = 300$ mm

$$\text{Approx. } A_s = M_n/(f_y z) = 0.1294/400 \times 0.3$$
$$= 1078 \times 10^{-6} \text{ m}^2 = 1078 \text{ mm}^2$$
$$\text{Min. } A_s = (1.4/f_y)bd = 1.4 \times 1 \times 0.315/400$$
$$= 1103 \times 10^{-6} \text{ m}^2 > 1078 \times 10^{-6} \text{ m}^2$$
$$A_s \geqslant 1103 \text{ mm}^2$$

USE #20 at 250 mm for toe (1200 mm²/m)

$$\ell_d = 0.8(0.019 \times 300 \times 400/\sqrt{20}) \text{ for wide spcg.} = 408 \text{ mm}$$
$$\leqslant \text{toe length less 50 mm end cover.} \qquad \textbf{O.K.}$$

(j) Assembly Problems* at Junction of Stem, Heel, and Toe; Revision of Stem

The stem height was based on an assumed 400 mm base thickness, whereas the design above uses 525 mm for the heel and 400 mm for the toe. The way the junction between the three members is handled influences the stem height and stem design. For this reason the *final* stem thickness and stem A_s would normally be left until after the heel and toe designs were completed.

The junction proposed by the author is shown in Fig. 9.15a and his reasoning will be clear as the various checks are made.

First, provision is usually made for a shear key between the stem and base. The key is usually made by embedding a beveled timber in the top of the footing as shown in Fig. 9.15b.† When the timber is removed and the stem is cast in place, the 40 mm vertical face provides an allowable bearing (Code 10.16.1) of (1×0.04) $0.70(0.85 \times 20) = 0.476$ MN/m of wall compared to the approximately 0.164 MN/m horizontal force on the stem (Fig. 9.9). The distribution of shear force on the horizontal section of the key at the top of footing is uncertain. If it is taken as parabolic, as for a homogeneous rectangular beam, the equation is

$$V_n \geqslant V_u/\phi = (vbh)2/3$$
$$v = 1.5(0.164/0.85)/1 \times 0.15 = 1.93 \text{ MPa}$$

The allowable shear in such a case is also not too definite. It is somewhat

---

*See Reference 7, especially pp. 1244–1246.

†Some designers prefer that the key be cast as a projection on top of the base that extends up into the stem. Some now question the need for a shear key.

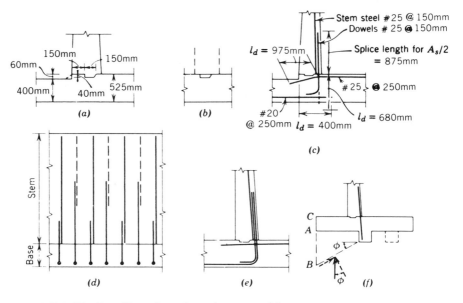

FIGURE 9.15   Details at junction of stem and base.

similar to the shear permitted between stirrups, which the Code limits to roughly $0.8\sqrt{f_c'} = 3.57$ MPa. As a practical matter, the bending moment on the stem gives $N_c = N_t = A_s f_y = 3044 \times 10^{-6} \times 400 = 1.218$ MN/m of compression pressing down on top of the base, which with a strength reduction factor of $\phi = 0.90$ could be taken as the design compression of 1.096 MN/m producing more than adequate friction to resist the 0.164 MN/m sliding force. The shear strength in front of the key must also be maintained, especially when the toe is of lesser thickness. Under this compression and with a little more area it seems to present no problem.

If heel and toe were of equal thickness, the cantilever stem height would clearly be measured from the top of the base. In the present case it is less specific. It will be quite safe to calculate stem thickness at the level of the top of the toe and steel area at the level of the heel steel; or one might calculate both at the top of the heel and put in a fillet between stem and toe to deepen the stem, as shown by the dotted line in Fig. 9.15a. The first appears the simpler from the construction viewpoint and will be used.

For compression, the height of stem is 5.5 m minus 0.4 m for a net of 5.1 m, but the loading is only on the height above the heel which is 5.5 m minus 0.525 m for a net of 4.975 m. The stem thickness was based on 5.1 m for both, which is close enough.

For tension, the height of stem is 5.5 m minus the heel d of 0.4 m, which is 5.1 m, the same as used originally. No change in the stem cross section is necessary.

Since the stem steel cannot be readily supported in position while the base concrete is placed, it is customary to provide dowels (or stub bars) in the base equal to the stem steel, #25 at 150 mm. These must project a full anchorage length into the base and a full splice length (Sec. 7.17) above the base, as shown in Fig. 9.15c. The anchorage length into the base, for a spacing $\geqslant 150$ mm, is

$$\ell_d = 0.8[0.019 \times 500 \times 400/\sqrt{20}] = 680 \text{ mm}^*$$

This length can be obtained vertically below the stem if the bar is curved into the toe. A large radius bend is more effective than a standard hook, but there should be more limitation on anchorage around a bend. For a #25 the limitation of $f_h = \zeta\sqrt{f_c'} = 45\sqrt{20} = 201$ MPa on a standard hook (Code 12.5.1) is not unreasonable. This dowel length into base is satisfactory.

The splice lap in the stem must be 1.3 ℓ_d if not over half of the bars are spliced at one level, as in Fig. 9.15d, but 1.7 ℓ_d if all bars are spliced at one point.

USE staggered splices (min. lap = $1.3 \times 680 = 884$ mm)

with 900 mm lap

The possibility of eliminating the splice on half the bars by extending half the dowels as high in the stem as they are needed for moment will be checked later.

The #25 bars in the heel are top bars (more than 300 mm of concrete below them) and are spaced more than 150 mm apart.

$$\ell_d = 1.4 \times 0.8(0.019 \times 500 \times 400/\sqrt{20}) = 952 \text{ mm}$$

Anchor the #25 bars 975 mm, as shown in Fig. 9.15c. The compression from the stem across part of the length merits some recognition, but what numerically this should be is unknown.

The bars in the toe must be anchored 400 mm.

It should be obvious, as a matter of detailing, that some of the stem dowel steel might be replaced by tow steel bent up as shown in Fig. 9.15e. This would probably save some steel and certainly would tie the toe well to the wall. The detailing then is facilitated by using the same spacings for toe and stem steel.

(k) Key Against Sliding

Resistance against sliding as provided in Sec. 9.5e assumed a key into the soil to take the difference between the coefficient of friction of 0.62 and 0.50, or 0.12 $\Sigma(W + D) = 0.12 \times 309.8 = 37.1$ kN/m. This requires only a

---

*ℓ_d could be taken directly from Table B.3, Appendix B.

limited amount of depth into the soil, and 300 mm appears adequate, as sketched in Fig. 9.15f. The key is usually placed sufficiently below the stem in order to use it for anchorage of the stem dowels, but it may be more effective when somewhat more to the rear. The maximum effect of such a shift cannot be more than to mobilize the passive resistance of the soil over the depth AB in Fig. 9.15f, where ϕ is the friction angle for soil on soil. The key should be designed as a bracket and the details will not be shown here.

(l) Temperature and Shrinkage Steel

Longitudinal bars are used to space the moment bars and to provide for shrinkage and temperature stresses; both horizontal and vertical bars are needed on exposed faces. The wall should be placed in short lengths, not to exceed 9 m, to reduce shrinkage stresses.

The Code contains the following provisions for walls, not relating particularly to retaining walls:

10.15.2 Minimum ratio of vertical reinforcement area to gross concrete area shall be:
  (a) 0.0012 for deformed bars not larger than #15 and with a specified yield strength not less than 400 MPa, or
  (b) 0.0015 for other deformed bars, or
  (c) 0.0012 for welded wire fabric (smooth or deformed) not larger than W31 or D31.
10.15.3. Vertical reinforcement shall not be spaced farther than 3 times the wall thickness, nor 500 mm.
10.15.5. Minimum ratio of horizontal reinforcement area to gross concrete area shall be:
  (a) 0.0020 for deformed bars not larger than #15 with a specified yield strength not less than 400 MPA, or
  (b) 0.025 for other deformed bars, or
  (c) 0.0020 for welded wire fabric (smooth or deformed) not larger than W31 or D31.
10.15.6. Horizontal reinforcement shall not be spaced farther apart than 3 times the wall thickness nor 500 mm.

For Grade 400 bars #15 or smaller, this leads to areas on *each* face:

	Vertical, 0.0006 bh	Horizontal, 0.0010 bh
$b = 1000$ mm, $h = 300$ mm	180 mm²/m	300 mm²/m
400 mm	240	400
525 mm	315	525

In the exposed face of the retaining wall stem the author would prefer a little more than this and possibly less on the side against the earth. In the base that will be covered, no bars are needed on the compression faces of heel and toe, and the main function of the longitudinal bars is as spacers. The following are chosen:

Stem: exposed face, horizontal,

upper 2.5 m #15 at 500 mm = 400 mm^2/m

below 2.5 m #15 at 300 mm = 667 mm^2/m

exposed face, vertical, #15 at 500 mm = 400 mm^2/m

rear face, horizontal, #15 at 500 mm = 400 mm^2/m

Toe: longitudinal spacers in bottom, #15 at 500 mm = 400 mm^2/m

Heel: longitudinal spacers in top, #15 at 500 mm = 400 mm^2/m

Temperature and shrinkage steel is shown in Fig. 9.16, with the exposed face steel in the foreground, as seen most clearly at the bottom of the illustration, adjacent to the form panel that is being lifted into place.

FIGURE 9.16   Reinforcing in wall stem with the front form being lifted into place. Frequently the rear or front face form is first erected and the reinforcement is next placed. (Courtesy Texas Highway Department.)

(m) Drainage

Since walls are usually not designed against water pressure, the design must provide complete drainage of backfill by French drains, drains through the wall, or the like. A porous backfill, such as gravel, should be provided directly behind every wall to allow the water to reach these drains.

SELECTED REFERENCES

1. Whitney Clark Huntington, *Earth Pressures and Retaining Walls*, John Wiley and Sons, New York, 1957.
2. Karl Terzaghi and Ralph B. Peck, *Soil Mechanics in Engineering Practice*, John Wiley and Sons, New York, 2nd ed., 1968.
3. *Standard Specifications for Highway Bridge*, AASHO, Washington, 10th ed., 1969.
4. "Retaining Walls and Abutments," *AREA Manual*, Vol. 1, Chapter 8, Part 5, Chicago, 1953.
5. B. K. Hough, *Basic Soils Engineering*, Ronald Press, New York, 1957.
6. Wayne C. Teng, *Foundation Design*, Prentice-Hall, Englewood Cliffs, New Jersey, 1962.
7. I. H. E. Nilsson and A. Losberg, "Reinforced Concrete Corners and Joints Subjected to Bending Moment," *Proc. ASCE, Jour. Struct. Div.*, V. 102, ST6, June 1976, p. 1229.
8. R. B. Peck, W. H. Hanson, and T. H. Thornburn, "*Foundation Engineering*," John Wiley & Sons, New York, 2nd ed., 1974.

PROBLEMS

PROB. 9.1.

(a) Investigate under service load conditions the stability, sliding resistance, and foundation soil pressure of the wall of Fig. 9.17. Surcharge = 10 kPa, soil weight = 16 kN/m^3, $K_A = 0.30$, concrete weight (including reinforcing) = 24 kN/m^3, coefficient of friction of concrete on soil = 0.40, allowable soil pressure = 145 kPa. Ignore uncertain back pressure from soil in front of toe.

(b) Same as (a) except use load factors of Sec. 9.5 and a permissible ultimate soil pressure of 290 kPa.

PROB. 9.2. Design (a) stem, (b) heel, and (c) toe for the wall of Fig. 9.17 assuming all the foundation soil pressures and sliding resistances found in

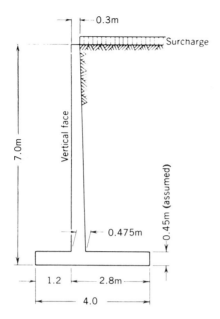

FIGURE 9.17   Retaining wall for
Probs. 9.1 and 9.2.

Prob. 9.1 are used as satisfactory. Use factored loading, $f'_c = 30$ MPa Grade
400 bars. (Use toe no thinner than heel.)

PROB. 9.3.   Redesign the retaining wall of Sec. 9.5 for horizontal pressure based on $K_A = 0.27$, $f'_c = 30$ MPa, allowable foundation soil pressure
under factored loading $= 240$ kPa, and a *minimum* base length.
(*a*) Stem design.
(*b*) Minimum base length.
(*c*) Heel design.
(*d*) Toe design with thickness based on stresses, but not less than thickness of heel.
(*e*) Sketch the detail of the joint where the stem, heel, and toe come
together.

PROB. 9.4.   Redesign the retaining wall of Sec. 9.5 changing only the
following data from those originally used:
Omit surcharge.
Limit foundation soil pressure under factored loading to 290 kPa.
Use minimum length base.
Coefficient of friction of concrete on soil $= 0.40$, of soil on soil $= 0.55$.
Avoid toe thinner than heel.

PROB. 9.5. Assume that the retaining wall of Sec. 9.5 is to be made without any toe projection, that is, as an L-shaped wall. What length of heel will be necessary to keep the service load resultant at the third point of the base? What will then be the maximum foundation pressure (ignoring the change in heel thickness that will probably be necessary): (*a*) At service load? (*b*) With load factors?

PROB. 9.6. Would it be feasible to design the retaining wall of Sec. 9.5 without any heel slab, that is, as an L-shaped wall? (Two problem aspects.)

10
CONTINUOUS BEAMS AND ONE-WAY SLABS

10.1 TYPES OF CONSTRUCTION

Most reinforced concrete members are statically indeterminate because they are parts of monolithic structures. However, it would convey the wrong impression to ignore the many forms of precast concrete construction which are taking a significant part of the present market. These vary from statically determinate floor channels, double-T sections, and other shapes to systems where precast units are combined with cast-in-place girders or a cast-in-place slab to form essentially monolithic construction or composite construction. Often the precast members are also prestressed.

The most usual form of building construction consists of a slab cast monolithically with a beam-and-girder floor framing that carries the floor load to the columns. The plan view of such a floor in Fig. 10.1a indicates that such slabs are supported on all four sides and should properly be designed with two-way steel by the methods of Chapter 12.

When the slab is more than twice as long as it is wide, it is usually designed as a one-way slab continuous over the beams, but with special negative moment steel added across the girders as required by Code 8.10.5.* When such a slab is designed for a uniform load the supporting beams are also designed for a uniform load,† this process in effect ignoring

---

*Because they find the one-way steel arrangement simpler, many engineers actually design one-way slabs for all panels except those that are nearly square.

†As computers come more into the design process, such oversimplifications are less often used.

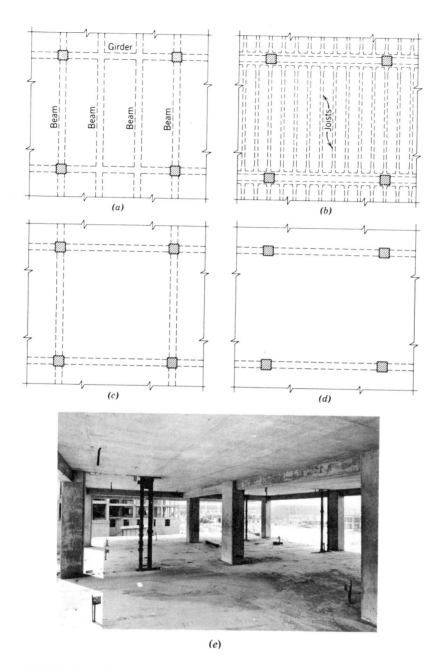

FIGURE 10.1 Typical floor beam framing. (*a*) Beam-and-girder layout. (*b*) Joist construction. (*c*) Two-way slab and beams. (*d*) One-way slab and beams. (*e*) One-way construction. (Courtesy Portland Cement Assn.)

the portion of the slab load that goes directly to the girder from the end of the panel. For the beams, the uniform load assumption is on the safe side since it overestimates the beam load. For the girders, the corresponding assumption is to place the beam reactions on the girder as concentrated loads and to add the uniform load situated directly over the girder. This assumption is not on the safe side for the girder design, particularly for maximum shear. More realistic tributary loading areas are shown in Fig. 16.1b. The student might investigate the difference in maximum girder shear between the two loadings for a girder simply supported.

Sometimes, for long spans, closely spaced beams with a very thin slab are used. This so-called joist construction (Fig. 10.1b) is facilitated by the availability of removable pans that are used as forms between the joints.

When no beams are used except those between columns, as in Fig. 10.1c the slabs are definitely supported on all four sides and both beams and slabs should be designed as discussed in Chapters 12 or 16.

In light construction the beams at times are run in only one direction, as shown in Fig. 10.1d and e. In this case true one-way slabs result, except at the beams supporting the walls at the ends. The slab band construction of Sec. 16.3 is really this same type of construction utilizing shallow beams, often quite wide.*

Two slab-type floors with beams only at outside walls and around large openings are economical, with the flat plate floor used for light loads and the flat slab floor almost always used for very heavy loads. The (patented) lift slab construction is a flat plate floor without any beams, which is cast at grade and provided with special collars around the columns for lifting into place. Flat slabs and flat plates are presented in detail in Chapter 15, including mention of the two-way joist slab commonly referred to as the waffle slab (Fig. 15.3).

10.2 INTERACTION BETWEEN PARTS OF THE STRUCTURE

In all these various types of construction the designer is faced with a highly indeterminate type of structure, a structure in three dimensions that cannot be precisely analyzed as a planar structure.[1-3] More specifically, the intermediate beams of Fig. 10.1a cannot be analyzed exactly without considering the vertical deflection and torsional stiffness of the girders as well as the stiffness of the columns. Likewise, the beams framing into the columns have moments which are influenced by the column joint rotations and hence by any torsion present in the girders.†

---

*The spandrel beams on the outside are often omitted with light live loads.
†Torque stiffness drops sharply after first diagonal cracking occurs. See Sec. 6.11 discussion.

The Code (8.5.3.1) permits "any reasonable assumptions for computing relative flexural and torsional stiffnesses," but the assumptions must be "consistent throughout analysis." This may include ignoring torsion in many situations, but Chapter 16 points out the importance of torsion in designing two-way slabs and flat slabs and plates.

Designers commonly utilize approximate methods of analysis, but they become better designers when they understand the nature of their approximations and how exact or how crude these may be. This chapter assumes that the reader is familiar with ordinary moment distribution procedures and the plotting of continuous member moment and shear diagrams.

The ordinary conventional assumption is that analysis in two dimensions is adequate for most designs. This assumption fits in better with structures designed for uniform floor loads than for those with moving concentrated or wheel loads. It may be reasonably assumed that, in carrying a uniformly distributed load, any one beam gets little help from its neighbor, because this neighboring beam is probably also fully loaded.* Likewise, in a one-way slab each strip may be assumed to carry the load directly above it when the entire slab is loaded.

When moving concentrated loads are involved. each slab strip may carry a different moment and the load on a single beam may depend greatly on the stiffness of the slab and the adjoining beams. The problem of moving concentrated loads is discussed in Chapter 17.

Moment diagrams for continuous beams normally show negative moments over supports and positive moments near mid-span, as indicated in Fig. 10.2a. The presence of columns changes the picture only slightly, typically by making the moment at opposite faces of the column slightly different. The moment diagram in any span may be considered as the sum of two parts, one the simple beam moment diagram for that span, and the other the moments across the span due to the negative support moments, as diagrammed for the center spans, Fig. 10.5a,b. Since these negative moments at supports are influenced by loads on *any* span, it follows that the moment at each point is influenced by every load on every span. Loading patterns for maximum moment are thus a major consideration.

The typical shear diagram (Fig. 10.2b) differs only slightly from that for a series of simple spans. In any span the shear diagram for a given loading

---

*This statement is not strictly true, of course. In Fig. 10.1a, with equal spacing of beams, the beam framing directly into the column is stiffer because its end joints are stiffer, whereas the girder provides only a yielding support for the neighboring beam. The beam framing into the column thus tends to carry the greater load and to some extent relieves the intermediate beam. It might be noted, however, that at the ultimate or collapse load, for beams of equal size, each beam would probably be carrying almost exactly the same load, since the load distribution changes after some yielding takes place.

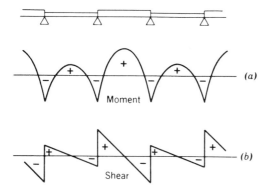

FIGURE 10.2   Typical M and V diagrams for a continuous beam.

consists of the simple beam shear plus a constant correction which will be designated as the continuity shear V_c. The continuity shear is usually relatively small except in end spans.

10.3 THE GENERAL DESIGN PROBLEM FOR CONTINUOUS MEMBERS

Each span of a continuous beam or slab requires a separate design for negative and positive moment conditions, even though practice in the United States usually uses a constant depth member for any given span, often for all spans. Figure 10.3 shows the design conditions for continuous slabs, rectangular beams, and T-beams, in diagrammatic fashion. To indicate clearly that at this stage no consideration is being given to the arrangement of steel, the reinforcement is indicated only in the vicinity of maximum moment zones. Moment at A (on the left) is assumed to be the governing (maximum) negative moment in all cases. The letter t designates the tension face.

In the thinner slabs compression steel is not desirable because it would lie so close to the neutral axis that a slight displacement would make it ineffective. Hence Fig. 10.3a shows the maximum negative moment at A and a limiting percentage of steel determining the slab depth, with relatively underreinforced sections elsewhere. Thick slabs can be designed with compression steel, similar to rectangular beams, if desired, but this does not reduce slab cost.

Rectangular beams may be designed with compression steel at supports, as in Fig. 10.3b. Code 12.12 requires that part of $+A_s$ extend into the support.

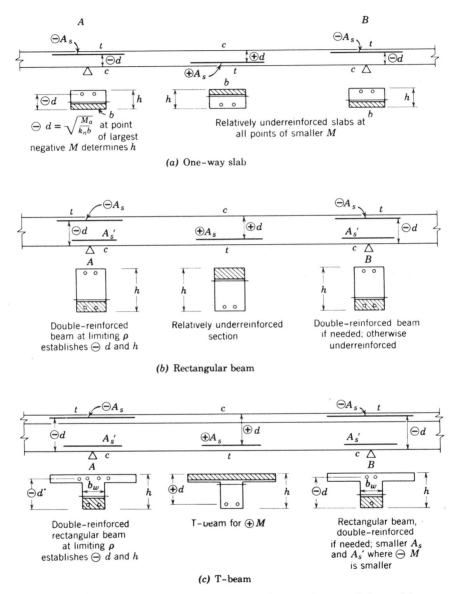

FIGURE 10.3 Strength design procedures for continuous slabs and beams.

T-beams become, in effect, inverted rectangular beams at the supports with only the web width b_w effective in compression, as indicated in Fig. 10.3c. This restricted compression zone is improved by using a double-reinforced section; this is recommended for ordinary use.

Shear (diagonal tension) rarely governs in ordinary one-way slab design, in spite of the fact that the allowable v_c is only $\frac{1}{6}\sqrt{f_c'}$. Stirrups in slabs are awkward but may be used. Continuous rectangular beams and T-beams normally require stirrups, but the web size for the continuous T-beams very rarely will be governed by shear instead of moment and only occasionally by the size needed for placing the bars.

10.4 LOADING PATTERNS FOR MAXIMUM MOMENTS

(a) Maximum Positive Moment

Elastic theory is specified for calculation of design moments except that some limit design modification is permitted.

Influence lines might be used to determine the loading arrangement, but the critical patterns for an elastic analysis are easy to deduce from a single load-deflection sketch, such as that in Fig. 10.4a in which deflections are greatly exaggerated. The usual carry-over moment idea leads directly to the moment diagram in Fig. 10.4b. It is observed that this loading produces positive moment near mid-span in all even-numbered spans. One concludes that if all even-numbered spans were loaded, each such load would increase the positive moments in the other loaded spans. This loading pattern is usually stated as follows:

> For maximum positive moment near the middle of a span, load that span and alternate spans on each side, as shown in Fig. 10.5a.

Load factors are applied to all loads. The dead load with its load factor is always considered to act* and is included on the moment diagrams shown.

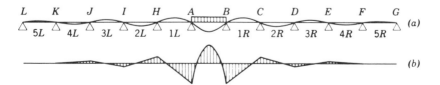

FIGURE 10.4   The influence of a single panel load on a continuous beam.

*Except that when it offsets the effect of wind or earthquake loading, the load factor is reduced.

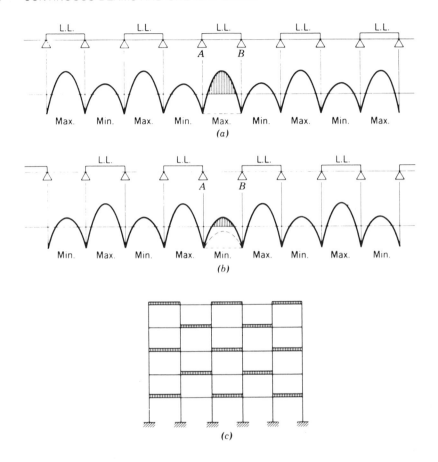

FIGURE 10.5 Loading for (*a*) maximum positive moment and (*b*) minimum positive moment at mid-span. (*c*) Checkerboard loading for multiple stories.

This one loading arrangement gives the maximum positive moments on all the loaded spans. The maximum positive moment on all other spans is given by loading only all those spans, as in Fig. 10.5*b*, taking off the live loads shown in Fig. 10.5*a*. Thus all maximum positive moments on all spans are determined by two load arrangements and two moment distributions. Extended to a multistory frame, these patterns become checkerboard arrangements of loading (Fig. 10.5*c*), still with only two patterns needed for the complete analysis.

(b) Minimum Positive Moment (or Maximum Negative Moment) Near Midspan

Since the minimum positive moment is simply the opposite extreme of loading from that for maximum positive moment on beam AB, all the live loads in Fig. 10.5a must be removed and live loads with their load factor must be placed on the other spans as in Fig. 10.5b. It should be noted that the two loadings already discussed for maximum positive moment, Fig. 10.5a and b give all the necessary minimum positive moments or maximum negative moments near midspan. The loading criterion is:

Omit live load on span considered; load adjacent spans and alternate spans beyond.

With smaller dead load moment, the midspan moment in AB could be negative, as indicated by the dashed curve in Fig. 10.5b, especially since the load factor for dead load is smaller than that for live load.

(c) Maximum Negative Moment at Left Support

The single panel loading of Fig. 10.4 shows that negative M is produced at

B by loading adjacent span to left

D by loading third span to left

F by loading fifth span to left

A by loading adjacent span to right

I by loading third span to right

K by loading fifth span to right

This can be summarized by the criterion:

For maximum negative moment at a given support, load adjacent spans on each side and alternate spans beyond.

Load factors must be applied to the loads. Applied to maximum moment at the left support A, this criterion results in the loading and moment diagram shown in Fig. 10.6a. Only the resulting moments adjacent to A are significant, since no others are either maximum or minimum moments. These moments are critical at the face of support, as discussed in Sec. 10.4g.

Adjustment of these moments from elastic analysis in the direction of limit design is considered in Chapter 11.

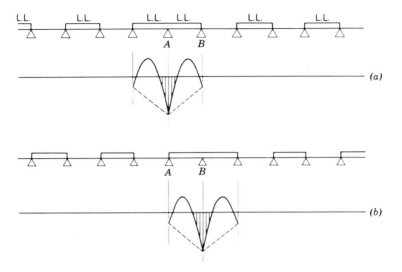

FIGURE 10.6   Loadings for maximum negative moment: (a) at left support A; (b) at right support B.

(d) Maximum Negative Moment at Right Support

The same criterion applies to this maximum moment as in the preceding case. The loadings are placed on spans adjacent to B, and alternate spans beyond, as shown in Fig. 10.6b. Unfortunately, the loading gives maximums only near the one support at B.

(e) Partial Span Loadings

It should be noted that both maximum positive and maximum negative moment conditions call for full load on the span under consideration. Maximum negative moment occurs with a very unsymmetrical moment diagram in the span involved. On the other hand, maximum positive moment results in a moment diagram nearly symmetrical about the middle of the span. Minimum positive moment at midspan (or maximum negative there when dead load is small) also gives a moment diagram nearly symmetrical, in this case a moment diagram based on dead load alone on the span in question.

In building frame design it is not customary to deal with partial span loads since they do not increase the principal design moments. In highway structures partial span loads would have a considerable influence on detailing the steel. Partial span loads are mentioned briefly in Sec. 10.6 in connection with maximum moment diagrams.

Some codes specify a single moving load to be placed anywhere on the structure as an alternate loading to care for special cases.

(f) Permissible Simplifications

The ACI Code encourages the use of a reduced or simplified frame in the analysis of buildings (Code 8.9) and specifies that moments be calculated by elastic analysis. One floor may be analyzed at a time with the far ends of columns taken as fixed. In calculating maximum negative moments the live load may be applied on only the two adjacent spans. This permits the use of simplified moment distribution procedures such as the two-cycle procedure given in Reference 1. These methods, in the author's opinion, are quite reasonable for ordinary structures, except that they give beam and column moments at the exterior columns that tend to be considerably too small. The load factors, on both dead and live load, must not be overlooked for any design.

(g) Moment at Face of Support

Moment distribution generally implies moments calculated on the basis of spans taken center to center of supports. Since such calculations treat the support reaction as though it were concentrated at a point, the resulting moment diagram within the support width is entirely imaginary. This is of no concern since both theory and tests show that the critical moment is at the *face* of the support (Code 8.7.3).

Many engineers establish the design moment at the face of support from the calculated moment at the center line of support, simply by deducting $Va/2$, where V is the shear and a is the column width, as illustrated in Fig. 10.7a. This is almost the same as scaling the moment value from the moment diagram, because the small length of uniform load within the column width (Fig. 10.7b) would cause little change in moment.

The author prefers the more conservative correction recommended by the old Joint Committee Specification, namely, a reduction taken as $Va/3$. The reasoning behind this value is as follows. The support stiffens the end of the beam much as would a haunch. A calculation considering this increased end stiffness would lead to an increased negative moment at the center of the column, as indicated by M_1' in Fig. 10.7c. Instead of calculating M_1' (which might be roughly $Va/6$ larger than M_1), an approximate equivalent is obtained by applying a smaller correction, $Va/3$, to the original calculated M_1 value. The design moment is then $M_1 - Va/3$, as in Fig. 10.7c.

The extra stiffness at the support also leads to smaller positive moments. However, the correction would be only about $Va/6$. Most engineers consider this correction less certain than that to the face of the column.

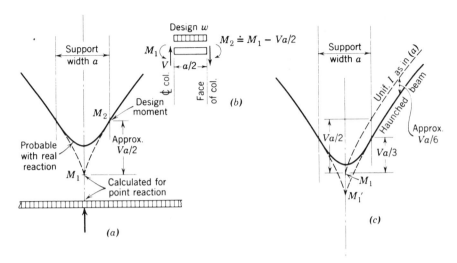

FIGURE 10.7 Moment at face of support. (a) Without correction for increased stiffness at support. (b) Free-body diagram establishing procedure in (a). (c) Recommended procedure recognizing increased stiffness at support.

They recognize the lesser accuracy of positive moment calculations, always sensitive to relative column stiffness values, and simply use the original positive moments without any correction.

(h) Validity of Elastic Analysis in Strength Design

Philosophically, the use of elastic theory for moments and the use of the very nonelastic methods of section analysis are not consistent. This leads design thinking toward some kind of inelastic frame analysis as discussed in Chapter 11. Code 8.4 permits some redistribution in this direction. Although the elastic theory may seem incompatible with strength design, it errs only by giving moments that are too large and hence too safe. This chapter is based wholly on moments obtained by the elastic theory.

10.5 MOMENT COEFFICIENTS

Any given moment can be expressed as a moment coefficient times $w\ell_n^2$, where w is the total design load (including load factors) and ℓ_n is the clear span. The maximum moment coefficients will be largest when the ratio of live load to dead load is large and when the column or other joint restraint is relatively small. Negative moment coefficients may also be large when adjacent spans are longer or more heavily loaded than the span in question.

Based on uniform live loads not greater than three times the dead load

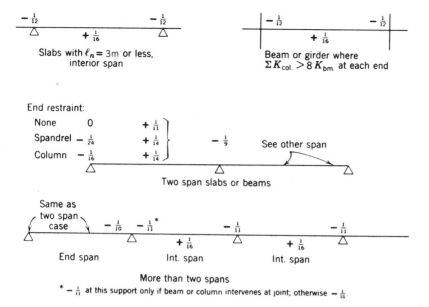

FIGURE 10.8 ACI Code (8.33) moment coefficients for nearly equal spans and live load less than three times the dead load. $M = $ (coef.) $w\ell_n^2$.

and on span lengths "approximately equal (the larger of two adjacent spans not exceeding the shorter by more than 20%)," the Code committee has established by analysis certain reasonable moment coefficients to use for maximum moment calculations. Code 8.3.3 tabulates these coefficients and Fig. 10.8 presents them in diagrammatic fashion.

The same Code section lists two shear values:

In end members at face of first interior support	$1.15\ w\ell_n/2$
At face of all other supports	$w\ell_n/2$

When the ratio of dead to live load is particularly large, some economy can be achieved by calculating the moments by more accurate methods. Also some economy can be achieved by less approximate analysis and the redistribution discussed in Chapter 11.

The use of moment coefficients should be restricted to rather standard conditions unless one uses detailed general tables.

10.6 MAXIMUM MOMENT DIAGRAMS

To determine the best arrangement of the reinforcing it is necessary for the designer to have in mind a clear picture of the extreme range of moments all along the beam or slab. The major part of such a diagram can be assembled from the several moment diagrams for critical maximum moments already illustrated in Figs. 10.5, 10.6, and 10.7. For a typical interior span with equal spans and equal live loads, these moment curves are drawn to larger scale in Fig. 10.9a–d and grouped together in Fig. 10.9e.

Recent frame studies have pointed out a moment condition not shown in Fig. 10.9 and one that is easy to overlook. When the floor system is light and the lower-story columns (rather than a shear wall) resist high wind or earthquake shears, the sum of the moments from the column above and

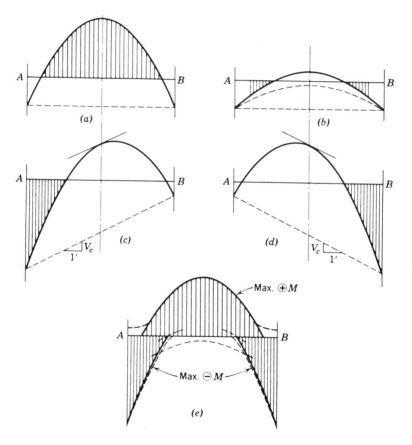

FIGURE 10.9 Moment diagrams for maximum moment. (*a*) Positive moment at mid-span. (*b*) Negative moment near midspan; the dotted curve would show with a relatively large live load or long adjacent span. (*c*) Negative moment at left support. (*d*) Negative moment at right support. (*e*) Composite maximum moment diagrams.

the column below add at the joint and rotate it and the attached beams. This increases the negative moment in one of the beams enough to make it the weakest member of the system. Hence, the designer should be alert to combinations of live load with wind or earthquake to produce the governing *beam* negative moment in spite of the appropriate reduced load factors.

By loadings exactly opposite to those for maximum negative moment, some small positive moment can often be obtained over supports as suggested by the dashed lines. Some partial span loadings may also increase the negative moments slightly, as indicated by dashed lines, but these are not very significant changes.

A very significant fact about the maximum moment diagrams is that, over a considerable portion of the beam, the moment may change from positive to negative, or the reverse, as the loading in adjacent spans is varied. Consequently, the designer must provide tension steel in both top and bottom over this zone.

Definite locations for some of the inflection points on Fig. 10.9 are calculated in connection with the bending of bars in Fig. 10.11 and in Sec. 10.14.

10.7 MAXIMUM SHEARS

The loading for maximum shear at a support is the same as for maximum negative moment there. Hence the end shear can always exceed the simple beam shear by an amount equal to the continuity shear V_c. On interior spans the value of V_c will be in the order of 3 to 12% of the simple beam shear, with 8% a fair average value. On end spans it can run as large as 20% or more of the simple beam shear. This large V_c on the end span is additive only on the half beam adjacent to the first interior support. For nearly equal spans, the Code (8.3.3) suggests 15% extra shear in end members at the first interior support and no addition elsewhere. The author prefers to make a nominal addition in interior spans as well as the substantial addition in the end span.

10.8 DESIGN OF CONTINUOUS ONE-WAY SLAB

Design a continuous one-way slab supported on beams at 3.7 m on centers, ACI Code moment coefficients (Sec. 10.5 above), dead load of 1 kPa (plus slab weight), live load of 10.5 kPa, weight of concrete = 24 kN/m^3, f'_c = 20 MPa, Grade 400 steel with ρ limited to 0.18 f'_c/f_y. Assume the beam stem is 300 mm wide.

No particular merit is associated with ρ of 0.18 f'_c/f_y except that (1) design constants have already been developed in Fig. 3.8, and (2) anything near this value will be more economical than the use of the maximum ρ from Table 3.1. The maximum ρ is uneconomical for slabs or beams and will frequently lead to thin sections with deflection problems; it should be avoided except where conditions do demand an extreme design.

Solution

The moment coefficients shown in Fig. 10.8 indicate that the negative moment at the first interior support is the maximum, at $0.10\,w\ell_n^2$. The slab at this point, will be designed for the arbitrary limiting steel ratio, which leads to $k_n = 0.160\,f_c'$ (Fig. 3.8) = 3.2 MPa. Slab weight will be based on assumed h of 150 mm.

$$w_\ell = 10.5 \quad \times \quad 1.7\text{(L.F.)} = 17.85 \text{ kPa}$$
$$w_d = \quad 1.0 \quad \times \quad 1.4\text{(L.F.)} = \quad 1.40$$
$$\text{Slab wt.} = \quad 0.125 \times 24 \times 1.4 \quad = \quad 4.20$$
$$\text{Total } w = \overline{23.45} \text{ kPa}$$

$\ell_n = 3.7 - 0.3 = 3.4$ m clear span

$M_u = 0.10 \times 23.45 \times 3.4^2 = 27.11$ kN·m/m width of slab

$M_n = M_u/\phi = 27.11/0.9 = 30.12$ kN · m/m $= 30.12 \times 10^{-3}$ MN · m/m

$k_n b d^2 = 3.2 \times 1\, d^2 = 30.12 \times 10^{-3}$

$d = \sqrt{0.009\,41} = 0.097$ m $= 97$ mm

Code 7.7.1c specifies 20 mm clear cover, if not exposed to weather.

$$h = d + d_b/2 + 20 = 97 + 8 + 20 = 125 \text{ mm for } \#15 \text{ bars}$$
$$\text{USE } h = 125 \text{ mm}, d = 97 \text{ mm for } \#15 \text{ bars}$$

Code 9.5.2 gives a deflection warning (Table 3.2, Sec. 3.8) if $h < \ell/28 = 3400/28 = 121$ mm < 125 mm O.K. This design cannot use the savings from the limit design idea of Code 8.4 because approximate moment coefficients have been used.

All other sections have less moment and hence all steel can be designed assuming z slightly greater than the $0.89\,d$ from Fig. 3.8, say $z = 0.90\,d$. (For very small moments it may be desirable to check a and z for a more economical value.) Noting that A_s and M_n are each per foot of slab width,

$$A_s = \frac{M_n}{f_y z} = \frac{M_n \text{ in MN·m/m}}{400 \times 0.9 \times 0.097} \text{ m}^2/\text{m} = 28.64 \text{ kN · m/m}$$

The calculations are tabulated in Table 10.1 Code 7.12.2 for Grade 400 bars requires $\rho = 0.0018$ for shrinkage and temperature, as a minimum. (For beams Code 10.5 makes the minimum $\rho = 1.4/f_y$ which is $\rho = 0.0035$ for Grade 400 bars.) The required areas and the spacing of bars are given in Table 10.1 as though top and bottom steel were to be totally separate as they are in Fig. 10.10a.

For the layout of Fig. 10.10a the requirements are well matched by the bars specified at the bottom of the table.

TABLE 10.1   Calculation of Slab Steel

	Exterior Span		First Int Sup.	Typical Interior	
	Ext. End	Middle		Middle	Support
M coef.	$-\frac{1}{24}$	$+\frac{1}{14}$	$-\frac{1}{10}$	$+\frac{1}{16}$	$-\frac{1}{11}$
$M_n = \text{coef.} \times 23.45 \times 3.4^2/0.9$	-12.6	$+21.5$	-30.1	$+18.8$	-27.4 kN·m/m
$A_s = 28.64\, M_n$	361	616	862	538	785 mm²/m
Min. pos. $A_s = 0.0018\,bh$	—	225	—	225	— mm²/m
Spacing #10 bars at:	250	150	100	175	125 mm
Spacing #15 bars at:	500	300	200	350	250 mm
USE #10 at:	250	150	—	175 mm	—
Plus #15 at:	—	—	200	—	250 mm

Bar areas are taken from Table C.2.

Some designers prefer to use some bent-up bars, feeling that this positions the top steel more exactly. The arrangement of Fig. 10.10b is then the most common bending pattern. The top bars are actually all in one layer and the bottom bars in one layer, but they are sketched separately to indicate the patterns more clearly. Bending up half the bars provides negative moment reinforcement equal to the average positive moment reinforcement in adjacent spans and added straight bars can make up the deficit. This requires #10 at 300 mm over the first interior support and #10 at 350 mm at all typical interior supports.

The bend points and stop points for bars in slabs involve almost exactly the same considerations as for any other continuous member. These considerations are discussed in considerable detail in Secs. 10.13 through 10.16 in connection with the design of a continuous T-beam.

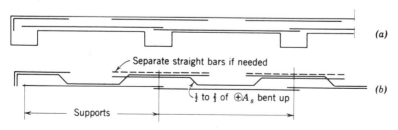

FIGURE 10.10   Arrangement of slab steel. (a) Straight bars alone. (b) Some bars bent.

The student should note that temperature steel is required parallel to the beams, at least in the amount specified in Code 7.12. The temperature bars are used as spacers for both top and bottom steel, tied to the underside of top bars and to the top of bottom bars. In addition, bar supports or chairs should be provided to hold the steel at proper levels.

Slab shear stresses could easily have been computed, but they do not control on one-way slabs of ordinary span.

The design of rectangular beams is discussed in Sec. 10.20.

10.9 CONTINUOUS T-BEAM DESIGN—GENERAL

The principles of design for a continuous T-beam can be discussed most easily in terms of a numerical example. Much of the remainder of this chapter (through Sec. 10.20) is devoted to various aspects of such design in terms of a typical interior span as follows:

> A typical interior panel of a continuous T-beam of 6 m clear span with 375 mm square columns is to be designed to carry the slab designed in Sec. 10.8. Beams are 3.65 m on centers, $h_f = 125$ mm, $w_\ell = 10.5$ kPa, $w_d = 4.0$ kPa (including slab weight but excluding stem weight), $f'_c = 20$ MPa, Grade 400 bars, and moment coefficients calculated by moment distribution, when expressed in terms of total load and clear span ($w\ell_n^2$), are as follows: -0.091, $+0.072$, and for minimum positive moment at midspan, -0.010.

There are many different choices of b_w and d for the web of a T-beam, any one of which might be a good design for specific conditions. A choice of d might be on the basis of stiffness against deflection, after the fashion of Table 3.2 of Sec. 3.8, especially for long spans. This would set $\ell/21$ as a probable minimum overall depth. The depth may be chosen to match some other member that is more critical in strength or stiffness or it may be related to depths needed for ducts. This member, for example, might well match the depth required for the exterior span with its larger moments and shears, if of the same span.

It would seem logical to choose depth on the basis of economy, remembering that A_s decreases with depth, whereas concrete and formwork are more expensive as depth increases. However, the economy of the structural frame may rarely agree with over-all building economy since greater depths for beams mean heavier beams (usually) and hence heavier columns and footings, higher exterior walls with expensive finish, more steps in a story height and hence larger stair wells, higher elevator lifts, and so on. The stem width may on occasion be chosen to fit into a wall.

One can almost say that the choice of web b_w and d is an arbitrary choice. Four primary conditions must be satisfied. (1) The capacity of the inverted rectangular section that carries negative moment (Fig. 10.3c) must be adequate for flexure with a "reasonable amount" of compression steel. (2) The shear capacity must be adequate, but such a wide range can be carried by stirrups that this provision is rarely restrictive. (3) There must be width for the required number of reinforcing bars, but the possibility of using higher strength steels, of bundling the bars, and of spreading the negative moment steel into the flange makes this much less restrictive than formerly. (4) There must be enough stiffness to keep deflections within proper limits, but the use of compression steel can stiffen considerably a shallow beam (Code 9.5.2.5).

In this example, deflection is assumed not to be critical on a heavily loaded 6 m span.* Depth will be considered on the basis of some arbitrary requirements for negative moment. The given moment coefficients indicate that positive moment steel will be approximately 75 to 80% of the negative moment tension steel, since z will not be greatly different for positive and negative moment. If half of the positive moment steel is bent up in the fashion indicated in Fig. 10.13b in Sec. 10.15c, the other half will continue in the bottom part of the beam into the column and thereby almost automatically provide compression steel in the amount $A_s' = 0.5 \times 0.8\,A_s = 0.4\,A_s$, where A_s is the negative moment tension steel. By lapping this compression steel from two adjacent spans as in Fig. 10.17b in Sec. 10.16 A_s' could become approximately $0.8\,A_s$.

A long-standing rule in continuous beam design has been to run at least one-fourth the positive moment steel into the support. Now it is further specified (Code 12.12.2) that, if the beam is part of a frame resisting lateral load, this one-fourth be anchored such that its full f_y in tension be available at the face of support. (The objective is greater ductility or energy absorption in case of moment reversal from unexpected loadings such as explosion, earthquake, unusual settlement, etc.) This reinforcement is obviously the minimum automatically available for A_s' at face of support.

The designer thus has considerable freedom in selecting the approximate A_s' to use, and estimates such as the first above can only be approximate at this stage. (For example, the positive moment steel may work out to be five bars, which means either 40 or 60% bent up, instead of half.) Here, for the first step, it is assumed that deflection probably does not control and that A_s' will be $0.4\,A_s$. The possible range of k_n values for moment will be explored.

For $f_c' = 20$ MPa, $f_y = 400$ MPa, Table 3.1 shows $k_n = 5.25$ MPa, $\rho = 0.0162$,

---

*The deflection calculation of Sec. 8.5c indicates a possible need for more consideration of deflection here. If the beam supports or is attached to partitions, the resulting deflection is barely satisfactory.

and $a/d = 0.38$. Assume that $A_{s2} = A'_s$.

$$A_{s2} = A'_s = 0.4\,A_s = 0.4(A_{s1} + A_{s2}) = 0.4(0.0162\,bd + A'_s)$$
$$A'_s - 0.4\,A'_s = 0.006\,48\,bd, \qquad A'_s = 0.0108\,bd$$

With $d - d'$ estimated as $0.88\,d$,

$$M_n = 5.25\,bd^2 + 0.0108\,bd \times 400 \times 0.88\,d$$
$$= (5.25 + 3.80)\,bd^2 = 9.05\,bd^2$$

Thus bd^2 can be reduced by, at least, 50% with this small A'_s, if one desires. The resultant small d would certainly not be economical if A'_s ran full length of a member, but here the need is only at extreme ends of the span. This small section might also get more into deflection problems or, with its larger A_s, into bar spacing problems.

This short analysis confirms the earlier statement that the choice of web can be quite arbitrary, with experience and judgment quite helpful in making the choice. On this basis, and recalling that minimum size is rarely the best solution, the author arbitrarily picks an intermediate value of $k_n = 7$ (near the middle of the range) for his starting point. This may prove either a wise or unwise decision; the choice can only be evaluated after the design is further advanced.

10.10 CONTINUOUS T-BEAM DESIGN—NEGATIVE MOMENT SECTION

Size will first be established for moment as a double-reinforced rectangular beam.

$$LL = 10.5 \times 3.65 \times 1.7 \qquad\qquad\qquad = 65.15 \text{ kN/m}$$
$$Slab + DL = 4.0 \times 3.65 \times 1.4 \qquad\qquad = \underline{20.44}$$
$$85.59$$
$$\text{Estimated stem wt.} = 3 \times 1.4 \qquad\qquad = \underline{\;\;4.20}$$
$$w = 89.79 \text{ kN/m}$$

$$M_u = -0.091 \times 89.79 \times 6^2 = 294.2 \text{ kN} \cdot \text{m}$$
$$M_n = 294.2/0.9 = 326.8 \text{ kN} \cdot \text{m} = 0.327 \text{ MN} \cdot \text{m}$$

This moment could be reduced by the redistribution provisions of Code 8.4 and Chapter 11, but will first be used as would be necessary when only approximate moments are available.

$$\text{Reqd. } b_w d^2 = M_n/k_n = 0.327/7 = 0.046\,69 \text{ m}^3$$
$$\text{If } b_w = 300 \text{ mm}, \quad d = \sqrt{0.156} = 0.395 \text{ m} = 395 \text{ mm}$$
$$b_w = 250 \text{ mm}, \quad d = \sqrt{0.187} = 0.432 \text{ m} = 432 \text{ mm}$$

The latter is nearer the usual d/b ratio of 1.5 to 2.5, although the narrow width may give a tight detailing situation. Add 40 mm cover + 10 mm stirrup + 10 mm ($= d_b/2$) = 60 mm.

$$\text{Min } h = 432 + 60 = 492 \text{ mm for over-all beam height.}$$
$$\text{USE } b_w = 250 \text{ mm, } d = 500 - 60 = 440 \text{ mm}$$

Check shear before calculating A_s. The shear is critical at d from support and an extra 8% of the simple beam shear* will be allowed for the continuity shear. (More generally, to fix the stem size, the *end* span shear 1.15 $w\ell_n/2$ would be used to permit uniformity in size for exterior and interior spans.)

$$V_u = 89.79(1.08 \times 3 - 0.44) = 251.4 \text{ kN}$$
$$V_n = 251.4/0.85 = 295.8 \text{ kN} = 0.296 \text{ MN}$$

$v = V_n/b_w d = 0.296/0.25 \times 0.44 = 2.69 \text{ MPa} < (\tfrac{1}{6} + \tfrac{2}{3}) \sqrt{f'_c} = 3.73 \text{ MPa}$ **O.K.**

Stem wt. (Fig. 10.11b), $w_s = 0.25(0.500 - 0.125) \times 24$

$$= 2.25 \times 1.4 \, LF = 3.15 \text{ kN/m} \qquad \textbf{O.K.}$$
$$M_n = 0.327 \text{ MN} \cdot \text{m}$$

From Table 3.1, $k_n = 5.25$ MPa, $p = 0.0162$, $a/d = 0.38$

$$M_{n1} = 5.25 \times 0.25 \times 0.44^2 = 0.254 \text{ MN} \cdot \text{m}$$
$$M_{n2} = 0.327 - 0.254 = 0.073 \text{ MN} \cdot \text{m} \qquad d - d' = 440 - 60 = 380 \text{ mm}$$
$$A_{s1} = 0.0162 \times 250 \times 440 = 1782 \text{ mm}^2$$
$$A_{s2} = 0.073/400 \times 0.38 = 0.480 \times 10^{-3} \text{ m}^2 = \overline{480 \text{ mm}^2}$$
$$\text{Neg. } A_s = \overline{2262 \text{ mm}^2}$$

Considering displaced concrete, effective $f'_s = 400 - 0.85 \times 20 = 383$ MPa

$$A'_s = 0.073/383 \times 0.380 = 0.502 \times 10^{-3} \text{ m}^2 = 502 \text{ mm}^2$$

At midspan (Fig. 10.11j) the minimum positive moment is actually negative, $-0.010 \, w\ell_n^2$, requiring, closely enough by proportion,

$$A_s = 2262(0.010/0.091) = 249 \text{ mm}^2$$

---

*In Fig. 10.11h,

$$V_c = \frac{(0.091 - 0.048)w\ell_n^2}{\ell_n} = 0.043 \, w\ell_n$$

which is 8.6% of the simple beam shear at the end and 10 to 11% of that at a distance d from the support.

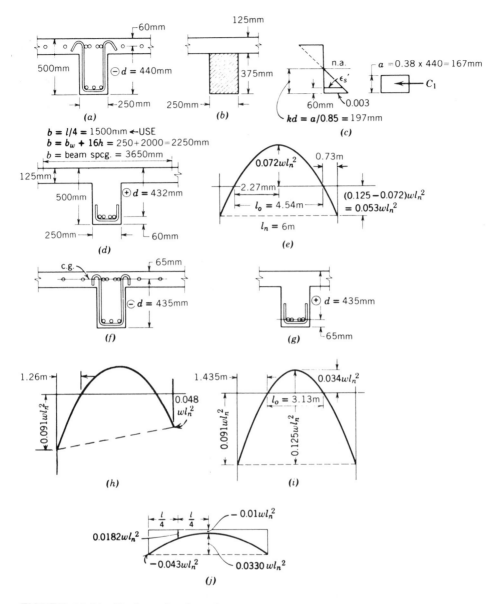

FIGURE 10.11 Design sketches for continuous T-beam design. (a) Assumed depth for negative moment. (b) Estimated weight, shaded area. (c) Strain triangle for f_s' calculation. (d) Assumed depth for positive moment. (e) Location of PI with maximum positive M. (f) Actual depth to top steel as selected. (g) Actual depth to bottom steel as selected. (h) "Exact" location of PI with maximum negative M. (i) Approximate location of PI with maximum negative M. (j) Minimum positive moment near mid-span.

284

with $1-\#20$ adequate (300 mm^2). At the quarter point of span, $M_u = -0.0182 \, w\ell_n^2$, and

$$A_s = 2262(0.0182/0.091) = 452 \text{ mm}^2$$

The $2-\#20(600 \text{ mm}^2)$ are adequate back to about 2 m from the support.

10.11 CONTINUOUS T-BEAM DESIGN—POSITIVE MOMENT DESIGN

The over-all depth of 500 mm already chosen for negative moment fixes the effective depth at mid-span. For one layer of steel (Fig. 10.11d),

Positive $d = 500 - 40$ cover $- 15$ stirrup $- 13$ (for $d_b/2$) $= 432$ mm

Trial $z = 432$ mm $- h_f/2 = 370$ mm, or $0.9 \, d = 389$ mm. Use larger value.

$$M_u = +0.072 \times 89.79 \times 6^2 = 232.7 \text{ kN} \cdot \text{m}$$
$$M_n = 232.7/0.9 = 258.6 \text{ kN} \cdot \text{m}$$

Trial $A_s = \dfrac{M_n}{f_y z} = \dfrac{258.6 \times 10^{-3}}{400 \times 0.389} = 1.662 \times 10^{-3} \text{ m}^2 = 1662 \text{ mm}^2$

The compression area is probably limited to less than the depth of the 150 mm wide flange and thus acts as in a wide rectangular beam.

$$a = 1662 \times 400/0.85 \times 20 \times 1500 = 26 \text{ mm}$$
$$z = 432 - \tfrac{1}{2} \times 26 = 419 \text{ mm}$$
$$A_s = 258.6 \times 10^{-3}/400 \times 0.419 = 1.543 \times 10^{-3} \text{ m}^2 = 1543 \text{ mm}^2$$

A further cycle could change only the last digit. Since a is so small, the beam is obviously in the underreinforced classification.

This steel should be checked against the minimum for positive moment steel as required by Code 10.5, that is, against $1.4 \, b_w d/f_y = 1.4 \times 250 \times 432/400 = 378 \text{ mm}^2$. Bars must be so arranged to maintain this steel area throughout the entire positive moment length, but this is no problem since it is here essentially the one-quarter of the positive moment steel which the Code requires to be continued into the column in all continuous beams.

10.12 CONTINUOUS T-BEAM DESIGN—CHOICE OF BARS

The size of bars used for positive moment may be limited by the specified ℓ_d at the point of inflection (Sec. 7.13a, Code 12.12) at least as regards the bars continuing toward the column. For these n bars of individual area A_b

the available $\ell_d = M_n/V_u + d$. At the P.I. which is located in Fig. 10.11e at 2.27 m from mid-span,

$$M_n/V_u = f_yz/V_u = nA_b \times 400 \times 0.419/89.79 \times 10^{-3} \times 2.27 = 822\ nA_b$$

if nA_b is half of the maximum required positive moment steel (required steel being 1543 mm²),

$$M_n/V_u = 772 \times 10^{-6} \times 822 = 0.635\ \text{m, available}\ \ell_d = 635 + 432 = 1067\ \text{mm}$$

For #25, $\ell_d = 0.019\ A_bf_y/\sqrt{f'_c} = 0.019 \times 500 \times 400/\sqrt{20} = 850$ mm.
This permits the use of #25 bars.

Bars will now be selected and detailed, using bent bars. The detailing will be revised in Sec. 10.15e,f for straight bars.

	$-A_s$	A'_s	$+A_s$
Reqd, mm²	2262	502	1543
USE 4-#20 Bent	$= 1200\ \text{mm}^2 \leftarrow$	——— 2-#20 Bent	$= 1600\ \text{mm}^2$
4-#20 Straight	$= 1200\ \text{mm}^2$ 2-#25 str. $\leftarrow$2-#25 Straight	$= 1000\ \text{mm}^2$	
	$\overline{2400\ \text{mm}^2}$ $= 1000\ \text{mm}^2$		$\overline{1600\ \text{mm}^2}$

The schematic bending arrangement of the steel selected is the same as that shown for the slab in Fig. 10.10b.

For the positive moment steel, 1-#20 + 1-#25 bars will be bundled* in each corner of the stirrups since there is not room to space four bars singly in the clear space between stirrup legs which is $250 - 2 \times 40$ cover $- 2 \times 15$ stirrup $= 140$ mm. With bars bundled as above, the clear space between bundles is $140 - 2(25 + 20) = 50$ mm. The spacing for bundled bars in Code 7.6.6.5 is the same as for a single bar of the *combined* area. In this case the area is $300 + 500 = 800$ mm² with equivalent diameter of $\sqrt{800 \times 4/\pi} = 32$ mm; hence the space between bars is ample, since aggregate will probably be 20 mm maximum size.

The crack control provisions should be considered here, but this will be delayed until after all the bars are tentatively selected in order not to confuse the basic selection process.

The negative moment steel can be placed in one layer by bundling the four bent-up #25 bars into two bundles adjacent to the stirrups and placing the other two straight bars in the slab. The use of #25 bars increases the estimated d for all steel by 3 mm, a little more with the #15 bars at the top, but the change appears negligible (Fig. 10.11f, g).

The compression steel (at support) might be detailed as 2-#20 from each

*Note that bundled bars require the use of stirrups around them. If shear did not call for stirrups all the way across the span, as is the case in Sec. 10.17, the bundling here would have required them anyway.

side, each caring for the stresses on the entering face of the column. Since this means developing all four bars into (and actually beyond) the far face of the column, it appears simpler to run 1-#20 from each side far enough to make it also effective on the far face of the column. This leaves 1-#20 bottom bar to be cut off short of the column on each side.

The crack control provision of Sec. 8.13 (Code 10.6.4) is

$$z = f_s \sqrt[3]{d_c A} \lessgtr 30 \text{ MN/m (for interior exposure)}$$

The positive moment bars will be considered first. This equation was not designed for bundled bars. Whether to count each bundle as one or two bars is not definite; the use here as one bar is on the conservative side. Try the two bars stacked vertically as suggested above, instead of horizontally as shown in Fig. 10.11g, since this is on the safe side with A larger (Fig. 8.15). Although it may at first sound contradictory, d_c will be taken from the center of outermost bar (65 mm) because crack width spreads from the outermost bar (Fig. 8.15) to a width dependent upon the free cover. The suggested $0.6 f_y$ will be used for f_s.

$$A = 2(40 \text{ cover} + 15 \text{ stirrup} + 25) \times \tfrac{1}{2} \times 250 = 20\,000 \text{ mm}^2 = 0.020 \text{ mm}^2$$
$$z = 0.6 \times 400 \sqrt[3]{0.065 \times 0.020} = 240 \times 0.11 = 26.4 \text{ MPa} < 30 \text{ MPa}$$

The bars are thus satisfactory for either orientation of the bundle.

For the negative moment bars, with the bundles in the horizontal plane and the bars spread over the flange width of 1 500 mm, $d_c = 65$ mm, $A = 2 \times 65 \times 1500/6 = 32\,500 \text{ mm}^2$, again counting a bundle as one bar.

$$z = 0.6 \times 400 \sqrt[3]{0.065 \times 0.0325} = 240 \times 0.128 = 30.72 > 30 \text{ MPa}$$

In consideration of the conservative treatment of the bundled bars, the arrangement is approved.

It is not uncommon for the designer to increase b_w arbitrarily or even change d to make the steel used space properly or to modify the required A_s to fit available bar sizes closer.

A cheaper beam would result if advantage were taken of the redistribution provisions of Code 8.4. Where the moments have been established by elastic theory this is recommended.

10.13 CONTINUOUS T-BEAM DESIGN—GENERAL REQUIREMENTS FOR BENDING BARS

The fabricator must detail each bar, but the designer can be satisfied to locate bend points and cutoff points reasonably accurately. Many offices use rules such as "bend up half the bottom steel at the quarter point of clear span," but more exact procedures are presented here to indicate more adaptable methods needed in many cases.

The author recommends a very conservative attitude toward bar detailing. There has been much ineffective detailing of what would otherwise have been good designs. The final member is not better than its details.

Bend points may be governed by:

1. Moment requirements
2. Development lengths.
3. "Arbitrary" requirements of Code 12.11.3, 12.12.1, 12.12.2 (Sec. 4.11d)
4. Use of bent bars as stirrups.

The several "arbitrary" specification requirements will first be summarized. Code 12.13.3 requires that at least one-third of the total reinforcement for negative moment be extended beyond the extreme location of the point of inflection by the greater of: (1) $\ell_c/16$; (2) beam depth d; (3) 12 d_b. Code 12.12.1 requires that at least one-fourth of the positive reinforcement in continuous beams shall extend 150 mm into the support. Both of these, in the words of the old Joint Committee Specification, are "to provide for contingencies arising from unanticipated distribution of loads, yielding of supports, shifting of points of inflection, or other lack of agreement with assumed conditions governing the design of elastic structures." Code 12.12.2 goes further to add that when the member is part of a "primary lateral load resisting system" the extension of the bottom steel covered in 12.12.1 must anchor it to develop its full-yield stress in tension at the face of the support, that is, be much more than 150 mm. This requirement is to provide some ductility in the event of stress reversal from wind, earthquake, or explosion.

Code 12.11.3 also requires that every bar, whether required for positive or negative reinforcement, be extended the depth of the beam or 12 bar diameters beyond the point at which it is no longer needed to resist flexure. This requirement, discussed in Sec. 7.11, the author designates as ℓ_a and uses as a shifted moment diagram, but the Code itself seems to be less strict and to leave much to the designer's discretion. Also, since the Code uses simply the word "extended," it seems to imply that the extension might exist entirely as a bent bar. The author prefers to consider at least half of it ($\frac{1}{2}\ell_a$) as a relaxed requirement for a shifted moment diagram requiring one-half of the extension to remain straight, but relaxed because a bent bar remains partially effective in flexure while in the tension half of the effective depth.

Arrangements satisfying moment requirements, development lengths, and the "arbitrary" requirements may sometimes be varied slightly to help out in development length demands elsewhere or in web reinforcement.

For example, if development of the bottom steel at the point of inflection proves difficult, it may be possible to shift the bend-up points toward the column and to keep more steel available at the P.I.; or the spacing of bend points may be shifted to make bent bars more useful for web steel.

10.14 CONTINUOUS T-BEAM DESIGN—MOMENT DIAGRAMS GOVERNING BAR BENDS

The positive moment diagram has already been established from Fig. 10.11e. For maximum negative moment the actual moment diagram is unsymmetrical and the maximum moment calculation alone does not provide sufficient data for the entire diagram. The diagram can be reasonably approximated by using a negative moment at the far end as slightly less than that accompanying the maximum positive moment, in this case say $-0.048\,w\ell_n^2$ instead of the $-0.053\,w\ell_n^2$ value shown in Fig. 10.11e. Figure 10.11h would be almost exact insofar as the location of PI (point of inflection) on the left is concerned. The approximate PI is often calculated for the assumed symmetrical diagram of Fig. 10.11i, which will be used here.[*]

$$\tfrac{1}{8}\,w\ell_0^2 = 0.034\,w \times 6^2$$
$$\ell_0 = 6\sqrt{0.272} = 3.13 \text{ m}$$

Hence the PI is 1.435 m from the column. The further approximation of using a triangle for the negative moment section of the parabola is reasonable when the PI is not too near midspan, say, outside the middle third of the span.

The minimum positive moment at mid-span is also part of a symmetrical diagram, but in this case the simple beam moment diagrams is that for dead load alone, which can be expressed as follows in terms of total load w, that is, $20.44 + 4.20 = 24.64$ kN out of 89.79 kN total.

$$\text{Simple beam } M_s = 0.125\,w_d\ell_n^2 = 0.125\,\frac{24.64}{89.79}\,w\ell_n^2 = 0.0343\,w\ell_n^2$$

The diagram is sketched in Fig. 10.11j. The maximum moment diagrams are assembled in Fig. 10.12a.

---

[*]More accuracy is not difficult and should usually be used for end spans. The moment diagram is a simple span moment added to the diagram for the two end moments (and their equilibrating shears). The continuity shear V_c is the slope of a straight line joining the two end moments. For a uniform beam load this V_c will be the only shear at midspan. The maximum positive moment will then be a distance V_c/w off center and increased over the centerline moment by the area of the small shear triangle of height V_c and length V_c/w. The resulting diagram is a symmetrical parabola about this positive moment point.

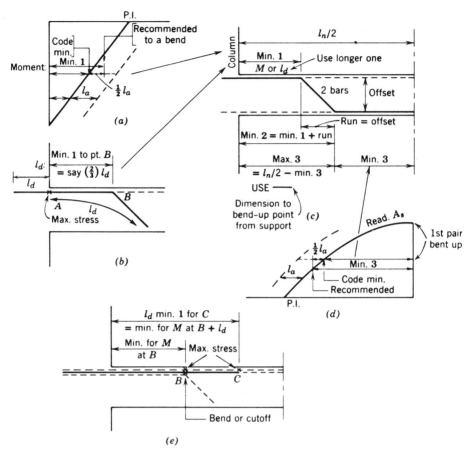

FIGURE 10.12 Schematic for data assembly.

As is usual in the office unless it is a special case, the A_s required diagram is based on the total steel used.

10.15 CONTINUOUS T-BEAM DESIGN—BENDS AND STOP POINTS FOR TENSION BARS

(a) Purpose of Discussion

It is not the intent of this section to consider simply the details of bar detailing. Rather, experience has shown that no other type of problem so clearly brings to the fore the fundamentals of how reinforced concrete works and how both flexure and bar development length affect each other.

(b) General Assembly of Data

The data can be arranged in many possible patterns, but several pieces of data are interrelated in each decision; and a regular repeating pattern of presentation is desirable. The author's form is shown schematically in Fig. 10.12, for one pair of bars bent first.* The minimum distance to the bend-down point is determined by the larger of two requirements:

1. Moment length plus $\frac{1}{2}\,\ell_a$, as shown in Fig. 10.12a, the second $\frac{1}{2}\,\ell_a$ being available in the bend length itself.
2. The development needs, shown in Fig. 10.12b. The author assumes two-thirds of ℓ_d is permissible here because the bent portion also helps to develop the bar. (The ℓ_d into the support is also necessary.)

These moment length and development length values are tabulated in Fig. 10.12c, immediately above the bar sketch; the larger is "Min.1." Since bend points are usually dimensioned at the level of the bottom bars, "Min.2" is next found from "Min.1" by adding the run of the bar, equal to the offset of the bar. The bottom bar can be bent up at "Min.3" (Fig. 10.12d) given by the moment length plus $\frac{1}{2}\ell_a$ from midspan or at "Max.3" from the column = $\ell_n/2$ – Min.3. The values of "Max.3" and "Min.2" determine whether the bend can be made and the leeway open to the designer in choosing the final dimension.

If Fig. 10.12 were based on a first cutoff point instead of bend point, only three changes would be necessary (but the stirrups of Code 12.11.5 would be necessary; see Sec. 5.16).

1. The moment length in Fig. 10.12a,d would include the full ℓ_a length (not $\frac{1}{2}\,\ell_a$), as a Code requirement.
2. The development length would be the full ℓ_d.
3. Only "Min.1" would determine the top cutoff and only "Min.3" the bottom cutoff.

For any succeeding bar cutoffs one other change is involved, as in Fig. 10.12e. The ℓ_d value is to be measured from point B, the *theoretical* cutoff for the last bar stopped or bent. The point B does not shift if the bar is run farther because of ℓ_d or some other reason. (If the bar runs on to C, the maximum stress point is eliminated and the column face takes its place.)

This form is expandable to as many bars or cutoffs as are necessary. Development lengths are obviously needed for this example as follows.

---

*The student might note that "first" relates to the nearness to the point of maximum moment for that steel. If only one pair is bent, the first bar bent down is also the first bar bent up. If two bars are bent singly, the first bent down matches the last one bent up and the second bent down matches the first bent up.

Bottom #20, $\ell_d = 0.019\, A_b f_y/\sqrt{f'_c} = 0.019 \times 300 \times 400/\sqrt{20} = 510$ mm

Bottom #25, $\ell_d = 0.019 \times 500 \times 400/\sqrt{20} = 850$ mm

Top #20, $\ell_d = 1.4 \times 510 = 714$ mm

(c) Author's Preferred Arrangement of Bars

The author's preferred arrangement consists of 2 − #25 bars bent as a pair (or singly) and all other bars straight and continuing at least to the nominal PI as sketched in Fig. 10.13b. Since no bars are cut off inside the PI, the extra stirrups of Code 12.11.5 are not needed.

The bent bars are calculated in Fig. 10.13 (based on the ideas of Fig. 10.12a to d); and at the bottom bar level a maximum of 1.51 m results. The next bottom stop point is for one #25 bar, established solely from the minimum from mid-span. Since the size of this bar was originally selected in Sec. 10.12 to match development length needs, it is automatically satisfactory if extended d beyond the PI.

The top straight bars are next considered. One-third of the total of 2400 mm², say 3-#20 = 900 mm², must extend past the PI (which goes with that negative moment diagram) at least $\ell_n/16$, d, or 12 d_b, or d (in this case 0.44 m). One ot these will necessarily be spliced at midspan with a lap of 1.7 $\ell_d = 1.7 \times 0.714 = 1.21$ m, with the 1.3 factor not acceptable because more than 50% of the bar is needed at midspan, that is, $f_s > f_y/2$. The other #25 bar must wait until the development length requirements are fixed by what is done with neighboring bars.

The initial decision to carry straight bars to the nominal PI automatically goes beyond any moment requirement. Hence, moment lengths are not all shown. For this case, arrangement becomes primarily a matter of staggering the bar cutoffs. For the second 2-#20 bars the critical stress point is at 0.32 m where the first 2-#20 are bent down and the #20 ℓ_d of 0.714 m is added to give an ℓ_d requirement of 1.0 m, well inside the PI. These bars are stopped at the nominal PI, at 1.5 m. The third pair of 2-#20 bars has the same requirements, and the cutoff is simply staggered 0.25 m to 1.75 m. The 1-#20 must go to the nominal PI plus 0.44 m to satisfy the arbitrary extension, or 1.44 + 0.44 = 1.88 m; or for the dead load moment diagram 1.49 + 0.44 = 1.93 m. The small stress peak where the last #20 bars are cut off can be ignored. The flat dead load moment diagram demands also the extension and center splicing of the last 1-#20 bar.

(d) Bent Bars with Straight Bars Cut Off by Shifted Moment Diagram

This arrangement will require extra stirrups under Code 12.11.5 over the bars cut off inside the PI. These stirrups are not computed here.

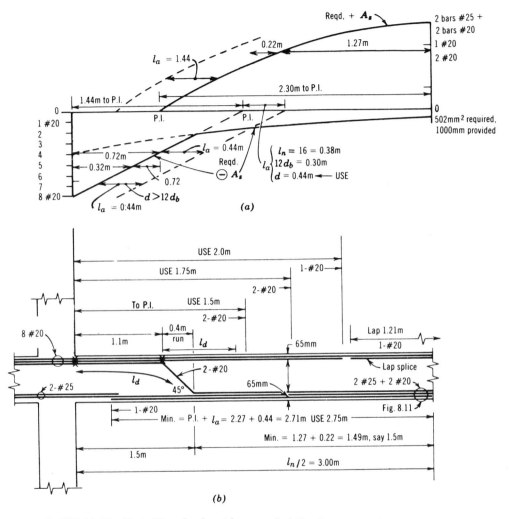

FIGURE 10.13 Detailing for bent bars and shifted moment diagram.

The bend point (Fig. 10.14) for the first 2-#20 bars is the same as in c above. The second pair of 2-#20 have an ℓ_d requirement of the 1.0 m moment length just noted for the bars cut plus an ℓ_d of 0.714 m beyond, a total of say 1.75 m, with the moment requirements much less.

The development length of the last #20 is 1.44 + 0.714 m = 2.154 m, say 2.25 m.

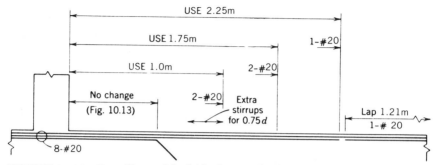

FIGURE 10.14 Detailing with shifted moment diagram but with minimum cutoff points.

(e) All Bars Straight and Cut Off by Shifted Moment Diagram

The final results are tabulated in Fig. 10.15. Only the pair of #20 bottom bars, and the last two pairs of #20 bars are changed from Fig. 10.14.

(f) All Bars Straight Without Shifted Moment Diagram Idea

The change that occurs in this case compared to (e) is that all lengths based on ℓ_d are relaxed by assuming that the bar stress is maximum at the basic moment diagram rather than the shifted diagram. This would make the moment requirement of Fig. 10.15 the larger requirement in Fig. 10.16

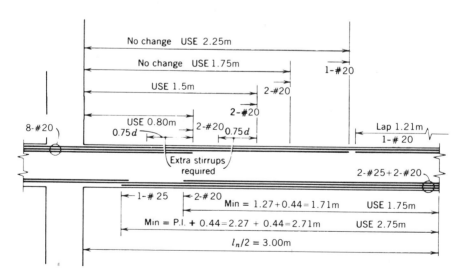

FIGURE 10.15 Detailing straight bars with shifted moment diagram.

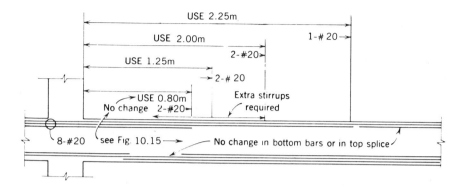

FIGURE 10.16 Detailing straight bars without shifted moment diagram.

and thus lower the lengths for the last two pairs of #20 bars to 1.25 m and 0.80 m, respectively.

(g) Bent Versus Straight Bars

A major advantage of bent or offset bars is that they do not lower the shear strength of the member as do bars simply cut off and stopped. They also help to keep top steel at the proper level and probably tend to reduce placement errors. However, practice has moved largely toward straight bars, probably because of cost considerations in placement. If bars are bent, it is easier to satisfy moment requirements with bars bent singly; but it is easier to satisfy development lengths needed with bars bent as pairs.

10.16 CONTINUOUS T-BEAM DESIGN—STOP POINTS FOR COMPRESSION STEEL

The compression length of the #25 straight bottom bars is also established by both a moment and a development requirement.

Since only two bars are required at each column face, bars from *both* beams are not required (Fig. 10.17c), although it proved convenient in this design to lap one bar from each beam. Since both bars are then needed at both sides of the column, the extension must go through the column and then beyond the far side enough to develop the bar there in compression, as in Fig. 10.17b; the length for moment may also control.

For compression development length Sec. 7.16 gave the relation for

$f_c' < 30.6$ MPa

$\ell_d = 0.24 f_y d_b \sqrt{f_c'} \geq 200$ mm
For #25, $\ell_d = 0.24 \times 400 \times 25/\sqrt{20} = 537$ mm

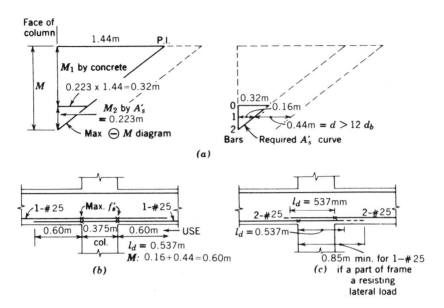

FIGURE 10.17 Detailing of compression steel.

For moment the length needing some help from M_2 is easily established in Fig. 10.17a from the negative moment diagram. In this case, M_{n2} was found in Sec. 10.10 to be 0.073 MN·m out of a total of 0.327 MN·m, or $M_{n2} = (0.073/0.327)M = 0.223 M$. This indicates *all* compression steel could theoretically be omitted at 0.223×1.44 (distance to PI) = 0.32 m from the column. Half the steel could be stopped at $\frac{1}{2} \times 0.32 = 0.16$ m *plus* the arbitrary requirement of d or 12 d_b (0.44 m controlling) for a total of 0.60 m. This exceeds ℓ_d and controls the bar length, as sketched in Fig. 10.17 b.

The detail of A_s' if both bars required are furnished from the adjacent beam is shown in Fig. 10.17c. Since compression steel in this case extends as far as negative moment, there is no need to consider the moment diagram at all. (Within the column the column compression provides good development conditions and even if the column were very wide the negative moment in the middle of the column would in effect be in a deeper beam.) The ℓ_d already calculated gives the distance the bars must extend *into* the column, which in this case happens to go a little beyond the far face of column.

Finally, attention is called to the special requirements of Code 12.12.2, already mentioned in Sec. 10.13. If the beam is part of a primary lateral load resisting frame, that is, if the beam must hold a column carrying lateral shear from, say, a wind loading on the frame, one-fourth of the required positive moment steel (1-#25 bar in this beam) must be anchored into the column for its full f_y in tension. For tension

$$\ell_d = 0.019 \, A_b f_y / \sqrt{f'_c} = 0.019 \times 500 \times 400 / \sqrt{20} = 850 \text{ mm}$$

This would call for a change in Fig. 10.17c but would not change Fig. 10.17b. However, notice the special limitation this places on bottom bar sizes in the case of a beam stopping at an exterior column, already discussed in Sec. 7.14.

10.17 CONTINUOUS T-BEAM DESIGN—STIRRUPS

Since bundled bars were selected in Sec. 10.12, stirrups are required all across the middle of the beam to satisfy Code 7.6.6.2. However, stirrups are usually considered a matter of shear design.

Under maximum moment loading the beam is subject to an end shear equal to the simple beam shear $w\ell_n/2 = 89.79 \times 6/2 = 269.37$ kN plus a continuity shear, again assumed as $0.08 \, w\ell_n/2$ or 21.55 kN, acting both at the end and at midspan as shown dashed in Fig. 10.18a. The shear at midspan will be greater when live load is removed from the left half of the span. This loading will produce a smaller continuity shear, which will be neglected. This simple beam shear at midspan is $65.15 \times 3 \times 1.5/6 = 48.86$ kN. The solid line in Fig. 10.13a will be used as a maximum shear diagram for stirrup design. With $b_w = 250$ mm and $d = 435$ mm at the end, Fig. 10.11f, the critical shear is at the distance d from support.

$$\text{At support } v_0 = \frac{290.92 \times 10^{-3}}{0.85 \times 0.25 \times 0.435} = 3.15 \text{ MPa}$$

$$\text{At midspan } v_{10} = \frac{48.86 \times 10^{-3}}{0.85 \times 0.25 \times 0.435} = 0.53 \text{ MPa}$$

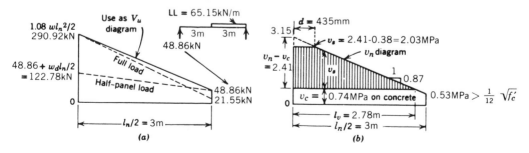

FIGURE 10.18 Shear diagrams for stirrup calculations. (a) V_u diagram. (b) v_v diagram and v_s (shaded).

The slope of the v diagram is $(3.15 - 0.53)/3 = 0.87$ MPa/m. The critical shear at 435 mm from support is $3.15 - 0.87 \times 0.435 = 2.77$ MPa. The unit shear curve is plotted Fig. 10.18b. Stirrups are needed for the ℓ_v length where v is in excess of $\frac{1}{6}\sqrt{f'_c}$; also for all the remainder of the length since everywhere $v > \frac{1}{2}v_c = \frac{1}{12}\sqrt{f'_c} = 0.37$ MPa. The stirrup requirement over bundled bars is thus automatically satisfied; also the requirement for stirrups over A'_s bars under Code 7.12.5.

If bars were cut off in a tension zone, *extra* stirrups would be needed for $0.75 \, d$ over each cut bar as indicated in Figs. 10.14, 10.15, and 10.16. Code 12.11.5 waives these stirrups only where shear is equal to or less than two-thirds the allowable. These stirrups must provide an extra $\rho_n f_y$ of 0.4 MPa with a resulting spacing not exceeding $\frac{1}{8} d \, \beta_b$, where β_b is the proportionate part of the bars cut off at the particular section.

Except for such special extra stirrups, a design with similar data is worked out in detail in Sec. 5.13.

10.18 CONTINUOUS T-BEAM DESIGN—PLACEMENT OF STIRRUPS

The proper placement of stirrups in continuous beams presents a problem. The best anchorage of stirrups would call for the hooks to be in compression concrete, which near the supports would be the bottom concrete. Since construction is simpler with the open end of the stirrup turned up, the matter of anchorage has generally been ignored. Furthering this easier placement is the requirement that the compression (bottom) steel be tied as specified in Code 7.11.1. One stirrup must extend completely around all longitudinal A'_s bars. The ties, over the length where A'_s is needed must satisfy the requirements for a column, that is, be "so arranged that every corner (bar) and alternate . . . bar shall have lateral support provided by the corner of a tie . . . and no bar shall be further than 150 mm clear on either side of such a laterally supported bar." Closed stirrups, like column ties, would be excellent but these have limited usage except where torsion may be present because they complicate bar placement. The chief objection to these relates to difficulty in dropping reinforcement into place inside the ties.

10.19 CONTINUOUS T-BEAM DESIGN—DEFLECTION

Deflection under service load is the critical case and this calculation is somewhat toward an elastic analysis. The deflection calculations for nearly this same beam are given in Sec. 8.5c.

10.20 CONTINUOUS RECTANGULAR BEAMS

The design differences between continuous rectangular beams and continuous T-beams are all minor. The foregoing design of the T-beam can also serve as a model for continuous rectangular beam design. Other than the obvious minor difference in designing the positive moment steel for a rectangular beam and a somewhat greater congestion of negative moment steel (possibly two layers required), attention must be called to the special requirements for lateral supports as given in Code 10.4.1. Freestanding rectangular beams lack the extra stiffness and strength of the T-beam in resisting torsion.

10.21 SPANDREL AND OTHER L-BEAMS

A spandrel beam is one supporting an exterior wall, although occasionally the problems (without the name) may be associated with an interior beam adjacent to a stairwell, shaft, or other interior opening. The spandrel carries more dead load because of the exterior wall weight, but often is lighter than an interior beam because of live load coming from a reduced floor area.

The L-beam, with flange on only one side, is an unsymmetrical section. As such, Code 8.10.3 limits the flange counted for flexure more strictly than in a T-beam, that is, to the least of $\ell_n/12$, $6\,h_f$, or half the clear distance to the next beam. However, in flexure, an L-beam cannot act as an unsymmetrical section because the slab prevents the normal lateral deflection which would occur when freestanding.

The lateral deflection is actually prevented only at the level of the slab. With the slab loaded, a secondary rotation about an axis on the intersection of the center lines of slab and web does occur and causes some lateral deflection of the web. This results from the slab deflection and rotation at the beam junction, and it introduces torsion into the beam.

There is a current trend, with which the author disagrees, to consider only the rectangular web of the L-beam in computing torsional resistance (author's comments in Sec. 6.14d). The slab is both a loading and a resisting portion of the system; a strength calculation that ignores the strength aspect is oversimplified and wasteful.

How to establish the design torsion is still the major uncertainty, in the author's opinion. Elastic analysis tends to overemphasize torsional demands. After cracking, a rapid decrease in torsional stiffness occurs. For the additional rotation necessary to develop the resisting stirrups, torsion will build up at only 5 to 10% of the rate it does under initial rotations. It is, therefore, difficult to project the initial elastic calculations to a reasonable design basis. The alternate Code design method summarized in Sec. 6.12 looks like a good approach if member torque is not essential to equilibrium.

10.22 END SPANS AND IRREGULAR SPANS

End spans always involve negative moments smaller at the outer end and larger at the inner end, with the points of inflection and point of maximum positive moment shifted toward the outer support. Irregular spans and loads also result in maximum moment diagrams that are less symmetrical than those used in this chapter.

Proper detailing of end spans and irregular spans requires a better knowledge of unsymmetrical moment diagrams, but no additional reinforced concrete theory. Where approximations are deemed proper, the designer should be more conservative than where more exact moment requirements are known.

10.23 JOINTS IN FRAMES

Beam-column joints in frames constitute a major area for reinforced concrete designers. Even in braced frames they become three-dimensional with reinforcing bars in columns competing for space with bars from the beams, usually beams in at least two directions, each with reinforcement running into the columns at two levels. In unbraced frames many joints must be designed for reversal of beam moments.

Horizontal and vertical shearing forces within many joints (as in Fig. 23.6) called for closed ties or stirrups within the joints. Earthquakes and reversal of beam moments produce horizontal shears in joints that complicate design. The designer at the same time must be concerned with erection problems and about leaving room for some concrete in the joint.

An adequate discussion of joints is not feasible here. Instead the author would recommend Reference 5 (especially Sec. 13.8) and Reference 6 (emphasizing corner joints), both of which are excellent in this area.

SELECTED REFERENCES

1. "Continuity in Concrete Building Frames," Portland Cement Association, Chicago, 3rd ed.
2. Phil M. Ferguson, "Analysis of Three-Dimensional Beam-and-Girder Framing," *Jour. ACI, 22,* Sept. 1950; *Proc.,* 47, p. 61.
3. R. H. Wood, "Studies in Composite Construction: Part I, The Composite Action of Brick Panel Walls Supported on Reinforced Concrete Beams;

Part II, The Interaction of Floors and Beams in Multi-Story Buildings,"
National Building Studies, *Research Papers No. 13* (1952) and *22*
(1955), Her Majesty's Stationary Office, London.

4. *Design Handbook*, V. 1, ACI Sp-17(73), Detroit, 1973.
5. R. Park and T. Paulay, *Reinforced Concrete Structures*, John Wiley &
Sons, New York, 1975.
6. I. H. E. Nilsson and A. Losberg, "Reinforced Concrete Corners and
Joints Subjected to Bending Moment," *Proc. ASCE, Jour. Struct. Div.*,
V. 102, ST6, June 1976, p. 1229.

PROBLEMS

PROB. 10.1. The reduced frame of Fig. 10.19*b* should be used for the
analysis of the second floor of the bent of Fig. 10.19*a* since standard

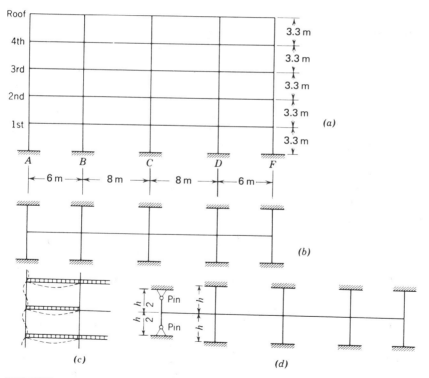

FIGURE 10.19 Analysis of building frame. (*a*) The frame considered. (*b*)
Ordinary reduced frame for calculation of moments on a single floor. (*c*)
Loading for maximum negative moment at exterior end of beam; also
nearly maximum for exterior column moment. (*d*) Improved form of
reduced frame for maximum moment at exterior joint.

coefficients are not applicable. Assume all the beams have $I = 10.4 \times 10^9$ mm^4, the 400 mm columns below the floor have $I = 2.2 \times 10^9$ mm^4, and the 350 mm column above the floor have $I = 1.3 \times 10^9$ mm^4. Each beam carries a dead load of 15.3 kN/m and a live load of 31.4 kN/m. With factored loads:

(a) Calculate the maximum negative moment for the beam CD at C and correct this to the design moment at the face of the column. (Note that symmetry about C is equivalent to a fixed end for moment distribution purposes.)

(b) Calculate the maximum negative design moment at D of beam CD.

(c) Calculate the maximum positive moment for CD; also the minimum positive moment.

(d) Locate the several points of inflection that are useful in detailing steel.

(e) Compare the points of inflection in (d) with those which would be obtained if each moment diagram were assumed to be symmetrical about midspan, as in Fig. 10.11i.

PROB. 10.2. In Prob. 10.1 calculate the corresponding maximum moments in span DF and the points of inflection needed. (For this end span the use of a symmetrical moment diagram is scarcely valid.)

PROB. 10.3. In the frame of Prob. 10.1 calculate the maximum bending moment on column D, using factored loads. Simulate the worst loading condition, reversed curvature, as in Fig. 10.19c,d. (Increasing that column's stiffness in Fig. 10.19b by a factor of 1.5 gives the same result.)

PROB. 10.4. Repeat Prob. 10.3 for the exterior column F, using a reduced frame similar to the reduced frame of Fig. 10.19c,d.

In addition, note that loads on successive floors as in Fig. 10.19c give a reverse bending condition in the exterior columns which is almost equivalent to a reduced frame similar to that of Fig. 10.19d. Recalculate column D moments and negative moment at D in beam CD, using factored loads and this reduced frame.

PROB. 10.5. Design an interior span of a continuous one-way slab supported on beams 4.2 m on centers using moment coefficients of Sec. 10.5 (Code 8.3.3), live load of 7 kPa, no dead except slab weight, $f'_c = 20$ MPa, Grade 300 steel, beam stems 275 mm wide, cover over center line of steel of 30 mm (as a simplification). Carry the design through the choice and spacing of bars (same size bars for both positive and negative moment). Draw a lengthwise section of slab and sketch the bars in place, assuming no bars bent up.

PROB. 10.6. Assume the steel found in Table 10.1 for a typical interior span of the slab of Sec. 10.8 is to be arranged as shown in Fig. 10.10b, with half of the bottom bars bent up and extra straight top bars added over the support. (This involves a new choice of steel at the support.) Calculate all bend points and stop points (for a typical interior span).

PROB. 10.7. A continuous rectangular beam is to carry a uniform live load of 45 kN/m plus 15 kN/m dead load and its own weight over equal 6 m spans. Supports may be considered of negligible width (knife edges). Design a typical interior span through the choice of beam size and the choice of reinforcing. Assume maximum $V = 1.05\ w\ell_n/2$, $f'_c = 30$ MPa, Grade 400 steel. Limit k_n to 9 MPa which means some compression steel at the worst section. Show a cross section at support and at midspan with detailed spacing of bars; also an elevation of beam showing the schematic arrangement (bending) of bars. Exact bend points, and so forth, are not a part of this problem.

PROB. 10.8. Design an intermediate span of one of a series of continuous T-beams spaced 3.3 m on centers to carry a 125 mm slab, live load of 8.4 kPa, and its own weight over a 6.6 m clear span, using $f'_c = 20$ MPa, Grade 300 steel, ACI moment coefficients, $V = 1.08\ w\ell_n/2$. It is suggested that the stem size be based on using $k_n = 7$ MPa at the support and a stem width of 300 mm be first tried. Stem weight may be assumed 12 kPa (without revision). Sketch the cross section at the support and at midspan and also the trial arrangement of longitudinal steel.

PROB. 10.9. Assume that an interior span of a continuous beam with uniform live load has been designed for moment coefficients of -0.093 and $+0.067$, based on using the clear span of 7 m. The steel used is 12-#20 for negative moment, 9-#20 for positive moment, and 5-#20 for compression steel at support. Assume columns 400 mm square, $b_w = 400$ mm, $d = 475$ mm, $d - d' = 410$ mm, $f'_c = 30$ MPa, $f_y = 300$ MPa, and $f'_s = 248$ MPa. The M_2 couple represents 0.33 of the maximum negative moment. Arrange the bars to be bent up, using a single bent bar nearest midspan and then a pair of bent bars nearer the support. For location of points of inflection, the simplifying assumption of Fig. 10.11i may be used. Sketch the arrangement of steel and detail all bend and stop points. (If any bends fail to work out satisfactorily, note this fact and compromise lengths as seems best, but do not revise the bending scheme.) Show the results on a sketch similar to Fig. 10.13. Assume minimum positive moment does not control bar lengths. Assume the beam is cast with a 100 mm slab and is part of a frame braced by a shear wall.

PROB. 10.10 A typical interior span of continuous beam of 6.5 m clear span supported by 450 mm square columns in a braced frame and designed for uniform load moment coefficients of $-\frac{1}{12}$ and $+\frac{1}{16}$ (based on clear span) requires 9-#25 for negative moment, 6-#25 for positive moment, and 6-#25 for compression steel at the support. If $M_2 = 0.54$ of maximum negative moment, $f'_c = 20$ MPa, $f_c = 300$ MPa, $f'_s = 193$ MPa, and the offset distance between top and bottom steel is 350 mm, sketch the steel arrangement using straight bars and locate stop points for bars. For detailing bar lengths take d as 425 mm. Assume midspan minimum positive moment as $+0.015\ w\ell_n^2$ with the factored dead load equal to $w/3$. Record the results as in Fig. 10.15. (A 125 mm slab may be assumed with beam $b_w = 350$ mm).

PROB. 10.11.   An interior 5.9 m clear span of continuous T-beam with a 125 mm slab is loaded at its third points with concentrated loads P such that the uniform load may be neglected in establishing the shape of the moment diagram. Columns are 400 mm square and a primary part of the lateral load resisting system. The maximum moments, including an allowance for uniform load, are $-0.25\ P\ell_n$ and $+0.17\ P\ell_n$. The negative moment is broken down into $M_1 = 0.42\ M$ and $M_2 = 0.58\ M$ for purposes of design. The minimum midspan positive moment is $0.050\ P\ell_n$. Factored dead load is $P/3$. For bar lengths, d may be taken as 450 mm and $d - d'$ as 375 mm. The required number of #20 bars is 10 for negative moment A_s, 6 for compression steel A_s', and 6 for positive moment steel. $f_c' = 30$ MPa, $f_y = 400$ MPa, $f_s' = 345$ MPa. Arrange and detail the steel attempting to bend at least two bars up from the bottom. Record the results as in Fig. 10.13. (If any bends fail to work out satisfactorily, note the fact and compromise lengths as seems best, but for this problem stay with two bars up in a web 375 mm wide.)

PROB. 10.12.   Same as Prob. 10.11 except straight bars are to be used and detailed.

11
LIMIT DESIGN

11.1 TERMINOLOGY

Limit design and plastic design are often considered as synonymous terms. A distinction is made here and in most discussions of reinforced concrete. In steel design, the term *plastic design* includes not only the change in the pattern of moments beyond the yield point but also the increased resistance of a cross section after its extreme fiber reaches the yield point. In reinforced concrete, *limit design* is used only to refer to the changing moment pattern; ultimate strength design already includes the increase in strength of a given cross section after some stress reaches the yield point.

For reinforced concrete, the frame can be designed under any *fixed* loading to have strengths closely matching the moment diagram, with the concrete tailored to fit by cutting off and the bending of bars. In such a case the limit design concept adds nothing because capacity is reached simultaneously at a number of points and the elastic theory analysis is fully adequate to predict collapse.

However, when several pattern loadings are considered possible on a continuous flexural member, the sum of the maximum positive moment and the average of the two maximum negative end moments will total more than the simple beam moment. In such a case proper limit design might reduce the member cost. Slab design, on the basis of extensive testing, uses the limit design idea in Chapter 12 without the name or the usual limit design calculations.

11.2 LIMIT DESIGN FOR REINFORCED CONCRETE

When yielding starts, deflections increase sharply and repeated loading would introduce an element of fatigue. Hence under working load conditions yielding is certainly rather undesirable. On the other hand, as a reserve against final failure or collapse, the strength between yielding and failure has proved to be quite significant for structural steel, and the fact that this stage involves larger deflections does not seem too important.

Limit design is significant because it points out that (1) a statically indeterminate member or frame cannot collapse as the result of a single yielding section, and (2) between first yielding and final frame failure there normally exists a large reserve of strength.

Limit design was not recognized in the ACI Code until 1963 and then only in a rather minor way. In the 1971 and 1977 Codes under the title of redistribution of negative moments (Code 8.4) the provisions of 1963 are relaxed somewhat, but not expanded. (The beams carrying two-way slabs, discussed in Chapter 16, might be considered as now under a special form of limit design, but it is not so designated.) Nonelastic design conditions have long been recognized in averaging the moments across a footing or across fairly wide strips of two-way slabs, flat plates, and flat slabs. This is a kind of limit design applied transversely, but it is not the usual limit design concept applied lengthwise to change the given moment diagram itself.

There are several reasons why the ACI Code can afford to appear even to be a little timid in the area of limit design.

First, in terms of economics, the situations for steel and for concrete are not comparable. With steel the efficient wide flange section has only a nominal increase in strength between M_y and M_p, these being the moment at first yield of steel and the fully plastic moment. With essentially the constant strength of a steel beam along the span, a change which equalizes several maximum moments is particularly helpful for most types of loading.*

It is only because reinforced concrete frames are designed for many alternate loadings that reserve strength exists somewhere under almost any specific loading. Although such reserve strength can very often be utilized to advantage by limit design, it is not proportionately as important as in steel construction.

A second reason why limit design is not pushed so strongly for reinforced concrete is that there is more uncertainty as to how concrete acts under all conditions of overload. It does not have the uniform ductile character exhibited by steel—unless certain design restrictions are imposed. Some of these behavior patterns are not yet sufficiently documented to make complete limit design acceptable to all designers.

---

*Not, for instance, with a fixed midspan load on a fixed-end beam that under elastic conditions has equal positive and negative moments.

11.3 MOMENT-CURVATURE RELATIONS

The basis for limit design lies in the inelastic behavior of materials at high stresses, that is, their ability to sustain a given yield moment while a considerable increase in local curvature develops. In a statically indeterminate frame this means that a given local section that tends to be overloaded yields and, in effect, refuses to accept more moment, but does not fail. Instead it forms what is called a "hinge" and thereby forces responses to further loading onto sections less fully stressed. A study of limit design must start by considering the relation between bending moment and the resulting curvature of a short length of the member, as shown in Fig. 11.1a for steel and 11.1b for reinforced concrete. The curve for steel can be approximated with considerable accuracy by two straight lines, as shown in Fig. 11.2a. The curves for reinforced concrete can, a

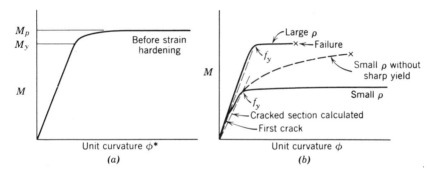

FIGURE 11.1 Moment curvature. (a) Structural steel. (b) Reinforced concrete. (*There is no real distinction between ϕ as used here and $d\theta$ used in Chapter 8.)

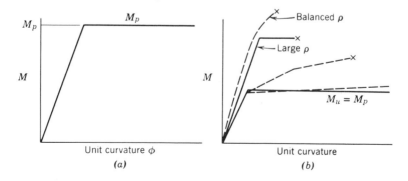

FIGURE 11.2 Conventional moment-curvature relations. (a) Structural steel. (b) Reinforced concrete. Compare Fig. 11.1b.

little less accurately, be approximated by two straight lines or three straight lines as in Fig. 11.2b. For small percentages of sharply yielding steels, such as ordinary Grade 300 steel, the approximation is quite satisfactory. With a large percentage of the same steel, the yielding actually usable would be too short (before failure) to modify the ultimate behavior of the frame significantly. With a balanced beam, such as shown in the upper curve of Fig. 11.2, the inelastic behavior of the concrete in compression makes for some small nonproportional curvature increases at high loads, but does not provide the ductility necessary for significant modification of moments in the frame. For reinforcing steels without a sharp yield point, typical of some of the present-day high strength steels, the behavior might call for three straight lines as shown in Fig. 11.2b. Although behavior with such steel may be quite favorable, it is much more difficult to handle this case mathematically and it will not be considered further in this discussion.

It should be noted that either axial tension or axial compression in the member substantially modifies the $M-\phi$ curve for either steel or reinforced concrete, but more so for the latter.

11.4 FIXED-END BEAM WITH HINGES, AS AN EXAMPLE

The principal behavior that makes limit design interesting can be illustrated with a fixed-end beam under increasing uniform load as shown in Fig. 11.3. This will be discussed for the simplified two line curves of Fig. 11.2. Moments build up proportionately until moment diagram 1 shows the negative moment as M_p and the positive moment roughly half as large. Under further loading M_p cannot increase, but the end element is forced to turn through a larger angle ϕ to permit the beam to deflect more and build up the necessary added resistance elsewhere. Thus the moment diagram

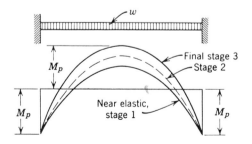

FIGURE 11.3 Development of moment with increasing load, for idealized moment curvature.

would go through increased stages such as 2 until it finally developed a positive moment M_p as shown by diagram 3. At this point it could carry no more added load because it would be on the verge of developing into a mechanism, with hinges at each end and at the middle of the span.

The increase in load between diagrams 1 and 3 is 33% if M_p is the same for positive and negative moments, as in a rolled steel beam. However, in reinforced concrete the M_p for positive moment would usually be much less than for negative moment. Thus this redistribution of moments, that constitutes limit design as it relates to reinforced concrete, is of less economic significance for reinforced concrete than for steel. Nevertheless, a reinforced concrete beam designed under elastic theory normally is not fixed ended and is designed separately for different loads to produce maximum positive moment and maximum negative moment, as illustrated by Fig. 10.9e. Under either loading there is surplus moment capacity, in the one case at the supports, in the other at the mid-point. Thus, even in reinforced concrete, there is some room for limit design concepts.

11.5 CODE PROVISIONS FOR LIMIT DESIGN

Code provisions for limit design are under the title "redistribution of negative moments in continuous . . . members," Code 8.4. Redistribution is permitted *only* if the moments at supports are calculated by elastic analysis and if the ratio of steel ρ (or $\rho - \rho'$) is less than 0.5 ρ_b where

$$\rho_b = \frac{0.85\,\beta_1 f_c'}{f_y} \times \frac{600}{600 + f_y}$$

Thus ρ_b is defined such that it is independent of any A_s' which may be present.

The negative moments may be decreased (or increased) by a percentage defined as

$$20\,[1 - (\rho - \rho')/\rho_b]$$

Restricted in this way the problem of necessary hinge rotation at the ends of the beam is very easily handled by the very long flat $M-\phi$ curve that applies to the beam with small percentages of reinforcement. The curves in Fig. 11.4 for various grades of steel in the beam (indicated in the upper left of the figure) relate the theoretical possible change in moment to the steel used in the beam. The permissible change in moment under the Code is bounded by the two straight lines.

The next section reconsiders the continuous beam design of Sec. 10.10 and changes that might be made by redistribution of its moments.

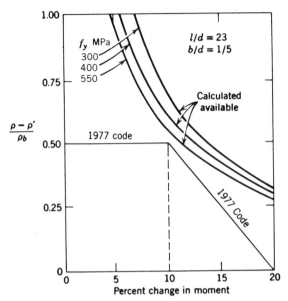

FIGURE 11.4 Moment redistribution allowed and studies of minimum rotation capacities available. (Modified from ACI Code Commentary.)

11.6 REDISTRIBUTION APPLIED TO BEAM OF CHAPTER 10

The beam chosen in Chapter 10 starting at Sec. 10.10 is compact and rather small (250 mm web and h of 500 mm overall), just stiff enough for deflection and, simply as a structural member, on the expensive side because of the area of steel required. If beam weight is not critical, this member could be made eligible for redistribution of moments simply by widening its web to 350 or 375 mm. Beam weight would go up 50% to 70% and steel would not be greatly reduced by this process, but ρ would be reduced by the extra width to around the 0.5 ρ_b threshold required for redistribution.

If clearances permit, it would be better to increase b_w and d without changing their ratio so much. To this end, one could note that k_n (or k_m) in Table 3.1 is based on a ρ of 0.75 ρ_b. The use of k_n about 70% to 72% of the tabulated value (allowing for increased internal lever arm) would lead to ρ approximately 0.5 ρ_b. Alternatively, consideration of the usual required presence of 25% of the positive moment reinforcement at the support would mean a ρ' of roughly 20% of ρ and would permit a k_n of around 85% of the tabulated value, say, 4.5 MPa for k_n instead of the 5.25 MPa tabulated and the 7 MPa used in the earlier design. The first decision toward redistribution would be to accept the size this k_n leads to as an appropriate beam.

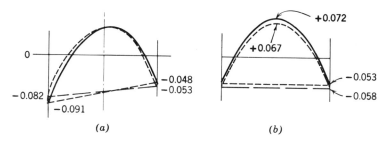

FIGURE 11.5   Design moments (solid lines) redistributed (dashed lines). (*a*) Maximum negative moment at left. (*b*) Maximum positive moment.

Balanced reinforcement, needed in assessing redistribution, is accurately available from Table 3.1, simply as 4/3 of the maximum ρ permitted there.

For a member chosen on the above basis (with ρ and ρ' to be verified later), at least a 10% reduction in negative moments is permissible by redistribution, a little more if $\rho - \rho'$ proves to be less than 0.5 ρ_b. From the negative moment diagram of Fig. 10.11*h* and the positive moment diagram of Fig. 10.11*e*, the modified diagrams of Fig. 11.5 (moment coefficients shown but omitting the wl_n^2 which goes with each) are obtained. In Fig. 11.5*a* the negative moment at the left is decreased to $-0.082\ wl_n^2$, approximately 90% of the original $-0.091\ wl_n^2$. At the same time the moment on the right is similarly increased to $-0.053\ wl_n^2$, leaving a positive moment at midspan almost unchanged, $+0.053\ wl_n^2$, and less than the positive moment in Fig. 11.5*b*. The opposite hand of the same diagram would be appropriate for the right support of the beam.

In a similar fashion the negative moment in Fig. 11.5*b* can be increased 10% to $-0.058\ wl_n^2$, which reduces the positive moment coefficient from 0.072 to 0.067. This remains larger than the coefficient in Fig. 11.5*a* and hence is permissible; in fact there is still the spread between 0.067 and 0.053 which could be reduced if the redistribution could exceed the 10% used here.

The result of these moment changes is nearly a 10% reduction of all A_s values and the elimination of any A_s' requirement as such. Should the resulting $\rho - \rho'$ values at the support be less than 0.5 ρ_b, the reduction could be greater, as limited by the spread in positive moment values in the above paragraph. In that connection some allowance should be made for the slightly higher positive moment off center in Fig. 11.5*a*.

The positive moment reinforcement ρ is subject to the 0.5 ρ_b limit just discussed. This ρ should be lower than at the support unless A_s' there was large.

11.7 BENDING AND STOPPING BARS

The Code, in taking the first step toward limit design, has not been specific with regard to details. We hope the joint ACI-ASCE Committee studying limit design will recommend details concerning bar lengths and development needed as the moment diagrams shift.

It is not the Code intent, as the author visualizes it, to reduce the steel requirements all across the span. Its intent is to reduce the peak moments and to reduce steel concentrations across columns where beam and girder steels must cross each other. The reduced steel of Sec. 11.6 will not modify the moments of Fig. 10.11e and h except in the last 10% increment of load. Cracking away from points of maximum moment should not be encouraged by reducing bar lengths in addition to total steel area. Hence for bar bending the author recommends using the basic moment diagrams of Fig. 10.11e and h with only the minor modifications shown in Fig. 11.6a and d. The required A_s curves then become those of Fig. 11.6b and d and their use involves little difficulty since the peak dotted values are easily calculated, either in terms of bars or bar areas, from the ratios 0.072/0.067 and 0.091/0.082. These required A_s curves are directly comparable to those of Fig. 10.13a.

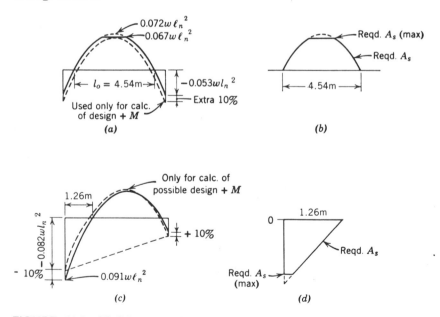

(a)

(b)

(c)

(d)

FIGURE 11.6 Modification of design moments toward limit design. (a) Reduced positive moment for design. (b) Required A_s diagram for positive moment reinforcement. (c) Reduced negative moment for design. (d) Required A_s diagram for negative moment reinforcement.

11.8 PRESENT STATUS OF LIMIT DESIGN THEORY

Aside from bounding the redistribution problem more closely and liberalizing it within these bounds, there has been little progress toward general agreement on the way limit design for reinforced concrete should develop. There has been some considerable progress in basic concepts, but limit design in unbraced frames still appears improbable in the near future.

If one considers only frames sufficiently braced (without restricting normal joint rotation) that lateral deflection of the joints is eliminated, this restricted limit design may be reasonably near practical application.[5] Such frame restrictions make frame failure as a whole meaningless and failure of individual members the basic concept. The continuity that surrounds the member at service load gradually fades away as hinges form (Sec. 11.9) first at one end and later at the other end of the member (Sec. 11.10). Thus this problem gradually reduces to one of individual member collapse. Frame action after first yielding is discussed in Sec. 11.11 where it is pointed out that it is not necessary to trace these intermediate steps even where complete frame collapse is involved.

11.9 THE PLASTIC HINGE IDEA

If ϕ designates the angle change that develops over a unit length of the member, the M–ϕ curve or the relation between M and ϕ is shown in the simplest or ideal case in Fig. 11.1a and idealized in Fig. 11.2a. It can be similarly idealized for reinforced concrete members as in Fig. 11.2b.

The significant part of these M–ϕ curves for limit design lies beyond the first yield moment, in the zone where the plastic moment M_p remains unchanged over a very large range in ϕ values. When a point along the beam develops a moment M_p, it will act as a plastic hinge. This means it will continue to resist the moment M_p, but further loading will result only in an increased angle change ϕ rather than an increased moment. If an ordinary structural hinge is thought of as frictionless, a plastic hinge may be considered as a "rusty hinge" having a definite, but limited, resistance to rotation. Thus, upon further loading of the member, the moment *changes* produced elsewhere on the member are the same as though a real hinge existed at the plastic hinge point.

The concept of a hinge at a point is mathematically convenient, but in real frames (steel or concrete) a plastic hinge is spread over some length of top and bottom beam surfaces, a length normally in the order of the depth of the member.[6] Occasionally, as in a symmetrical third-point loading of a beam, the yielding region can occur over all the middle third of the beam, a very generalized form not too closely resembling the simple plastic hinge idea.

11.10 THE COLLAPSE MECHANISM

In limit design as developed for steel frames, the ultimate strength considered is that which brings either the frame or the individual member to the verge of failure or collapse, assuming perfect plastic hinge action. To fit the present restricted idea of braced frames of reinforced concrete, consideration here is first directed to the simpler individual beam members.

A member cannot collapse under moment loading (except by buckling) until enough actual hinges or plastic hinges form to transform it into a mechanism. A cantilever beam thus collapses when a single hinge forms at the support, but a simple or continuous beam span must have three hinges to collapse, as indicated in Fig. 11.7. For the simple beam this requires the formation of only one plastic hinge because the supports furnish two real hinges. The entire moment diagram is statically determinate for this case. The total deflection consists of two parts, as indicated in Fig. 8.14 with the added deflection that occurs after M_p develops consisting of a simple triangle, just as though the two segments were plane and not curved by the earlier loading.

For a single span of the usual continuous beam three plastic hinges are

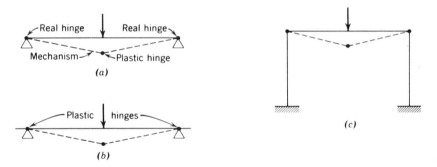

FIGURE 11.7 Local collapse mechanism. (*a*) Simple span. (*b*) Continuous span. (*c*) Frame member.

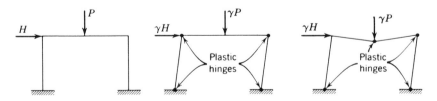

FIGURE 11.8 Mechanisms for collapse of frame as a whole, under load factor γ.

required for collapse; and these hinges can greatly modify the moments from their elastic analysis values.

A frame may collapse as a whole without any individual member developing the three hinges, provided the frame as a whole develops enough hinges to act as a mechanism, as in the middle sketch of Fig. 11.8.

11.11 FRAME ACTION BETWEEN FIRST YIELDING AND COLLAPSE

In Sec. 11.2 it was stated that yielding under working loads were objectionable, although this does not mean that such a condition is totally prohibited under all circumstances. Generally, however, a study under working loads involves analysis on an elastic basis.

As overloads are applied, one or more of the moments might be expected to pass the M_y value and reach the M_p value.* However, it would be quite an unusual frame which would have its maximum moments so equalized that all the plastic hinges required to form a mechanism would develop at any one loading stage and thus lead to an immediate collapse. The usual pattern is for a single highly stressed section (or several) to develop first M_y and then M_p, with more load necessary before some other plastic hinge forms; a number of such loading increments might be required to produce all the plastic hinges necessary for a mechanism.

Although it might be interesting to trace out this sequence of loads and moment diagrams, it would be lengthy process. Fortunately it is an unnecessary process. For any given loading on a given frame, the collapse pattern can be established entirely independently of the original elastic moment pattern and of the intervening elastic-plastic stages.

The ratio of this collapse loading to the service loading is the available safety factor γ, or in reinforced concrete it might be better to say the available load factor γ. This brings out one limitation inherent in the usual limit design approach. *This available load factor is meaningful only as it relates to that particular loading used* and it implies that all portions of that loading are considered only as increasing proportionately.

This limitation is important enough for elaboration through a specific example. If the frame of Fig. 11.8 were analyzed and γ were established as 3, it would mean that if all loads were increased in proportion, the frame would not collapse until it carried loads of $3H$ and $3P$. This would not mean it could carry $3H$ alone or $3P$ alone or $3H$ along with P or any other combination except $3H$ and $3P$. If the load factor for H alone is needed, this requires a separate analysis.

---

*In steel construction M_p is definitely greater than M_y; in reinforced concrete M_p is definitely M_u and M_y need not be differentiated in the calculations.

The student should also note that the collapse load is far beyond the elastic range; hence the effect of the two loads H and P may be far different from the sum of their individual effects.

11.12 LOWER BOUND OR LIMIT ON LOAD FACTOR

A lower bound or lower limit on the real load factor is determined from a moment diagram, *any* moment diagram that is consistent with the given loads, that is, *any* such moment diagram that satisfies statics.

The fundamental idea may be very simply indicated in terms of the elastic moment diagram. Let M_p represent the plastic moment (different values for different members or for different parts of members) and M_s represent the corresponding service load moments from the moment diagram. If M_p/M_s is determined for every member, the smallest ratio found is a lower bound on the true load factor. Why? Up to value M_p the idealized M–ϕ diagram assumes elastic action, which means the moment is directly proportional to the load. Hence the ratio M_p/M_s indicates how much the load may be increased without inelastic action and before any readjustment of the moment diagram shape occurs. This is a *lower* limit on how much the load may be increased because the moment diagram will probably shift to a more favorable shape as this critical section begins to act as a plastic hinge. For further loading the critical moment at the hinge remains constant at M_p, whereas other parts of the moment diagram increase toward the M_p values.

Greenberg and Prager have proved mathematically that any moment diagram satisfying statics may be used as the basis of a lower bound. Normally it is easier to develop an "arbitrary" moment diagram than to establish the elastic moment values.

As one tries different possible moment diagrams, one discovers that higher values of this *lower* bound are obtained when more individual ratios of M_p/M_s become equal to this bound. The moment diagram can thus be "equalized" toward values establishing the true load factor. In complex cases it is desirable finally to assume plastic hinges at these high moment points and check such a solution by the energy-mechanism procedure (Sec. 11.13) which fixes an *upper* bound, just to verify the solution.

If this approach is applied to the beam of Fig. 11.9a, with all M_p values 135 kN · m, one might start with a simple beam moment diagram as in (b) with $M_{max} = 90 \times 2.4 \times 3.6/6.0 = 130$ kN · m and then try to develop three peaks of moment having the same ratio of M_p/M_s. This is easily accomplished by shifting the axis as in (c), making M_p/M_s uniformly $135/65 = 2.08$ at each peak moment value. This 2.08 is the true load factor possible. If M_p for positive moment were 80 kN · m and M_p for negative moment remained 135 kN · m, the moment diagram of (d) would be the proper one and the load factor would be 1.67.

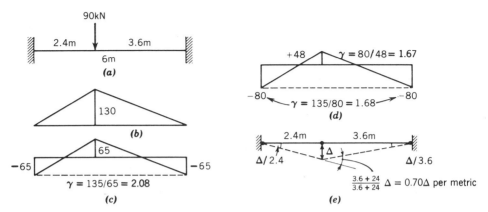

FIGURE 11.9   Establishing possible load factor, γ. $(a\text{-}b)$ Beam and simple span M. $(c\text{-}d)$ Adjusting M diagram. (e) Energy approach.

11.13 UPPER BOUND OR UPPER LIMIT ON LOAD FACTOR

If a frame is on the verge of collapse, it must have developed a sufficient number of plastic hinges to change it into a mechanism. As the mechanism starts to collapse, the loads move in such a way as to contribute energy to the system while rotations occur at the plastic hinges that absorb energy. For a specific given or assumed mechanism a token deflection establishes specific angle changes at all plastic hinges. The required external load to maintain this movement can be established by equating internal energy absorbed at the plastic hinges to the external energy input by the loads.

The true load factor for the frame can be no larger than the ratio of P (collapse) obtained for this particular mechanism to P (working load); it may be smaller for some other (more probable) mechanism. The mechanism-energy approach thus establishes an upper bound or upper limit on the value of the true load factor.

If all possible mechanisms are investigated the lowest load factor will be the true one; but it is desirable to check this load factor against the lower bound discussed in the Sec. 11.12. For this purpose it should be noted that the moments are all statistically determinate when M_p values are used at each plastic hinge in a mechanism.

Consider again the beam of Fig. 11.9a, with positive moment $M_p = 80\,\text{kN} \cdot \text{m}$, negative moment $M_p = 135\,\text{kN} \cdot \text{m}$. From the discussion of Sec. 11.12 the necessary hinge locations here are obviously at the three possible peaks on the moment diagram, as in Fig. 11.9e. The energy input from the load is $90\,\text{kN} \times \Delta$. Energy absorption at hinges $= 135(\Delta/3.6 + \Delta/2.4) + 80 \times 0.70\,\Delta = 150.5\,\Delta\,\text{kN} \cdot \text{m}$. This leads to the load factor possible as $150.5\,\Delta/90\,\Delta = 1.67$. It is not necessary to check other solutions even though this is an upper bound theorem, because the pattern is unique.

In entire frames the hinge location is not so obvious. The student might wish to imagine the hinge at midspan instead of the load and see how it leads to a higher and hence invalid γ.

11.14 BASIC CONCEPTS ESSENTIAL TO LIMIT DESIGN AS SUMMARIZED BY FURLONG

(a) Assigned Limit Moments

Furlong proposes[5] limits within which he feels it feasible to design totally without an elastic frame analysis. He limits himself to frames braced to avoid joint deflection laterally. His paper has taken the limit design concept out of the realm of involved theory and injected a practical design approach. Whether the profession is ready to depart this far from elastic analysis remains to be seen. Some such approach appears necessary if a later Code is to go beyond the moment redistribution now authorized. The direct approach appeals to the author as having much merit compared to the present "patch-up" of elastic analysis. Furlong's proposal at least presents pertinent data and a challenge and encourages examination of the basic problem of whether limit design for reinforced concrete can be made worthwhile.

(b) Flexural Ductility

All limit design analyses require a material that will sustain M_p through a large angle of rotation in order that the simple hinge concept be feasible. Solutions are more complex if a sloping line, as shown dotted near the bottom of Fig. 11.2b, is required and quite complex if the three-line-dotted shape in the middle of that figure is required. Ductility here is defined as the ratio of the ultimate ϕ_u at the end of the horizontal line to the initial ϕ_e at the start of the horizontal section. Values of 4 to 6 are desirable and Reference 5 relates available ductility to variation in either $\rho - \rho'$ or $M_u/(\phi b d^2)$. The method of A. L. L. Baker[3] in general frames uses a group of simultaneous equations to find the related values of needed hinge rotation at each joint to assure that local failure of a hinge will not lower the frame capacity as usually computed.

(c) Ratio of M_y to M_p

The ratio of M_y/M_p lies between 0.94 and unity for singly reinforced members made with reinforcement having a long flat yield section in their stress-strain curve. This implies to the author that the use of $M_p = M_u$ will usually be feasible without involving M_y much, if at all, in the calculations.

(d) Yielding at Service Loads

Furlong has shown that reasonable limits are possible to assure no yield under service loads in load patterns which elastic analyses usually recognize.

SELECTED REFERENCES

1. Lynn S. Beedle, *Plastic Design of Steel Frames*, John Wiley and Sons, New York, 1958.
2. Michael Rex Horne, "A Moment Distribution Method for the Analysis and Design of Structures by the Plastic Theory," *Jour. Inst. Civil Engrs.*, 3, Part III, Apr. 1954, p. 51.
3. Report of Institution Research Committee on Ultimate Load Design, A. L. L. Baker, "Ultimate Load Design of Concrete Structures," *Proc. Inst. Civil Engineers (London)*, 21, Feb. 1962, p. 399.
4. M. Z. Cohn, "Rotational Compatibility in the Limit Design of Reinforced Concrete Continuous Beams," *Flexural Mechanics of Reinforced Concrete*, SP-12, ACI/ASCE, Detroit, 1965, p. 359.
5. Richard W. Furlong, "Design of Concrete Frames by Assigned Limit Moments," *Jour. ACI*, 67, No. 4, April 1970, p. 341.
6. Alan H. Mattock, "Rotational Capacity of Hinging Regions in Reinforced Concrete Beams," *Flexural Mechanics of Reinforced Concrete*, SP-12, ACI/ASCE, Detroit, 1965, p. 143.

12
TWO-WAY SLABS ON STIFF BEAMS

12.1 ONE-WAY AND TWO-WAY SLABS

One-way slabs were discussed in connection with continuous beams in Chapter 10. Such slabs differ from beams only in the large width-to-depth ratio of the slabs, which results in less important shear stresses in one-way slabs than in beams. Such slabs need transverse steel, usually an arbitrary amount to care for shrinkage and temperature stresses, and sometimes a computed amount to assist in distributing concentrated live loads on the slab. In the latter case some two-way action is being called for.

Some simple two-way slab arrangements are given in Fig. 10.1a and 10.1c. The slab of Fig. 10.1c supported by beams between columns should be designed by the methods of Chapter 16 as a part of a frame made up of columns and beam-slab combinations between columns. This is especially true if it involves at least a three-span series and qualifies for the direct design method. Code Chapter 13 covers such cases in detail.

Code 13.3.1 authorizes design of a slab system "by any procedure satisfying conditions of equilibrium and geometrical compatibility, if shown that the design strength at every section is at least equal to the required strength considering Sec. 9.2 and 9.3*, and that all serviceability conditions, including specified limits on deflections, are met." Designs under the yield line method (Chapter 13) or strip method (Chapter 14) can be made to meet this requirement.

The methods of Code Chapter 13 are relatively complex, especially

---

*Sec. 9.2 and 9.3 cover load factors and ϕ factors.

320

where the conditions for the direct design method are not met. Such conditions (among others) involve a minimum of three continuous spans in each direction, successive span lengths not differing by more than one-third the longer span, and columns not offset more than 10 percent of the span (in the direction of offset). Provisions there also do not seem to apply to beams intermediate between columns, as in Fig. 10.1a.

The author feels that numerous two-way slabs supported on stiff beams or walls, where column interaction is not significant, justify the use of the slightly less economical slab design methods given in the 1963 and earlier Codes. These were based on moment coefficients from elastic analysis. Whether used today or not, some of the inherent behavior patterns on which these were based give valuable insight into slab behavior for the many irregular slab cases the designer faces, including slabs with openings as well as slabs of irregular shape.

This chapter is limited to the behavior of two-way rectangular slabs supported on stiff beams or walls and the semielastic analysis method of the 1963 and earlier Codes. The author believes this still is valuable design background.

12.2 ELASTIC ANALYSIS—MATHEMATICAL APPROACH

Two-way slabs, even single panel simply supported, require a three-dimensional approach for analysis. Such slabs are rarely statically determinate in their internal moments and shears. They are the most highly indeterminate of all ordinary structures.

Usually slabs have been analyzed as flat thin plates made of a homogeneous elastic material that has equal strength and stiffness in every direction, that is, an isotropic material. On this basis solutions for simple cases can be established from partial differential equations by the use of advanced mathematics. Westergaard was a pioneer in such analysis in this country.

With differential equations, approximations must generally be introduced to handle practical cases. The use of difference equations and various computer techniques has led to reasonably accurate solutions for some problems. Theoretical boundaries thus established are valuable in assessing the merits of shorter and less accurate analyses. Today computer solutions can handle most cases, but these solutions are not always available for everyday design.

12.3 TYPICAL MOMENT PATTERNS

Since the student cannot really analyze a slab except by arbitrary code provisions, he or she should have a clear picture of the physical action of

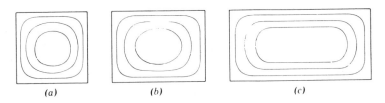

FIGURE 12.1 Approximate contours for two-way simply supported. (*a*) Square slab. (*b*) Oblong slab. (*c*) Long rectangular slab.

the slab. With a little imagination one should be able to visualize the general deflected shape taken by a uniformly loaded slab. A simply supported square slab will deflect into a saucerlike shape; and unless the corners are held down they will actually rise a little off the supports. An oblong slab will take a platterlike shape. A very long narrow slab will take a troughlike shape except near the ends. For fixed edges, there must also be a transition zone around the edges in which the slope gradually turns downward from the horizontal edge tangents. If contours are roughly sketched, as in Fig. 12.1 for the simple supports, they give more than a clue to the moment pattern.

In a square slab, simply supported, a strip across the middle cuts the greatest number of contours and has the sharpest curvature and largest moment. In a long narrow slab only the short strips have significant curvature and bending over most of the panel length. The long center strip is essentially flat and without moment except near its ends.

In the continuous slab all slab strips in each direction have negative moments near the supports and positive moments near midspan, as shown in Fig. 12.2*a*. The long center strip in a long narrow slab is an exception, or a special case, in that its positive moment occurs not at the center but at the point where the strip starts to curve upward; the moment over its mid-span length is small or zero (Fig. 12.2*b*) and the short center strips act almost exactly as one-way slabs.

Mathematical analysis shows that the negative moment on such a *long* strip is nearly the same regardless of the long span length. It is almost the

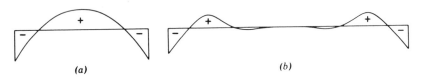

FIGURE 12.2 Typical moment diagrams for continuous slabs. (*a*) Short-span strip. (*b*) A very long strip, say, length three times width.

same as it would be for a square panel having the short span dimensions. One method in the 1963 Code* found it appropriate to express both long span and short span moments in terms of coefficients times the *short span* squared.

12.4 APPROXIMATE ANALYSES

Approximate analyses of slabs are usually somewhat crude, certainly so from the standpoint of the theoretical person or mathematician. Section 12.3 has indicated that the slab in each direction acts somewhat as a one-way slab. But the perpendicular slab strips of Fig. 12.3a are not independent in action. They share in carrying the load and thus each has a smaller moment than a one-way slab. They must deflect the same total amount; hence their relative stiffness becomes a factor in establishing the load and the moment each must carry. With slab thickness a common factor, the longer span is the more flexible and carries the smaller moment and load.

The use of a *single* slab strip in each direction is obviously a crude analogy. The short-span element across the middle of the panel deflects a much greater amount than a parallel strip alongside the edge beam; the edge strip can deflect little more than the beam. A better analysis would result if the slab were considered as several strips in each direction, held to common deflections at their intersections, as in Fig. 12.3b. A still better approximation would consider even more strips and torsional stiffness of these strips as well as their bending stiffness. Since three strips each way would give nine intersections, it should be noted that the labor of such an analysis increases at least with the square of the number of strips considered. Many simultaneous equations must be solved, one for each intersection, or many successive approximations must be tried for a solu-

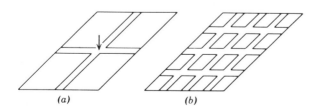

(a)   *(b)*

FIGURE 12.3   The strip idea for two-way slabs. (a) Single intersecting strips. (b) Multiple intersecting strips.

*See text Sec. 12.8.

tion. With enough strips, results closely equivalent to those obtained from the partial differential equation solution are found, but the labor is excessive except on the computer.

Although elastic analyses of homogeneous isotropic plates provide much information that is helpful, they are exact solutions only for the assumed conditions. They usually do not recognize the fact that reinforcing steel is made lighter near the panel edges than at the center and that short span steel is heavier than long span steel. Nor do they usually recognize the changes in relative stiffness which result as cracked sections develop from the bending moment.

12.5 INELASTIC CONSIDERATIONS IN SLAB DESIGN

In a 1926 paper Westergaard recommended moment coefficients that gave considerable weight to the nonelastic readjustments in slab moments which take place before failure. In recognition of these favorable readjustments, his recommended coefficients were established at 28% below strictly elastic values. The percentage reduction corresponds in a way to a similar reduction which had been accepted for flat slabs, but for two-way slabs it gave more recognition to maximum moment loadings. The 1963 ACI Code requirements, including flat plates, fundamentally stemmed from Westergaard's recommendations, although work by Marcus and by vanBuren and diStassio was also recognized.

Design practice and codes by no means attempt to design for the real distribution of bending moment existing across the slab under elastic conditions. This moment varies from element to element across a strip. If a square slab is considered as subdivided into 1-m strips in a given direction, the center strip obviously has the sharpest curvature and largest moment. Curvature and moment on adjacent strips decrease gradually until alongside the edge beam the deflection, curvature, and moment all approach zero.

Even if this were a one-way slab under variable loading it would not collapse when the maximum moment raised the steel stress in a single narrow strip beyond its yield point. This one strip would then simply become more flexible and leave more of any added load to be carried by adjacent elements less highly stressed. Before real failure could occur, all elements would have to be yielding in some fashion. (Chapter 13 discusses the yield-line analysis and collapse conditions for the ultimate strength of slabs.)

12.6 INFLUENCE OF SLAB SUPPORTS ON THE DESIGN

If a two-way rectangular slab panel is supported on stiff beams running between corner columns, part of the load on this panel is carried N-S to the beams running E-W, and part E-W to the beams running N-S. The shorter (and hence stiffer) slab spans carry the larger unit load and larger moments. The longer slab strips carry less unit load and less moment. If the slabs were supported on bearing walls instead of beams and columns, the distribution of the slab moments would completely solve the problem.

At the other extreme, if there were no supporting beams or walls, the slabs would be supported only by the columns, this constituting the flat plate floor discussed in Chapter 15. There the slab would have to be a two-way slab; it would not be enough to carry the load N-S because it would still require that the load be carried E-W to accumulate it at the columns. In fact essentially *all* of the load must be carried N-S and *all* of it also must be carried E-W. If shallow beams were added, the total moment on N-S strips would be shared with the N-S beams; and the *total* N-S moment would be the same as without the beams, but the slab would carry less. These are the concepts developed for the general case of two-way slabs in Chapter 16.

With stiff beams between columns the slab and beam design can be separately calculated. However, it is still true that 100 percent of the total load must be carried N-S, some by the N-S slab strips, some by the N-S beams that pick up the reactions from the E-W slab strips. Likewise, the E-W slab and beams must carry 100 percent of the total load over the E-W span.

12.7 EFFECT OF THE DISTRIBUTION OF REINFORCEMENT

If in one direction the designer fails to provide adequate strength, that slab will yield locally when its limit is reached; but it can still shift the overload to the perpendicular strips if these (and their supporting beams) have the necessary reserve strength.

This readjustment in stiffness and bending moment indicates that the steel does not have to be placed exactly in accordance with the elastic moment requirements. One can say that *almost* any arbitrary arrangement of steel, in sufficient quantity to carry the total load, will be developed before the slab completely fails. The strip method for designing slabs in Chapter 14 moves in that direction.

At working loads, however, a poor distribution of slab steel may result in local yielding of the steel, large cracks, and increased deflection. For desirable action under working loads, steel should be placed at least roughly in accord with the moments existing at the service load; and these moments are more nearly the elastic analysis moments.

Engineering practice uses a uniform spacing of steel over the center strip (one-half the panel width in a square panel) and reduces the steel towards the edges of the panel. This uniform center strip steel is designed to take the *average* rather than the maximum moment on the center strip,

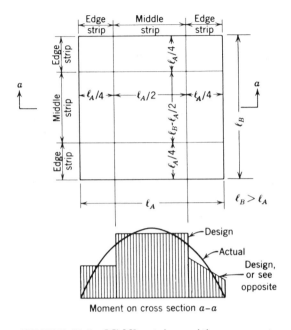

Moment on cross section a–a

FIGURE 12.4 Middle strip positive moment, slab simply supported. Comparison of actual and design moments. This subdivision of ℓ_B fits the 1977 two-way slab system; Secs. 12.8 and 12.9 use $\ell_B/4$ wherever $\ell_A/\ell_B \geq 0.5$ (Fig. 12.6).

as shown in Fig. 12.4 for a simply supported slab. Thus the method takes considerable cognizance of nonelastic behavior as failure approaches.

Despite the fact that in practice approximate moments are used and the steel is designed for average moments rather than maximum, slabs have proved one of the most trustworthy structural elements and have stood up well under great abuse and overload.

12.8 VISUALIZING THE EFFECT OF REDUCED RESTRAINT

Designers of beams are accustomed to the exterior support situation where a light end support results in a large distribution of the fixed end moment back to the beam, with probably half of this distribution carried back to the first interior column joint. A slab support, even with little restraint, is rarely such that it can rotate uniformly over a considerable width;[1] mathematically this would have to be discussed in terms of summing some very unequal restraints and rotations. This is a problem of accuracy, but not one of visualization. One can accept a distribution of fixed end moment

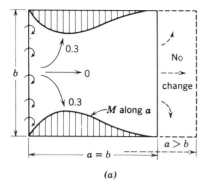
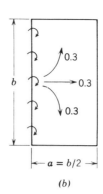

FIGURE 12.5 Carry-over moment factors in continuous slabs on stiff beams.

back to the slab as realistic. However, slabs differ greatly from beams in their carry-over moments. In a square panel with stiff beams *essentially none of this distribution carries back to the opposite beam support.* In a square panel the moment distributed back to the slab goes roughly equally to the two adjacent supports (90 degrees from the rotating edge); not only does it go to the perpendicular slab elements, but most of it goes to those parallel elements in the near half of the slab. Distribution to the slab is thus a more localized operation and this means pattern loadings are important chiefly where they are local patterns. The distribution is roughly indicated in Fig. 12.5a for a square panel and assumes *very stiff* beams.

Increasing dimension *a* to make a rectangular slab influences the distribution very little. If *a* is reduced to *b*/2, a three way distribution results roughly equal to all three beams, again for very stiff beams.

This three-dimensional behavior is significant because with stiff beams it means even free rotation of one edge is a very local matter, not very important except to the positive moments on slabs in both directions and to the negative moments added on the parallel elements nearby the rotating edge. Carry-over moments do not reach into more remote slabs in the same way they do in beams. This makes pattern load moments very much less important in slabs.

12.9 RECOMMENDED TWO-WAY SLAB DESIGN FOR SIMPLER CASES

As indicated in Sec. 12.1, the author recommends slightly less economical designs made under older Codes for many slab layouts, including most of those not meeting the direct design requirements* of Chapter 16. These simpler designs would include many slabs not significantly interlocked with general frame action, those of less than three spans, slabs supported on walls or steel beams (without composite construction), floor slabs in low rise buildings, slabs serving as roofs over tanks or as sidewalls for structures below grade, and the like. The only addition included here from the 1977 Code is the minimum slab thickness from Code 9.5.3.1, simplified by the use of the curves of Fig. 15.14 from Sec. 15.11.

Method 2 of the 1963 Code is the simplest and possibly the best of the three 1963 methods for designing two-way slabs supported on beams. It should still be useful in estimating what should be done with all those slabs not really large enough to justify a slab systems approach.

As in the 1977 Code, two-way slabs are divided into strips as indicated in Fig. 12.4. Although partially in an old notation of S for short span, Fig. 12.6 shows more clearly the design strips from the 1963 Code as used here. For this purpose Table 12.1 is reproduced here from the 1963 Code (Appendix A), with 1977 notation, along with about a page of that Code to indicate its limitations and its general use:

(a) *Limitations*—These recommendations are intended to apply to slabs (solid or ribbed), isolated or continuous, supported on all four sides by walls or beams, in either case built monolithically with the slabs.

ℓ_s = length of *short* span for two-way slabs. The span shall be considered as the *center-to-center* distance between supports or the clear span plus twice the thickness of slab, whichever value is the smaller.

w = total uniform load per unit area

A two-way slab shall be considered as consisting of strips in each direction as follows:

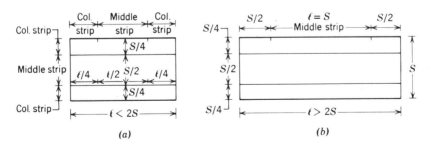

(a)   (b)

FIGURE 12.6   Column and middle strips for 1963 ACI Method 2. (a) Long span < 2S. (b) Long span > 2S. (The S is the old notation, no longer used.)

*Partially listed in fourth paragraph of Sec. 12.1.

TABLE 12.1 Method 2 Moment Coefficients

| | Short Span | | | | | | Long Span, All Span Ratios |
| | Span Ratio, Short/Long | | | | | | |
Moments	1.0	0.9	0.8	0.7	0.6	0.5 and less	
Case 1—Interior panels							
Negative moment at—							
Continuous edge	0.033	0.040	0.048	0.055	0.063	0.083	0.033
Discontinuous edge	—	—	—	—	—	—	—
Positive moment at midspan	0.025	0.030	0.036	0.041	0.047	0.062	0.025
Case 2—One edge discontinuous							
Negative moment at—							
Continuous edge	0.041	0.048	0.055	0.062	0.069	0.085	0.041
Discontinuous edge	0.021	0.024	0.027	0.031	0.035	0.042	0.021
Positive moment at midspan	0.031	0.036	0.041	0.047	0.052	0.064	0.031
Case 3—Two edges discontinuous							
Negative moment at—							
Continuous edge	0.049	0.057	0.064	0.071	0.078	0.090	0.049
Discontinuous edge	0.025	0.028	0.032	0.036	0.039	0.045	0.025
Positive moment at midspan	0.037	0.043	0.048	0.054	0.059	0.068	0.037
Case 4—Three edges discontinuous							
Negative moment at—							
Continuous edge	0.058	0.066	0.074	0.082	0.090	0.098	0.058
Discontinuous edge	0.029	0.033	0.037	0.041	0.045	0.049	0.029
Positive moment at midspan	0.044	0.050	0.056	0.062	0.068	0.074	0.044
Case 5—Four edges discontinuous							
Negative moment at—							
Continuous edge	—	—	—	—	—	—	—
Discontinuous edge	0.033	0.038	0.043	0.047	0.053	0.055	0.033
Positive moment at midspan	0.050	0.057	0.064	0.072.	0.080	0.083	0.050

329

A middle strip one-half panel in width, symmetrical about panel center line and extending through the panel in the direction in which moments are considered.

A column strip one-half panel in width, occupying the two quarter-panel areas outside the middle strip.

Where the ratio of short to long span is less than 0.5, the middle strip in the short direction shall be considered as having a width equal to the difference between the long and short span, the remaining area representing the two column strips.

The critical sections for moment calculations are referred to as principal design sections and are located as follows:

For negative moment, along the edges of the panel at the faces of the supporting beams.

For positive moment, along the center lines of the panels.

(b) *Bending moments*—The bending moments for the middle strips shall be computed from the formula

$$M = (\text{Coef.})w\ell_s^2$$

The average moments per unit width in the column strip shall be two-thirds of the corresponding moments in the middle strip. In determining the spacing of the reinforcement in the column strip, the moment may be assumed to vary from a maximum at the edge of the middle strip to a minimum at the edge of the panel.

Where the negative moment on one side of a support is less than 80 percent of that on the other side, two-thirds of the difference shall be distributed in proportion to the relative stiffness of the slabs. (This ends the extract from Code.)

Note first that all coefficients imply application to unit width slab strips and these can lead to the moments per unit width or per strip width as the designer chooses; the coefficients represent the average moment over a strip width. Second, note that long strip moments defined in terms of the coefficients are to be multiplied by the square of the *short* span. Third, the coefficients specifically assume the supporting beams or walls are "built monolithically"; if not so built on one edge, that edge provides less restraining moment and positive moments in both directions should increase a bit.

The method envisions a pattern type of loading; the coefficients thus provide more moment resistance than the M_o of the 1977 Code Chapter 13 would require. However, where ρ (or $\rho - \rho'$) is not greater than $0.5 \rho_b$ the negative moments may be partially redistributed as indicated in Code 8.4.

The 1963 Code minimum slab thickness was 3.5 in. (90 mm) or the clear slab perimeter divided by 180, whichever was larger. The 1977 Code is somewhat stricter, using the same three limiting equations for slabs on beams and those on columns only, with three variables to define the exact case. A more complete discussion is given in Sec. 15.11; and Fig. 15.14 from that section summarizes the requirements, using the following symbols:

ℓ_n = clear span in *long* direction, face to face of supports

α_m = average ratio of beam to slab stiffness on the four supporting faces, usually equal to 2 or more for "stiff" beams

β = ratio of slab *clear* spans, the longer divided by the shorter. (Note that this is the inverse of the Table 12.1 ratio.)

β_s = ratio of the sum of the lengths of continuous edges to total perimeter of slab panel

This figure will be used in the example that follows: (1) starting vertically from α_m (used as 2 if not given), (2) to one of the two interior panel ($\beta_s = 1.0$) curves or an interpolated curve dependent on the β ratio of the two slab spans, and finally (3) going horizontally to the left to read maximum usable ℓ_n/h, which establishes the minimum h. Note that the plotted ℓ_n/h is for $f_y = 300$ MPa and the figure title states the simple correction for $f_y = 400$ MPa: divide the ℓ_n/h by 1.10.

The lengthy equations given early in Sec. 15.11 simplify when (1) α_m is taken as 2 or more (for a stiff beam), (2) a numerical f_y is introduced, and (3) an interior span slab is considered (which means $\beta_s = 1$). For $f_y = 300$ MPa and $\alpha_m = 2$, Code Eq. 9-12 becomes $h \not> \ell_n/36$ and Code Eqs. 9-10 and 9-11 become the same: $h \geq \ell_n/(36 + 10\,\beta)$. For $f_y = 400$ MPa, these equations are changed only by a coefficient of 1.07 multiplying the ℓ_n value in the numerator.

For exterior spans $\beta_s < 1$ reduces both denominators and increases the required h, as Fig. 15.14 indicates.

At an exterior corner of the slab, special reinforcement is called for in Sec. 13.4.6. The discontinuity in two directions requires an increase in the reinforcement in this corner of the slab.

12.10 EXAMPLE OF TWO-WAY SLAB DESIGN—STIFF BEAMS

Design the interior bay two-way slab on stiff beams for a bay 5 m by 6 m to centerline of beams with service loads of 3.5 kPa (including slab weight) and live load of 5 kPa, $f_c' = 20$ MPa, Grade 400 bars, 20 mm clear cover on slabs. Assume beams stiff enough to qualify slab for design by Method 2 of 1963 Code Appendix.

Solution

Minimum thickness for deflection control.

Assume beams 300 mm wide. Slab spans = center to center of beams or clear span plus twice slab thickness = say 5 m and 6 m.

Assume h in 1963 Code = (slab perimeter)/180 = 2(5 + 6) × 1000/180 = 122 mm,

TABLE 12.2 Slab Calculations for Sec. 12.10

For #15 bars: $h = 130$	Short Span		Long Span	
	108 mm	108 mm	87 mm	87 mm
$w\ell_s^2/\phi = 0.372$ Mn · m/m	**Neg. M**	**Pos. M**	**Neg. M**	**Pos. M**
Middle Strips				
M coefficient	0.046	0.034	0.033	0.025
M_n = (coef.) 0.372 MN · m/m	0.0171	0.0126	0.0123	0.0093
Min. $d = \sqrt{M_n/k_n} = 1000\sqrt{M_n/5.25}$ mm	58	—	48	—
For $h = 130$ mm, $d = 110 - 8$	102	102	—	—
or $d = 110 - 23$	—	—	87	87
$A_s = M_n/400 \times 0.9\,d$				
$\quad = M_n \times 10^9/360\,d$ mm²/m	466	343	393	297
0.0018 bh min. $= 0.0018 \times 1000 \times 130$	234	234	234	234
Spacing of #15 mm	400	500	500	600
Max. spcg. $= 3h = 390$ mm	350	350	350	350
*USE #15 at spacing:	350 mm	350 mm	350 mm	350 mm
$\quad$ Strip width	3 m	1.5 m	2.5 m	1.5 m
*USE in strip total #15 bars	9	4	7	4
Edge Strips, each side				
$\quad$ Strip width		1.5 m		1.5 m
$M_n = 0.5(2/3)(M_n$ on mid-strip) is not critical, because maximum spacing rules				
*USE #15 at spacing:		350 mm		350 mm
*USE in strip total #15		4		4

*Alternate forms of listing.

332

say, 125 mm. This is more than the 90 mm minimum of the present Code 9.5.3.1c. However, the present Code also requires a formula check using three equations. For this example the equivalent graphs in Fig. 15.14 will be used.

Assume $\alpha_m \gtrless 2$ for stiff beams

$$\beta = \text{ratio of clear spans} = (6000 - 300)/(5000 - 300) = 1.21$$

$\beta_s = 1$ for interior panel (all edges continuous)

From Fig. 15.14, entering with $\alpha_m = 2$, read:

For $\beta = 1$ $\ell_n/h = 46$
    $\beta = 2$ $\ell_n/h = 56$ $\beta = 1.21,\ \ell_n/h = 46 + 0.21(56 - 46) = 48.1$

Correct this reading for $f_y = 400$ MPa: $\ell_n/h = 48.1/1.07 = 45.0$
Minimum $h = \ell_n/45.0 = 5\,700/45.0 = 126.7$ mm, say 130 mm.
Moment calculations.

$$w_u = 1.4\,w_d + 1.7\,w_\ell = 1.4 \times 3.5 + 1.7 \times 5 = 13.4 \text{ kPa}$$

All moment coefficients in Table 12.1 are based on the use of the short span ℓ_s for moments in both directions.

$$M_u = (\text{coef.})w_u\ell^2_s = 13.4 \times 5^2 = (\text{coef.})335.0 \text{ kN} \cdot \text{m/m}$$
$$\text{Reqd. } M_n = M_u/\phi = (\text{coef.})\ 335.0/0.9 = (\text{coef.})\ 372.2 \text{ kN} \cdot \text{m/m}$$

Table 12.1 shows moment coefficients for span ratio of $5/6 = 0.83$ as follows:

Short span neg. M 0.046   Long span neg. M 0.033
Short span pos. M 0.034   Long span pos. M 0.025

Table 12.2 is convenient, although some values have been omitted as obviously not governing and others could have been so classified. The short span moments call for the largest d. For $k_n = 5.25$ MPa (Table 3.1 or B. 5 in Appendix B)

$$d \text{ for neg. } M = \sqrt{M_n/k_nb} = \sqrt{0.046 \times 372.2 \times 10^{-3}/5.25 \times 1} = 0.0571 \text{ m} = 57 \text{ mm}$$
$$h \text{ for neg. } M = 57 + 20 \text{ cover} + 8(\text{for } d_b/2) = 85 \text{ mm} < 130 \text{ mm}$$
$$\text{USE } h = 130 \text{ mm} \qquad d = 130 - 20 - (0.5\,d_b \text{ or } 1.5\,d_b)$$
$$= 110 - (0.5\,d_b \text{ or } 1.5\,d_b)$$

For positive moment in the middle strip, the short span would use the 0.5 d_b and in the long span the 1.5 d_b for the perpendicular bars. In a square slab, the average of 1.0 d_b is often used for both directions because the beams can usually carry the slight difference in loading this can introduce. The negative moment on the edge strips also requires overlapping bars, and this would also apply to negative moments on middle strips if continuous top bars were used; these are not used here.*

---

*Slab reinforcing bars are flexible and the projecting end of a bent bar can easily be forced into a plane one bar diameter above or below that which it occupies at the bend point. For this reason many designers feel it unnecessary to be as careful as the author about theoretical bar interference. The author's reasoning is as follows. When bars must be

The calculations in Table 12.2 indicate that another solution with #10 bars would have been possible with some saving in the total steel used, but more field labor would be used in placing the extra bars (possibly 60 to 80 percent more). Normally only one of the optional steel listings shown would be calculated. The more usable is probably the total bars per strip. The spacing of bars in edge strips would be more logical if it started adjacent to the middle strip with bars at the middle strip spacing, gradually increasing to a maximum of three times that spacing at the edge of slab, an average of 1.5 times the middle strip spacing. The **maximum bar spacing limitation prevents this with the #15 bars.**

Shear in two-way slabs on beams would govern only for extremely heavy loads since the beams provide reactions on all four sides. The load distribution on the beams is usually taken from the contributing areas shown in Fig. 12.7. Shear, if it should control the slab would be shear at a distance d from the face of support, in **this case for the short strip at $\frac{1}{2}(5 - 0.3 - 0.102) = 2.30$ m from midspan.**

$$V_u = 13.4 \times 2.30 = 30.82 \text{ kN} \cdot \text{m/m}$$

$$V_n \geq V_u/\phi = 30.82/0.85 = 36.26 \text{ kN} \cdot \text{m/m}$$

$$v_c = V_n/bd = 36.26 \times 10^{-3}/1 \times 0.102 = 0.35 \text{ MPa} < \tfrac{1}{6}\sqrt{f'_c}$$

12.11 FREELY SUPPORTED SLAB CORNERS

Where a slab is simply supported on masonry walls or steel beams, a problem arises at the corners if the slab is not securely fastened down.

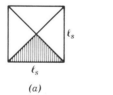
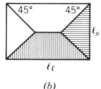

FIGURE 12.7 Slab load areas to beams. (a) Square slab. (b) Rectangular slab.

displaced because of interference, it requires close field inspection to see that the correct group of bars is displaced. The construction worker frequently has no knowledge of the designer's desires in this respect. Proper placement is more likely to result if bars are detailed in such a way that the field worker does not have to readjust the steel levels.

The most critical portion of the long-span top steel is over the middle section of the short beams. At this point the top steel could easily be brought up one bar diameter closer to the top surface. Beyond each side of this section, approximately between the quarter and one-eighth points, the short span top steel could be depressed enough to put the long-span top steel in this more favorable position. Designers must decide whether (1) they want to assign this readjustment to the field, (2) they want to use the larger d near the center width and a smaller d near the edges, or (3) they wish to follow the very conservative approach of using the smaller d for all the long-span negative steel, as in this book.

Such corners rise when the slab is loaded, rising far enough to create a conspicuous horizontal crack in the masonry walls in which they may be embedded. This is especially the case with roof slabs carrying parapet walls. Such slabs should be anchored down (to a substantial mass of masonry, for example) and have reinforcing steel provided for the restraining moment thus developed.

The exterior corner of any slab needs special treatment if the support is stiff, either a beam or wall. Code 13.4.6 gives a detailed specification covering slabs monolithic with moderately stiff concrete beams.

SELECTED REFERENCE

1. C. P. Siess and N. W. Newmark, "Rational Analysis and Design of Two-Way Concrete Slabs," *Jour. ACI*, 20, Dec. 1948; Proc. 45, p. 273.

PROBLEMS

PROB. 12.1   Design a 6.4 m square (c-c of beams) interior panel of a two-way slab on beam stems 300 mm wide (beams assumed to give $\alpha_m = 2$), $f'_c = 30$ MPa, Grade 400 steel, $w_d = 5.8$ kPa (including slab weight), $w_\ell = 2.9$ kPa. Slab thickness must satisfy deflection requirements (close of Sec. 12.9 and Fig. 15.1) and this could determine slab thickness.

PROB. 12.2.   For the data of Prob. 12.1 assume a structure only two bays wide. Design a typical slab for this layout.

PROB. 12.3.

(*a*) Pick a slab thickness based on the Code deflection requirements (close of Sec. 12.9) for an isolated two-way slab supported by tie-beams 300 mm wide on top of masonry walls (assumed stiff enough to qualify for $\alpha_m = 2$). The panel is 4.9 m by 5.5 m center to center of beams, $f'_c = 30$ MPa, Grade 400 bars, $w_d = 4.8$ kPa (including slab weight), $w_\ell = 2.4$ kPa.

(*b*) Check the depth for moment and design the reinforcement on the basis of Table 12.1.

PROB. 12.4   Design an interior slab panel 6.4 m by 5.5 m, otherwise like that of Prob. 12.1 for $w_\ell = 9.6$ kPa without any change in w_d.

PROB. 12.5   A structure consists of a series of transverse bays each extending laterally 6.4 m center to center between longitudinal beams 300 mm wide. If transverse beams 300 mm wide are 5.8 m center to center, loads are $w_d = 5.8$ kPa including slab weight and $w_\ell = 9.6$ kPa, $f'_c = 20$ MPa Grade 400 bars, design the two-way slab, assuming $\alpha_m = 2$.

13
YIELD-LINE THEORY
FOR SLABS

13.1 YIELD-LINE THEORY AS A DESIGN GUIDE

The yield-line theory is not specifically recognized by the ACI Code, but designs are now being made on this basis by good engineers, under the special systems provisions of Code 1.4. Such applications seem particularly advantageous for irregular column spacing.[14, 15]

The yield line is introduced here because it helps the engineer to think about failure patterns and to visualize the ultimate behavior of slabs made with simple reinforcement patterns. For these the method is straightforward, reasonably simple in its concepts, and emphasizes the lines of highest stress. It is adaptable to irregular cases. The engineer can trace slab behavior and visualize some of the effects of moving to less simple reinforcement patterns, cutting off bars, and the like. The strip method discussed in the next chapter is even more usable as a design method where slabs are not supported directly on columns, but designers are apt to be embarrassed by their freedoms unless they have some background such as the yield-line method presents.

Yield line analysis is a limit design method. Designers, until they have a good code to guide them, should assume a liberal load factor, because slab deflection is quite large before failure. Alternatively, some of the ultimate strength might be discounted because of large deflections, just as engineers discount that portion of structural steel strength that lies between the first yield point and the ultimate. The yield line analysis deals with moment alone; it does not assure adequate strength in diagonal tension, but diagonal tension in two-way slabs is a problem only in special cases. On the other hand, in flat plates the diagonal tension (punching

336

shear) around the columns is quite often the weakest element of strength.

Tests have closely verified the yield-line analysis. The calculated load normally underestimates the actual test results. Whether the excess arises from the flat arch action or membrane action, it appears that engineers may use the yield-line method with confidence; it gives a conservative estimate of moment strength.

The real concern of designers using this method will be to establish that their slabs will be entirely satisfactory at working loads. The Danish code has for some time permitted a form of ultimate strength design for slabs. It is the author's understanding that design for ultimate strength alone does not necessarily lead to acceptable slabs. The deflection and stiffness of such slabs may not be satisfactory. Evidently these specific matters must be investigated or limited by establishing maximum ℓ/h ratios.

13.2 BASIC IDEAS OF YIELD-LINE THEORY

(a) Angle Changes at Yielding

The yield-line theory is a form of limit design (Chapter 11) and yields an ultimate strength. Slabs are normally underreinforced, with much less than balanced steel on a strength basis. As a result, on progressive loading the steel reaches its yield-point stress before the slab reaches its ultimate strength. As the steel yields, the center of compression on the cross section moves nearer the face of the slab until finally a secondary failure in compression takes place, at a moment only slightly greater than the yield-point moment.

For a one-way simple span slab, the increase in moment after the steel starts to yield is not large, in the order of 5 to 10%, but the further angle change ϕ occurring at the point of maximum movement is quite large in comparison with the "elastic" angle change. Relative values can be shown as an $M-\phi$ curve, as in Fig. 11.1b.

In a statically indeterminate slab these extra angle changes permit (or cause) significant modifications of the resisting moments and shears. Yielding at one point in such a slab marks the gradual beginning of larger deflections but by no means marks the end of reliable load capacity.

(b) Yield Lines as Axes of Rotation

Yielding under increasing load progresses to form lines of yielding. Until yield lines are formed in sufficient numbers to break up the slab into segments that can form a collapse mechanism, additional load can be supported. To act as hinges for a collapse mechanism, yield lines must usually be straight lines although the fan pattern discussed under (c) has one curved yield-line boundary.

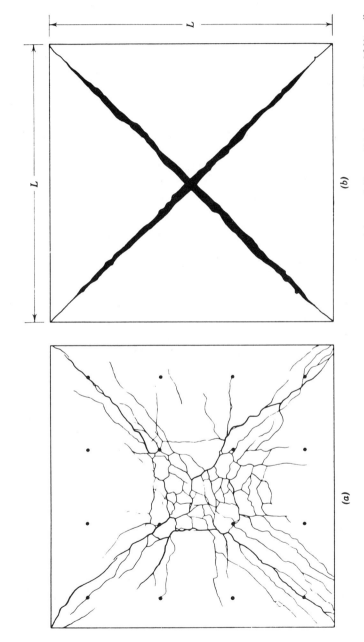

FIGURE 13.1 Yield lines in a square slab on stiff beams. (a) Actual cracking at failure. (b) Assumed "yield lines" or "fracture lines." (These figures are copied by permission from *Reinforced Concrete Review*.[5])

It would be a more accurate description to say that yielding zones develop on the tension face over narrow bands as in Fig. 13.1a and the yield line is an idealization in which all the angle change is considered on a line at the center of the yielding bands, as in Fig. 13.1b. Yield lines are axes of rotation for the movements of the several parts of the final mechanism. Yield lines form at lines of maximum moment, but this action is not restricted to those maximum moment lines originally developed under initial "elastic" conditions.

(c) Interrelationships Between Axes of Rotation

The supports of a slab determine some of the axes of rotation of the several slab segments. In general, each support line constitutes an axis of rotation and each separate column support constitutes a pivot point, that is, a point on an axis of rotation. For example, a one-way continuous slab in a given span must fail by the development of yield lines at each support acting together with an intermediate yield line dependent on the loading, as indicated in Fig. 13.2a. This might be compared to the local collapse mechanism of Fig. 11.7 in limit design discussion.

Consider a continuous slab with a skewed support b as shown in Fig. 13.2b. With a yield line over two adjacent supports such as a and b, one segment rotates about a and one about b; and some common yield line *between* a and b joins the two plate segments rotating about these axes. Hence the common yield line must lie on an axis through O_1 at the intersection point of a and b extended. Which of the possible dashed axes through O_1 will develop depends on the reinforcement and type of loading on the span. Likewise, for collapse in span bc, the third yield line must pass through O_2.

In Fig. 13.2c a slab is shown supported on two adjacent sides and a column. The general pattern of failure will be as indicated, but the axis through the column is at an unknown angle and the yield-line intersection point in the slab depends on the loading and on whether the supported edges are simple supports or are lines of negative moment resistance.

Around a concentrated load a circular yield line of negative moment (Fig. 13.2d) tends to form, with radial yield lines of positive moment, like spokes of a wheel. When supports are nearby, only circular segments instead of full circles may form, as in Fig. 13.2f. These portions are called fans and consist of multiple segments similar to the one shown in Fig. 13.2e. These are discussed in Sec. 13.7.

(d) Segments as Free Bodies To Verify Yield-Line Locations

The free body represented by each collapsing segment must be in equilibrium under (1) its applied loads, (2) the yield moments on each yield line, (3) the reaction or shear on support lines, and at times (4) correction forces.

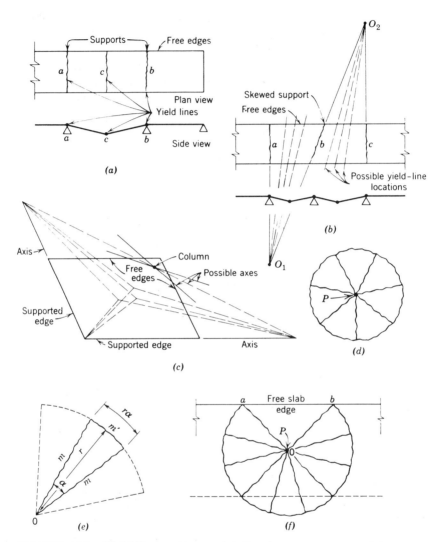

FIGURE 13.2 Yield-line patterns. (*a*) Continuous one-way slab, right supports at *a* and *b*. (*b*) Continuous slab with a skewed support at *b*. (*c*) Slab supported on two adjacent edges and a column. (*d*) Fan pattern. (*e*) A fan segment. (*f*) Fan pattern near free edge.

Since the yield lines form at lines of maximum moment, neither shear nor torsion is typically present at positive moment yield lines.* Torsion can be present in special cases and then it must be represented by correction

*Lenschow and Sozen[12] show that both torsion and a lowered moment resistance may exist at a yield line when $m_x \neq m_y$.

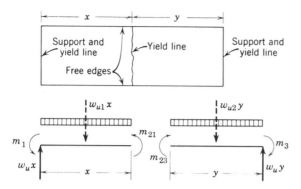

FIGURE 13.3 Free body diagrams for verification of yield-line location.

forces called nodal or knot forces, as discussed in Sec. 13.2g. The yield-line moments thus usually establish the load the segments can support, as indicated in Fig. 13.3. For a uniformly loaded slab, the location of the intermediate yield lines must subdivide the slab into segments each of which support the same uniform ultimate load w_u. If, for given values of yield moments m_1, m_2, and m_3 and a given trial location of the yield line, the calculated w_u values for the separate segments differ, the yield line must lie in a different position. The real location must be such as to reduce the segment size where the calculated w_u is smaller and to increase segment size where the calculated w_u is larger. The correct location is a unique location for a given loading and type of collapse pattern.*

In terms of design it may be convenient to take the known quantity as the load times the desired load factor. Then; with the segments as the free bodies, the required yield moments may be calculated. At a given yield line the moment thus found for one segment must match that from the adjoining segment, that is, m_{21} must be the same as m_{23}; otherwise the yield line is incorrectly located.

When only one yield line remains to be located, an algebraic equation can be solved for a governing dimension, but in more complex cases this may not be as simple as a cut-and-try procedure.

Wood[7] has raised some serious questions about the universal applicability of the correction forces discussed in Sec. 13.2g for use in equilibrium solutions. This clouds the use of the equilibrium process in novel cases, but the virtual work procedure presented in Sec. 13.2e is not subject to any limitations and may be confidently used in all cases.

---

*However, if a different pattern is possible, this solution says nothing as to which pattern is the governing one. In complicated cases the latter demands the energy approach, at least for verification of findings.

(e) Virtual Work or Energy-Mechanism Criterion

If a slab has been reduced to a mechanism by yield lines acting as hinges, a known additional deflection of any specific point in the mechanism (by geometry) establishes the additional deflections at all points, along with the additional angle changes at the yield lines or hinges. For such a slab deflection, the loads also deflect and thereby contribute energy to the mechanism, whereas the hinges resist movement and absorb energy from the system. Thus, for a given mechanism, a given loading imparts enough energy to develop specific yield moments in the hinges. No smaller yield moment will be adequate to resist these loads. Some other mechanism may represent a more probable failure pattern; if so, it demands a larger yield moment to balance the given load. Thus the worst mechanism, and the real mechanism which actually forms, is the one that requires the largest resisting yield moment. Any trial mechanism or yield-line pattern establishes a lower bound or lower limit on the required yield moment.

The same procedure can be used with given yield moments to establish the ultimate load. Any given mechanism establishes an upper bound or upper limit on the collapse load. The real collapse load is the smallest that can be found from all possible mechanisms.

The energy calculations and the segment equilibrium conditions are alternative procedures. In ordinary cases either can be used to check the other. Each establishes the critical dimensions for a given pattern. Neither automatically compares this pattern with others that could be more critical. Hence, neither provides a lower bound on the load capacity (nor an upper bound on the needed moment capacity) and Wood emphasizes the need for these bounds that are available only for a few cases.

Each method has its advantages. The equilibrium method is efficient and points the way to the most critical dimensions; the energy method requires more groping. But in novel situations, especially around openings, the equilibrium method demands some correction forces [see subsection (g)] that may not be available from simple methods or may even be overlooked. Hence only the energy method is always dependable, as already mentioned at the close of subsection (d).

The energy-mechanism approach indicates that a solution is not very sensitive to small changes in the same yield-line pattern, but one must be very careful not to overlook the critical form of the pattern.

(f) Yield Moment on Axes Not Perpendicular to Reinforcing

When the yield moments in two perpendicular directions are equal and no twisting moment (torque) exists, the yield moments in all directions are equal. This condition of isotropic reinforcement simplifies the problem very considerably.

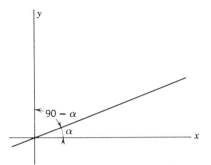

FIGURE 13.4   Moment axes.

Hognestad[2] has shown (following Johansen's demonstration[1,2,4]) that, when the reinforcement in one direction differs from that in the perpendicular direction by some constant ratio, the slab dimensions can be modified to permit analysis as an isotropic slab. Such cases will not be considered here.

If m_x represents the yield moment about the x-axis and m_y about the y-axis, the yield moment about an axis at angle α with the x-axis (Fig. 13.4) will be

$$m_\alpha = m_x \cos^2 \alpha + m_y \cos^2 (90 - \alpha) = m_x \cos^2 \alpha + m_y \sin^2 \alpha$$

When $m_x = m_y = m$, this leads to $m_\alpha = m$ as stated at the beginning of this section.

Wood[7] refers to this relationship as the "square yield criterion" and states that it can be true only when applied to a point where there is a state of "all-round uniform moment," that is, uniform in all directions. He reports that experiments indicate the slab load capacity may be increased as much as 16% by the increased m capacity on a diagonal. He concludes that the square yield criterion normally used is only an approximation for the more complicated actual criterion; but this simple criterion is the one generally used.

On the other hand, Lenschow and Sozen[12] rather conclusively prove by test and theory that there is no appreciable change in the resisting m in the case just discussed. The increased capacity observed in the English tests must have an explanation different from the one Wood suggests.

(g) Correction Forces Where Yield Line Intersects a Free Edge

Correction forces, usually called edge forces, nodal forces, or knot forces, are a necessity for general solutions by equilibrium methods. They appear most certainly when a yield line intersects a free edge at other than a 90° angle. These forces are substitutes for either torsion or twisting moments along assumed yield lines and for normal shears existing at negative moment yield lines. The correction forces are applied in the corner of segments, normal to the plane of the slab, in the magnitude $m_t = m \cot \alpha$, as shown in Fig. 13.6. Since these correction forces act between segments or between a segment and its reaction, they are of the nature of internal forces except when a segment is considered separately as a free body. Hence they completely disappear from any virtual work or energy equations applied to the entire slab. These forces can be totally ignored except in equilibrium calculations (and in reaction calculations).

These forces have been "established" in different manners by different authors. The writer agrees with Wood's evaluation that each of these proofs is lacking in rigor and open to some question. Yet, as Wood shows by various examples, the method works successfully in many instances and it avoids the necessity of differentiation for a minimum value. Hence it has merit even though it is an uncertain tool in new situations. It would seem wise to say that no equilibrium solution can now be considered entirely satisfactory by itself without a check by the energy method.

One approach to the value of the correction force is to say that the use of a straight line as the yield line can only be an approximation at an edge. The reasoning is that a single yield line, to satisfy statics, must intersect a free edge or a simply supported edge at 90° to avoid the necessity of torsional or twisting moments on the yield line. Such twisting moments are not normal on a true maximum moment line. Hence, near an edge,

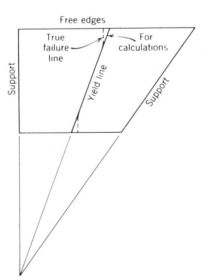

FIGURE 13.5 The curved yield line at a free boundary.

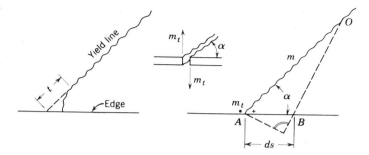

FIGURE 13.6 The shear load m_t at boundary when straight
yield lines are used. (Reprinted from *Jour. ACI.*[3])

yield lines must turn as shown by the dotted lines of Fig. 13.5. If one
insists on using the false "straight-line" yield line, one must consider the
torsional moment $m_t = m \cot \alpha$ in equilibrium equations, and this is most
simply done by using correction forces at the tip of the segments. The
following "proof" and Fig. 13.6 are taken directly from Hognestad's ACI
paper[3]:

> The magnitude of m_t may be established by considering the equilibrium of
> the infinitesimal triangle AOB shown in Fig. 13.6 in which AO is a finite
> length and AB is infinitesimal. Neglecting differentials of higher order, the
> moment in the section OB must equal the moment m in the yield line OA as
> m is a maximum value. Since the bending moment is zero along $AB = ds$, the
> total moment acting on the triangle AOB is found by vector addition
>
> $$m(\overline{AO} + \overline{OB}) = m\overline{AB} = m\,\overline{d}s$$
>
> Equilibrium of moments about OB then gives
>
> $$m\,ds \cos \alpha = m_t\,ds \sin \alpha, \qquad \text{or } m_t = m \cot \alpha$$
>
> differentials of higher order again being neglected. It should be noted that m_t
> acts down in the acute corner. These boundary conditions were first intro-
> duced into the yield-line theory by Johansen in 1931.

Other more detailed "proofs" start from a consideration of intersecting
yield lines anywhere. These deduce the same result plus two more useful
conditions:

1. If all intersecting yield lines have positive moments (or all negative
    with no positive), the number of intersecting lines is unlimited, as the
    "fan" of Fig. 13.2*d*, and no knot force (correction force) may be needed
    at the intersection where only three such yield lines intersect; but knot
    forces will sometimes be needed when *more than three intersect*.

2. If all intersecting lines are not of one kind, only three different directions are possible. Of the six half lines resulting, any one half line may be omitted in a particular case, *or* two half lines may be omitted if they are either halves that form a single line or are not adjacent half lines.

Wood notes that the value of $m_t = m \cot \alpha$ exceeds m when the angle α is less than $45°$ and that such a value would be impossible. However he points out that, in accepting an admittedly approximate or simplified yield line, one may not be limited to exactness as to such an angle. Some solutions giving essentially true results involve smaller angles.

When an entire slab is considered, this downward shear on one segment and the upward shear on the adjacent segment are internal and mutually offsetting forces, and hence do not show in the over-all energy equation.

(h) Corner Pivots and Corner Yield Lines

Supported slab corners, as in Figs. 13.11 and 13.12, introduce the problem of what might be called localized yield-line patterns. Hognestad[3] reports: "According to Johansen it is most expedient in practical design to disregard the corner levers and then later apply corrections, for which he has developed general equations and tabulated the most common cases." In this brief treatment of the subject, corner patterns will be evaluated as any other regular pattern. Such discussion will be deferred until an example not needing this corner analysis has been considered.

13.3 SLAB EXAMPLE, ONE-WAY STEEL

In Fig. 13.7a consider that the yield moment at a is $m_a = -18 \text{ kN} \cdot \text{m/m}$ width, at b is $m_b = -22 \text{ kN} \cdot \text{m/m}$, and for positive moment is $m_c = +13 \text{ kN} \cdot \text{m/m}$. By the yield-line method, calculate the ultimate uniform load the slab will support on a 3.6 m span.

Solution

The slab capacity is independent of adjacent panel conditions except as these are reflected in the values of m_a and m_b. The one-way slab can be very simply solved by limit design procedures, with identical results, but the object here is to introduce yield-line procedures.

Because only one yield-line pattern is possible, an algebraic solution is simple, based on the free body diagrams of Fig. 13.7b.

$$\Sigma M_A = -18 - 13 + wx^2/2 = 0, \qquad w_A = 31 \times 2/x^2 = 62/x^2$$
$$\Sigma M_B = +13 + 22 - w(3.6 - x)^2/2 = 0, \qquad w_B = 35 \times 2/(3.6 - x)^2$$

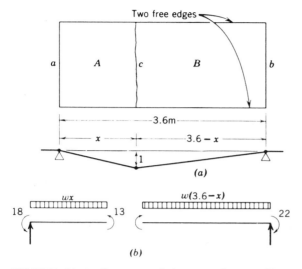

FIGURE 13.7 One-way slab example. (*a*) Failure mechanism. (*b*) Free body diagrams.

Since w_A is to be equal to w_B,

$$62/x^2 = 70/(3.6 - x)^2$$

The quadratic could be solved algebraically, but a solution more typical of the general possibilities of this method would be by trial.

If $x = 1.8$ m $19.14 < 21.60$

$x = 1.7$ m $21.45 > 19.39$

$x = 1.75$ m $20.24 = 20.45$ approx.

Ultimate $w = 20.35$ kPa

The virtual work or energy-mechanism solution is also entirely feasible, using Fig. 13.7*a*. For a 1 m strip the energy input for a unit deflection at the center yield line is

$$0.5 \, wx + 0.5 \, w(3.6 - x)$$

The work done on the slab at the yield lines or hinges is $m\theta$, where θ is the angle of rotation. At the center yield line the total angle change depends upon the combined rotation of A and B. However, the most convenient treatment is to calculate the work there as the separate amounts due to A and B respectively. The work at the yield lines is then:

Due to rotation of A: $(18 + 13)(1/x)$

Due to rotation of B: $(13 + 22)[1/(3.6 - x)]$

Equating energy input to energy consumption,

$$0.5\,wx + 0.5\,w(3.6 - x) = 31/x + 35/(3.6 - x)$$
$$1.8\,w = 31/x + 35/(3.6 - x)$$

For the needed minimum w, $dw/dx = 0$

$$1.8(dw/dx) = -31/x^2 - 35(-1)/(3.6 - x)^2 = 0$$

This is the same algebraic equation solved in the other approach, which gave $x = 1.75$ m. If this value is substituted in the energy equation,

$$1.8\,w = 31/1.75 + 35/1.85 = 36.63$$
$$w = 20.35 \text{ kPa}$$

For the more usual slab with a more complex yield-line pattern, a group of partial derivatives would replace dw/dx. This type of solution might not be practical. However, it is also practical to try different patterns, in this case different x values, in the energy equation: $1.8\,w = 31/x + 35/(3.6 - x)$.

Try $x = 1.8$ $w = 17.22/x + 19.44(3.6 - x) = 20.3667$ kPa

     $x = 1.7$ $w = 20.9294$ kPa

     $x = 1.75$ $w = 20.3481$ kPa

     $x = 1.76$ $w = 20.3493$ kPa

     $x = 1.74$ $w = 20.3482$ kPa

Therefore $x = 1.75$ m

This solution indicates, as is usually the case, that the calculated w by the virtual work or energy approach is not too sensitive to the exact yield-line location. A cut-and-try approach to the problem is thus feasible.

The calculation of w for the several segments by the equilibrium method is more sensitive, but it has the advantage of indicating more definitely the needed shift in assumed yield-line location. (The equilibrium equation for a segment is actually an identity with an energy equation written for that *single* segment.) The shifts with the energy method are more in the nature of groping one's way toward the correct solution rather than observing the needed change.

13.4 SLAB ON NONPARALLEL SUPPORTS, TWO-WAY STEEL

The slab of Fig. 13.2b illustrates two further points of procedure and brings up a possible practical complication. In addition, it serves to emphasize that moments are vector quantities and as such may need to be resolved into components. The ultimate uniform load will be calculated on the basis of the dimensions of Fig. 13.8a, $m_a = -18$ kN · m/m, $m_b = -22$ kN · m/m, and $m_c = 13$ kN · m/m both longitudinally and transversely.

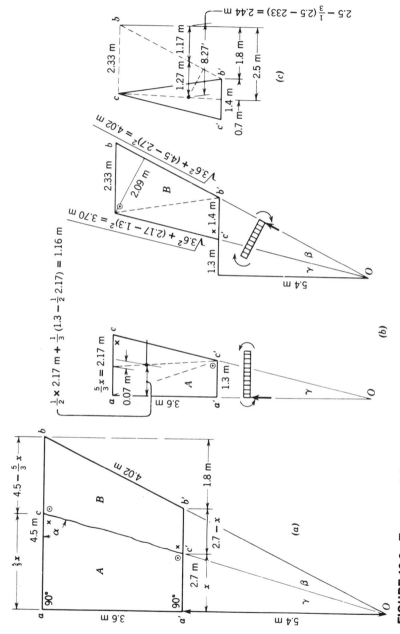

FIGURE 13.8 Two-way slab on nonparallel supports. (*a*) Slab layout. (*b*) Detailed dimensions of segments *A* and *B*. (*c*) Subdivision of Segment *B*.

349

Solution

Since yield lines for negative moment will occur over each support, the corresponding axes of rotation intersect at O. The intermediate positive moment yield line (extended) must also pass through O and can be defined in terms of an unknown angle at O or by the dimension x, the length of one side of the A segment.

This solution requires the yield moment m_c along the inclined yield line between A and B. With equal m_α values longitudinally and transversely, m_c is the same for all orientations. This follows from the equation of Sec. 13.2f:

$$m_\alpha = m_x \cos^2 \alpha + m_y \sin^2 \alpha = m_c(\cos^2 \alpha + \sin^2 \alpha) = m_c$$

If the transverse reinforcement were lighter, as would often be the case, the yield moment would have a specific m_α value for each assumed slope of yield line. The variable m_x would constitute an extra complication* in the solution, which would make general equations rather difficult. However, successive trials for different values of x would be feasible. Solution by trial may constitute the simpler approach even for the given isotropic case because of the somewhat involved geometry for moment arms and angles.

The second new procedure arises from the fact that the middle yield line crosses two free edges at other than a 90° angle. This requires a correction (Sec. 13.2g) in the form of a downward shear m_t at the acute angles and an upward shear m_t at the obtuse angles.

$$m_t = m \cot \alpha = m(x/5.4) = mx/5.4$$

These are marked on the plan view in Fig. 13.8a by an x for a downward force and by a dot within a small circle for an upward force. These shear forces influence an analysis by segments but not an energy equation. The first trial will be made with segments.

Assume $x = 1.3$ m, giving the dimensions in Fig. 13.8b. The moment about aa' of the yield moments acting along cc' is

$$m_c(\text{length } cc') \cos \gamma = m \times 3.6$$

$$m_t = m_c x/5.4 = 13 \times 1.3/5.4 = 3.130$$

$$\Sigma M'_{aa} = -18 \times 3.6 - 13 \times 3.6 - 3.13 \times 1.3 + 3.13 \times 2.17$$
$$+ \tfrac{1}{2} w \times 3.6 \times 1.3 \times \tfrac{1}{3} \times 1.3 + \tfrac{1}{2} w \times 3.6 \times 2.17 \times 1.16 = 0$$
$$-64.8 - 46.8 - 4.07 + 6.79 + 1.014 w + 4.531 w = 0$$
$$w_A = 108.88/5.545 = 19.64 \text{ kPa}$$

*Hognestad presents Johansen's proof showing that the slab dimensions can be modified to give a solution based on the simpler isotropic case. Assume the reinforcement in the transverse direction leads to a value of μm across longitudinal sections compared to m for the longitudinal strips. The simpler isotropic case ($m_x = m_y = m$) can be used if the length in the transverse direction and the size of any concentrated loads are first divided by $\sqrt{\mu}$, any uniform load w remaining unchanged. This provides a relatively simple solution for the problem but it appears that in a general case the ratio μ would have to be the same for both positive and negative moments in a given direction.

The yield moment along $cc' = m$(length cc'), which gives a moment about axis $bb' = m$(length cc')$\cos \beta$. The angle β in triangle $c'Ob'$ is

$$\beta = \tan^{-1} 2.7/5.4 - \tan^{-1} 1.3/5.4 = 26°34' - 13°32' = 13°2'$$

$$\cos \beta = 0.974 \qquad \cos(\gamma + \beta) = 0.894$$

An alternate procedure would be to obtain the moments for cc' as components about the x- and y-axes and then to take the components of these about bb'.

The slab load on segment B will be subdivided into triangular areas of load as in Fig. 12.8b,c. Triangle $bb'c$ has an area $2.33 \times \frac{1}{2} \times 3.6 = 4.19$ m^2 which indicates an altitude perpendicular to bb' of $4.19 \times 2/4.02 = 2.09$ m. For triangle $cc'b'$, the centroid will be on the median at 2.44 m to the left of b. Thus the horizontal distance between the centroid and bb' is 1.27 m. The arm about $bb' = 1.27 \cos(\gamma + \beta) = 1.27 \times 0.894 = 1.14$ m.

$$\Sigma M'_{bb} = 13 \times 3.70 \times 0.974 + 22 \times 4.02 - 3.13 \times 1.4 \times 0.894 + 3.13 \times 2.33 \times 0.894$$
$$- \tfrac{1}{2} \times 2.33 \times 3.6 \, w \times \tfrac{1}{3} \times 2.09 - \tfrac{1}{2} \times 1.4 \times 3.6 \, w \times 1.14 = 0$$
$$46.85 + 88.44 - 3.92 + 6.52 - 2.922 \, w - 2.873 \, w = 0$$
$$w_B = 137.89/5.795 = 23.80 \text{ kPa} > w_A = 19.64 \text{ kPa}$$

The indication is that x was taken a little too large, but the difference between w_A and w_B is quite small and an answer of ultimate $w = 22$ kPa would be quite close. Normally one would not expect results of a trial to be this near to a correct answer; this choice of x benefited from the somewhat similar analysis of the slab in Sec. 13.3. On the other hand, the individual trial calculation should not be as long as here shown. The rather involved geometry was worked out in detail; graphical determination of some of the dimensions might be preferable, especially for initial trials.

The same problem will be solved from the energy-mechanism or virtual work approach, still using Fig. 13.8. Here also the geometry and angles must be carefully determined. Use $x = 1.3$ m as before and assume a unit vertical movement at c'. Point c then deflects $9/5.4 = 1.667$ units (in proportion to the distance from O). Part B rotates through an angle $1/1.4 \times 0.894 = 0.799$/m, and the yield hinge at bb' absorbs energy of

$$E_1 = 22 \times 4.02 \times 0.799 = 70.66 \text{ kN} \cdot \text{m/m}$$

For the yield hinge at cc', it is convenient to work with the x- and y-components of these moments and the respective components of the rotation angle. For rotation of B:

$$E_2 = 13 \times 3.6/1.4 + 13(2.17 - 1.3)/9 = 34.69$$

For the rotation of A, the energy at cc' is:

$$E_3 = 13 \times 3.6/1.3 = 36.00$$
$$E_4 = 18 \times 3.6/1.3 = 49.85$$

The total energy absorbed is $70.66 + 34.69 + 36.00 + 49.85 = 191.20$ kN $\cdot$ m/m.

The same load triangles will be used as before. Triangles $bb'c$ at its centroid deflects $\frac{1}{3}(1.667) = 0.555$.

$$E_5 = \tfrac{1}{2} \times 3.6 \times 2.33 \, w \times 0.555 = 2.238 \, w$$

Triangle $cc'b$, $E_6 = \tfrac{1}{2} \times 3.6 \times 1.4 \, w \times 1.27/1.4 = 2.286 \, w$

Triangle $aa'c$, $E_7 = \tfrac{1}{2} \times 3.6 \times 1.3 \, w \times \tfrac{1}{3} = 0.780 \, w$

Triangle acc', $E_8 = \tfrac{1}{2} \times 3.6 \times 2.17 \, w \times 1.16/1.3 = 3.485 \, w$

The total energy available from loads is $2.328 \, w + 2.286 \, w + 0.780 \, w + 3.485 \, w = 8.879 \, w$

$$8.879 \, w = 191.20$$
$$w = 21.53 \text{ kPa}$$

An ultimate load of 22 kPa is a good answer, the reliability of this answer being judged more on the basis of the equilibrium calculation for parts A and B than on the energy equation. It must be kept in mind that the energy equation always gives loads at least equal to and generally greater than the true ultimate load; that is, *errors are always on the unsafe side*.

To indicate the effect of a small error in locating the center yield line, the energy calculation will be repeated for $x = 1.35$ m instead of the 1.30 m used above. (The equilibrium calculation indicated the true value was *less* than 1.30 m). The dimensions are shown in Fig. 13.9.

$$\cos(\gamma + \beta) = 3.6/4.02 = 0.895$$

Energy absorbed:

$$
\begin{aligned}
22 \times 4.02/1.35 \times 0.895 \qquad\quad &= 73.20 \\
13 \times 3.6/1.35 + 13(2.25 - 1.35)/9 &= 35.97 \\
13 \times 3.6/1.35 + 13(2.25 - 1.35)/9 &= 35.97 \\
18 \times 3.6/1.35 \qquad\qquad\quad &= \underline{48.00} \\
& \;\,193.14 \text{ kN} \cdot \text{m/m}
\end{aligned}
$$

Energy from loads:

$$
\begin{aligned}
bb'c: \;\; &\tfrac{1}{2} \times 3.6 \times 2.25 \, w \times 0.555 &&= 2.248 \, w \\
cc'b: \;\; &\tfrac{1}{2} \times 3.6 \times 1.35 \, w \times 1.17/1.35 &&= 2.106 \, w \\
aa'c: \;\; &\tfrac{1}{2} \times 3.6 \times 1.35 \, w \times \tfrac{1}{3} &&= 0.810 \, w \\
acc': \;\; &\tfrac{1}{2} \times 3.6 \times 2.25 \, w \times 1.17/1.35 &&= \underline{3.510 \, w} \\
&&& \;\, 8.674 \, w
\end{aligned}
$$

$$w = 193.14/8.674 = 22.27 \text{ kPa}$$

This differs very little from the better solution above.

13.5 SQUARE PANEL, IGNORING CORNER EFFECT

Find the ultimate uniform load that a continuous two-way slab 5 m square can carry if the yield moment is 13 kN · m/m for positive moment and 18 kN · m/m for negative moment, equal in both directions.

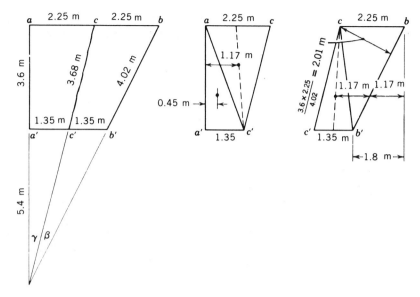

FIGURE 13.9   Dimensions for another trial solution of slab of Fig.13.8.

Solution

The yield pattern is established by symmetry in this case, with triangular segments rotating about each edge, as shown in Fig. 13.10.

When triangle *abo* is considered, each diagonal carries a positive moment of 13 kN · m/m, since the slab is isotropic with equal resistance at any angle. The component about *ab* of the moments on the diagonals is equal to the moment *m* times the projected length *ab*. Hence the equilibrium equation for this segment becomes:

$$5(13 + 18) - \tfrac{1}{2} \times 5 \times 2.5 \, w \times 0.833 = 0$$

$$w = 155/5.21 = 29.75 \text{ kPa}$$

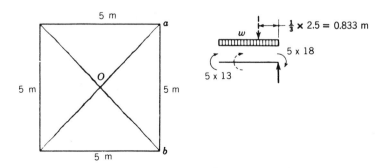

FIGURE 13.10   Two-way slab supported on all four sides.

Except for possible corner effects (Sec. 13.6), this is an exact solution and there is no need for trial solutions. The energy equation would serve just as well. For a unit deflection at o:

$$5(13 + 18)/2.5 = \tfrac{1}{2} \times 5 \times 2.5 \, w \times \tfrac{1}{3}$$

$$62 = 2.083 \, w$$

$$w = 29.76 \text{ kPa}$$

13.6 CORNER EFFECTS

A simply supported square slab at a corner may not follow the simple yield-line pattern used in Sec. 13.5. If the corners are not fastened down, they will rise off the supports as the slab is loaded and the diagonal yield line will split or divide into two branches to form a Y, as in Fig. 13.11a. This forms an additional corner segment with yield moments on only two faces and with support only on the two points where the yield lines cross the boundary. This condition is called a corner pivot, since the segment pivots about these two points.

If the corners are held down, but are not specially reinforced, similar yield lines form but in addition a corner crack opens along the pivot line as in Fig. 13.11b. If the corner is specially reinforced for negative moment, this adds a yield moment m' across this boundary of the triangular segment which increases its capacity and moves the junction of the Y farther from the corner. With a large enough m', the triangular segment fails to form and the simple diagonal yield line into the corner is correct without modification.

In continuous slabs, the corners have top steel which provides an m' for the corner segment. Whether the Y-pattern or the straight diagonal yield line forms depends upon the amount of this reinforcement.

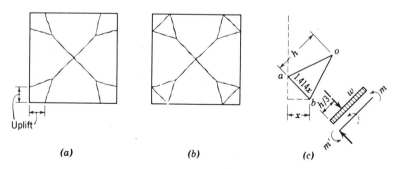

(a)   (b)   (c)

FIGURE 13.11 Corner segments. (a) When corners are not fastened down. (b) When corners are held down but not reinforced for negative moment. (c) Equilibrium of corner segment.

Consider the free body formed by one of these corner segments, as in Fig. 13.11c. The positive moment m along sides ao and ob adds vectorially to give a positive moment m on the width ab, when m is the same in each direction. If no knot force acts at the apex of segment,

$$\Sigma M_{ab} = -1.414\ xm' - 1.414\ xm + 0.5(1.414\ xhwh)/3 = 0$$
$$wh^2/6 = m + m'$$
$$h = \sqrt{6(m + m')/w}$$

The length h is not dependent on x or the width of the segment and increases with the ratio of the sum of positive and negative yield moments to w. If, in a given slab, the moment-load ratio establishes a distance h much less than the length of the diagonal along which it lies, this corner pattern will control; that is, it does reduce the slab capacity. If the calculated h is large enough to push the Y-intersection beyond the limits of its particular diagonal, it means that the corner segment does not form. Intermediate values of h indicate that the corner element controls but has a smaller effect on the ultimate load or required moment strength of the slab.

With a simple supported square slab of side dimension a the ultimate moment is $wa^2/24$ if no corner segment forms. This is the true condition with the corner held down and with m' made equal to m. With $m' = 0$, or with the corners free, the corner pivots produce corner segments which increase the ultimate moment to $wa^2/22$. In the square panel the maximum effect of the corner segments is thus slightly less than 9%.

13.7 CIRCULAR SEGMENTS OR FANS

Under a concentrated load the negative moment yield line tends to form somewhat in a circle around the load with positive moment yield lines as spokes or radial lines, as shown in Fig. 13.12d. If a segment of this circular area is isolated as in Fig. 13.12e, the total load P at O results in a load $(\alpha/2\pi)P$ to be carried by the particular segment from O to its circular boundary. The segment will be small enough to be considered as a triangle, which is similar to the corner segment of Fig. 13.11c, except for the loading. Then ΣM about the (circular) chord becomes:

$$(m + m')r\alpha - (\alpha/2\pi)Pr = 0$$
$$P = 2\pi(m + m')$$

It is interesting to note that the load P is thus not a function of r and hence the slab is equally subject to failure at any surrounding circle. This is not quite true when the load is applied over a finite area and the weight of the slab is considered, but it points out the need for both positive and

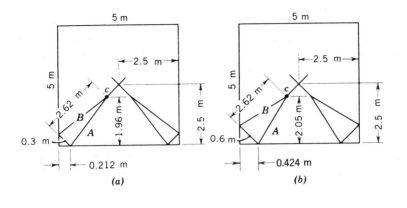

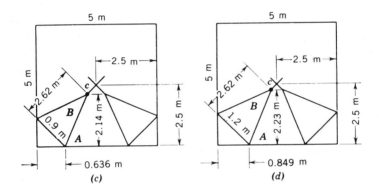

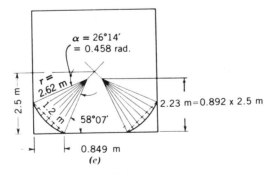

FIGURE 13.12   Corner segment trials.

negative moment resistance over the full area of a slab subjected to heavy concentrated loads.

Generally, the segments tend to have a large radius, and consideration of the equilibrium equation indicates that the radius might vary from segment to segment with only minor changes in the strength, resulting in a form resembling that of Fig. 13.15g. When a slab has a free edge nearby, as in Fig. 13.2f, it is weaker on the triangle abO because this has no m' and the circle will develop such as to make this triangle large. If a second free edge were on the other side of the load as shown dotted, a similar triangle would also develop on that side instead of the segments shown. On the other hand, if the edges were restrained, providing an m', the failure might be similar to Fig. 13.15h, although the corner segments might be more like Fig. 13.12e.

Wood[7] points out an interesting case, that of a simple cantilever slab with a concentrated load in the corner. The failure section is more probable as a segmental fan plus edge triangles as at A in Fig. 13.15i. These indicate the need for m' resistance in both directions and without reduction near the corner, since any radius gives the same resistance. Even without the fan concept, as at B, a similar conclusion follows for negative moment (alone) since the moment zP must be carried by a strip $2z$ wide. The fan, however, indicates the need also for positive moment steel.

13.8 SQUARE PANEL, CONSIDERING CORNER EFFECTS

The effect of the slab corners on the ultimate moment on the square slab of Sec. 13.5 will now be investigated by basic principles.*

In Sec. 13.5 w was found to be 29.75 kPa, and this will not be seriously changed, say, not below 27 kPa based on the 9% correction of Sec. 13.6. With this assumed w, summation of moments for the corner segment, as above, gives:

$$m' + m = wh^2/6$$
$$13 + 18 = 27\,h^2/6$$
$$h = \sqrt{31 \times 6/27} = 2.62 \text{ m}$$

The full diagonal length is $\frac{1}{2} \times \sqrt{2} \times 5 = 3.54$ m. Hence it is probable that the corner segment does form. Since the edge does not have $M = 0$, M_t edge shears are not necessary in this example.

Try the failure pattern of Fig. 13.12a, using the energy method with a

---

*Reference 3 develops three simultaneous equations for the square slab case that establish the corner segment dimensions algebraically.

center deflection of one unit. Since the pattern repeats, only one set of areas A and B will be included.

The deflection of point c is $1.96/2.5 = 0.784$. Energy input from the load is:

$$E_A = \tfrac{1}{2} \times 5 \times 2.5\,w \times \tfrac{1}{3} - 2 \times \tfrac{1}{2} \times 0.212 \times 1.96\,w \times \tfrac{1}{3} \times 0.787 = 1.974\,w$$
$$E_B = \tfrac{1}{2} \times 0.3 \times 2.62\,w \times \tfrac{1}{3} \times 0.787 = 0.103\,w$$

Energy absorbed in hinges:

$$E_A = (18 + 13)(5 - 2 \times 0.212)/2.5 = 56.742$$
$$E_B = (18 \times 13) \times 0.212 \times 0.787/2.62 = 1.974$$
$$1.974\,w + 0.103\,w = 56.742 + 1.974$$
$$w = 58.716/2.077 = 28.3\ \text{kPa} < 29.76\ \text{kPa of Sec. 13.5}$$

This comparison proves that the corner segment does form.

A larger corner segment may drop the ultimate w lower. Try the increased segment of Fig. 13.12b. The deflection of c is $2.05/2.5 = 0.82$. As for the above trial,

$$1.845\,w + 0.215\,w = 51.485 + 4.114$$
$$w = 55.599/2.06 = 27.0\ \text{kPa} < 28.3\ \text{kPa}$$

Since this governs over the preceding calculation; further trials were made for the conditions of Fig. 13.12c and d which gave 26.1 kPa and 25.4 kPa respectively.

The corner triangles might be replaced by a small fan as shown in Fig. 13.12e. Then for the quarter panel the energy imput is:

$$E_A = \tfrac{1}{2} \times 5 \times 2.5\,w \times \tfrac{1}{3} - 2 \times \tfrac{1}{2} \times 0.849 \times 2.23\,w \times \tfrac{1}{3} \times 0.892 = 1.520\,w$$
$$E_B = \tfrac{1}{2} \times 2.62^2 \times 0.458 \times \tfrac{1}{3} \times 0.892 \qquad\qquad = 0.467\,w$$

Energy absorbed in hinges:

$$E_A = (18 + 13)(5 - 2 \times 0.849)/2.5 = 40.945$$
$$E_B = (m + m')0.892\,\alpha = (18 + 13)0.892 \times 0.458 = 12.665$$
$$1.520\,w + 0.467\,w = 40.945 + 12.665$$
$$w = 53.610/1.987 = 26.98\ \text{kPa}$$

The original h calculation is still about right. This has not exhausted all the possibilities, but when one considers the basic approximation shown in Fig. 13.1, great mathematical refinement seems out of place.

13.9 RECTANGULAR SLABS

Investigate the ultimate load capacity of 3.6 m by 6 m slab continuous on all edges, which has a yield moment of 13 kN · m/m for positive moment and 18 kN · m/m for negative moment, both uniform in each direction.

Solution

Symmetry dictates a yield line at the middle, parallel to the long side, which merges with corner diagonals symmetrical about unknown points O and O' as shown in Fig. 13.13a. The corner effect with the Y-form on the corner diagonals is

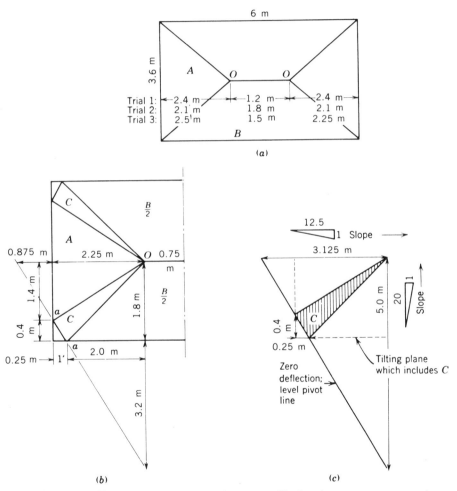

FIGURE 13.13 Rectangular slab analysis. (a) Neglecting corner segments. (b) With trial corner segments. (c) The rotation of corner segment.

also possible, but this will be treated as a later modification of the simple pattern.

The first trial dimensions shown in Fig. 13.13a will be considered first in terms of the equilibrium of segments A and B.

Segment A:

$$-3.6(18+13) + \tfrac{1}{2} \times 3.6 \times 2.4 \, w \times \tfrac{1}{3} \times 2.4 = 0$$
$$w_A = 111.6/3.456 = 32.29 \text{ kPa}$$

Segment B:

$$-6(18+13) + 1.2 \times 1.8 \, w \times \tfrac{1}{2} \times 1.8 + 2 \times \tfrac{1}{2} \times 2.4 \times 1.8 \, w \times \tfrac{1}{3} \times 1.8 = 0$$
$$w_B = 186/4.536 = 41.00 \text{ kPa} \gg w_A$$

Try a smaller A segment as noted in Fig. 13.13a for the second trial. These dimensions lead to $w_A = 42$ kPa and $w_B = 36$ kPa. A third trial with point O at 2.25 m from the end gives $w_A = 36$ kPa and $w_B = 38$ kPa. This is reasonably close and will be checked by the energy relationship, using half of the panel for convenience and a unit deflection along OO'.

$$
\begin{aligned}
A: \ & \tfrac{1}{2} \times 3.6 \times 2.25 \, w \times \tfrac{1}{3} && = 1.35 \, w \\
B: \ & 1.5 \times 1.8 \, w \times \tfrac{1}{2} && = 1.35 \, w \\
& + 2 \times \tfrac{1}{2} \times 2.25 \times 1.8 \, w \times \tfrac{1}{3} && = 1.35 \, w \\
\hline
& && \overline{4.05 \, w}
\end{aligned}
$$

Alternatively, visualize the total slab deflections as a volume somewhat similar to an inverted pyramid. This volume times w measures the energy imput.

$$
\begin{aligned}
2.25 \text{ m end as half pyramid:} \ & 2.25 \times 3.6 \, w \times \tfrac{1}{3} = 2.7 \, w \\
0.75 \text{ m center as wedge:} \ & 0.75 \times 3.6 \, w \times \tfrac{1}{2} = 1.35 \, w \\
\hline
& \overline{4.05 \, w}
\end{aligned}
$$

Energy absorbed:

$$
\begin{aligned}
A: \ & 3.6(18+13)/2.25 \ = \ 49.6 \\
B: \ & = 6(18+13)/1.8 \ = 103.3 \\
\hline
& \overline{152.9}
\end{aligned}
$$

$$w = 152.9/4.05 = 37.76 \text{ kPa}$$

For the location of O the third trial dimensions will be assumed correct enough.

The corner segment situation will now be investigated. For equilibrium of the corner segment,

$$-18 - 13 + \tfrac{1}{6} wh^2 = 0, \qquad h = \sqrt{186/w}$$

If $w = 37.76$ kPa, $h = 2.219$ m

This compares with the diagonal length of $\sqrt{2.25^2 + 1.8^2} = 2.881$ m. It appears that there will be a significant corner segment, probably one extending most of the way to point O.

The corner segment shown in Fig. 13.13b will be analyzed by the energy approach, using the deflection at O as unity. At aa the x- and y-components of the moments and rotations, as diagramed in Fig. 13.13c, are used for calculating the energy absorbed by the hinges. Only half the panel (one segment each of A, B, and two of C) will be used.

Energy input by loads:

$$B: \ 1.5 \times 1.8 \, w \times \tfrac{1}{2} \qquad\qquad = +1.35 \, w$$
$$+ \, 2 \times \tfrac{1}{2} \times 2.0 \times 1.8 \, w \times \tfrac{1}{3} \ = +1.20 \, w$$
$$A: \ 2 \times \tfrac{1}{2} \times 1.4 \times 2.25 \, w \times \tfrac{1}{3} \ = +1.05 \, w$$
$$C: \ 2 \times \tfrac{1}{2} \times 0.25 \times 1.8 \, w \times \tfrac{1}{3} \ = +0.15 \, w$$
$$+ \, 2 \times \tfrac{1}{2} \times 0.4 \times 2.25 \, w \times \tfrac{1}{3} \ = +0.30 \, w$$
$$-\, 2 \times \tfrac{1}{2} \times 0.25 \times 0.4 \, w \times \tfrac{1}{3} \ = \underline{-0.03 \, w}$$
$$4.02 \, w$$

Somewhat simpler, energy imput from the deflection volume:

2.25 m end as half pyramid:* $(2.25 \times 3.6 - 2 \times \tfrac{1}{2} \times 0.25 \times 0.4)w \times \tfrac{1}{3} = 2.67 \, w$

0.75 m center as wedge: $\qquad (0.75 \times 3.6) \, w \times \tfrac{1}{2} \qquad\qquad = \underline{1.35 \, w}$
$$4.02 \, w$$

Energy absorbed by hinges:

$$B: (6 - 2 \times 0.25)(18 + 13)/1.8 \ \ = 94.72$$
$$A: \ 2 \times 1.8(18 + 13)12.25 \qquad\ \ = 49.60$$
$$C: \ 2 \times 0.25(18 + 13)0.25/5.0 \ \ = \ \ 0.78$$
$$+ \, 2 \times 0.4(18 + 13)0.25/3.125 = \ \ \underline{1.98}$$
$$147.08$$

$$w = 147.08/4.02 = 36.59 \text{ kPa} < 37.76 \text{ kPa}$$

A number of variations, shown in Fig. 13.14, where computed by the energy method, partly to be certain the worst case was found, partly to study the rather minor differences in w that resulted. It appears that an ultimate load of about 36 kPa is correct, possibly with corner fans.

13.10 INFLUENCE OF STRENGTH OF EDGE BEAMS

Tests of slabs in England have indicated that the failure pattern of Figs. 13.10 or 13.11 occurs with stiff supporting beams but the pattern can be altered by varying the size of beams. If the beams in one direction are very

---

*Where the corner elements go to a common intersection, the net area of the base as used here is convenient. Where the corner elements stop short, the "gross" pyramid for the total height less small corner triangular pyramids of lesser height is convenient as a concept.

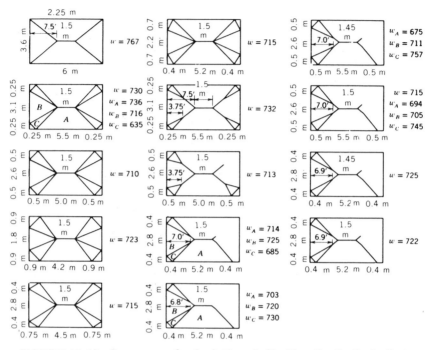

FIGURE 13.14 Summary of calculations indicating the limited effect of corner segments.

flexible, the slab fails almost as does a one-way slab and the yield line lies directly across the middle of the span and continues through the light beams. Then the strengths of the beams and the slab add in the evaluation of the *m* and *m'* moments. Obviously, at some particular beam stiffness, it would be a matter of chance whether the diagonal failure or the midspan failure line develops.

In tests of multiple panels of two-way slabs at the University of Illinois[10] a positive moment yield line developed near mid-span in the exterior panels parallel to the outside edges and the failure extended across the interior and spandrel beams that framed perpendicular to the outside edge. In such a case slab and beam strengths are additive in establishing the maximum load capacity. However, it should be pointed out that, although the calculated failure load for this mode of failure closely matched the actual failure load, the structure did not fail in the mode the yield-line method would have predicted. Failure of the individual slabs was indicated by theory at some 25% lower load.

The example problems in this chapter have assumed that beams furnished stiff supports of such strength that the slab would fail first.

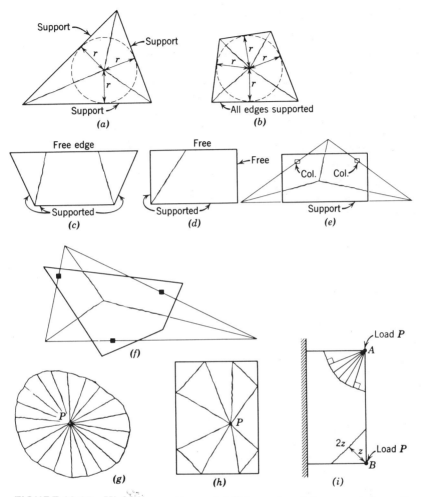

FIGURE 13.15 Yield-line patterns. (*a*) Triangular slab supported on all sides. (*b*) Polygonal slab circumscribed around a circle, all edges supported. (*c*) Slab supported on three of four sides. (*d*) Slab supported only on adjacent sides. (*e*) Slab supported on one side and two columns. [(*a*) to (*e*) adapted from Ref. 4.] (*f*) Slab supported on three columns. (*g*) Yield pattern around a concentrated load. [(*f*) and (*g*) adapted from Ref. 3, ACI.] (*h*) Yield pattern from an unsymmetrical concentrated load on a rectangular slab. (Adapted from Reference 6, Chamecki.) (*i*) Corner load on cantilever slab.

13.11 OTHER YIELD-LINE ANALYSIS

A few yield-line patterns for other cases may be helpful as suggestions. Figure 13.15a shows a triangular slab simply supported and carrying uniform load. Johansen[4] shows for isotropic conditions that the yield lines intersect at the center of the inscribed circle and result in $m = wr^2/6$, where r is the radius of the circle. He extends this case to show the same moment in any polygon shape that circumscribes a circle, as in Fig. 13.15b. It is assumed that any of these might need to be investigated for a possible corner pivot or segment. Johansen also shows free edge slabs as shown in Fig. 13.15c,d,e.

Hognestad[3] shows the slab of Fig. 13.15f. He also analyzes a radial pattern which at times develops under concentrated loads, Fig. 13.15g, as already discussed in Sec. 13.7.

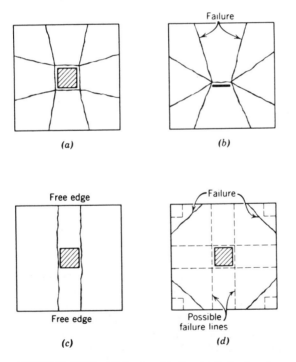

(a) *(b)*

(c) *(d)*

FIGURE 13.16 Slabs supporting column loads. (From Reference 9, ACI.) (a) Simply supported on all edges, corners free. (b) Eccentric column load treated as a line load, simply supported on all edges, corners free. (c) Simply supported on two opposite edges. (d) Supported at corners only.

With a single concentrated load in a rectangular panel Chamecki[2] shows eight triangular segments radiating from the load P whether P is centered or off center from both axes as in Fig. 13.15h. He develops equations to locate all key dimensions for the simple support case when the corners are anchored down and when they are free to rise. He also shows the condition necessary to eliminate the corner segments.

Elstner and Hognestad[9] show the yield patterns of Fig. 13.16 for slabs carrying a center load in the form of a column stub cast monolithically with the slab. Figure 13.16a shows a slab simply supported and corners free; Fig. 13.16b is the same except that an eccentric column load is considered as a line load. The case of simple supports on two opposite sides only is shown in Fig. 13.16c and simple support on four corners only in Fig. 13.16d; in the latter case the dashed lines represent an alternate yield-line pattern.

13.12 SLABS WITH OPENINGS

One advantage of the yield-line method is that slabs with openings can be analyzed. The designer must, however, be alert to limitations on the effective usage of the equilibrium method.

When two positive moment yield lines meet at the corner of an opening, the knot forces are *not* zero. In fact, shearing forces will often be transferred from one element to another.

For example, the load of 33 kPa and the yield lines shown in Fig. 13.17a were calculated as the governing case from the energy approach using $m = 13$ kN $\cdot$ m/m and $m' = 18$ kN $\cdot$ m/m and neglecting corner effects. The equilibrium method would lead to the idea that the height of triangle A

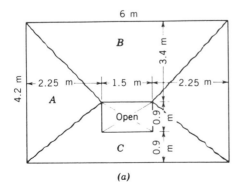
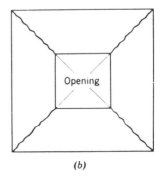

FIGURE 13.17 Yield-line patterns in slabs containing openings. (a) Large knot forces present. (b) No knot forces.

should be 2.3 m on the basis of the equation for h in Sec. 13.6. However, a net downward knot force at the apex makes a shorter height of triangle the correct one. The student can use equilibrium of each segment to find these knot forces, after having w established. For this slab anyone visualizing the deflection of segment B would quickly see from the way a slab deflects that some support for B is required at the two corners of the opening. This is what causes the knot forces there.

A second problem also involves the equilibrium method. The discontinuity at the 90° corner of the opening prevents any effective use of the ordinary edge forces m_t. It will be obvious that a single yield line going symmetrically into the corner, as in Fig. 13.17b, has two obtuse angles and that the nominal upward forces in each corner could not represent real internal forces (because they would not be "equal and *opposite*"). Symmetry considerations indicate that this must be the correct yield-line pattern (ignoring corner effects) and that there can be no knot forces at the opening. Thus one must conclude that knot forces around an opening may be absent or may be present (as in Fig. 13.17a), but are not likely to be of the nominal magnitude m_t. Thus the energy method is essentially mandatory for such cases.

13.13 REINFORCEMENT AT VARIABLE SPACING AND/OR WITH BARS CUT OFF

Taylor *et al.*,[13] tested a series of simply supported square slabs of several thicknesses but with the chief variable the arrangement of reinforcement. Specimens included uniform spacing as the yield line theory uses most

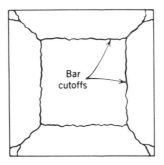

FIGURE 13.18 Test yield line pattern on simply supported slabs having bars cut off according to moment.

conveniently, variable spacing strips as most commonly used in practice, and 50 to 70% of the bars cut off to the moment requirements.

Variable spacing showed only a minimal advantage, if any, since the larger area of reinforcement near the middle decreased the internal lever arm. Variable spacing gave a slightly stiffer slab, with the cracks appearing first near the corners rather than near the center; it did not increase strength.

Some economy of material results from stopping some bars short of the supports, and ultimate design loads were still exceeded. However, the excess load capacity was less than when bars were full length and these slabs deteriorated rapidly after forming a center square of positive moment yield lines at the points where bars were cut off, as in Fig. 13.18.

13.14 FLAT PLATE FLOORS

Flat plate floors directly supported on columns are used for many buildings designed for light or moderate live road. The Code design for these is covered here in Chapter 15. However, an equilibrium analysis, using in part yield line concepts, has been found desirable for such floors carried by irregularly spaced columns. Reference 14 covers this approach and Reference 15 relates it also to the strip method of Chapter 14.

SELECTED REFERENCES

1. K. W. Johansen, *Pladeformler*, Polyteknish Forening, Copenhagen, 2nd ed., 1949.
2. K. W. Johansen, *Paldeformier; Formelsamling*, Polyteknish Forening, Copenhagen, 2nd ed., 1954.
3. Eivind Hognestad, "Yield-Line Theory for the Ultimate Flexural Strength of Reinforced Concrete Slabs," *Jour. ACI*, 24, No. 7, Mar. 1953; *Proc.*, 49, p. 637.
4. K. W. Johansen, "Yield-Line Theory," English translation, Cement and Concrete Assn., London, 1962.
5. F. E. Thomas, "Load Factor Methods of Designing Reinforced Concrete," *Reinf. Conc. Review*, 3, No. 8, 1955, pp. 540, 544.
6. Samuel Chamecki, *Calculo No Regime de Ruptura, Das Lajes de Concreto Armadas em Cruz*, Curitiba, Parana, Brazil, 1948, 107 pages.
7. R. H. Wood, *Plastic and Elastic Design of Slabs and Plates*, Ronald Press, New York, 1961.
8. L. L. Jones, *Ultimate Load Analysis of Reinforced Concrete Structures*, Interscience Publishers, New York, 1962.

9. Richard C. Elstner and Eivind Hognestad, "Shearing Strength of Reinforced Concrete Slabs, Appendix 1," *Jour. ACI, 28*, No. 1, July 1956; *Proc.*, 53, p. 55.

10. W. L. Gamble, M. A. Sozen, and C. P. Siess, "Measured and Theoretical Bending Moments in Reinforced Concrete Floor Slabs," Univ. of Illinois *Civil Engineering Studies Structural Research Series* No. 246, June 1962.

11. Telemaco van Langendonck, *Charneiras Plásticas em Lajes de Edifícios*, Associacão Brasileira de Cimento Portland, São Paulo, Brazil, 1966, 81 pages.

12. Rolf Lenschow and Mete A. Sozen, "A Yield Criterion for Reinforced Concrete Slabs," *Jour. ACI, 64*, No. 5, May 1967, p. 266.

13. R. Taylor, D. R. H. Maher, and B. Hayes, "Effect of the Arrangement of Reinforcement on the Behavior of Reinforced Concrete Slabs," *Magazine of Concrete Research*, 18, No. 55, June 1966, p. 85.

14. Frederick P. Wiesinger, "Design of Flat Plates with Irregular Column Layout," *Jour. ACI, 70*, No. 2, Feb. 1973, p. 117.

15. Frederick P. Wiesinger, "Yield Line Method—Strip Method—Segment Equilibrium Method," *ASCE Preprint 2502*, ASCE Natl. Structural Convention, April, 1975.

PROBLEMS

PROB. 13.1. In Figs. 13.2a and 13.7 calculate the ultimate load by yield-line procedures if:

(a) The positive moment capacity m_c is increased to $+17$ kN $\cdot$ m/m, m_a and m_b remaining unchanged at -18 kN $\cdot$ m/m and -22 kN $\cdot$ m/m respectively.

(b) The negative moment capacity m_a is decreased to -12.5 kN $\cdot$ m/m, m_b and m_c remaining at -22 kN $\cdot$ m/m and $+13$ kN $\cdot$ m/m respectively.

PROB. 13.2. Recalculate the ultimate load for the slab of Sec. 13.4 and Fig. 13.8 if the right support has only a 1.2 m skew, that is, if b' in Fig. 13.8a is moved 0.6 m to the right.

PROB. 13.3. If $m = +12.5$ kN $\cdot$ m/m and $m' = -16$ kN $\cdot$ m/m, find the ultimate load which can be carried by a 6 m square slab continuous on all sides. Consider first without corner effect and then establish effect of corner segments.

PROB. 13.4. Investigate the ultimate load capacity of a 4.5 m by a 6 m slab continuous on all sides and having a yield moment of 13 kN $\cdot$ m/m for positive moment and 20 kN $\cdot$ m/m for negative moment.

PROB. 13.5.   Assume the rectangular slab of Sec. 13.9 had a 0.9 m square hole in the middle of the span (centered each way). Find the ultimate uniform load by yield-line method: (*a*) neglecting corner effect; (*b*) with corner effects.

PROB. 13.6.   Assume the rectangular slab of Sec. 13.9 has no support at the upper 6 m side (Fig. 13.13). Find the ultimate uniform load by yield-line method.

14
STRIP METHOD FOR SLAB DESIGN

14.1 DEVELOPMENT OF STRIP METHOD

A method for *designing* slabs, published by Hillerborg in 1956 and 1959 in Swedish,* has received more attention in recent years because of study and tests at the Building Research Station in England. Wood and Armer[1,2,3] in 1968 reported a critical analysis of the method and their own tests of typical slabs designed by this method. They found (mathematically) that a design made by the strip method and reinforced exactly according to the moments found (without averaging across a band, but with reduction of steel as moment decreased) was an *exact* solution rather than just a lower-bound solution of the problem.† Such a slab did develop the load capacity introduced into the design.

The strip method of design gives the designer wide freedom of choice in the design approach. Hence many different solutions for a given slab design are possible. Obviously not all solutions will be of equal economy. Wood‡ points out that a design using moments approaching those from elastic analysis is an efficient design and to be preferred.

The strip method is simplest for slabs on simple supports, but continuity can be handled on a basis similar to limit design. The suitability of the method to slabs with openings is a strong point in its favor.§ The most

*Now in English; see Reference 6.
†Note that the yield-line method is an *upper* bound procedure.
‡Wood's name will be used in this chapter for easy reference to the joint work of Wood and Armer.
§The author accepts the strip method with some enthusiasm because it formalizes a very approximate method that designers have been using for many years, that of designing by their "feel" for the way the load was most apt to be transferred to the supports.

difficult slabs for this method are slabs supported on columns (Sec. 14.8). For such a case, Hillerborg developed what Crawford[4] calls the advanced strip method, using a rectangular element carrying load in two directions to a support at one corner of the element. Although the mathematical basis for this type of element could not be proved by Wood, a test result for a slab design using this method was satisfactory. Hillerborg states that the advanced strip method leads to a more economical design with a simple reinforcement pattern than does the substitute type of element Wood suggests. The discussion here will stay with the simpler strips that Wood endorsed. (The author leans heavily on work by Wood in this area.)

Wood calls attention[1] to the normal use of A_s varying in accord with the strip moments, in contrast to the complexities inherent in a yield-line analysis when m (or A_s) is not constant. If the reinforcement is cut off where it is not needed, the slab designed by the strip method does not fail by sharply developed yield lines, but by "yield in nearly all directions at failure, . . . somewhat like a plastic hammock." Hence yield-line concepts are only marginally useful with this method. The yield-line method deals with "rigid plate" rotations; the strip method tends to eliminate rigid plate failure at ultimate.

14.2 THEORETICAL BASIS

The equilibrium equation for slabs is

$$\frac{\delta^2 M_x}{\delta x^2} + \frac{\delta^2 M_y}{\delta y^2} - 2\frac{\delta^2 M_{xy}}{\delta x \delta y} = -w$$

where the bending moments, M_x and M_y, and the twist moment M_{xy} follow Timoshenko's notation and w is the load per unit area on the slab. Hillerborg designs the slab to make M_{xy} unnecessary, that is, he assumes $M_{xy} = 0$ and then apportions the load to $\delta^2 M_x/\delta x^2$ and to $\delta M_y/\delta y^2$ as he wishes, usually at a particular spot wholly to one or to the other. This particular apportionment is more of a convenience than a necessity, however.

Loads in a particular area are assigned to particular slab strips, as the next section will illustrate, and continuity of the resulting moments and shears must be carefully maintained. Apparent discontinuity in torque or deflection may be disregarded, but a discontinuity in moment or shear is not permitted. Both elastic and plastic analysis concepts are permissible in evaluating moments on strips, but both Wood and Hillerborg note that under service load the slab behavior is more nearly in the elastic range. Hence elastic concepts are valid and acceptable even though the uses suggested below ignore relative deflections with apparent unconcern.

14.3 SIMPLY SUPPORTED RECTANGULAR SLAB

A rectangular slab is adequately represented by the simplified concept of a grid of strips in the x- and y-directions, that is, 90° apart.

Consider that boundaries, called lines of stress discontinuity (or discontinuity lines), are set up as indicated in Fig. 14.1a, creating an area 1 and two areas 2. These discontinuity lines indicate the designer's decision to carry all the load in areas 2 in the x-direction on x-strips and all load in area 1 in the y-direction on y-strips. The discontinuity lines are *not* yield

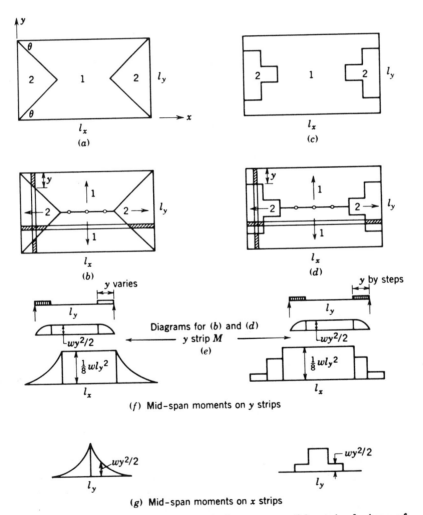

FIGURE 14.1 Layouts and details for two possible strip designs of rectangular slab simply supported.

lines, and the designer is free to choose the angle θ. If the designer chooses 90°, he or she will design a one-way slab; the result will be adequate strength, early cracking along the y-supports, and an overall solution approaching the absurd.

Wood suggests that discontinuity lines might be taken as sketched in Fig. 14.1c to suit bands of reinforcement, since in limit analysis one is not restricted to the use of a single straight line in a quadrant.

As in any flexural member, a load anywhere on a strip produces a shear along the entire strip. However, it is convenient to add a zero shear line on the x-axis as shown in Fig. 14.1b and d and think of the y-strip as carrying all the load above this zero shear line to the upper slab support as indicated by the arrows.

In either layout the central y-strips are simple one-way slab strips under a uniform load or such other distribution of load as may exist. The y-strips running through an area 2 are unloaded in that area and loaded only in the two area 1 end portions, as indicated by the shaded areas. Likewise x-strips in Fig. 14.1b are all unloaded except near the supports, but this is not quite true in Fig. 14.1d. With square discontinuity lines the x-strips near the top and bottom are totally unloaded, and the load pattern changes by substantial steps where the strips enter area 2.

The moment in each strip defines the necessary steel and even indicates where some may be cut off. In Fig. 14.1b the required steel requires a variable spacing; in Fig. 14.1d bands of steel are indicated. For the variable spacing Hillerborg actually substituted a weighted average; Wood shows that the lower bound concept is then not satisfied, although membrane action may bridge the gap.

14.4 CONTINUITY IN RECTANGULAR SLABS

Since slab design by the strip method is a form of limit design, the ratio of negative to positive moment on a strip is not rigidly fixed. The use of moments approximating elastic analysis assures a satisfactory service load response, but this is not mandatory for strength. Wood suggests the assumption of a point of inflection (PI) for the middle y-strips in Fig. 14.2a at about 0.2 ℓ_y from the support, which is close to the elastic condition. Nearly the same result would be obtained by using the negative moment as $-(1/12)w\ell_y^2$; but for skewed slabs Hillerborg and Wood use the initial selection of the PI locations as the key to the desirable slab strips (Sec. 14.7).

For y strips that pass through area 2 (Fig. 14.1b) and thus are loaded symmetrically over the y lengths only on their ends, the best assumption for the negative moment or PI is not so clear. Wood, following Hillerborg, breaks the PI trace and takes it diagonally to the corner, as shown at the top of Fig.

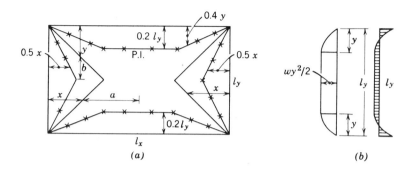

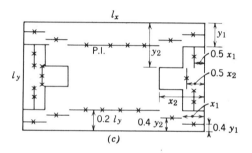

FIGURE 14.2 Points of inflection for continuity. (*a*) Typical layout. (*b*) *M* on *y* strip, simple span and continuous span. (*c*) Proposed points of inflection.

14.2*a*, making the distance from support to PI 0.4 *y*. If one considers the simple span moment on these strips, as in Fig. 14.2*b*, the maximum simple beam moment is $wy^2/2$, with this moment constant across area 1. The use of the diagonal PI line leads to negative moments here that are substantially less (down 50% or more*) compared to the elastic analysis values for fixed ends, but this is permissible with limit design concepts.

For the *x*-strips a similar treatment of PI lines at 0.4*x* from the support establishes negative moments relatively *much* less than elastic analysis

---

*The values vary strip to strip and this is near that for the middle strip of the group. The elastic analysis for fixed ends shows that the PI distance from support becomes $y(1 - 0.82\sqrt{y/\ell_y})$ and the negative moment is $M_s(0.67 + b/y)/(1 + b/y)$, where M_s is the maximum simple span moment, *y* is the distance to the discontinuity line, and *b* is the remaining distance to midspan, that is, $b = (\ell_y/2) - y$. In these terms the same expressions can be applied to the *x*-strips, with distances *x*, *a*, and ℓ_x replacing *y*, *b*, and ℓ_y.

values (in some cases down by 70%). Although the matter deserves further study in terms of the implied degree of hinging and relative costs, the author (in the absence of specifications by Wood) recommends the use of the PI line at $0.5x$ as possibly more realistic, Fig. 14.2c.

14.5 DESIGN OF RECTANGULAR SLAB

A 4.5 m by 6 m rectangular slab with restrained edges at beams is designed below for a factored load w of 14 kPa. The design moments are computed without the ϕ factor included. Once the slab thickness is determined for moment or deflection (shear rarely controls a slab on beams), the required A_s calculated from $M/(\phi f_y z)$ takes the same distribution as the moments plotted in Fig. 14.3.

Discontinuity lines are arbitrarily chosen as shown in Fig. 14.3a and PI locations are selected as discussed in the section above.

Moments will be calculated first for each y-strip. The loadings are sketched to the right in Fig. 14.3b. Here the moments will be calculated as for a simple span between PI points and as a cantilever carrying these reactions plus its own uniform load. The moments on the x-strips are sketched in Fig. 14.3c and on y-strips in b.

Strip 1-1   No loading

Strip 2-2   Pos. $M = +\frac{1}{2} \times 14 \times 0.45^2 = +1.418$ kN · m/m
Neg. $M = -14 \times 0.45 \times 0.45 - \frac{1}{2} \times 14 \times 0.45^2 = -4.253$

Strip 3-3   Pos. $M = +\frac{1}{2} \times 14 \times 1.05^2 = +7.718$
Neg. $M = -14 \times 1.05 \times 1.05 - \frac{1}{2} \times 14 \times 1.05^2 = -23.153$

Strip 4-4   Pos. $M = +\frac{1}{2} \times 14 \times 0.54^2 = +2.041$
Neg. $M = -14 \times 0.54 \times 0.36 - \frac{1}{2} \times 14 \times 0.36^2 = -3.629$

Strip 5-5   Pos. $M = +\frac{1}{2} \times 14 \times 1.08^2 = +8.165$
Neg. $M = -14 \times 1.08 \times 0.72 - \frac{1}{2} \times 14 \times 0.72^2 = -14.515$

Strip 6-6   Pos. $M = +\frac{1}{8} \times 14 \times 2.7^2 = +12.758$
Neg. $M = -14 \times 1.35 \times 0.9 - \frac{1}{2} \times 14 \times 0.9^2 = -22.680$ kN · m/m

These moments are summarized in Fig. 14.3d and e into diagrams of negative moment crossing the face of support and positive moment crossing midspan, which determine the amount and distribution of reinforcing steel. If this arrangement appears awkward or seems to have excessive local demands, the designer can modify the discontinuity lines in such a way as to shift the requirement in the desired direction. The several moment diagrams sketched in Fig. 14.3b and c are useful for establishing bar cutoff points.

The minimum temperature steel may well control where computed moments are small, certainly in strip 1-1 close to the long edge beams.

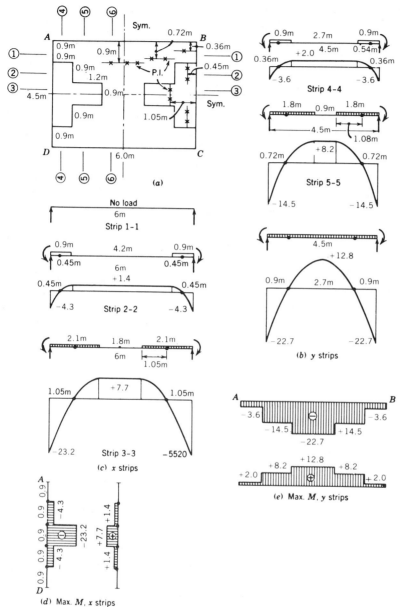

FIGURE 14.3 Design moments for a rectangular slab.

The loadings on the beams supporting the slabs are most logically established from the strip reactions.

14.6 DESIGN OF A RECTANGULAR SLAB WITH HOLE

The symmetrical slab of Fig. 14.4a, restrained at all edge beams illustrates the rerouting of loads when strips are interrupted. Since slabs are normally considerably under-reinforced, it is possible to use certain strips near the opening (Fig. 14.4b) as small beams simply by increasing the local reinforcement. If the opening is so large that even extra slab steel is inadequate to care for the moment, a real beam is needed around one or more sides of the opening, quite probably spanning to the edge beams.

The assumed "beam strips" are shown dotted around the opening and the assumed discontinuity lines and PI lines are added in Fig. 14.4c. Only the PI line between the long side of the opening and the long support beam requires special discussion. This 1.2 m long strip in the y-direction could be designed as a cantilever from DC, but it appears that this might be ignoring excessive deflection at the opening. Consider instead a discontinuity line at 0.9 m from the support, which will terminate the cantilever and leave a 0.3 m strip 1-1 alongside the opening in the x direction. Strip 2-2 in the y-direction will pick up the reaction from strip 1-1. Assume this strip 2-2 to be 0.6 m wide, since strips 1-1 alone would add 50% to the strip 2-2 load if only a 0.3 m strip were used, and additional loads come from strips 6-6 and 7-7. Strip 7-7, because of the opening, is terminated on strip 2-2; it seems logical to treat strip 6-6 in the same manner.

The design moments will be calculated for a total design (factored) load of 14 kPa. The strip loadings and moment diagrams are shown in Fig. 14.4d, along with a final sketch that assembles the maximum negative and positive design moments on both x- and y-strips. The supporting calculations are as follows.

Cantilever strips near strip 1-1, 0.9 m length, 1.8 m wide
  Neg. $M = -\frac{1}{2} \times 14 \times 0.9^2 = -5.67$ kN $\cdot$ m/m

Strip 1-1 with simple span 2.4 m to center of strip 2-2, 0.3 m wide
  Pos. $M = +14 \times 0.9 \times 1.2 - \frac{1}{2} \times 14 \times 0.9^2 = +9.45$ kN $\cdot$ m/m
  Reaction on strip 2-2 $= 14 \times 0.9 = 12.60$ kN/m

Strip 6-6 spanning 2.4 m to centre of strip 2-2, 0.6 m wide
  Reaction* on strip 2-2 $= 14 \times 0.3 \times 0.15/2.1 = 0.30$ kN/m
  Pos. $M < +0.30 \times 1.95 = +0.59$ kN $\cdot$ m/m
  Neg. $M = +0.30 \times 2.4 - 14 \times 0.6 \times 0.30 = -1.80$ kN $\cdot$ m/m

---

*From ΣM about PI using "simple span" as freebody.

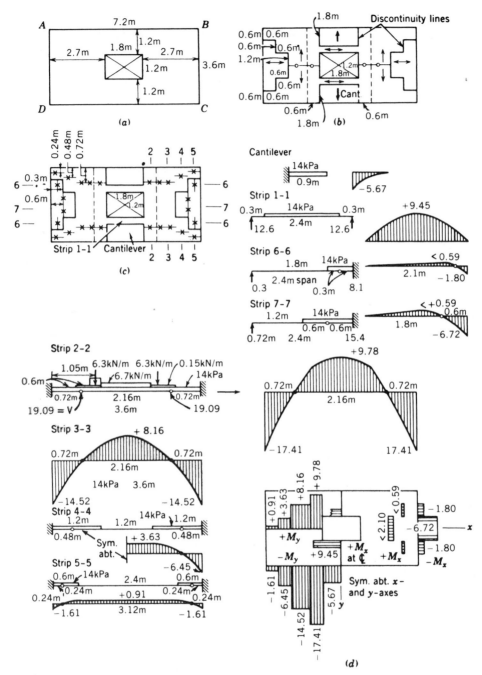

FIGURE 14.4 Design based on strip method. (*a*) Layout. (*b*) Discontinuity lines. (*c*) Inflection point lines. (*d*) Design moments.

378

Strip 7-7 spanning 2.4 m to center of strip 2-2, 1.2 m wide
   Reaction on strip 2-2 $= 14 \times 0.6 \times 0.3/1.8 = 1.40$ kN/m
   Shear at point of inflection $= 14 \times 1.2 - 4.67 = 12.13$ kN/m
   Pos. $M < +1.40 \times 1.5 = +2.10$ kN $\cdot$ m/m
   Neg. $M = +1.4 \times 2.4 - 14 \times 1.2 \times 0.6 = -6.72$ kN $\cdot$ m/m

Strip 2-2 spanning 3.6 m, 0.6 m wide
   The reactions from 1-1, 6-6, and 7-7 are distributed over the two strips,
   each 0.6 m wide, total width 1.2 m
   Total load on each strip 2-2:
   $3.6 \times 14 + 0.6 \times 12.60/1.2 + 1.2 \times 1.40/1.20 + 1.2 \times 0.30/1.20 = 58.40$ kN/m
   Shear at point of inflection: $29.20 - 0.72 \times 14 - 0.12 \times 0.30/1.2 = 19.09$ kN/m

Pos. $M = 19.09 \times \frac{1}{2} \times 2.16 - \frac{1}{2} \times 14 \times 1.08^2 - \dfrac{12.6}{1.2} \times 0.3(0.6 + 0.15)$

$\qquad - \dfrac{0.30}{1.2}(0.6 - 0.12) \times (0.6 + 0.24) - \dfrac{1.4}{1.2} \times 0.6 \times 0.3 = +9.78$ kN $\cdot$ m/m

Neg. $M = -19.09 \times 0.72 - \frac{1}{2} \times 14 \times 0.72^2 - \dfrac{0.30}{1.2}$

$\qquad\qquad \times 0.12 \quad (0.60 + 0.06) = -17.41$ kN $\cdot$ m/m

Strip 3-3 spanning 3.6 m, 0.9 m wide
   Pos. $M = +\frac{1}{8} \times 14 \times 2.16^2 = +8.16$ kN $\cdot$ m/m
   Neg. $M = -14 \times 1.08 \times 0.72 - \frac{1}{2} \times 14 \times 0.72^2 = -14.52$ kN $\cdot$ m/m

Strip 4-4 spanning 3.6 m, 0.6 m wide
   Shear at point of inflection $= 14 (1.2 - 0.48) = 10.08$ kN/m
   Pos. $M = +10.08 \times 0.72 - \frac{1}{2} \times 14 \times 0.72^2 = +3.63$ kN $\cdot$ m/m
   Neg. $M = -10.08 \times 0.48 - \frac{1}{2} \times 14 \times 0.48^2 = -6.45$ kN $\cdot$ m/m

Strip 5-5 spanning 3.6 m, 0.6 m wide
   Shear at point of inflection $= 14 \times (0.6 - 0.24) = 5.04$ kN/m
   Pos. $M = +5.04 \times 0.36 - \frac{1}{2} \times 14 \times 0.36^2 = +0.91$ kN $\cdot$ m/m
   Neg. $M = -5.04 \times 0.24 - \frac{1}{2} \times 14 \times 0.24^2 = -1.61$ kN $\cdot$ m/m

The maximum moments assembled in Fig. 14.4d look reasonable in distribution and no modification of the strip arrangement is indicated as desirable. The maximum moments fix the slab depth and shear may be checked from the strip loads if that appears possibly to control. The reinforcement should be arranged in bands corresponding to the strips used and this calculation is simple. The short y-strip steel should be placed at the greater depth and for the x-strips this d must be reduced by one bar diameter. Minimum temperature or spacing steel will control where the indicated moments are very small. The moment diagrams for the strips give the basic data for required length of bars, subject to the usual Code requirements for arbitrary extensions beyond the theoretical cutoff points.

Deflection *at service load* must be considered in checking serviceability. In any actual design the service load is available, and it should be on the safe side to use the strip service load moments with EI based on the cracked section. The more detailed calculations of Sec. 8.5 may be used to interpolate between cracked and uncracked sections. Although long, that method considers the usual intermediate slab condition and is less severe than the use of the cracked section. In fixing M_a the designer must be clear as to whether this is calculated per strip width or per unit width; then must evaluate the accuracy to be expected as a result of uncertainty as to what is the best M_a to use.

14.7 NONRECTANGULAR SLABS

The strip method is not limited to rectangular strips supported on beams on all four sides. A few layouts, largely those already in the literature, should indicate the possibilities.

A free edge, as in Fig. 14.5a, can be considered as a strong strip with extra reinforcement and can even pick up a limited amount of load from cross strips.

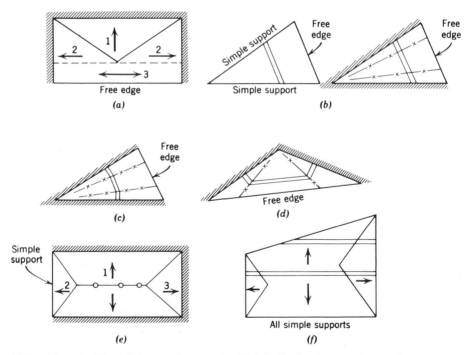

FIGURE 14.5 Varied types of supports. Hatched edges are restrained.

Where triangular slabs exist with a free edge opposite an acute angle, strips can span the short way between supported sides as in Fig. 14.5b. The same can be done if the supported edges are restrained, but Hillerborg suggests strips that change direction, as in Fig. 14.5c. With an assumed PI line, the mid-length is like a simple span, and the negative moment region acts like a cantilever to pick up the simple span reaction plus its own load. The obvious torsion implications are not a concern in this method. However, the cantilever and mid-strip must have different widths to care for the geometry of the layout. A similar idea shows in Fig. 14.5 d.

When fixed and simply supported edges occur nearby, elastic concepts are helpful. In Fig. 14.5e, for example, area 2 should be smaller than area 3.

14.8 LAYOUT AROUND COLUMNS SUPPORTING THE SLAB

The treatment of flat plates around the columns makes a difficult problem when one uses the strip method. Hillerborg's 1959 publication dealt particularly with this problem and resulted in a recommended type-3 rectangular element carrying load in two directions from zero shear lines and delivering the reaction to one corner (the column). This is obviously a concept much more involved than that of the strip. Wood was not able to prove the complete mathematical basis for this element, but he reported the resulting reinforcement looked reasonable. Armer reported[2] a test that was quite satisfactory.

Wood[1] developed a substitute prodedure using the type-3 rectangular element with computer calculated tables leading to the moment coefficients. Armer designed one test specimen that substituted for the type-3 elements strong slab strips in the x- and y-directions to carry load from slab strips directly to the column. Test results on all of these variations proved satisfactory. Wood has also suggested a wide "beam band" over the columns that could be strengthened at the column by local (short) strong bands across the column, these apparently acting in flexure much like steel shearhead reinforcement in flat plates.

Reference 5 is an interesting approach to flat plates supported on columns at irregular spacings.

Reference 6 is an excellent overall coverage of the strip method.

SELECTED REFERENCES

1. R. H. Wood and G. S. T. Armer, "The Theory of the Strip Method for Design of Slabs," *Proc. Institution of Civil Engineers*, 1968, Vol. 41 (October), p. 287.

2. G. S. T. Armer, "Ultimate Load Tests of Slabs Designed by the Strip Method," *Proc. Institution of Civil Engineers*, 1968, Vol. 41 (October), p. 313.

3. G. S. T. Armer, "The Strip Method: A New approach to the Design of Slabs," *Concrete*, Vol. 2, No. 9, Sept. 1968, p. 358.

4. Robt. L. Crawford, "Limit Design of Reinforced Concrete Slabs," *Proc. ASCE, Jour. Eng. Mech. Div.*, Oct. 1964, EM5, p. 321.

5. Frederick R. Wiesinger, "Design of Flat Plates with Irregular Column Layout," *Jour. ACI*, Proc. 70, Feb. 1973, p. 117.

6. Arne Hillerborg, "Strip Method of Design," Viewpoint Publications, Cement and Concrete Association, Wexham Springs, Slough, SL3 6PL, England, 1975.

15
FLAT PLATES AND
FLAT SLABS

15.1 SCOPE OF CHAPTER

Earlier chapters have dealt with slabs primarily supported on beams, in most cases where slab and beam design would be separate steps. This chapter deals primarily with slabs that are totally supported directly on columns. Waffle slabs and slab band construction are included only for identification purposes.

This chapter gives design examples of flat plates, both interior and exterior panels, and an interior panel of a flat slab with drop panel and column capital. These designs are in accord with the slab design methods of Code Chap. 13, but they appear more simplified than those of the next chapter because the combined beam-slab interaction is minimized in the absence of beams. Chapter 16 relates to the general and more involved case of beam and slab interaction. The sequence in this chapter is as follows:

15.2 FLAT PLATES, FLAT SLABS, AND RELATED TYPES—IDENTIFICATION

A flat plate is a concrete slab reinforced in two directions so that it brings its loads directly to supporting columns. Most generally it is used for light loadings and the spandrel beams shown in Fig. 15.1 are omitted. Beams may be used around stairs or other large openings in the slab.

The flat slab is an older type developed primarily for heavy loadings and longer spans (Fig. 15.2) typically using the flared column capital and often the thickened slab around the column, called a drop panel. The waffle slab (a perpendicular joist-like system in effect) shown in Fig. 15.3 may be designed as a flat slab suitable for long spans.

For heavy service live loads, that is, over 5 kPa, flat slabs have long been recognized as the most economical construction.

The slab band type of floor thickens the slab into bands of greater depth, as in Fig. 15.4. Although the most usual pattern results in one-way transverse slabs of variable depth supported on wide shallow beams or bands in the longitudinal direction, some engineers have used such thickened sections in end panels much like a drop panel around a flat slab column. Such construction partakes somewhat of the nature of flat slabs.

FIGURE 15.1 Flat plate with shallow spandrel beam. The spandrel is more often omitted to leave a totally flat ceiling. (Courtesy Portland Cement Assn.)

FIGURE15.2 Typical flat slab construction, with drop panel. (Courtesy Portland Cement Assn.)

Two-way slabs supported on all sides by wide shallow beams also involve some flat slab action. When the slab proper occupies only the middle half of the panel, the Code recognized it as a paneled ceiling form of flat slab.

Flat plates, being thin members, are not economical of steel, but they are economical in their formwork. Since formwork represents over half the cost of reinforced concrete, economy of formwork often means over-all economy. Since the 1950s flat plate floors have proved economical in tall apartment house construction. Reduced story height resulting from the thin floor, the smooth ceiling, and the possibility of slightly shifting column locations to fit the room arrangement all seem to be factors in the over-all economy.

FIGURE 15.3 Waffle slab. (Courtesy Portland Cement Assn.)

FIGURE 15.4   Slab band construction. (Courtesy Portland Cement Assn.)

15.3 HISTORICAL DEVELOPMENT

(a) Early Slabs as Empirical Construction

Flat slabs were built and sold many years before an adequate analysis was available.[3] The originator was an American, C. A. P. Turner. He built from intuition as to how slabs would function and applied a proof load to the completed slab to satisfy the owner.

Numerous flat slab structures were load-tested during the period from 1910 to 1920.* These slabs performed well under test loads. The difficulty in closely correlating test results with the moments that statics shows must be present is mentioned in (c).

(b) The Statics of a Flat Slab

In 1914 J. R. Nichols[4] showed that statics required a total of positive and negative moments equal to

$$M_0 = \frac{1}{8}W\ell\left(1 - \frac{2}{3}\frac{c}{\ell}\right)^2$$

where W is the total uniform panel load, ℓ is the span, and c is the

---

*In 1947 Professor J. Neils Thompson and the author ran such a test on a flat slab constructed about 1912. Although it cracked badly, its strength was surprising.

diameter of the column capital. This conclusion follows from an analysis of a half panel.

Consider the interior panel of Fig. 15.5a, loaded with a unit load w and surrounded with similar panels equally loaded. The straight boundaries of the slab are all lines of symmetry, indicating that they are free from shear and torsion. Hence all the shear and torsion must be carried around the curved corner sections which follow the column capital.

Likewise, if the slab is subdivided along the middle of the panel, this line is a line of zero shear and torsion. Thus the free body of Fig. 15.5b,c is subject to a downward load W_1 acting at the centroid of the loaded area, an equal upward shear W_1 acting on the curved quadrants, the total

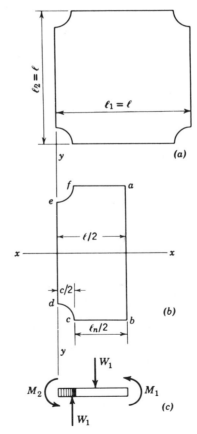

FIGURE   15.5 Equilibrium conditions indicate $M_1 + M_2$ is established by statics.

positive moment M_1 acting on the middle section ab, and the total negative moment M_2 acting about the y-axis on section $cdef$. This assumes the bending moment around the column capital is uniformly distributed, which means no torsional moments exist there. Moments also exist about the x-axis on efa and dcb, but these do not enter into $\Sigma M_y = 0$.

$$W_1 = w(\ell^2/2 - \pi c^2/8) = (\ell^2 - \pi c^2/4)w/2$$

The moment of this load about axis y-y is:

$$\frac{w\ell^2}{2} \times \frac{\ell}{4} - \frac{\pi c^2 w}{8} \times \frac{2c}{3\pi} = \frac{w\ell^3}{8} - \frac{wc^3}{12}$$

If the upward shear W_1 is considered uniformly distributed around the quadrants cd and ef, the resultant acts at a distance c/π from the y-axis. Equilibrium of moments about the y-axis then gives:

$$-M_1 - M_2 + \frac{w\ell^3}{8} - \frac{wc^3}{12} - \frac{w}{2}\left(\ell^2 - \frac{\pi c^2}{4}\right)\frac{c}{\pi} = 0$$

$$M_1 + M_2 = M_0 = \frac{w\ell^3}{8} - \frac{wc^3}{12} - \frac{wc\ell^2}{2\pi} + \frac{wc^3}{8} = \frac{w\ell^3}{8}\left(1 + \frac{c^3}{3\ell^3} - \frac{4c}{\pi\ell}\right)$$

$$M_0 = \frac{w\ell^3}{8}\left(1 - \frac{2}{3}\frac{c}{\ell}\right)^2$$

The value of $[1 - \frac{2}{3}(c/\ell)]^2$ approximates the longer parenthesis reasonably well.*

This statics solution tells nothing about how this total moment is distributed between positive or negative moment or how either varies along the slab width. The solution also neglects possible torsional moments around the column capital that will act if the tangential bending moments there are not uniform in their distribution.

When flat slab panels are not square, the long span produces the larger moment. The student should contrast this with the two-way slab supported on all four sides where the short slab strips carry the larger load and larger moment. In such a slab the long slab strips carry less load and less moment, but it should be noted that the long beams have to carry heavy slab reactions and large moments. Two-way slab moments in both directions are reduced because the beams help to carry the moment. In the flat slab, in contrast, the *full* load must be carried in *both* directions by the slab alone.

---

*The error is 0.5% low for $c/\ell = 0.1$, 0.5% high for $c/\ell = 0.2$, 1.3% high for $c/\ell = 0.25$, and 2.0% high for $c/\ell = 0.30$.

(c) Correlation of Statics with Tests

In early tests (mechanical) strain measurements were necessarily over considerable lengths and indicated average rather than maximum steel stresses. This led to an overly optimistic evaluation of the reserve strength, even when attempts were made to allow for this problem. Also tests were not to failure and thus reflected less tension cracking than would develop at ultimate. Thus early test evaluations led to a total M_0 of $0.09\, w\ell_1^2\ell_2\,[1-(2/3)(c/\ell_1)]^2$, which was accepted until the change in the 1971 Code, in spite of the apparent conflict with statics.

The behavior of large areas of this type of construction has proven that the static analysis fails to give the total picture. There is a horizontal thrust, a form of arching that, under favorable conditions, greatly strengthens interior panels and even adds some strength to edge panels. Tests in the 1960s have also shown this reserve strength, to such an extent that the Code now uses $M_0 = w\ell_2\ell_n^2/8$, which bases the moment on clear span ℓ_n between square supports*. This looks in Fig. 15.8a like true statics and it is an adaptation (reduction) of Nichol's value. It is admittedly still short of the full static moment, since all of the reaction is not on the near face of the column, but this M_0 is larger than in earlier codes.

15.4 THEORETICAL DISTRIBUTION OF MOMENT

Westergaard developed a theoretical slab analysis that accompanied Slater's test analyses.[1] He established the distribution of positive and negative moments which would exist throughout a flat slab when it was considered as an isotropic plate. His work, roughly checked by Slater's test analyses, formed the basis for the specified subdivision of the total M_0 into positive and negative moments and for the further subdivision of these into moments on the two design strips set up in the specifications. Figure 15.6a shows his calculated distribution of negative moments on strips crossing the column center line for $c/\ell = 0.15$ and 0.25. Around the column capitals the moment in a radial direction is considered constant at $-0.223\, w\ell^2$ for $c/\ell = 0.15$ and at $-0.143\, w\ell^2$ for $c/\ell = 0.25$. These values are nearly four times the maximums shown between columns. The positive moments on strips crossing the middle of the span are shown in Fig. 15.6b. For a strip running along the line between column centers, the moment diagram is shown in Fig. 15.7a and for a strip running along the middle of the panel in Fig. 15.7b. It will be noted that both these strips have a form of moment diagram very similar to that of a one-way slab, with positive moment in the middle of the span and negative moments at the ends.

*If supports are not square or rectangular replace them (for this value of ℓ_n) with square columns of equal area.

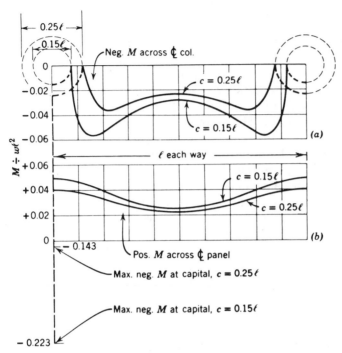

FIGURE 15.6 Theoretical moments on vertical strips (*a*) crossing column centerlines and (*b*) crossing midspan of panel (Poisson's ratio zero). (Adapted from Reference 1, ACI.)

As a result of extensive analytical work and quarter scale tests on multipanel slabs[4,5,6] at the University of Illinois, and a three-quarter scale test by the Portland Cement Association,[10] the distribution of moments actually used since the 1971 Code (Fig. 15.8*b*) has been slightly modified from the 1963 Code values. The 1971 values now apply to cases of slabs with monolithic cast beams (if framing directly into columns) as well as to flat slabs. An interesting study and evaluation of the Code procedures is given in Reference 12.

15.5 CODE DIRECT DESIGN RULES—BASIC REQUIREMENTS FOR FLEXURE

When layouts are relatively simple, a direct design procedure is given in Code 13.6, limited to slabs meeting these limitations:

1. A minimum of three spans each way, directly supported on columns.

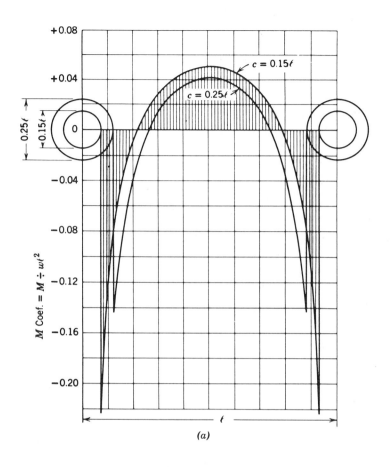

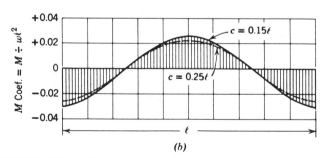

FIGURE 15.7 Theoretical bending moments (Poisson's ratio zero). (Adapted from Reference 1, ACI.) (*a*) Moment diagram for strip along center line of columns. (*b*) Moment diagram for strip along middle of panel.

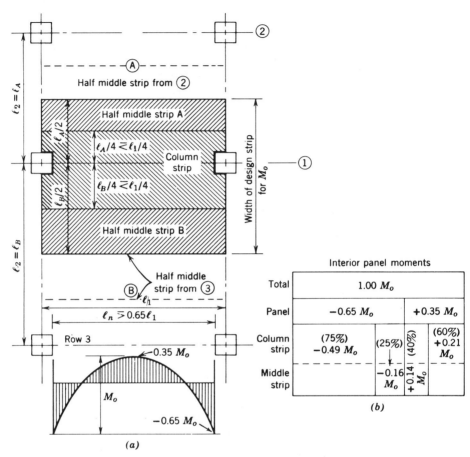

FIGURE 15.8 Calculation of M_0 and its subdivision, for an interior panel (without beams) in x-direction; similay in y-direction except for half column strip at exterior.

2. Rectangular panels with the long span not more than twice the short span.
3. Successive spans not differing by more than 1/3 of the longer span.
4. Live load not more than 3 times dead load.

Columns may be offset from either axis by up to 10% of the span in the direction of the offset.

The ACI Code Commentary has many helpful sketches, tables, and discussions on two-way slab systems.

The minimum thickness rules of Code 9.5.3.1 appear very complex but, for the case of no beams, the interior panels are always governed by the minimum of 120 mm when without drops, 100 mm with drops, or by Code Eq. 9-10 and 9.12 which then become the same* and specify the thickness as:

$$h = \ell_n(800 + f_y/1.5)/36\,000 = \ell_n(0.8 + f_y/1500)/36$$

This equation reduces to $h = \ell_n/36$ for Grade 300 bars and $h = \ell_n/33.75$ for Grade 400 bars, where ℓ_n is the clear span.

For exterior and corner panels *without* edge beams the thickness must be increased 10% above those just stated (Code 9.5.3.3).[†]

Only bending moments for the interior panels will be discussed first, to show the basic procedures without the slight complications arising from evaluation of the effective restraints at the exterior columns (discussed in Sec. 15.9).

The basic moment calculation, $M_0 = (\frac{1}{8})\,w\ell_2\ell_n^2$, is for a panel centered on a column line, that is, for a half panel from panel A and a half panel from panel B in Fig. 15.8a. This moment calculation, centering on the column lines in each direction, uses moments based on the clear span ℓ_n and the average width of $0.5(\ell_{2A} + \ell_{2B})$. For interior panels (with or without beams) $-0.65\,M_0$ is considered negative moment and $+0.35\,M_0$ is positive moment, as in Fig. 15.8a.

The moments, as already noted earlier, are larger close to the columns than elsewhere. While they vary greatly, it is safe to average them over a considerable width for a design. For this purpose the Code defines a column strip and a middle strip.

A column strip is a width (Fig. 15.8a) of $\ell_{2A}/4$ plus $\ell_{2B}/4$ (but not more than $\ell_1/2$ total) adjacent to the centerline of the columns. The remainder of the area used in Fig. 15.8 makes up two half middle strips, one for panel A and one for panel B. In design the two adjacent half strips at the column are treated as one design strip and the two adjacent half middle strips as the other design strip. In uniform panels each strip is $\ell_2/2$ in width.

Without beams (which means $\alpha_1 = 0$ in the Code notation), the column strip negative moment is 75% of the $-0.65\,M_0$, that is, $-0.49\,M_0$, the other $-0.16\,M_0$ going to the two half middle strips, as indicated in Fig. 15.8b. Likewise the column strip is assigned 60% of the $+0.35\,M_0$ moment, or $+0.21\,M_0$, and the $+0.14\,M_0$ remaining goes to the two half middle strips. For uniform panels, the two half middle strips are alike, giving $-0.16\,M_0$ and $+0.14\,M_0$ for the total middle strip moments.

The designer is permitted to modify any design moment by 10%, if all of M_0 is still assigned.

The system for exterior panel moments is quite similar, except that less

---

*$\alpha_m = 0$ when no beams; $\beta_s = 1$ for interior panel. See Sec. 15.11 for the general discussion of the h equations.
[†]As used in Sec. 15.13a.

negative moment goes to the exterior, and more to the interior column line; the positive moment is correspondingly increased (Sec. 15.12), much as for typical exterior beam panels. The detailed discussion of this operation is delayed because the exterior slab restraint involves not only the exterior column but also torsion in the edge beams or, in their absence, in the slab strip. The details of this now would tend to cloud the simple structure of the direct design method.

15.6 CODE SHEAR REQUIREMENTS AND BAR DEVELOPMENT

(a) Shear

Although shear requirements were adequately discussed in Chapter 6, shear is introduced here again for emphasis. Although the yield line theory shows that flexural capacity can be obtained by several arrangements of reinforcing, no such freedom exists in caring for shear around columns in the case of flat plates and flat slabs. In either, shear is *the critical element* of design. Nearly any test slab, if loaded to collapse, will show a final shear failure or, around an exterior column, a combined shear and torsion failure.

On interior columns it is possible to use crossed steel shearheads to strengthen the slab (up to 75% more v_u) in shear (Code 11.11.4); but a shearhead modification of an exterior column to the same specification *does not function* and is wasted. Around interior columns it is also possible to use closed stirrups effectively (up to 50% more v_u but with v_c decreased to $\frac{1}{6}\sqrt{f_c'}$), if the bars or stirrups are carefully developed.

Around exterior columns, especially when spandrel beams (edge beams) are omitted or when the slab extends only to the outside of the column (less than all the way *around* the column), careful design for shear *and torsion* is absolutely essential. An overhang of the slab provides a great improvement. Many flat plate structures have been built without extending the slab beyond the column, but the author is still prejudiced against this construction detail. A very conservative approach is fully warranted. Attention is specifically directed to Code 13.6.3.5.

> Edge beams or the edges of slabs shall be proportioned to resist in torsion their share of the exterior negative factored moments.

The provisions of Code 13.3.4.2 about the transfer of part of the moment by an eccentricity of shear are also in the right direction. The designer must be alert to the fact that the shear transfer to the exterior column is (nearly always) the weakest point in the slab.

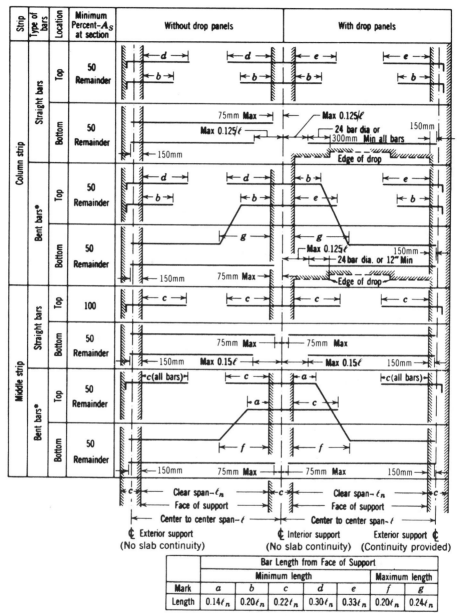

FIGURE 15.9 Minimum length of slab reinforcement, slabs without beams (From ACI Code, Chap. 13). Note: Not adequate for lateral loads in unbraced frames. (See Code 12.12.1 regarding extending reinforcing into supports.) *Bent bars at exterior supports may be used if a general analysis is made.

(b) Development of Reinforcement in *Braced* Frames

The Code includes Fig. 15.9 which indicates the requirements for extend-ing some of the positive moment bars into the support area, to within 75 mm of the centerline of supports. The left half of this figure is for slabs of uniform thickness; the right half is for slabs thickened around the column by a drop panel (Fig. 15.2 and Sec. 15.14). Of course, bars must be of such a size that they can be developed within the lengths shown (or else be extended farther), but this is not commonly a serious problem.

Note that Fig. 15.9 is not adequate in an unbraced frame.

15.7 DESIGN OF FLAT PLATE—INTERIOR PANEL

Design by the direct method a 5.5 m by 6 m flat plate interior panel for a service dead load, including its own weight, of 5 kPa* and a live load of 3 kPa. Assume 375 mm square columns, $f_c' = 30$ MPa, Grade 400 steel, 20 mm cover because not exposed to the weather.

Solution

Normally the slab thickness used would be determined by the thickness of an exterior panel. However, since a transition in depth can often be accommodated, the design will be developed here as if this panel stood alone in the design.

Depth probably will be determined by shear around the column or by the thickness needed against deflection. The minimum thickness (Code 9.5.3.1) is 120 mm or

$$h = \ell_n(800 + f_y/1.5)/36\,000 = \ell_n(0.8 + f_y/1500)/36 = \ell_n/33.75$$
$$= (6000 - 375)/33.75 = 168 \text{ mm, say, } 175 \text{ mm}$$

$$\text{Dead load} = 1.4 \times 5 = 7.0$$
$$\text{Live load} = 1.7 \times 3 = \underline{5.1}$$
$$w = 12.1 \text{ kPa}$$

Shear must be checked on a section enclosing the column at $d/2$ from the column face. For shear, the average depth of the two steel layers seems logical and safe, say, $d = 175 - 20 - 0.50$ bar $= 147$ mm for a #15 bar. The width of each side of this section thus becomes $375 + 147 = 522$ mm.

$$V_u = 12.1(6.0 \times 5.5 - 0.522^2) = 396 \text{ kN}$$
$$V_n \geq V_u/\phi = 396/0.85 = 466 \text{ kN}$$
$$\text{Allowable } v_c = \tfrac{1}{3}\sqrt{f_c'} = \tfrac{1}{3}\sqrt{30} = 1.83 \text{ MPa}$$
$$d_v = V_n/v_c b = 466 \times 10^{-3}/1.83 \times 4 \times 0.522 = 0.122 \text{ m} < h = 175 \text{ mm above}$$
$$\text{USE } h = 175 \text{ mm}$$

---

*This should include floor finish, partition allowance, and the like.

For moment the steel in the long direction will be nearest the top or bottom faces of the slab, that for the short span next inside, making:

$$\text{Trial } d_l = 175 - 20 - d_b/2 = 147 \text{ mm for long span}$$
$$d_s = 175 - 20 - 1.5\, d_b = 132 \text{ mm for short span}$$

both assuming #15 bars. Note that in a square panel the smaller d would control, *not* the average d, for steel in *both* directions, unless supervision were close enough to justify different steel in the two directions. The use of average d for A_s calculations (as in slabs of Chapter 12) is *not* safe here. The small bar size was selected here because the depth for deflection seems to be governing over stress criteria and this points to a lightly reinforced slab.

$$M_{ol} \text{ (long span)} = 0.125 \; w\ell_2\ell_n^2$$
$$= 0.125 \times 12.1 \times 5.5 \; (6.0 - 0.375)^2/1000$$
$$= 0.263 \text{ MN} \cdot \text{m}$$
$$M_{os} \text{ (short span)} = 0.125 \times 12.1 \times 6.0 \; (5.5 - 0.375)^2/1000$$
$$= 0.238 \text{ MN} \cdot \text{m}$$

Next check depth for moment by comparing $k_n = M_n/bd^2$ to the allowable of 7.89 MPa from Table 3.1. The highest moment is the long span negative moment in the column strip (over the column), Fig. 15.8b.

$$-M_u = -0.75(0.65\, M_o) = -0.75 \times 0.65 \times 0.263 = -0.128 \text{ MN} \cdot \text{m}$$
$$-M_n = M_u/0.9 = -0.143 \text{ MN} \cdot \text{m*}$$

The strip width is half the transverse panel length, 2.75 m.

$$\text{Reqd. } k_n = M_n/bd^2 = 0.143/2.75 \times 0.147^2 = 2.40 \ll 7.89 \text{ MPa}$$

This indicates such a greatly underreinforced slab that a check in the short direction is not needed.

A tabular form, as in Table 15.1, expedites this type design and organizes the results in a manner easily available to the detailer. The author finds the use of nominal M_n convenient since it eliminates ϕ from all subsequent calculations, but this is optional. The lever arm z of the internal couple will be estimated as 0.95 d, a little high in the usual range of 0.90 d to 0.95 d because the slab is much thicker than moment requires. This value typically need not be modified for the various moments; but z could be checked where any close decision is to be made in fixing the bar spacing. Here, since the student probably has little "feel" for the proper z, a check will be made at the worst section, that just used in checking k_n.

$$A_s = M_n/f_y z = 0.143/400 \times 0.95 \times 0.147 = 2.56 \times 10^{-3} \text{ m}^2 = 2560 \text{ mm}^2$$
$$a = N_n/(0.85\, f_c'b) = 2.56 \times 10^{-3} \times 400/0.85 \times 30 \times 2.75 = 0.0146 \text{ m}$$
$$z = 147 - \tfrac{1}{2} \times 14.6 = 140 \text{ mm} = 0.95\, d$$

---

*It is optional whether M_u or M_n should be distributed. The author prefers to insert the ϕ as early as possible.

TABLE 15.1 Steel Calculations for Flat Plate Panel of Sec. 15.7.

	Long Span				Short Span			
	Column Strip (2750 mm)		Middle Strip (2750 mm)		Column Strip (2750 mm)		Middle Strip (3250 mm)	
	Negative	Positive	Negative	Positive	Negative	Positive	Negative	Positive
Distribution of M_{on}	$-0.49\,M_{on}$	$+0.21\,M_{on}$	$-0.16\,M_{on}$	$+0.14\,M_{on}$	$-0.49\,M_{on}$	$+0.21\,M_{on}$	$-0.16\,M_{on}$	$+0.14\,M_{on}$
$M_{on} \geq M_u/0.9$		292 kN·m				264 kN·m		
M_n, kN·m	−143	+61	−47	+41	−129	+55	−42	+37
d, mm	147	147	147	147	132	132	132	132
zf_y	56 000	56 000	56 000	56 000	50 000	50 000	50 000	50 000
A_s reqd., mm²	2 560	1 090←	840	730	2 580	1 100←	840←ᵃ	740
Min. $A_s = 0.0018 \times 175\,b$	—	866	866←	866←	—	866	1 024	1 024←
Straight Bars								
No. of #15	13	6	5		13	8	5	6
Aver. spcg., mm	200	450	Use #10		200±		Use#10	Use #10
Max. spcg, mm	—	350	350		—	350←	350	350
No. of #10			9	8			9	11
Aver. spcg., mm			300±	350			350	300±
Alternate-Bent & Str.								
Bent	6-#15	3-#15	8-#10	4-#10	8-#15	4-#15	10-#10	5-#10
Straight	7-#15	3-#15	1-##	4-#10	5-#15	4-#15	—	6-#10

ᵃNeg. A_s + half pos. A_s can be made effective for temp.

398

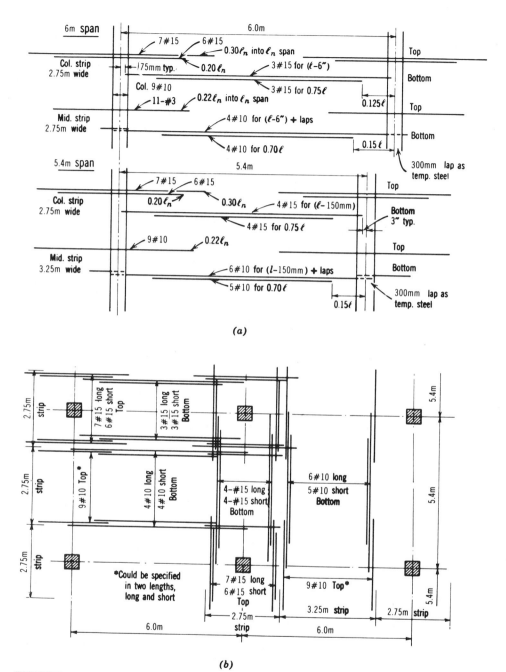

FIGURE 15.10 Steel for flat plate floor. (*a*) Symmetrical schematic arrangement as in elevation. (*b*) Plan showing bands of reinforcing. Lengths of bars would often be listed instead of simply "long" or "short."

399

Thus $0.95\,d$ is satisfactory here and will be on the safe side elsewhere (where smaller moments lead to still smaller a values).

This area of steel also indicates 13-#15 bars (2600 mm²) at a spacing of 2750/13 = 212 mm, which is reasonable for the worst spot. Bars will usually be specified by total number in the strip rather than by exact spacing, as 13-#15 at 200 mm ± , thus also letting the field worker know that a uniform spacing close to 200 mm will need little rearrangement to cover the strip properly.

In Table 15.1 the cross-section sketches are helpful to emphasize the steel arrangement. Solid circles represent the bars being designed. The zf_y constant used must be multiplied by 10^{-6} to obtain the result in mm². Minimum temperature and spacing steel (Code 7.12.2) is also shown, with the ratio for Grade 400 bars being 0.0018 based on the area of the *total slab thickness*. This steel includes both top and bottom steel where both exist, as in a negative moment region. Although in such a region the main steel is top steel for negative moment, half of the positive moment steel is required to be run into the support region and stays effective for temperature cracking resistance. If the temperature needs *at* the support appear critical these bottom bars can be lapped instead of stopping them 75 mm short of the column center line as indicated in Fig. 15.9. Finally, the maximum spacing for slabs is fixed at $2\,h$ which is 750 mm and this governs in several places.

Two arrangements of bars are shown in the table, the first for all bars straight as shown schematically in Fig. 15.10a. These would usually be shown on the drawings as bands, somewhat as in Fig. 15.10b. There for simplification the listings have been shown almost on two separate panels, one for the steel in each direction. Where spans are unequal and steel must be shown in both directions on many

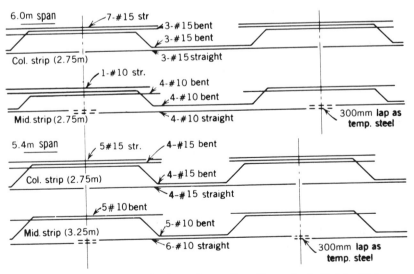

FIGURE 15.11 Pattern for bent bars, the alternate in Table 15.1.

panels, it is possible to superimpose as Fig. 15.10b shows along one column centerline.

The second arrangement in Table 15.1 is for bent bars, as sketched in Fig. 15.11.

The foregoing concludes the primary or main design calculations for this example. However, there are four special aspects of design that might influence or change the decisions already made. The following aspects deserve more generalized comment than would have been appropriate within the design itself:

Sec. 15.8   Minimum stiffness of columns.
Sec. 15.9   Effective column stiffness in a slab system.
Sec. 15.10   Concentration of reinforcement over the column.
Sec. 15.11   Code thickness equations—comments on their use.

15.8 MINIMUM STIFFNESS OF COLUMNS

Slab moments, especially for slabs without beams, are sensitive to pattern loadings.[5] Provision for the total M_o moment based on full loading of all panels still leaves overstress possible under pattern loadings. The Code accepts a possible overstress to 33% under these conditions. To stay within this limit when the unfactored dead load is less than twice the unfactored live load, the system requires either a specific minimum size column or, with a lesser column, an increase in the positive moments above those indicated by M_o.

The minimum column is defined by a minimum required stiffness ratio

$$\alpha_{min} = \Sigma K_c / \Sigma (K_s + K_b)$$

where K_c, K_s, and K_b are the flexural stiffness of columns (above and below the slab), slabs and beams (at the joint), respectively. In Table 15.2 α_{min} is tabulated for various values of β_a, the ratio of dead to live load (at service load level), panel proportions ℓ_2/ℓ_1, and relative beam stiffness values α. For dead load of twice the live load, α_{min} is zero, but α_{min} increases as β_a decreases.

As an example, consider the slab design of Sec. 15.7 where $\beta_a = 5/3 = 1.67$, which is less than the specified ratio of 2 and requires investigation. $\ell_2/\ell_1 = 6/5.5 = 1.09$. $\alpha = 0$ (no beams). From the table:

$$\beta_a = 2.0, \text{ any } \ell_2/\ell_1, \ \alpha = 0, \ \alpha_{min} = 0$$
$$\beta_a = 1.0, \ \ell_2/\ell_1 = 1.11, \ \alpha = 0, \ \alpha_{min} = 0.74$$

Interpolating: $\beta_a = 1.67$, $\ell_2/\ell_1 = 1.11$, $\alpha = 0$, $\alpha_{min} = 0.24+$, say 0.25

$$\alpha_{min} = 2 \, K_c / 2 \, K_s = K_c / K_s = 0.25$$

The smaller $I_s = 5500 \times 175^3 / 12 = 2.456 \times 10^9 \text{ mm}^4$

For a slab span of 6 m, $K_s = 4 \, EI_s / \ell = 4 \, E \times 2.456 \times 10^9 / 6000 = 1.637 \times 10^6 \, E$

TABLE 15.2ᵃ Minimum Column, α_{min} Values

β_a	Aspect Ratio ℓ_2/ℓ_1	Relative Beam Stiffness, α				
		0	0.5	1.0	2.0	4.0
2.0	0.5–2.0	0	0	0	0	0
1.0	0.5	0.6	0	0	0	0
	0.8	0.7	0	0	0	0
	1.0	0.7	0.1	0	0	0
	1.25	0.8	0.4	0	0	0
	2.0	1.2	0.5	0.2	0	0
0.5	0.5	1.3	0.3	0	0	0
	0.8	1.5	0.5	0.2	0	0
	1.0	1.6	0.6	0.2	0	0
	1.25	1.9	1.0	0.5	0	0
	2.0	4.9	1.6	0.8	0.3	0
0.33	0.5	1.8	0.5	0.1	0	0
	0.8	2.0	0.9	0.3	0	0
	1.0	2.3	0.9	0.4	0	0
	1.25	2.8	1.5	0.8	0.2	0
	2.0	13.0	2.6	1.2	0.5	0.3

ᵃCopy of Code Table 13.6.10.

The varying I of Code 13.7.3.3, is not used although appropriate (not required for direct design method) if one had the related curves convenient.

$$0.25 = K_c/K_s = K_c/1.637 \times 10^6 \, E, \qquad K_c = 409.4 \times 10^3$$

For a story height of 3 m, $K_c = 409.4 \times 10^3 \, E = 4 \, EI_c/3000$

$$\text{Min. } I_c = 409.4 \times 10^3 \times 3000/4 = 307.0 \times 10^6 = t^4/12, \, t = 246 \text{ mm}$$

The 375 mm column used is thus more than adequate with this 175 mm slab. The design positive moments used remain acceptable without penalty.

If α_{min} could not be satisfied, an increased positive moment would be required as given by multiplying the nominal moment values by δ_s:

$$\delta_s = 1 + \frac{2 - \beta_a}{4 + \beta_a}(1 - \alpha_c/\alpha_{min}) \qquad \text{Code Eq. 13.5}$$

where all terms have already been defined except α_c, the actual column value.

15.9 EFFECTIVE COLUMN STIFFNESS IN A SLAB SYSTEM

Even if a flat plate were supported on very stiff columns, the slab spans would *not* be really fixed at the column. The slab strip framing directly into the column would be fixed over the column width, but the parallel strips farthest from the column would be restrained only by the next slab span, practically none at all by the column itself. The torsional restraint that the slab could transfer laterally to this strip would be nearly negligible. With a heavy transverse beam there could be more restraint, but on the average across the slab the restraint would still be less than directly at the column. This is a way of saying that insofar as the slab is concerned the slab restraint averages less than the column stiffness would suggest. Only a transverse beam infinitely stiff in torsion between two columns could make this difference negligible.

This reduced column effectiveness is a new concept insofar as the Code is concerned. The Code uses the term "the equivalent column" and describes its behavior in terms of flexibilities (the inverse of stiffnesses). For the complete exterior column (above and below), the specified flexibility is

$$1/K_{ec} = 1/(\Sigma K_c) + 1/K_t$$

where ΣK_c represents the sum of the stiffnesses of column above and below the joint and K_t represents the sum of the torsional stiffness of the slab or beam on each side of the column. The reason K_t and K_{ec} do not carry a Σ is that the Σ shows earlier as part of the K_t equation and K_{ec} definition.

The author prefers to think in terms of stiffness rather than flexibility, which would make the equation

$$K_{ec} = \frac{1}{1/(\Sigma K_c) + 1/K_t}$$

The effective column K_{ec} taking the place of ΣK_c is either more flexible or less stiff than the column would be alone. If K_t is small, say, only a slab, K_{ec} is *much less* stiff than ΣK_c.

The evaluation of K_t is based on simple (rather crude) assumptions and basic theory, with final adjustment by an empirical factor to fit test results. The final evaluation is taken as $\frac{1}{3}$ as stiff as Fig. 15.12 from the Code Commentary would suggest. This figure shows the transverse beam (or slab) loaded in torque by unbalanced slab moments that vary from zero at midspan to a maximum at the centerline of the column. The total torque (area of the load curve) is 0.5 for each half span and the shape of the torque curve is a second power parabola with an ordinate at the face of column

$$T_u = \frac{1}{2}\left(1 - \frac{c_2}{\ell_2}\right)^2$$

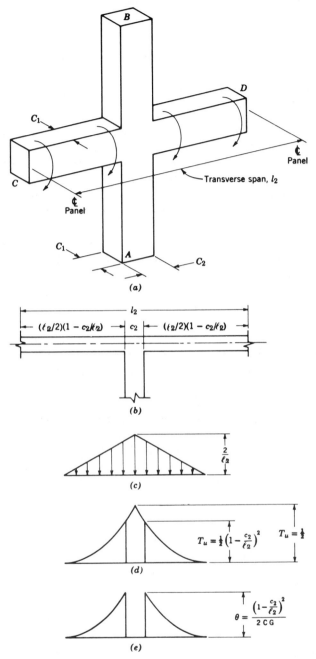

FIGURE 15.12 The equivalent column concept. (*a*) Simplified physical model. (*b*) Beam-column combination. (*c*) Distribution of torque or twisted load, based on a total of unity. (*d*) Twisting moment or torque diagram. (*e*) Distribution of twist angle per unit length. (Modified from ACI Commentary on Building Code.)

The unit rotation angle at any point is the torque divided by both G, the shear modulus, and C, the torque stiffness of the cross section. The integral of the area under this curve measures the difference between column rotation and midspan beam rotation,

$$\theta_t = (1/3)[(\ell_2/2)(1 - c_2/\ell_2)0.5(1 - c_2/\ell_2)^2]/CG$$
$$= (1/12)\ell_2(1 - c_2/\ell_2)^3/CG$$
$$= (1/6)\ell_2(1 - c_2/\ell_2)^3/CE, \text{ assuming } G = 0.5\ E.$$

One may think of the average θ for the slab as $\theta_t/3$ or he may think of using a factor of 1/3 to fit the equation to the test data. Fitting could be necessary to represent a nonlinear torque load, the difference between torque at the column face and at the center of column, and so on,

$$\theta_t = (1/18)\ell_2(1 - c_2/\ell_2)^3/EC^*$$

The stiffness is the ratio torque (used as 0.5 per arm) to this θ_t, or

$$K_t = 0.5/\theta_t = 9\ EC/[\ell_2(1 - c_2/\ell_2)^3]$$

for one arm, or twice this for the two arms in the symmetrical case. The Code indicates the doubling by writing the Σ sign as part of the definition of K_t

$$K_t = \Sigma 9\ EC/[\ell_2(1 - c_2/\ell_2)^3]$$

(Clarity, to the author, would be improved if the Σ were moved into the K_{ec} equation to show ΣK_t.)

The value of C in these equations is the necessary term to replace the polar moment of inertia when considering the torque of noncircular members:

$$C = \Sigma(1 - 0.63\ x/y)x^3y/3$$

where x and y are the small and large dimensions of the various rectangles making up the cross section, as shown in Fig. 15.13a. Where no beam stem is present, Code 13.7.5.1a defines the beam as a width of slab equal to the column width. Where a beam stem is present, the beam includes the stem plus the adjoining slab on *each* side (one side if a spandrel or edge beam) of width equal to $4\ h_s$ or, if smaller, the stem projection above or below the slab, as indicated in Fig. 15.13b.

For moment distribution under the equivalent frame method, or in moment formula evaluation under the direct design method, K_{ec} is the column stiffness used. It reflects the reduced fixity of slab because some middle strips of the slab are less restrained by abutting slabs than are

---

*C is similar in use to the polar moment of intertia used for a circular elastic member under torsion.

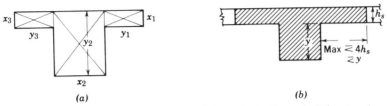

FIGURE 15.13 (*a*) Rectangles used in calculation of *C* for torsion.
(*b*) Flange in slab assumed as part of beam.

those close by the column. Where this creates a torque in the slab this torque is carried back to the column and the effect on the column is a bending moment at the joint equal to the sum of moment transferred directly from the slab to the column and that transferred indirectly through torque. But this total is less than if all the slab were directly framing into the column, which is what K_{ec} is intended to indicate.

15.10 CONCENTRATION OF REINFORCEMENT OVER THE COLUMN

Where moment must be transferred from slab to column, as at an exterior column, or from column to slab, as in resistance to wind or earthquake moments, it is desirable that this be accomplished as directly as possible. The Code (13.3.4.2) considers a direct transfer of moment possible only through reinforcement that passes through the column itself or within a width $1.5\,h_s$ on each side.

Tests have shown that the shear stresses are shifted by the general moment transfer taking place at the column, somewhat as indicated in Sec. 6.4 and as calculated in Sec. 15.13f. The Code Commentary suggests that 40% of the moment transferred should be represented by this eccentric shear (or torque) and that 60% should be transferred directly. If the column depth c_1 (parallel to ℓ_1) is greater than the transverse column width, Code 11.12.2.3 gives a formula that increases the percent to be transferred by eccentric shear or torque.

Some concentration of the column strip negative moment steel within the specified width may be necessary for the direct transfer of moment to the columns. If wind load moment is being transferred to the slab, extra moment steel may be required. Live load pattern moments or unequal spans also involve these transfer moments in interior panels. For equal spans the usual steel arrangement for the column strip will usually be adequate, as now demonstrated.

The Code (13.6.9.2) gives an equation defining the moment to be transferred to columns above and below the joint. For equal spans this equation simplifies to:

$$M = 0.08(0.5 \ w_1 \ell_2 \ell_n^2)/(1 + 1/\alpha_{ec}) = M_u{}^*$$

where w_ℓ is the live load and α_{ec} comes from K_{ec} just discussed in Sec. 15.9:

$$\alpha_{ec} = K_{ec}/\Sigma(K_s + K_b)$$

where subscript s is for slab and b for beam.

Thus far there has been no need to calculate K_{ec} numerically. Rather than do so now where emphasis is on moment transfer rather than on the equivalent column, the present calculation is based on data developed in Sec. 15.13b,c where for an exterior column of the same size, K_{ec} was $1.667 \times 10^6 \ E_c$, K_s for a 200 mm slab was $2.444 \times 10^6 \ E_c$, K_b was zero, α_{ec} was 0.682, and $1 + 1/\alpha_{ec} = 2.47$. At the typical interior column of Sec. 15.7 K_s must be based on a 175 mm slab and be doubled for two slabs instead of one at the exterior column. $K_s = 2.444 \times 10^6 \ E_c \ (175/200)^2 = 1.871 \times 10^6 \ E_c$.

$\alpha_{ec} = 1.667 \times 10^6 \ E_c/2 \times 1.871 \times 10^6 \ E_c = 0.445$

$1 + 1/\alpha_{ec} = 1 + 1/0.445 = 3.25$

$M_u = 0.08 \times 0.5(1.7 \times 3) \times 5.5 \times (6.0 - 0.375)^2/3.25 = 10.9 \ \text{kN} \cdot \text{m}$

Some 60% of this M_u should be carried within a 375 mm column width plus $3.0 \ h_s = 450$ mm which totals 0.875 m, or $M_n \geqslant 0.60 \times 10.9/0.9 = 7.3 \ \text{kN} \cdot \text{m}$. The original column strip design M_n was 143 kN · m on a 2.75 m width and this has already provided a capacity in the required width of $143 \times 0.875/2.75 = 45.5 \ \text{kN} \cdot \text{m}$. No concentration of reinforcing is necessary or even nearly so.

For unequal spans the Code formula adds to the numerator already used the difference between $0.08 \ w_d \ell_2 \ell_n^2$ for the long span and the corresponding quantity for the short span.

For the exterior column this requirement is more serious, as in Sec. 15.13(f). There 60% of the total negative moment must be carried within this section close to the column and special provisions must be made for bringing in the torque.

There will be no need to check the concrete in compression for this direct moment transfer if the designer is careful to keep the steel ratio less than $0.75_{\rho b}$ locally. The slab thickness provided against deflection should usually result in lower steel ratios.

---

*The author interprets this M as the external moment with no ϕ factor yet associated. For clarity, the subscript to give M_u is added.

15.11 CODE THICKNESS EQUATIONS—COMMENTS ON THEIR USE

The Code establishes minimum thicknesses of slab in two-way construction primarily by formula values, but with certain limiting considerations, especially minimum values (9.5.3.1):

For slabs without beams or drop panels	120 mm
For slabs without beams, but with drop panels satisfying Code 9.5.3.2	100 mm
For slabs having beams on all four edges with a value of α_m at least equal to 2	90 mm

The three Code equations, minimum values in Eqs. 9-10 and 9-11 and an upper limit in Eq. 9-12, are:

$$\text{Eq. 9-10: } h \geq \frac{\ell_n(800 + f_y/1.5)}{36\,000 + 5000\,\beta[\alpha_m - 0.5(1 - \beta_s)(1 + 1/\beta)]}$$

$$\text{Eq. 9-11: } h \geq \frac{\ell_n(800 + f_y/1.5)}{36\,000 + 5000\,\beta(1 + \beta_s)}$$

$$\text{Eq. 9-12: } h \not> \frac{\ell_n(800 + f_y/1.5)}{36\,000}, \text{ permissive, not mandatory}$$

All three equations have the common numerator that reduces to 1000 ℓ_n for Grade 300 steel and 1067 ℓ_n for Grade 400 steel*. Another format will be introduced before discussing the denominators.

The Portland Cement Association[2] treatment of these equations appears to offer better visualization of the thickness requirements. The three equations above are first restated with both numerator and denominator divided by 1000, as follows:

$$\text{Eq. 9-10: } h \geq \frac{\ell_n(0.8 + f_y/1500)}{36 + 5\,\beta[\alpha_m - 0.5(1 - \beta_s)(1 + 1/\beta)]}$$

$$\text{Eq. 9-11: } h \geq \frac{\ell_n(0.8 + f_y/1500)}{36 + 5\,\beta(1 + \beta_s)}$$

$$\text{Eq. 9-12: } h \not> \frac{\ell_n(0.8 + f_y/1500)}{36}, \text{ permissive, not mandatory}$$

*Note that this difference of 7% between the two grades of steel is smaller than the 7% difference for beams suggested with Code Table 9.5a reprinted here as Table 3.2 (Sec. 3.8). Slabs crack less than beams under service conditions where deflections are of greatest interest; the grade and quantity of steel makes little difference in the uncracked portions.

The numerators then all reduce to ℓ_n for Grade 300 bars and 1.067 ℓ_n for Grade 400 bars. The last equation, for example, can be restated for Grade 300 bars as a required $\ell_n/h \nless 36$, or ℓ_n/h need not be lower than 36; or for Grade 400 bars that ℓ_n/h need not be lower than $36/1.067 = 33.75$. Likewise the denominator of the other equations establish similar requirements, which are detailed after a brief discussion of the various ratios represented by the symbols in the denominators:

β = ratio of long to short *clear* spans = 1 for a square panel; above a ratio of 2, based on spans center-to-center of columns, the slab is to be designed as a one way slab or a more exact analysis is required (Code 13.6.1.2).

β_s = ratio of length of continuous edges to total perimeter of a slab panel = 1 for an interior panel and never less than 0.5 *for the direct design method.*

α_m = average ratio of flexural stiffness of beam section to that of the slab. Slab width to the center of adjacent panels is used, that is, for uniform panels ℓ_2. The subscript m calls for the average for all beams around the panel. The value of α varies from 0 for an interior flat plate upward.

Interior panels of flat slabs or flat slabs have no beams and hence beam stiffness ratios α_m and α are zero. With continuity on all sides β_s is unity. Thus the denominator for Eq. 9-10 is the same for interior panels as that of Eq. 9-12 in the absence of beams. For a square exterior panel without an edge beam β_s becomes 0.75 and β becomes unity, leading to a numerator for Eq. 9-10 of only $36 + 5(-.25)2 = 33.5$, which permits Eq. 9-12 to be used as a lesser limit if desired. These equations are plotted for Grade 300 bars in Fig. 15.14, with only the ℓ_n/h value of 36 significant in the absence of beams. For Grade 400 bars the 36 becomes $36/1.067 = 33.75$.

Where beams are used (Chap. 16), the curves of Fig. 15.14 are very useful.

Two special cases are treated as modifications of the Code values of thickness. (1) With drop panels at the columns meeting the requirements illustrated in Sec. 15.14, a thickness 10% smaller than the equation values may be used. (2) In edge panels the formula thickness is too shallow unless edge beams having a stiffness ratio $\alpha \geq 0.80$ are provided; otherwise the required thickness is 10% more than the formula values, or 10% more than the special drop panel provision requires.

Finally, it is permitted, in lieu of the formulas, that deflections may be calculated and the immediate deflection under the live load then be limited to maximum values given in Code Table 9.5b which is not reproduced here. The listed permissible deflection limits vary with structure usage and are strictest where attached nonstructural elements may be damaged by deflections.

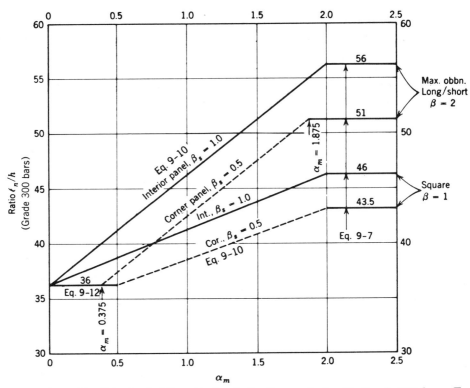

FIGURE 15.14 Required ℓ_n/h ratio for deflection control with Grade 300 bars. For Grade 400 divide ratio by 1.067. (Recase from Reference 2, PCA.)

15.12 FLAT PLATES AND FLAT SLABS—EXTERIOR SPANS

For strips parallel to the exterior wall, the methods for interior panels are available with little change. For the perpendicular strips running to the exterior a number of problems arise: the unsymmetrical negative moments, the combined shear and torsion problem at the exterior column, the torsion reinforcement often required in the slab or spandrel beam, and the direct transfer of moment to the exterior column (Sec. 15.10).

The Code gives formulas for the assignment of M_o into positive and negative moments, using functions of $\alpha_{ec} = K_{ec}/[\Sigma(K_s + K_b)]$, with the ratio of equivalent stiffness of exterior column to the stiffness of the slab (or slab and beam). These multipliers for M_o are:

Interior negative moment	$0.75 - 0.10/(1 + 1/\alpha_{ec})$
Positive moment	$0.63 - 0.28/(1 + 1/\alpha_{ec})$
Exterior negative moment	$0.65/(1 + 1/\alpha_{ec})$

TABLE 15.3 Distribution of Column Strip Negative Moment at Exterior. (Reproduced from Code 13.6.4.2)

ℓ_2/ℓ_1		0.5	1.0	2.0
$(\alpha_1\ell_2/\ell_1) = 0$	$\beta_t = 0$	100	100	100
	$\beta_t \geq 2.5$	75	75	75
$(\alpha_1\ell_2/\ell_1) \geq 1.0$	$\beta_t = 0$	100	100	100
	$\beta_t \geq 2.5$	90	75	45

Note that the denominator $(1 + 1/\alpha_{ec})$ shows in each case. The sum of the average negative moment and the positive moment is still M_o.

The break of moments into column strip and middle strip moments at the interior support is the same as for interior spans, 75% and 25% for negative moment and 60% and 40% for positive moment.

At the exterior support, for the same concrete in beam and column $(E_{cb}/E_{cs} = 1)$, Table 15.3 shows the portion going to the column strip is a function of $\beta_t = C/2I_s$, where C is the torsional value defining the edge beam and I_s is for a slab width equal to the beam span center-to-center of supports. The 2 factor comes from a ratio of E/G and the assumption that $G = 0.5\,E$. It might be debated whether to use $C = 0$ when no exterior spandrel beam is present. When calculating C in the determination of K_{ec}, a slab strip the width of the column is specified to be used. It would appear consistent to do the same here. Actually, unless C is large, nearly all the moment finally goes to the column strip and it is usually just as practical an answer to assume 100% to the column strip and skip the calculation unless a real beam is present.

The following section covers the basic calculations for an exterior panel matching the interior panel of Sec. 15.7.

15.13 DESIGN OF FLAT PLATE—EXTERIOR PANEL

Design by the direct method an exterior panel flat plate slab 5.5 m wide parallel to the free edge and 6 m center to center of columns perpendicular to this edge. Assume no beam,* no wall load (an omission for simplicity in the design example), and a slab extending to the outside face of the column, as in Fig. 15.15a. Use $f_c' = 30$ MPa for both slab and columns, Grade 400 steel, dead load (including slab

*The influence of an edge beam is discussed in Sec. 15.13g at the close of this section.

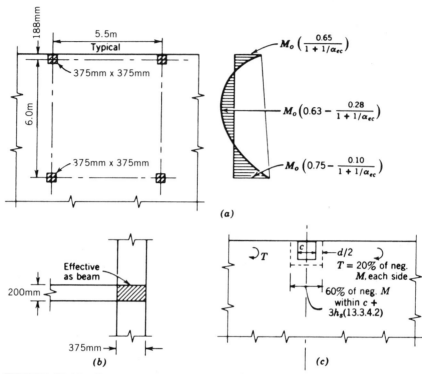

FIGURE 15.15   Exterior slab panel. (*a*) Layout and bending moments. (*b*) Slab section acting as beam in torsion. (*c*) Moment concentration over column.

weight) of 5 kPa and live load of 3 kPa. Assume all columns are 375 mm square with story heights of 3 m.

Solution

(a) Slab Thickness

Thickness for deflection will be considered first, following the comments of Sec. 15.11.

$$\ell_n = 6.0 - 0.375 = 5.625 \text{ m} \qquad f_y = 400 \text{ MPa}$$

β = ratio of *clear* spans = $(6.0 - 0.375)/(5.5 - 0.375) = 1.10$

$\alpha_m = 0$ (no beams)

β_s = ratio of length of continuous edge to total perimeter of panel

$\quad = (2 \times 6.0 + 5.5)/(2 \times 6.0 + 2 \times 5.5) = 0.761$

The curves of Fig. 15.14 will be used for the governing ℓ_n/h ratio. For $\beta_s < 1.0$, Eq.

9-10 generates a curve between the interior and corner panel curves, a curve that intersects the vertical axis ($\alpha_m = 0$) at something less than the Eq. 9-12 value of 36. Therefore, remembering the 1.067 ratio for Grade 400 bars, we calculate reqd. $h = \ell_n/(36/1.067) = 5625/33.75 = 167$ mm. Code 9.5.3.3 requires either a spandrel beam or an increase of 10% in these formula values.

$$h = 1.10 \times 167 = 183 \text{ mm, say } 200 \text{ mm}$$

The most critical design provisions are those for shear around the columns, but these cannot be calculated before the bending moments in the exterior panel perpendicular to the free edge are known. (An interior panel under the same conditions was found in Sec. 15.7 to require a d_v of only 123 mm.)

(b) Effective Exterior Column Stiffness

The equivalent exterior column stiffness must be found, which requires that the slab strip between exterior columns first be considered as a beam carrying torsion, as discussed in Sec. 15.9. This slab strip (Fig. 15.15b) is 375 mm wide by 200 mm deep, leading to the effective torsional inertia term C, or use Table 16.7 in Sec. 16.5.

$$C = (1 - 0.63 \, x/y)(x^3 y/3)$$
$$= (1 - 0.63 \times 200/375)(200^3 \times 375/3) = 0.664 \times 10^9 \text{ mm}^4$$

The torsional stiffness for the beam to one side of the column is

$$K_t = 9 \, E_{cs} C/[\ell_2(1 - c_2/\ell_2)^3]$$

where E_{cs} is simply E_c for the slab.

$$K_t = 9 \, E_c \times 0.664 \times 10^9/5500 \left(1 - \frac{375}{5500}\right)^3$$
$$= \frac{9 \times 0.664 \times 10^6}{5.5 \times 0.8090} E_c = 1.343 \times 10^6 \, E_c$$

For the two adjacent slabs, acting as transverse beams, $K_t = 2 \times 1.343 \times 10^6 \, E_c = 2.686 \times 10^6 \, E_c$.

The column above or below gives, for a *uniform* cross section:*

$$\tfrac{1}{2} K_c = 4 \, EI/\ell = 4 \, E_c \times \tfrac{1}{12} \times 375^4/3000 = 2.197 \times 10^6 \, E_c$$
$$K_c = 2 \times 2.197 \times 10^6 \, E_c = 4.395 \times 10^6 \, E_c$$

This leads to the equivalent column stiffness:

$$K_{ec} = \cfrac{1}{\cfrac{1}{2.686 \times 10^6 \, E_c} + \cfrac{1}{4.395 \times 10^6 \, E_c}} = 1.667 \times 10^6 \, E_c$$

(c) Slab Design Moments

For strips running perpendicular to the exterior edge, the design moments of Sec. 15.12 and Fig. 15.15a each contain a fractional multiplier that includes the denominator $1 + 1/\alpha_{ec}$, where α_{ec} by definition is $K_{ec}/\Sigma(K_s + K_b) = K_{ec}/K_s$ in this case.

*In the equivalent frame method the extra stiffness within the slab thickness is recognized.

$$I_s = \tfrac{1}{12} \times 5500 \times 200^3 = 3.667 \times 10^9 \text{ mm}^4$$
$$K_s = 4\, E_c I_s / \ell_1 = 4\, E_c \times 3.667 \times 10^9 / 6000 = 2.444 \times 10^6\, E_c$$
$$\alpha_{ec} = 1.667 \times 10^6\, E_c / 2.444 \times 10^6\, E_c = 0.682$$
$$1 + 1/\alpha_{ec} = 1 + 1/0.682 = 2.47$$

The design moments can now be established from the design load.
$$w_u = 1.4 \times 5 + 1.7 \times 3 = 12.1 \text{ kPa}$$

6 m Strips

$$M_o = 0.125 \times 12.1 \times 5.5(6.0 - 0.375)^2 = 263 \text{ kN} \cdot \text{m}$$
$$M_{on} \gtrless M_o/\phi = 263/0.9 = -292 \text{ kN} \cdot \text{m}$$
$$\text{Total neg. } M_n \text{ at int. col.} = [-0.75 - 0.10/(1 + 1/\alpha_{ec})]M_{on}$$
$$= -(0.75 - 0.10/2.47)292 = -207 \text{ kN} \cdot \text{m}$$
$$\text{Total positive } M_n = [0.63 - 0.28/(1 + 1/\alpha_{ec})]M_{on}$$
$$= (0.63 - 0.28/2.47)\, 292 = +151 \text{ kN} \cdot \text{m}$$
$$\text{Total neg. } M_n \text{ at ext. col.} = [0.65/(1 + 1/\alpha_{ec})]M_{on}$$
$$= -(0.65/2.47)292 = -77 \text{ kN} \cdot \text{m}$$

The interior negative moment is assigned to column strip (75%) and middle strip (25%) just as in an interior panel.

$$\text{Int. neg. } M_n \text{ on column strip} = -0.75 \times 207 = -155 \text{ kN} \cdot \text{m}$$
$$\text{Int. neg. } M_n \text{ on middle strip} = -207 + 155 = -52 \text{ kN} \cdot \text{m}$$

Likewise the positive moment is assigned 60% to the column strip and 40% to the middle strip, just as in an interior panel.

$$\text{Positive } M_n \text{ on column strip} = 0.6 \times 151 = 91 \text{ kN} \cdot \text{m}$$
$$\text{Positive } M_n \text{ on middle strip} = 151 - 91 = +60 \text{ kN} \cdot \text{m}$$

For the exterior negative moment, the proportion going to the column strip (Table 15.3) depends upon $\beta_t = E_{cb} C/(2\, E_{cs} I_s)$.

$$\beta_t = E_c \times 0.664 \times 10^9 / 2\, E_c \times 3.667 \times 10^9 = 0.0905$$

Interpolating in Table 15.3 of Sec. 15.12:

For $\beta_t = 0$, 100% goes to column strip

$\beta_t = 2.5$, 75% goes to column strip

$$\beta_t = 0.09,\ 100 - \frac{0.09}{2.5}(100 - 75) = 100 - 0.9 = 99.1\%$$

The difference between this and 100% is negligible and this is the reason for stating earlier that the difference could be ignored.

$$\text{Total exterior neg. } M_n \text{ on column strip} = -77 \text{ kN} \cdot \text{m}$$
$$\text{Exterior neg. } M_n \text{ on middle strip} = 0$$

This completes the design moments in the one direction, except for the following option.

Where unequal negative moments occur at a column line, as at the first interior column line, Code 13.6.3.4 requires the slab design be for the larger moment unless analysis is made to distribute the moments. The interior panel design in Sec. 15.7 found column strip negative moment M_n of -143 kN $\cdot$ m against -155 kN $\cdot$ m here and middle strip negative moment of -47 kN $\cdot$ m against -52 kN $\cdot$ m here. Design for the larger is no large penalty in this case.* Where significant, the unbalanced moment may be distributed to the slabs and column (using the K_{ec} values). For this case, with light columns and a small unbalance, the author would be inclined (1) to average the two panel values to use on both spans -149 kN $\cdot$ m for the column strip negative and -50 kN $\cdot$ m for the middle strip negative moments; (2) to neglect the resulting reduction in positive moments on the interior span; and (3) in the exterior span increase the positive column strip moment from 91 to 94 kN $\cdot$ m and increase the positive middle strip moment from 60 to 61 kN $\cdot$ m. The alternate procedure is to provide for the larger exterior panel negative moments on both sides of the first interior column line.

The strips parallel to the exterior are not different from an interior panel, except when a spandrel or edge beam is present. Such a beam, or in this case the slab since no beam is present, must be checked for torsion (as discussed in subsection g).

The design of reinforcement for the above strips would follow the method used for the interior slab in Table 15.1 in Sec. 15.7.

(d) Concentration of Negative Moment Reinforcement at Exterior Column

At the exterior column there is a necessary concentration of the column strip negative moment[5] as already discussed in Sec. 15.10. This requires a concentration of flexural reinforcement within a width equal to the column face plus $1.5\,h_s$ on each side for all moment to be transferred directly to the column; the remainder of the slab moment is transferred by torque, causing a resultant eccentricity of shear around the column on a width equal to the column face plus d, as indicated in Fig. 15.15c.

For a square column 60% of the exterior negative moment is considered transferred by moment, 40% by eccentric shear. (Note Code 11.12.2.3 for rectangular columns.)

$$\text{Direct moment transfer} = 0.60 \times 77 = 46 \text{ kN} \cdot \text{m}$$

$$\text{Assuming } d = 172 \text{ mm}, \ A_s = (46 \times 10^6/0.9)/400 \times 0.9 \times 172 = 825 \text{ mm}^2$$

This area, equivalent to 5#15 or 3#20, must be placed within a width of $c + 3\,h_s = 375 + 3 \times 200 = 975$ mm. This presents no serious problem, but may require a special note on the drawing as to this different spacing over the column.

---

*Note that the simplified "total dead load" used here has neglected the increased weight of the thicker exterior panel slab, in order to simplify the example; this is *not* a good practice.

(e) Minimum Column Stiffness

For the interior panel, minimum column stiffness was checked in Sec. 15.8. Since it is relative stiffness of column to slab that is checked, even a smaller column would satisfy the needed restraint at the exterior column. There, parallel to the exterior face, only a half panel width of slab is involved, which nearly doubles the relative column stiffness. In the other direction there is slab on only one face of the column, which has the same effect. The 375 mm square column is heavier than need be for this purpose.

(f) Shear at Exterior Column

Normally there would be some wall load, although for simplicity none is included in this text example; this would add a substantial extra shear. From Fig. 15.16a:

$$V_u = 12.1(3.0 \times 2 \times 2.75 - 0.535 \times 0.188 + 0.188 \times 2 \times 2.48) = 209 \text{ kN}$$
$$V_n \geqslant 209/0.85 = 247 \text{ kN}$$

From the basic ideas of Sec. 6.4, the centroid of shear resistance is at the centroid of the shaded areas of Fig. 15.16b.

$$A_c = 2 \times 460 \times 160 + 535 \times 170 = 238\,150 \text{ mm}^2 = 0.238 \text{ m}^2$$

From center of column,

$$\bar{x} = (2 \times 460 \times 160 \times 42 + 535 \times 170 \times 272)/238\,150 = 130 \text{ mm}$$

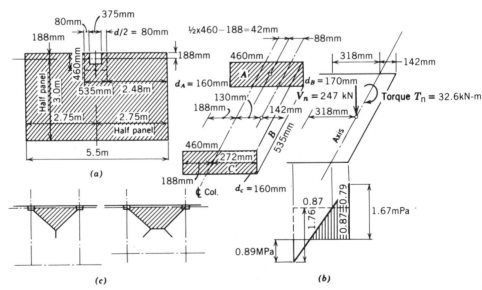

FIGURE 15.16 Shear considerations at exterior columns. (a) Shear load on column. (b) Shear and torque loading with resultant stresses. (c) Load to edge beam, when present.

The factored torque from both sides of the column is the 40% of the column strip negative moment not carried directly into the column.

$$\text{Col. strip neg. } M_u = \phi M_n = 0.9 \times 77 = 69.3 \text{ kN} \cdot \text{m}$$
$$T_u = 0.40 \times 69.3 = 27.7 \text{ kN} \cdot \text{m} \qquad T_n = T_u/\phi = 27.7/0.85 = 32.6 \text{ kN} \cdot \text{m}$$

The polar moment of inertia of these areas about a horizontal axis through the centroid and parallel to the center line of column is required. Using the area designations in Fig. 15.16b:

$$J_C = J_A = I_{xA} + I_{yA} \qquad\qquad J_B = A_B(460 - 188 - 130)^2$$
$$I_{xA} = \tfrac{1}{12} \times 460 \times 160^3 \quad = 157 \qquad\qquad = 142^2 A_B$$
$$I_{yA} = \tfrac{1}{12} \times 160 \times 460^3 \quad = 1298 \qquad\quad J_B = 170 \times 535 \times 142^2 = 1834 \times 10^6 \text{ mm}^4$$
$$+ 460 \times 160 \times 88^2 = \quad 570$$
$$J_A = \overline{2025} \times 10^6 \text{ mm}^4$$

$$J = 2 \times 2025 \times 10^6 + 1834 \times 10^6 = 5884 \times 10^6 \text{ mm}^4 = 5.884 \times 10^{-3} \text{ m}^4$$
$$v_{max} = V_n/A_c + T_n x_B/J$$
$$= 208 \times 10^{-3}/0.238 + 32.6 \times 10^{-3} \times 0.142/5.884 \times 10^{-3} = 0.874 + 0.787 = 1.67 \text{ MPa}$$
$$v_{min} = 0.874 - 0.787 \times 0.318/0.142 = -0.89 \text{ MPa}$$

The allowable maximum is $\tfrac{1}{3}\sqrt{f'_c} = \tfrac{1}{3}\sqrt{30} = 1.83$ MPa > 1.67 MPa   **O.K.**

If a wall had been included, as would be normal, it would have added substantially to the average shear but would have decreased the torque some since its weight would be to the left of the centroid.

(g) Torque on Slab

Torque might be calculated at a distance d from the column (Code 11.1.3.1) but will first be checked for the total torque to see if torque is serious.

$$T_n \text{ from above} = \tfrac{1}{2} \times 32.6 = 16.3 \text{ kN} \cdot \text{m}$$

For the specified slab strip equal to the column in width (Fig. 15.17b) and with torsion always on gross area (Sec. 6.5d), $\Sigma x^2 y/3 = 0.200^2 \times 0.375/3 = 5 \times 10^{-3} \text{ m}^3$.

$$v_{tn} = T_n/5 \times 10^{-3} = 16.3 \times 10^{-3}/5 \times 10^{-3} = 3.26 \text{ MPa}$$

The problem is serious. If one considers a parabolic torque diagram* similar to the one used in establishing K_t for the moment calculation, but with maximum at the *face* of column as shown in Fig. 15.17a, and takes $d = 200 - 20 - 15$ (col. strip bar) $- 8 = 157$ mm, T_n at a distance d from the column is:

$$T_n = 16.3(2.40/2.56)^2 = 14.3 \text{ kN} \cdot \text{m}$$
$$v_{tn} = 14.3 \times 10^{-3}/5 \times 10^{-3} = 2.86 \text{ MPa} = 0.52 \sqrt{f'_c}$$

The large torque shear stress indicates the basic problem of the slab without an

---

*This is not quite consistent with the 99.1% of negative moment to the column strip (subsection c), but is simple and on the safe side.

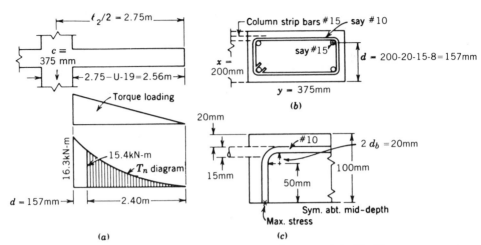

FIGURE 15.17 Torque provisions in beam element. (*a*) T_n diagram. (*b*) Cross section. (*c*) Development in vertical leg of closed tie.

edge beam.* Nor does the Code speak directly to the problem of how to design for torque quite close to a concentrated slab load, in this case the column reaction. Some two-way action exists on the inner face of the column but primarily one-way action exists near the outside column face. In this situation decisions must be somewhat arbitrary and there is no practical meaning in writing a third significant digit, although habit may make it easier that way.

If a beam existed in fact, one would here be calculating v_c and v_{tc} at a distance d from the column face, as here, but the criteria would be those for beam shear, critical on the inside face of the beam.

Since the vertical slab shear was calculated as some 52% of the permissible, the interaction of shear with torque will be considered relatively a normal one. If the vertical shear were zero, the allowable v_{tc} on the concrete under the 1971 Code 11.7.5 would be $v_{tc} = \frac{1}{5}\sqrt{f'_c}$. The 1977 Code Eq. 11.22 also used this allowable but, in its new force format, the $\frac{1}{5}$ is not so obvious. The permissible v_{tc} will be crudely estimated at about 50% of the pure torque allowable,† $v_{tc} = \frac{1}{2} \times \frac{1}{5}\sqrt{30} = 0.55$ MPa.

$$v_{ts} = v_{tn} - v_{tc} = 2.86 - 0.55 = 2.31 \text{ MPa}$$

$$A_t = [(v_{tn} - v_{tc})s\,\Sigma x^2 y/3]/(\alpha_t x_1 y_1 f_y)$$

$$\alpha_t = 0.66 + 0.33(y_1/x_1) = 0.66 + 0.33(375 - 2 \times 20 - 10)/(200 - 2 \times 20 - 15 - \tfrac{1}{2} \times 10)$$

$$= 0.66 + 0.33 \times 325/140 = 1.43$$

*This paragraph reflects the author's bias against slabs with neither an overhanging cantilever nor an edge beam. He feels that the 1977 Code does well to point to the problem. The state of the art, however, does not provide a documented design procedure. As in many design situations the engineer must visualize the slab behavior (deformation) and be overconservative in meeting it until more tests have been made of torsion reinforced slabs.
†The interaction with $\frac{1}{6}\sqrt{f'_c}$ for v_c seems inappropriate here with two-way bending and quite near a $\frac{1}{3}\sqrt{f'_c}$ allowable shear zone.

Reqd. $A_t = T_s s / \alpha_t x_1 y_1 f_y = v_{ts} \Sigma \frac{1}{3} x^2 ys / \alpha_t x_1 y_1 f_y$

$= 2.31 \times 5 \times 10^{-3} s / 1.43 \times 0.325 \times 0.140 \times 400$

$= 0.444 \, s \times 10^{-6} \, \text{m}^2 = 0.444 \, s \, \text{mm}^2$ (where s in in mm)

Max. spcg. in Code $= (x_1 + y_1)/4 = (325 + 140)/4 = 116 \, \text{mm}$

If $A_t = \#10 = 100 \, \text{mm}^2$, $s = 100/0.440 = 225 \, \text{mm} > \text{max. spcg.}$

USE #10 closed ties at 100 mm

Reqd. extra longitudinal steel $A_\ell = 2 \, A_t \, (x_1 + y_1)/s \, A_\ell$

$= 2 \times 0.486 \, s \, (325 + 140)/s = 452 \, \text{mm}^2$

$= 226 \, \text{mm}^2$ each top and bottom

With the above assumed v_{tc}, ties are needed wherever v_{tn} exceeds v_{tc} of 0.55 MPa or where T_n exceeds $0.55 \times 5 \times 10^{-3} \times 10^3 = 2.75 \, \text{kN} \cdot \text{m}$. With a parabolic T distribution rising to 16.3 kN · m at face of column.

$x_e^2/2.40^2 = 2.75/16.3$, $x_e = 0.98 \, \text{m}$ from midspan or 1.58 m from col.

Torsion reinforcement must be provided at least a distance $(d + b)$ beyond the point theoretically required $= 157 + 375 = 532 \, \text{mm}$. This increases the length from column face to $1.58 + 0.532 = 2.11 \, \text{m}$.

Use 19-#10 closed ties at 100 mm

3-#10 top and 3-#10 bottom bars extending 2.2 m from face of column

The latter steel is in addition to the usual flexural reinforcement.

Alternatively, the 1977 Code equations in terms of force instead of unit stresses are usable starting with $v_{tc} = 0.55 \, \text{MPa}$ as above. Although the $\frac{1}{5}\sqrt{f_c'}$ used in that calculation is not itself so obvious as a limit, it *is entirely* consistent with the newer code wording.

$T_c = v_{tc}(\Sigma x^2 y/3) = 5.5 \times 5 \times 10^{-3} \times 10^3 = 2.75$

$T_s = T_n - T_c = 14.3 - 2.75 = 11.55 \, \text{kN} \cdot \text{m}$

$s = A_t \alpha_t x_1 y_1 f_y / T_s$ $\alpha_t = 1.43$, as calculated above

For #10 ties, $A_t = 100 \, \text{mm}^2$

$s = 100 \times 10^{-6} \times 1.43 \times 0.325 \times 0.140 \times 400/11.55 \times 10^{-3} = 225 \times 10^{-3} \, \text{m} = 225 \, \text{mm}$

Max. Code spcg = 116 mm, as calculated above

USE #10 closed ties at 100 mm

The equations for longitudinal bars for torsion are not different from those of 1971. The value of required A_t/s is obtained using the required s of 225 mm

$A_\ell = 2 \, A_t(x_1 + y_1)/s = 2 \times 100(325 + 140)/225 = 413 \, \text{mm}^2$

$= 206 \, \text{mm}^2$ each top and bottom

The #10 tie in a 200 mm slab is on the verge of doubtful efficiency because of underdevelopment of its vertical leg. The hook (around the corner) is good (Code 12.5.1) for $43\sqrt{f_c'} = 236 \, \text{MPa}$. The clear vertical length between the start of hooks or bends (20 mm inside radius) is $200 - 2(20 + 10 + 20) = 100$ mm. If half of this 100 mm is available for development, the critical section being at midheight, as indicated in Fig. 15.17c, the straight development length is $100 - 20 - 10 - 20 = 50$ mm which can develop f_s as follows.

$$50 = 0.058 \, d_b f_s = 0.058 \times 10 \, f_s, \, f_s = 86 \text{ MPa}$$

The capacity of the hook is $43\sqrt{f_c'} = 236$ MPa.
This gives a total capacity of $236 + 86 = 322$ MPa.* The use of 100 mm spacing here instead of a theoretical 225 mm seems to give the required margin in strength required.

(h) Modifications in Design Where Edge Beam Exists

A separate example will not be developed for an exterior panel with an edge beam, but it is well to note the essential design differences in such a case.

The formula values for slab thickness will not then need the 10% increase if the edge beam has an α of at least 0.80.

The edge strip (half column strip) would be made up of a beam plus a slab, instead of a slab alone. The methods used in Secs. 15.13 and 16.3 are available for establishing the distribution of M_o in this half panel.

The equivalent exterior column stiffness found in (b) above is influenced by increased torsional stiffness of the beam, C in the edge beam case being based on the sum of two areas making up the inverted L section. The flange counted is limited to the smaller of 4 h_s or the projection of the beam above or below the slab (Code 13.2.4). The increased K_t results in a higher K_{ec}, α_{ec}, and negative moment assigned to the exterior column line, with minor changes in other moments. The relative torsional stiffness β_t of the beam would also modify (lower) the distribution of negative moment to the column strip, while increasing that to the middle strip.

The shear in this case would, in part, be brought into the column by the beam and, in part, directly by the slab. The load on the beams, including any direct wall load, could be assumed that bounded by 45 degree diagonals in the panel, as in Fig. 15.16c, plus a parallel centerline of panel if the exterior edge is the longer edge. Code 13.6.5.2 directs that where $\alpha_1 \ell_2 / \ell_1$ for the edge beam is less than unity this proportional part $(\alpha_1 \ell_2 / \ell_1)$ of the shaded area be used. The beam torque and shear should be analyzed (the shear at $\frac{1}{6}\sqrt{f_c'}$ allowable in combination with the allowable v_{tc}) at a distance d from the column, with closed stirrups provided for any excess.

The remainder of the column shear might, in the author's opinion, be assumed carried around the column outside the limits of the beam web, at a distance $d/2$ from the column faces. This shear would be limited to $\frac{1}{3}\sqrt{f_c'}$ unless the beam framed flush with the loaded face of the column and turned this into a one-way shear problem, critical at a distance d from the face. It would not be necessary to worry about a nonuniform distribution of this shear (in the nonflush case), although some variation undoubtedly would result.

*Since stirrups cannot be carelessly displaced in the direction of ℓ_d, the need for the 300 mm minimum is less here than for A_s bars. Code 12.2.5 excludes web reinforcement and lap splices from the 300 mm minimum ℓ_d requirement.

15.14 FLAT SLAB WITH DROP PANEL—DESIGN EXAMPLE

Design a typical interior bay of a flat slab to carry a dead load (including its own weight) of 5 kPa and a live load of 10 kPa over a 5.4 m by 5.9 m panel, center-to-center of columns. $f'_c = 20$ MPa, Grade 400 steel. Use a story height of 3.3 m, a column capital, and a drop panel.

Solution

A flat slab of this type (Fig. 15.18a) is particularly good for heavy manufacturing or warehouse loads. Shear can be controlled by the size of capital used and strength is more apt to control slab thickness than is deflection. However, deflection can be simply checked by calculating the required thickness h from the Code equations as soon as ℓ_n, the net span, is fixed.

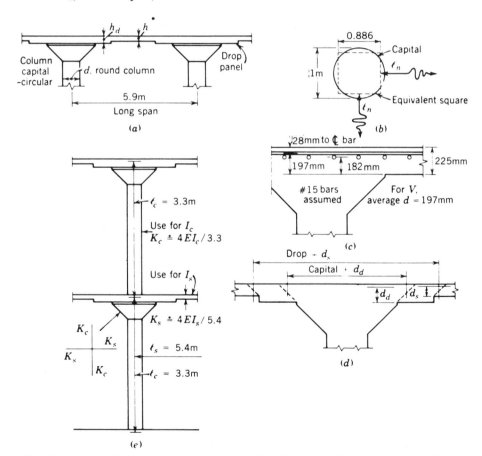

FIGURE 15.18 Flat slab construction with drop. (a) Nomenclature. (b) Net span l_n to face of equivalent square support. (c) Depth for shear. (d) Critical shear sections. (e) Calculation of minimum column, about center joint.

(a) Thickness for Deflection Control

The PCA equations discussed in Sec. 15.11 will be compared by evaluating their denominators, using $\alpha_m = 0$ because there are no beams and $\beta_s = 1$ because all panel edges are continuous.

$$\text{Eq. 9-10: } 36 + 5 \, \beta[\alpha_m - 0.5(1 - \beta_s)(1 + 1/\beta)$$
$$= 36 + 5(5.9/5.4)[0 - 0] = 36, \text{ same as for Eq. 9-12.}$$

Equation 9-12 cuts off any greater requirement and Eq. 9-10 permits no less; Eq. 9-11 is surplus in this case.

$$h_s = \ell_n(0.8 + f_y/1500)/36 = 1.067 \, \ell_n/36 = \ell_n/33.75$$

The net length must be estimated. Assume a column with a 1 m capital with an area of $\pi \times 1^2/4 = 0.786$ m^2. The equivalent side of a square having the same area $\sqrt{0.786} = 0.886$ m.

$$\ell_n = 5.9 - 0.886 = 5.0 \text{ m} \qquad h_s = 5000/33.75 = 149 \text{ mm}$$

With the drop panels specified in Code 13.4.7 (see (b) below) Code 9.5.3.2 lowers the requirement 10% to 134 mm, say 150 mm

(b) Depth for Flexure

$$\text{Dead load} = 1.40 \times 5 = 7$$
$$\text{Live load} = 1.70 \times 10 = 17$$
$$\overline{\phantom{\text{Live load} = 1.70 \times }24 \text{ kPa}}$$

$$\text{Long span } M_o = 0.125 \times 24 \times 5.4 \times 5^2 = 405 \text{ kN} \cdot \text{m} = 0.405 \text{ MN} \cdot \text{m}$$
$$M_{on} \geqslant M_u/\phi = 0.405/0.90 = 0.450 \text{ MN} \cdot \text{m}$$

$$\text{Total negative } M_n = -0.65 \, M_{on} = -0.65 \times 0.450 = -0.293 \text{ MN} \cdot \text{m}$$
$$\text{Col. strip neg. } M_n = -0.75 \times 0.293 = -0.219$$

The drop panel (9.5.3.2) should be at least $\ell_1/3$ by $\ell_2/3$, that is, in the long span direction 2.0 m and transversely 1.8 m. Its thickness h_d must be at least $1\frac{1}{4} \, h_s$ say in this case $1.25 \times 149 = 186$ mm say 200 mm, which seems light for this heavy a load. For strength design it is desirable to keep the steel ratio around 0.5 ρ_b* which means k_n slightly over two-thirds the value in Table 3.1 (for 0.75 ρ_b) which shows as 5.25 MPa. Try $k_n = 3.6$ MPa. The width in compression is that of the drop panel, 1.8 m.

$$\text{Reqd. } d_m = \sqrt{M_n/k_n b} = \sqrt{0.219/3.6 \times 1.8} = 0.183 \text{ m} = 183 \text{ mm}$$
$$\text{Drop } h_d = 183 + 20 \text{ cover} + 8 \text{ for } d_b/2 = 211 \text{ mm}$$

USE drop $h_d = 225$ mm, subject to shear check,

in panel 1.8 m by 2.0 m.

---

*For some extra ductility and usually also for overall economy.

(c) Shear at Column Capital

Use the assumed capital diameter of $1\,\text{m} + d = 1.000 + 0.183 = 1.18\,\text{m}$ to calculate V_u and the required capital for shear.

$$V_u = 24(5.4 \times 5.9 - \pi \times 1.18^2/4) = 738\,\text{kN}, \quad V_n \geq V_u/\phi = 869\,\text{kN}$$

Assume average $d = 225 - 20 - 8\,(=d_b) = 197\,\text{mm}$, as shown in Fig. 15.18c.
Shear perimeter (Fig. 15.18d) $= 2\pi(r + 0.5 \times 1.97) = 2\pi r + 619$

$$2\pi r + 619 = V_n/(v_c d)$$
$$= 0.869/\tfrac{1}{3}\sqrt{20} \times 0.197 = 2.960\,\text{m}$$
$$r = (2960 - 619)/2\pi = 373\,\text{mm}$$
$$2r = 746\,\text{mm vs. } 1000\,\text{mm assumed}$$

USE 0.9 m diameter column capital

The equivalent square (same area) $= 0.80\,\text{m}$ for use in ℓ_n.

(d) Shear Around Drop

This is critical at $d/2$ outside of the drop.

$$d = 150 - 20 - 8\,(=d_b) = 122\,\text{mm average}$$

Perimeter of shear section, if we keep drop dimensions, is one-third of center-to-center span each way; add $d/2$ each way to the critical section $= 2(2.0 + 0.122) + 2(1.8 + 0.122) = 8.088\,\text{m}$.

$$V_u = 24(5.9 \times 5.4 - 2.122 \times 1.922) = 667\,\text{kN}$$
$$v_c = 0.667/0.85 \times 8.088 \times 0.122 = 0.79\,\text{MPa}$$
$$\text{Allowable } v_c = \tfrac{1}{3}\sqrt{20} = 1.50\,\text{MPa} \quad \text{O.K.}$$

(e) Critical Moments

These must be recalculated since the size of capital has been changed

$$\text{Long span } M_o = 0.125 \times 24 \times 5.4(5.9 - 0.80)^2 = 421\,\text{kN}\cdot\text{m}$$
$$M_{on} = M_o\phi = 468\,\text{kN}\cdot\text{m}$$
$$\text{Short span } M_o = 0.125 \times 24 \times 5.9(5.4 - 0.80)^2 = 375\,\text{kN}\cdot\text{m}$$
$$M_{on} = M_o/\phi = 416\,\text{kN}\cdot\text{m}$$

The moment coefficients of Fig. 15.8b will be entered directly into Table 15.4.

Although the short strip moment is smaller than that for the long strip, the effective depth is also necessarily smaller by one bar diameter (because of crossing bars) than the depth with the long strip moment. Both depth requirements will be rechecked.

Long span col. strip neg. $M_n = -0.65 \times 0.75 \times 468 = -228\,\text{kN}\cdot\text{m}$
Short span col. strip neg. $M_n = -0.65 \times 0.75 \times 416 = -203\,\text{kN}\cdot\text{m}$

TABLE 15.4 Reinforcement for Flat Slab of Sec. 15.14

	Long Span (5.9 m) M_{on} = 468 kN·m				Short Span (5.4 m) M_{on} = 416 kN·m			
	Column Strip		Middle Strip		Column Strip		Middle Strip	
#15 bars assumed Strip width	Neg.	Pos.	Neg.	Pos.	Neg.	Pos.	Neg.	Pos.
M_n coef.	0.49	0.21	0.16	0.14	0.49	0.21	0.16	0.14
M_n = (coef.)M_{on} kN·m	−228	+98	−75	+66	−203	+87	−67	+58
d, mm	197	122	122	122	182	107	122	107
$A_s = \dfrac{M_n \times 10^6/400}{0.9\,d}$, mm²	3 229	2 231	1 708	1 503	3 114	2 259	1 526	1 506
Min. no. bars for $s = 2h$	9	9	9	9	9	9	11	11
No. of #15	17	12	9	8	16	12	8	8
Spacing, mm	150	225	300		175	225		
No. of #10				16			16	16
Spacing, mm				175			175	175

Reqd. long span $d = \sqrt{228 \times 10^{-3}/3.6 \times 1.8} = 0.188$ m $= 188$ mm

Min. $h_d = 188 + 20 + 8$ (for #15) $= 216$ mm < 225 mm above   **O.K.**

Reqd. short span $d = \sqrt{203 \times 10^{-3}/3.6 \times 2.0} = 0.168$ m $= 168$ mm

Min. $h_d = 168 + 20 + 1\frac{1}{2} \times 15 = 211$ mm < 225 mm   **O.K.**

If either h_d were greater than the 225 mm desired, the designer would have several choices:

1. Increase the drop panel thickness, weight, and moments.
2. Increase the drop width perpendicular to the strip, thus increasing the effective width b of strip (instead of d).
3. Increase column capital to reduce span and M_o.
4. Accept k_n several percent above the arbitrary 3.6 MPa value used, unless the ductility is particularly important. (Note that Code 8.4.3 permits moment redistribution only if $\rho \gtrless 0.5\,\rho_b$.)

USE drop panel 2.0 m in long span direction by 1.8 m in short span direction.

(f) Reinforcement Calculations

The remainder of the reinforcement calculations are made in Table 15.4. Since heavy loads and moments are involved the internal lever arm z is taken initially as $0.90\,d$ (compared to $0.95\,d$ in the earlier examples). The required A_s at the column in the long strip, the heaviest required, is 3229 mm^2 and will be used to check z. The concrete width used in the drop panel width of 1800 mm.

$$a = 3229 \times 400/0.85 \times 20 \times 1800 = 42 \text{ mm}$$
$$z = 197 - \tfrac{1}{2} \times 42 = 176 \text{ mm} = 0.89\,d$$

This z is about 1 percent low, but the excess in A_s used, offsets the 1 percent error in the calculation. In arranging these particular bars it would be better (except probably for the problem it adds to field supervision) to use the 17 bars spaced at a maximum of $2\,h_s = 300$ mm outside the 1.8 m wide drop panel and at a smaller spacing over and very near the drop to maintain the total number.

In the table, rather than estimate z individually for each case, the $0.90\,d$ was first used throughout, leading to tentative bar selections. The change in d resulting from the use of #10 bars is on the safe side.

Temperature steel calculations are not shown since a quick mental check of $0.0018\,bh$ in the short (wide) span middle strip indicated it could not control.

(g) Column Stiffness

Ratio of dead to live load, $\beta_a = 5/10 = 0.5$, is much below the factor of 2 required to make the above calculations acceptable without a special check on the column stiffness. $\ell_2/\ell_1 = 5.9/5.4 = 1.09$. Table 15.2 in Sec. 15.8 shows, fortunately with only the interpolation for this particular ratio and none for β_a, that $\alpha_{\min} = 1.7$.

The real column stiffness with the capital is greater than $4\,EI/\ell$. If I is based on minimum cross section, as usual, the coefficient for K of the column above is larger than 4 and in the column below (with the capital at the joint) *much* larger than 4, making more than half of the unbalanced moment at the joint go to the column below. The Code Commentary suggests more approximate methods are appropriate to the direct design method, both for the columns and slabs, but that similar simplifications should be used for both columns and slabs.

The slab stiffness is modified by both the drop panel and the column capital. Here the minimum I_s will be used, with the short span that is more restrictive in this case.

$$I_s = \tfrac{1}{12} \times 5900 \times 150^3 = 1659 \times 10^6 \text{ mm}^4$$

$$K_s = 4E_c \times 1659 \times 10^6/5400 = 1.229 \times 10^6 \, E_c$$

$$\alpha_{\min} = 1.7 = K_c/K_s = (2 \times 4 \, E_c I_c/\ell_c)/(2 \times 1.229 \times 10^6 \, E_c)$$

$$= [2 \times 4 \, E_c I_c/3300 \times 2 \times 1.229 \times 10^6 \, E_c = 986 \times 10^{-12} \, I_c$$

For a circular column of diameter d,

$$I_c = 1.7/986 \times 10^{-12} = 1724 \times 10^6 \text{ mm}^4 = \pi d^4/64$$

$$d^4 = 1724 \times 10^6 \times 64/\pi = 35\,118 \times 10^6 \text{ mm}^4$$

$$d = 432 \text{ mm}$$

This calls for a minimum 450 mm diameter column (or an increase in the positive moments already used for the slab design). In the strength design of this column the moments of Code 13.6.9.2 must be considered.

15.15 SHEAR REINFORCEMENT IN SLABS

The limited depth of slab makes the anchorage of shear reinforcement difficult and the anchorage requirements of Code 12.14 must be closely observed. The older ring type of wire reinforcement with inclined and inverted V-shaped wires welded around the perimeter has come into disfavor because its ability to pick up or develop the necessary stresses and again properly anchor them has been seriously questioned. As a result, the 1963 Code discontinued the crediting of shear reinforcement of bars, rods, or wires in slabs having a total depth of less than 250 mm. Since 1963, better tests have shown that *well anchored* bent bars and closed ties with bars in each corner can be made effective. The 1977 Code permits such shear reinforcement provided it takes care of all shear in excess of $\tfrac{1}{6}\sqrt{f'_c}$, thus leaving for concrete (in conjunction with shear reinforcement) only half of the usual v_c normally permitted around columns. Careful detailing and careful placement are both essential to this usage.

Because of these problems, shearheads of structural steel have also been further developed for slabs at interior columns.[11] Note that these are now only for interior columns. As now specified, *they do not work adequately at exterior columns*, but research in this area is making progress.

Shearheads consist of four crossing steel arms, welded together at a common level, to pick up both some shear and moment load from the concrete. Each shearhead arm may consist of a small wide flange beam or of two small channels turned back to back but spaced apart as much as the channel depth or more. These arms (totally within the slab thickness) pick up shear and moment beyond the column and bring the load to bearing on the column. The bottom flanges of the steel shapes are extended beyond the top flanges (with a sloping cut through the steel web) to pick up shear load that will exist low in the slab.

The critical section for shear on the concrete is thus removed to a larger perimeter farther than $d/2$ away from the column. The steel cantilever is considered effective in moving the critical shear section out from the column to a point three-fourths of its length from the column centerline. Between arms this critical section is considered a straight line joining the several three-quarter points on the arms, but never closer than $d/2$ to the face of the column.

The shearhead is a special type of construction that permits the use of a thinner slab where shear controls slab thickness at an interior column. It may be economical in some cases. Its detailed design is considered too specialized a problem for this book.

Finally, a repeated warning: shearheads are *not* for exterior columns at the present state of the art.

15.16 LIFT SLABS

Lift slabs are a flat plate type of slab cast at grade level and embedding steel shoes or collars that fit loosely around the columns. After the slabs are cured they are lifted by a patented jack system to the proper level where the shoes are welded to the steel columns. Although the Code makes no special attempt to provide for this type of design, the elastic method of flat slab analysis would appear to be proper for lift slabs. Since column stiffness is small, there seems to be no slab section that deserves to be treated as having a greatly increased moment of inertia. Loadings for maximum moment are also more significant in this case. A rigid joint between the collar and the column is an essential condition.

Because lift slabs do not qualify for empirical design procedures, total moment cannot be reduced to the M_o value. The collar stiffness may determine whether the critical section is at the center of the column or farther out.

The designer must consider carefully stresses caused by differential jack movements and make certain that in the field these do not exceed the limits considered in the office design.

15.17 CANTILEVERS FROM SLAB CONSTRUCTION

Lift slabs generally, and flat slabs and flat plates frequently, have the columns set back from the outside wall, thus causing the outermost section of the slab to act as a cantilever beyond the exterior columns. This is quite favorable to regular slab action. A proper overhang of the slab can provide a total negative moment that can almost eliminate the special problems associated with exterior slab panels.

The cantilever provides a total negative moment about the exterior columns that is statically determinate, but its distribution is not uniform. It is suggested that this distribution might be taken the same as that used to distribute negative moment between the column and middle strips in the elastic analysis method. If the cantilever projects the ideal distance, it can thus balance the typical interior slab negative moments. Longitudinal steel, as in any column strip, is needed to deliver the cantilever reaction ultimately to the columns.

It might also be noted that cantilever slabs have large deflections that become conspicuous after creep of concrete has taken place. The use of some compression steel solely to reduce deflections is often justified, as noted in Chapter 8.

SELECTED REFERENCES

1. H. M. Westergaard and W. A. Slater, "Moments and Stresses in Slabs," *ACI Proc.*, 17, 1921, p. 415.
2. *Notes on ACI 318-71 Building Code Requirements with Design Applications, Portland Cement Association*, Skokie, Illinois, 1972.
3. Mete A. Sozen and Chester P. Siess, "Investigation of Multi-Panel Reinforced Concrete Floor Slabs: Design Methods—Their Evolution and Comparison," *Jour. ACI*, 60, No. 8, Aug. 1963, p. 999.
4. J. R. Nichols, "Statical Limitations Upon the Steel Requirement in Reinforced Concrete Flat Slab Floors," *ASCE Trans.*, 77, 1914, p. 1670.
5. N. W. Hanson and J. M. Hanson, "Shear and Moment Transfer Between Concrete Slabs and Columns," *Jour. PCA Research and Development Laboratories*, 10, No. 1, Jan. 1968, p. 2.
6. D. S. Hatcher, M. A. Sozen, and C. P. Siess, "Test of a Reinforced Concrete Flat Slab," *Proc. ASCE*, 95, ST6, June 1969, p. 1051.
7. J. O. Jirsa, M. A. Sozen, and C. P. Siess, "Pattern Loadings on Reinforced Concrete Floor Slabs," *Proc. ASCE*, 95, ST6, June 1969, p. 1117.
8. W. L. Gamble, M. A. Sozen, and C. P. Siess, "Test of a Two-Way Reinforced Floor Slab," *Proc. ASCE*, 95, ST6, June 1969, p. 1073.

9. W. G. Corley and J. O. Jirsa, "Equivalent Frame Analysis for Slab Design," *Jour. ACI*, 67, No. 11, Nov. 1970, p. 875.

10. S. A. Guralnick and R. W. Fraugh, "Laboratory Study of a Forty-Five-Foot Square Flat Plate Structure," *Jour. ACI*, 69, No. 9, Sept. 1963, p. 1107.

11. W. G. Corley and N. M. Hawkins, "Shearhead Reinforcement for Slabs," *Jour. ACI*, 65, No. 10, Oct. 1968, p. 811.

12. Donald J. Fraser, "Equivalent Frame Method for Beam-Slab Structures," *Jour. ACI*, 74, No. 5, May 1977, p. 223.

PROBLEMS

PROB. 15.1.   Using $f'_c = 20$ MPa, Grade 400 steel, and no shear reinforcement, design (direct design method) an interior panel of a flat plate floor supported on colums 450 mm square spaced 5.4 m on centers each way. The slab is to carry a total dead load (including slab weight) of 5 kPa and live load of 2.5 kPa. Check minimum column stiffness for a story height of 2.7 m floor to floor.

PROB. 15.2.   Redesign the slab of Prob. 15.1 for the same data except with 6 m square panels. (For this problem neglect change in slab weight.)

PROB. 15.3.   Under the direct design method design a flat slab without a drop panel for a 6 m square interior panel using $f'_c = 30$ MPa and Grade 400 steel. Assume total dead load of 4 kPa and live load of 7 kPa. For the initial trial assume column capital $c = 1.35$ m. Story height is 3 m.

PROB. 15.4.   Redesign the flat slab of Prob. 15.3 using a drop panel, without changing the assumed design dead load.

PROB. 15.5.   Design a 5.7 m by 6.6 m interior panel of a flat slab without a drop panel, using the direct design method, $f'_c = 20$ MPa, Grade 400, a total dead load of 5 kPa, and a live load of 4 kPa. For the first trial use $c = 1.2$ m unless otherwise instructed.

PROB. 15.6.   Design the flat slab of Prob. 15.5 with a drop panel.

PROB. 15.7.

(*a*) Design an exterior panel (not a corner panel) of the flat plate described in Prob. 15.1, using the direct design method.

(*b*) What moment should be included in the exterior column design for this loading?

16
INTERACTION OF TWO-WAY SLAB SYSTEMS WITH BEAMS AND COLUMNS

16.1 TWO-WAY SLABS ON BEAMS—GENERAL

Code Chapter 13 on slab systems covers two-way slabs on beams that frame directly into columns, as well as flat plates and flat slabs. It provides a transition between the two extremes that will care for the range from slight stiffening of the slab to the provision of rather rigid beam supports. Being so broad in their coverage, the requirements are not simple to visualize in all their ramifications. The presentation introduces this design area by stages such as those already used for the flat plate and flat slab portions in Chapter 15.

At the beginning this type of slab will be introduced by a layout with beams stiff enough to minimize their contribution to slab deflection. The moment assignment tables from Code 13.6.4.1 and 13.6.4.4 are shown for reference as Table 16.1. For interior spans these assignment percentages

TABLE 16.1 Percentages of Moment Assigned to Column Strip

	Negative Moment			Positive Moment		
ℓ_2/ℓ_1	0.5	1.0	2.0	0.5	1.0	2.0
$\alpha_1\ell_2/\ell_1 = 0$	75%	75%	75%	60%	60%	60%
$\alpha_1\ell_2/\ell_1 \geq 1.0$	90%	75%	45%	90%	75%	45%

indicate that $\alpha_1 \ell_2/\ell_1 \geqslant 1$ is used as a break point. One might think of this as the point where beams carry essentially all the load the slab tends to deliver to them; below this $\alpha_1 \ell_2/\ell_1$ value lie small beams or deepened slab sections that simply stiffen the slab as an inseparable combination. Of course, $\alpha_1 = 0$ means no beams since α_1 is the ratio of beam stiffness to slab stiffness.

For convenience where $\alpha_1 \ell_2/\ell_1 \geqslant 1$, the term stiff beams is used here, although this stiffness is relative to the slab and not to beams in general. The beam values in the example of Sec. 16.2 indicate that the resulting "stiff" beam, viewed separately from a slab problem, might be considered just a usual beam.

Whatever the concept, at $\alpha_1 \ell_2/\ell_1 \geqslant 1$ the slabs and beams can be almost independently designed.* Where $\alpha_1 \ell_2/\ell_1 < 1$, the design must be a joint slab and beam problem starting from a trial combination of slab and stiffening beam, a problem where there are many independent design possibilities. For such combination design the designer must start with some assumed stiffening of the slab suiting the architectural demands or the designer's interest and verify its adequacy and the necessary reinforcement to be used, as is done in Sec. 16.3.

The ratio $\alpha_1 = I_b/I_s$ needs some special notice since I_b is for an arbitrary definition of a beam and I_s is for an entire panel width (not what is left after the beam is separated). The beam includes projections above or below the slab plus a flange equal on each side to the larger of these beam projections, but not greater than $4\,h_s$, as shown in Fig. 15.13 (Code 13.2.4). The value of I_s is that of all the slab, a width ℓ_2, as though no beam were present. Thus the same beam cross section on adjacent sides of a slab would lead to different α_1 values unless the slab were square ($\ell_1 = \ell_2$).

The next section designs a slab on stiff beams and Sec. 16.3 one employing small slab bands as stiffening.

16.2 TWO-WAY SLABS ON STIFF BEAMS—DESIGN EXAMPLE

Design an interior bay two-way slab on *stiff* beams for a bay 4.8 m by 6.0 m (to column centers) with service dead load (including slab weight) of 3.5 kPa and live load of 5 kPa, $f'_c = 20$ MPa, Grade 400 bars, 20 mm clear cover on slabs, 40 mm clear cover on beams, columns 375 mm square. Use direct design method of Code 13.6. Assume beams stiff enough to give $\alpha = $ beam I/slab $I = 1.00$ in the short direction and 1.25 in the long direction.

---

*α_1 is relative beam stiffness, discussed in the next paragraph. Since ℓ_2/ℓ_1 is limited to the range 0.5 to 2, an $\alpha_1 = 2$ will always give $\alpha_1 \ell_2/\ell_1 \geqslant 1$.

Solution

(a) Relative α_1 Values

For a given size of beam on all four sides such that α in the short direction is 1.0, the ratio $\alpha_1\ell_2/\ell_1 = 1.0 \times 6.0/4.8 = 1.25$ in the short direction. In the other direction I_s is only 4.8/6.0 as large, α_1 becomes 1.25, and $\alpha_1\ell_2/\ell_1 = 1.25 \times 4.8/6.0 = 1.0$. These α_1 values of 1.0 and 1.25 will be assumed here as the minimum to make $\alpha_1\ell_2/\ell_1 \geqslant 1$, although stiffer beams in either direction will not influence the slab part of the design.

(b) Slab Thickness

Based on the comments in Sec. 15.11:

$\ell_n = 6.0 - 0.375 = 5.625$ m in longer direction

β = long/short *clear* span ratio = 5.625/4.425 = 1.27

$\beta_s = 1$, continuous all around

α_m = average α for all beams around panel = $2(1.0 + 1.25)/4 = 1.12$

Compare the denominators of the h equations, using the PCA forms:

Eq. 9-10: $36 + 5\,\beta[\alpha_m - 0.5(1 - \beta_s)(1 + 1/\beta)$

$\qquad = 36 + 5 \times 1.27(1.12) = 43.1 > 36$

Eq. 9-11: $36 + 5\,\beta(1 + \beta_s) = 36 + 5 \times 1.27(1 + 1) >$ above; cannot govern.

Eq. 9-12 is an upper limit of no interest here, since Eq. 9-10 will call for a smaller thickness.

$\qquad$ Reqd. $h_s = (0.8 + 400/1\,500)\ell_n/43.1$

$\qquad\qquad = 1.067 \times 5.625/43.1 = 0.139$ m, say, 150 mm

This will be checked later for moment. Shear with this type slab with stiff beams is rarely critical.

(c) Design Moments

$$\text{Dead load} = 1.4 \times 3.5 = 4.9$$
$$\text{Live load} = 1.7 \times 5 \quad = 8.5$$
$$w = \overline{13.4}\ \text{kPa}$$

This does not include the stem weight of the beams.

Long Span

$M_o = 0.125 \times 13.4 \times 4.8 \times 5.625^2 = 254.4$ kN · m*

$M_{on} \geqslant M_u/\phi = 254.4/0.90 = 282.7$ kN · m

---

*The breakdown in moments can be in terms of M_{on} (ready for design) or in terms of M_o, for later conversion to M_n values.

Total negative moment $M_n = -0.65\,M_{on} = -0.65 \times 282.7 = -183.7\,\text{kN}\cdot\text{m}$

Total positive moment $M_n = 0.35\,M_{on} = 0.35 \times 282.7 = 98.9\,\text{kN}\cdot\text{m}$

Column strip:

$$\alpha_1 = 1.25 \text{ assumed}, \ell_2/\ell_1 = 4.8/6.0 = 0.8, \alpha_1\ell_2/\ell_1 = 1.0$$

For this $\alpha_1\ell_2/\ell_1$, Table 16.1 (from Code 13.6.4.1 and .4) shows moment assignment percentages to the column strip the same for negative and positive moment. Interpolations for ℓ_2/ℓ_1 of 0.8 and 1.25 give 81% and 67.5%:

ℓ_2/ℓ_1	0.5	0.8	1.0	1.25	2.00
$\alpha_1\ell_2/\ell_1$	90%	81%	75%	67.5%	45%

Code 13.6.5 assigns 85% of the column strip moment to the beam for these stiff beams.

Col. strip neg. $M_n = 0.81(-183.7) = -148.8\,\text{kN}\cdot\text{m}$

Beam neg. $M_n = 0.85(-148.8) = -126.5\,\text{kN}\cdot\text{m}$

Slab neg. $M_n = -148.8 + 126.5 = -22.3\,\text{kN}\cdot\text{m}$, or $-11.4\,\text{kN}\cdot\text{m}$ per half strip.

Col. strip pos. $M_n = 0.81 \times 98.9 = +80.1\,\text{kN}\cdot\text{m}$

Beam pos. $M_n = 0.85 \times 80.1 = +68.1\,\text{kN}\cdot\text{m}$

Slab pos. $M_n = 80.1 - 68.1 = +12.0\,\text{kN}\cdot\text{m}$, or $+6.0\,\text{kN}\cdot\text{m}$ per half strip

Middle strip (takes the remaining moments)

Mid. strip neg. $M_n = -183.7 + 148.8 = -34.9\,\text{kN}\cdot\text{m}$

Mid. strip pos. $M_n = 98.9 - 80.1 = +18.8\,\text{kN}\cdot\text{m}$

Short Span

$M_o = 0.125 \times 13.4 \times 6.0 \times 4.425^2 = 196.8\,\text{kN}\cdot\text{m}$

$M_{on} \geqslant 196.8/0.90 = 218.7\,\text{kN}\cdot\text{m}$

Total negative moment $M_n = -0.65 \times 218.7 = -142.1\,\text{kN}\cdot\text{m}$

Total positive $M_n = 0.35 \times 218.7 = +76.5\,\text{kN}\cdot\text{m}$

Column strip:

$$\alpha_1 = 1.0 \text{ assumed}, \ell_2/\ell_1 = 6.0/4.8 = 1.25, \alpha_1\ell_2/\ell_1 = 1.25$$

From the earlier interpolation from Table 16.1, $0.675\,M_n$ goes to the column strip and $0.325\,M_n$ to the middle strip. The final distributions, similar to those above for the long spans, are tabulated in Table 16.2 and used in Table 16.3. They are also shown on the slab layout in Fig. 16.1.

The middle strip negative moment on the short span is normally considered the most critical with regard to required slab depth for flexure. In this example the total negative moment on this strip is -46.2 compared to -34.9 on the long strip, but the respective widths (Fig. 16.1a) are 3.6 m and 2.4 m, making the worse moment per meter width that on the long span ($34.9/2.4 = 14.50\,\text{kN}\cdot\text{m/m}$ versus $4.62/3.6 = 12.8\,\text{kN}\cdot\text{m/m}$).

TABLE 16.2 Moment Assignment for Sec. 16.2

		Long Span		Short Span	
		Neg. M	Pos. M	Neg. M	Pos. M
	M_{on}	282.7		218.7	
Total		$0.65\,M_{on}$	$0.35\,M_{on}$	$0.65\,M_{on}$	$0.35\,M_{on}$
		$= -183.7$	$= +98.9$	$= -142.1$	$= +76.5$
		↓	↓	↓	↓
Col. strip		-148.8	$+80.1$	-95.9	$+51.6$
		↙ ↘	↙ ↘	↙ ↘	↙ ↘
Beam	-126.5		$+68.1$	-81.5	$+43.9$
Slab		-22.3	$+12.0$	-14.4	$+7.7$
Mid. strip		-34.9	$+18.8$	-46.2	$+24.9$

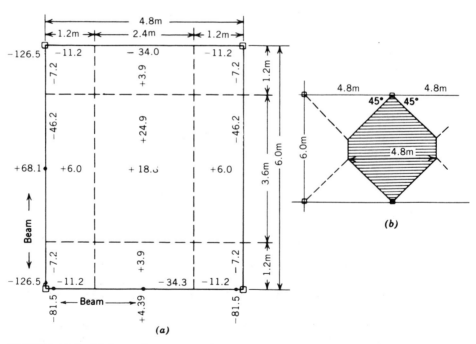

FIGURE 16.1 Slab on beams. (*a*) Distribution of moments to the strips and beams. (*b*) Load distribution to beams.

434

TABLE 16.3 Reinforcement for Slab of Sec. 16.2

Assume #10	Long Span		Short Span	
	Neg. M	Pos. M	Neg. M	Pos. M
Middle Strips				
Strip width	2.4 m		3.6 m	
M_{on}, kN · m	282.7		218.7	
M_n, kN · m	−34.9	+18.8	−46.2	+24.9
d, mm	125	125	125	115
$A_s = M_n \times 10^6/400$ $\times 0.95\, d$, mm²	735	396	973	596
$0.0018\, bh$, mm²	648	648	972	972
Min. no. bars, $s = 2h$	8	8	12	12
#10 bars(100 mm²)	8	8	12	12
Spacing, mm	300	300	300	300
Column Strips				
Strip width	Half strip = 1.2 m(including beam) = 1.1 m net		Half strip = 1.2 m(including beam) = 1.1 m net	
M_n, kN · m/half	−11.2	+6.0	−7.2	+3.9
d, mm	125	125	115	115
$A_s = M_n \times 10^6/400$ $\times 0.95d$, mm²	235	126	165	89
$0.0018\, bh$, mm²	297	297	297	297
Min. no. bars, $s = 2h$	4	4	4	4
#10 bars (100 mm²)	4	4	4	4
Spacing mm	300	300	300	300

435

For an effective depth $d = 150 - 20 - 5 = 125$ mm

$$k_n = M/bd^2 = 14.5 \times 10^{-3}/1 \times 0.125^2 = 0.93 \text{ MPa}$$

This is very much under the allowable k_n of 5.25 MPa in Table 3.1; the 150 mm slab is obviously satisfactory for moment.

USE slab $h = 150$ mm

(d) Check Minimum Column Stiffness for 3 m Story Height

The column stiffness requirement (already discussed in Sec. 15.8) is in the Code to protect slabs from pattern loading. The requirement should not be a problem where stiff beams are being used, but it will be checked anyway. The ratio of service dead to live load is $\beta_a = 3.5/5.0 = 0.7$, less than the criterion of 2, which causes the designer to check α_{min} from Table 16.2. In the two directions $\ell_2/\ell_1 = 6.0/4.8 = 1.25$ with $\alpha = 1.0$ and $\ell_2/\ell_1 = 4.8/6.0 = 0.8$ with $\alpha = 1.25$. Values of α_{min} increase as ℓ_2/ℓ_1 increases and decrease as the beam α increases. Hence the short span with $\ell_2/\ell_1 = 1.25$ and $\alpha = 1.0$ is the more critical for this check. Interpolating with $\ell_2/\ell_1 = 1.25$ and $\alpha = 1.0$ fixed:

$\beta_a = 1.0 \quad \alpha_{min} = 0$
$\beta_a = 0.5 \quad \alpha_{min} = 0.5$ $\qquad \beta_a = 0.70 \qquad \alpha_{min} = 0.30$

To use this, the stiffness of slab and beam must be developed for the short span.

$$I_s = \tfrac{1}{12} \times 6\,000 \times 150^3 = 1\,688 \times 10^6 \text{ mm}^4$$
$$I_b = \alpha I_s = 1.0\,I_s = 1\,688 \times 10^6 \text{ mm}^4$$

With slab and beams on both sides of the column and the column both above and below the joint:

$$\alpha_{min} = \Sigma K_c / \Sigma (K_s + K_b) = K_c/(K_s + K_b)$$

If the story height is 3 m

$$0.30 = (I_c/3.0)/(2 \times 1\,688 \times 10^6/4.8) = I_c/2\,110 \times 10^6$$
$$\text{Min. } I_c = 0.30 \times 2\,110 \times 10^6 = 633 \times 10^6 \text{ mm}^4 = h^4/12$$
$$\text{Min. } h = 295 \text{ mm}$$

The column size has not been specified, but should easily satisfy this minimum thickness.

(e) Slab Reinforcement

The positive moment bars in the middle strips do cross each other and require design for different depths. With moments (per unit width) so nearly the same it does not matter much whether long span or short span bars are given preferential location. Contrary to recommendations made in Chapter 12 design under Method 2 of the 1963 Code, the author will here place the long bars on the bottom, making:

$$d_\ell = 150 - 20 - 5 = 125 \text{ mm (for \#10 bars)}$$
$$d_s = 150 - 20 - 10 - 5 = 115 \text{ mm (for \#10 bars)}$$

Middle strip top bars in both directions may be in the top with $d = 125$ m, since the only interference is possibly near the quarter points of the sides. Column strip bottom bars will have to match what was done for middle strip bottom bars. Negative moments in the perpendicular column strips occur in the same area. Use long strip bars on top with $d = 125$ mm and short strip bars with $d = 115$ mm. The sketches and calculations in Table 16.3 complete the basic design of the slab.

The maximum bar spacing (13.4.2) is $2\,h_s = 300$ mm. The steel must also be adequate for temperature steel requirements, that is, $0.0018\,bh$.

(f) Beams

The moments are available for the beam designs except that the weight of stem must still be added. Although beams may be designed for strength, unless they provide $\alpha_1 \ell_2/\ell_1 \geqq 1.0$, the slab computations already made must be revised. This is equivalent to stating that the minimum stiffness represented by the assumed α values should be met. In this case this only means I_b in the short span of $1\,688 \times 10^6$ mm^4 or, say, a beam vertical projection below the 150 mm slab of about 225 mm over a 300 mm width.

The author calls attention to the fact that beam moments obtained in this way are lower than would be developed from conventional analysis. In effect this is a mild form of limit design for the beams comparable to the general slab design rules. As such, the effective percentage of reinforcement, in the author's opinion, should preferably meet the requirements of Code 8.4.3, that is, ρ or ρ-ρ' not exceeding $0.5\,\rho_b$, where ρ_b is that for a balanced member; this is lower than the $0.75\,\rho_b$ upon which Table 3.1 values of k_n are based. Code Chapter 13 is not written such as to require this limitation.

The shear load taken by the beams in this case must be that from the shaded area of Fig. 16.1b and any directly applied loads. If $\alpha_1 \ell_2/\ell_1 < 1.0$, the shear would be reduced proportionately with this ratio, with the remainder carried by the slab around the column.

(g) Exterior Panels

Design procedures for exterior panels parallel those of Sec. 15.13 for the flat plate, the column strip simply including beams and slab as above. There is one special provision (13.4.6) that requires special reinforcement, both top and bottom, in any *exterior* corner of the slab system to take care of a moment equal to the maximum positive moment per unit width of slab. The critical top reinforcement is parallel to the diagonal of the slab, and that on the bottom is at 90 degrees to that diagonal. Either diagonal bars or a rectangular grid of bars may be used, with the effective A_s/ft the maximum required in the slab for positive moment.

16.3 TWO-WAY SLAB WITH STIFFENED STRIPS—DESIGN EXAMPLE

Design an alternate interior panel slab for the conditions of Sec. 16.2, in this case considering the slab as stiffened only by strips 375 mm wide and 75 mm deeper than the slab running along the column centerlines each way.

Solution

(a) Slab Thickness

The α value for each beam can be established as soon as the slab thickness has been fixed. On the basis of h_s values already used in this chapter for deflection (150 mm in Sec. 16.2 for slab with stiff beams and 175 mm for a flat plate in Sec. 15.7 which happened to be 0.7 m wider than the present slab), 175 mm looks reasonable as an initial trial. Note that this requires an increase in dead load to 4.2 kPa, the initially specified 3.5 kPa not being adequate.

With a trial $h_s = 175$ mm the beam is then that of Fig. 16.2a. From the middle of the slab, y to the centroid becomes:

$$\bar{y} = -375 \times 75 \times 125/(525 \times 175 + 375 \times 75) = -29 \text{ mm}$$

$$\begin{aligned}
I_b: \tfrac{1}{12} \times 525 \times 175^3 &= 234 \times 10^6 \\
+ 525 \times 175 \times 29^2 &= 77 \times 10^6 \\
\tfrac{1}{12} \times 375 \times 75^3 &= 13 \times 10^6 \\
+ 375 \times 75(\tfrac{1}{2} \times 250 - 29)^2 &= 259 \times 10^6 \\
\hline
I_b &= \overline{583 \times 10^6} \text{ mm}^4
\end{aligned}$$

With $\ell_1 = 6.0$ m, $I_{s\ell} = \tfrac{1}{12} \times 4\,800 \times 175^3 = 2\,144 \times 10^6$ mm^4
With $\ell_1 = 4.8$ m, $I_{ss} = \tfrac{1}{12} \times 6\,000 \times 175^3 = 2\,680 \times 10^6$ mm^4

$$\alpha_{1\ell} = 583/2\,144 = 0.272 \text{ for long span}$$
$$\alpha_{1s} = 583/2\,680 = 0.218 \text{ for short span}$$

Average $\alpha_1 = 0.25$

The Code method is usable only if the relative beam stiffness in the two directions is within the ratio of 0.2 to 5.0 (Code 13.6.1.6).

$$(I_{b\ell}/\ell_\ell)/(I_{bs}/\ell_s) = (\alpha_{1\ell}/\alpha_{1s})(\ell_s/\ell_\ell)^2 = (0.276/0.221)(4.8/6.0)^2 = 0.80 \qquad \textbf{O.K.}$$

$\ell_n = 6.0 - 0.375 = 5.625$ m in the longer direction
$\beta = $ (long/short) clear span ratio $= 5.625/4.425 = 1.27$
$\beta_s = 1$, since continuous all around the panel

Compare the denominators of the h_s equations, using the PCA format (Sec. 15.11):

$$\text{Eq. 9-10: } 36 + 5\,\beta[\alpha_m - 0.5(1 - \beta_s)(1 + 1/\beta)]$$
$$= 36 + 5 \times 1.27 \times 0.25 = 37.6 > 36$$

The upper limit equation (9-12) is thus not useful here.

$$\text{Eq. 9-11: } 36 + 5\,\beta(1 + \beta_s)$$
$$= 36 + 5 \times 1.27(1 + 1) = \gg 37.7$$

Reqd. $h_s = (0.8 + 400/1\,500)5\,625/37.6 = 160$ mm < 175 mm
USE $h = 175$ mm

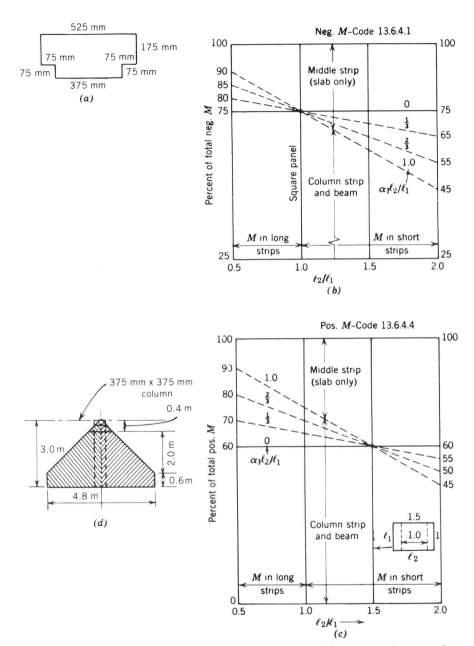

FIGURE 16.2 Slab of Sec. 16.3. (*a*) Beam used. (*b*) Moment assignment for negative moment. (*c*) Moment assignment for positive moment. (*d*) Shear load on beam.

439

(b) Shear Check

At this stage one can check the shear only in a rough fashion. If the slab is considered simply a flat plate, the critical shear section is a square around the column at $d/2$ beyond each face, with each side $375 + d = 375 + 140 = 515$ mm wide on each side of the column.

$$\text{Dead load} = 1.4 \times 4.2 = 5.88$$
$$\text{Live load} = 1.7 \times 5.0 = 8.50$$
$$w = \overline{14.38} \text{ kPa}$$

If concrete weighs 24 kN/m^3, the design stem weight $= 1.4 \times 24 \times 0.075 \times 0.375 = 0.95$ kN/m

$$V_u \doteq 14.38(6.0 \times 4.8 - 0.515^2) + 0.95 \times 2(6.0 + 4.8 - 2 \times 0.515) = 429 \text{ kN}$$
$$v_c = V_u/(\phi bd) = 429 \times 10^{-3}/0.85 \times 4 \times 0.515 \times 0.140 = 1.75 \text{ MPa}$$

The shear exceeds the allowable of $\frac{1}{3}\sqrt{20} = 1.49$ MPa, on the basis of the oversimplified concept of the flat plate. The designer has several choices. First the slab might be thickened, although it has not been proved that the beams with stirrups, in spite of their lower v_c, cannot handle v_u; the beams are rather shallow for effective stirrups. Second, the concrete strength might be increased to $f'_c = 30$ MPa which gives allowable $v_c = 1.83$ MPa. Third, the beams might be widened to make the 250 mm thickness effective all around the column, roughly a width of $375 + 175 = 550$ mm plus a little to keep the failure plane in the deeper elements, say, 650 mm. This same result could be accomplished by adding a triangle of concrete in each corner between beams to build up a kind of octagonal drop panel effect to resist shear. Fourth, a steel shearhead might be included within the slab depth. Tentatively, the design will proceed with f'_c to be changed to 30 MPa if that becomes necessary.

(c) Design Moments

Long Span

$$M_{o\ell} = 0.125 \times 14.38 \times 4.8 \times 5.625^2 = 273 \text{ kN} \cdot \text{m}$$

Total negative moment $M_u = -0.65 \times 273 = -178$ kN $\cdot$ m*

Total positive moment $M_u = 0.35 \times 273 = +96$ kN $\cdot$ m

Coefficients for moment distribution or assignment are given in Table 16.1 but for visualization have also been plotted in Fig. 16.2b,c.† For use, the following constants are required:

$$\alpha_1 = 0.28, \quad \ell_2/\ell_1 = 4.8/6.0 = 0.80, \quad \alpha_1\ell_2/\ell_1 = 0.22$$

---

*Although the author prefers going to M_n immediately, this example has not been revised to that format.

†See also further detailed discussion in Sec. 10.21.

Negative moment to the column strip is then $75 + (0.22/0.333)(0.2/0.5)5 = 75 + 1.3 = 76\%^*$ and positive moment to column strip is $60 + (0.22/0.333)(0.7 \times 10) = 65\%$. Code 13.6.5.1 for $\alpha_1\ell_2/\ell_1 > 1.0$ assigns $85 \, \alpha_1\ell_2/\ell_1$ percent of the negative or positive moment in the column strip to the beam. For $\alpha_1\ell_2/\ell_1$ between 1 and zero, Code 13.6.5.2 requires the designer to decrease the 85 percent to beam linearly with the decrease in $\alpha_2\ell_2/\ell_1$. These two conditions lead to:

Column strip negative moment $= -0.76 \times 178 = -135 \text{ kN} \cdot \text{m}$

  Beam neg. $M_u = -(0.85 \times 0.22)135 = -25 \text{ kN} \cdot \text{m}$

  Slab neg. $M_u = -135 + 25 = -110 \text{ kN} \cdot \text{m}$, or $-55 \text{ kN} \cdot \text{m}$ per half strip

Column strip positive $M_u = 0.65 \times 96 = +62 \text{ kN} \cdot \text{m}$

  Beam pos. $M_u = +(0.85 \times 0.22)62 = +12 \text{ kN} \cdot \text{m}$

  Slab pos. $M_u = 62 - 12 = 50 \text{ kN} \cdot \text{m}$, or $+25 \text{ kN} \cdot \text{m}$ per half strip

Middle strip neg. $M_u = -178 + 135 = -43 \text{ kN} \cdot \text{m}$

Middle strip pos. $M_u = 96 - 62 = +24 \text{ k-ft}$

The author notes that with this very light beam a distribution in proportion to web width alone would have given $-18 \text{ kN} \cdot \text{m}$ to the beam, which is over two-thirds of the $-25 \text{ kN} \cdot \text{m}$ found above; with the stiffer beam of Sec. 16.2 the 85% assignment led to $-114 \text{ kN} \cdot \text{m}$ to the beam.

Short span

$M_{os} = 0.125 \times 14.38 \times 6.0 \times 4.425^2 = 211 \text{ kN} \cdot \text{m}$

Total negative moment $M_u = -0.65 \times 211 = -137 \text{ kN} \cdot \text{m}$

Total positive moment $M_u = 0.35 \times 211 = +74 \text{ kN} \cdot \text{m}$

$\alpha_1 = 0.22, \ell_2/\ell_1 = 1.25, \alpha_1\ell_2/\ell_1 = 0.28$

For negative moment to the column strip Fig. 16.2b shows $75 - 2 = 73\%$ and for positive moment $60 + (0.28/0.333)0.7 \times 10 = 66\%$ (Fig. 16.2c)

Column strip neg. $M_u = -0.73 \times 137 = -100 \text{ kN} \cdot \text{m}$

  Beam neg. $M_u = -(0.85 \times 0.28)100 = -24 \text{ kN} \cdot \text{m}$

  Slab neg. $M_u = -100 + 24 = -76 \text{ kN} \cdot \text{m}$, or $-38 \text{ kN} \cdot \text{m}$ per half strip

Column strip pos. $M_u = +0.66 \times 74 = +49 \text{ kN} \cdot \text{m}$

  Beam pos. $M_u = +(0.85 \times 0.28)49 = 12 \text{ kN} \cdot \text{m}$

  Slab pos. $M_u = +49 - 12 = +37 \text{ kN} \cdot \text{m}$, or $19 \text{ kN} \cdot \text{m}$ per half strip

Middle strip neg. $M_u = -137 + 100 = -37 \text{ kN} \cdot \text{m}$

Middle strip pos. $M_u = +74 - 49 = +25 \text{ kN} \cdot \text{m}$

---

*This calculation, in effect, modifies the line for $\alpha_1\ell_2/\ell_1$ of $\frac{1}{3}$ in Fig. 16.2b to fit $\alpha_1\ell_2/\ell_1$ of 0.25. A value closer than a full percentage point is unrealistic. The curve equations are given in Sec. 16.8 along with some tabulated values in Table 16.8.

TABLE 16.4 Reinforcement for Slab of Sec. 16.3.

$h_s = 175$ mm generally, with 375 mm wide by 250 mm between columns

Middle Strips

	Long Span 2.4 m		Short Span 3.6 m	
Strip width	Neg. M	Pos. M	Neg. M	Pos. M
M_u, kN·m	-43	$+34$	-37	$+25$
$M_n = M_u/\phi$, kN·m	-48	$+38$	-41	$+28$
d, mm	150	150	150	140
$A_s = M_n \times 10^6/400$ $\times 0.95\,d$, mm²	842	666	719	526
0.0018 bh, mm²	756	756	1134*	1134
Min. no. bars, $s = 2\,h_s$	7	7	11	11
USE #10 (100 mm²)	9-#10	8-#10	8-#10 11-#10	12-#10
Spacing, mm	250	300	325	300

Column Strip

Long span

Strip width

Slab (each half strip)
Half strip = 1.2 m (incl. bm.)
= 1012 mm

Beam
$b_w = 375$ mm, $h = 250$ mm

	Neg. M	Pos. M	Neg. M	Pos. M
M_u, kN·m	-55	$+25$	-25	$+12$
$M_n = M_u/0.9$, kN·m	-61	$+28$	-28	$+13$

Slab (each half strip) — Half strip = 1.2 m (incl. bm.) = 1012 mm net
Beam $b_w = 375$ mm

	Slab		Beam	
	Neg. M	Pos. M	Neg. M	Pos. M
d, mm	147	147	212	197
$A_s = M_n \times 10^6/400 \times 0.95\,d$, mm²	1092	501	347	173
0.018 bh	318	318	169	169
Min. no. bars, s = 2h	3	3	1	1
USE bars	6-#15	6-#10	2-#15	2-#10
Spacing, mm	150	150	150	150

Column strip
Short span
Strip width

	Slab		Beam	
	Neg. M	Pos. M	Neg. M	Pos. M
M_u, kN·m	−38	+19	−24	+12
M_n, kN·m	−42	+21	−27	+13
d, mm	147	147	212	197
				0.27
$A_s = M_n \times 10^6/400 \times 0.95\,d$, mm²	751	376	335	173
0.0018 bh	318	318	169	169
Min. no. bars, s = 2h	3	3	1	1
USE bars	4-#15	4-#10	2-#15	2-#10
Spacing, mm	250	250	150	150

*Half of bottom bars count on this.

(d) Depth for Moment

The worst slab moment is $M_u = -55$ kN $\cdot$ m on a half column strip of width $\frac{1}{2}(2.4 - 0.375) = 1.013$ m in the long span where $d = 150$ mm $M_n \geqslant -55/0.9 = -61$ kN $\cdot$ m

$$k_n = 61 \times 10^{-3}/1.013 \times 0.15^2 = 2.68 \text{ MPa}$$

The allowable from Table 3.1 is 5.25 MPa

The worst beam section is 375 mm wide and carries $M_u = -25$ kN $\cdot$ m with d about $250 - 20$ top cover $- 10$ stir. $- 8 = 212$ mm. If beam moments were more important the normal beam cover of 40 mm would be more appropriate. $M_n = -25/0.9 = -28$ kN $\cdot$ m

$$k_n = 28 \times 10^{-3}/0.375 \times 0.212^2 = 1.66 \ll 5.25 \text{ MPa}$$

(e) Recheck Shear

A recheck is needed because the preliminary estimate left questions to be settled later. The load area contributing to the beam load is that shown shaded in Fig. 16.2d, but Code 13.6.8.2 reduces the load to $\alpha_1 \ell_2 / \ell_1$ parts where this ratio is less than 1.0. On the long beam, the critical shear is at d from face of column, $\frac{1}{2} \times 5.625 - 0.212 = 2.60$ m from midspan.

Gross V_u at critical section excluding beam weight
$$= (2.6 \times 4.8 - 2 \times \tfrac{1}{2} \times 2.0^2) \times 14.38 = 121.9 \text{ kN}$$

Effective $V_u = (\alpha_1 \ell_2 / \ell_1) 121.9 + \text{beam weight}$
$$= 0.22 \times 121.9 + 0.95 \times 2.60 = 29.3 \text{ kN}$$

$$v_c = 29.3 \times 10^{-3}/0.85 \times 0.375 \times 0.212 = 0.43 \text{ Mpa} < \tfrac{1}{6}\sqrt{25} = 0.83 \text{ MPa} \qquad \textbf{O.K.}$$

A quick mental check on the short beam with $\alpha_1 \ell_2 / \ell_1 = 0.28$ and less gross load shows it also has a low v_c. However, the remaining shear, around 270 kN, cannot be carried by the slab in the corners between the beams. (Allocating the low shear to the beams has increased the shear left for the remaining slab.)

With the beams working so inefficiently under the Code assignments, the check must return to the flat plate concept. The shear capacity, as a matter of logic, is not automatically reduced by deepening some portions 75 mm. The original calculation of $v_u = 1.75$ MPa > 1.49 MPa allowable still stands under this assumption.

The author is personally confident that the excess 0.26 MPa could be summed all around the critical section, and this excess assigned to stirrups in the four beams. Minimum stirrups ($\rho_n f_y$ of 0.33 MPa) would be ample, but stirrups in a 250 mm thickness lead to questions about stirrup development. If the irregular depth was created by steps on the tension face, instead of on the compression face, the greater effective depth would be available. The author is hesitant to count this extra depth here since diagonal cracks might break through the 175 mm depth before the 250 mm slab became fully effective. (Such a minimized resistance must be accepted where tests of unusual situations are not available.) The assumption is made here that specification of $f'_c = 30$ MPa is more acceptable than a change in slab thickness or stiffening thickness. With the change in concrete strength the allowable stress $\frac{1}{3}\sqrt{30} = 1.83$ MPa is greater than the calculated stress of 1.60 MPa.

$$\text{USE } f'_c = 30 \text{ MPa}$$

(f) Check on Minimum Column Size

A check on column stiffness is required where the ratio β_a, the ratio of service dead to live load, is less than 2. Here $\beta_a = 4.2/5.0 = 0.84$. Relative beam stiffness is 0.22 in the short direction where $\ell_2/\ell_1 = 1.25$, and 0.28 in the long direction where $\ell_2/\ell_1 = 0.8$. Table 15.2 (Sec. 15.8) shows that $\alpha_{\min}$ increases as ℓ_2/ℓ_1 increases and also as α decreases. Therefore the short span controls for a square column. Assume a column height of 3.0 m. With a beam stiffness ratio α of 0.22 and a length ratio of 1.25, Table 15.2 leads to:

$$\beta_a = 1.0 \qquad \alpha_{\min} \quad 0.6$$
$$\beta_a = 0.5 \qquad \alpha_{\min} = 1.5 \qquad \beta_a = 0.84 \qquad \alpha_{\min} = 1.0$$

$$\alpha_{\min} = \Sigma K_c/(\Sigma K_s + \Sigma K_b) = (I_c/\ell_c)/(I_s/\ell_s + I_b/\ell_b)$$

The values of I_s and I_b were calculated in (a) above.

$$1.0 = (2\, I_c/3.0)/(2 \times 2\,680 \times 10^6/4.8 + 2 \times 593 \times 10^6/4.8) = 2\, I_c/3.0 \times 1\,364 \times 10^6$$

Min. $I_c = 2\,046 \times 10^6$ mm^4 Min h for a square column = 396 mm

The column is too small.

$$\text{USE a column 400 mm square.}$$

(g) Reinforcement design

The calculations are summarized in Table 16.4 with no new problems except the extra calculation of beam steel and a possible alternate decision as to whether the 75 mm deepening of the slab really makes a beam with 40 mm cover required or just a deepened slab with 20 mm cover. The assigned moments are so low for positive moment it, in fact, makes no difference here. The table treats it as a slab type of element for positive and negative moment.

16.4 USE OF EQUIVALENT FRAME METHOD OF SLAB ANALYSIS

For slab systems not meeting the direct method requirements of Sec. 15.5 (Code 13.6.1) the design moments must be computed by the equivalent frame method;[2] elsewhere it is an optional method.

As in an analysis of beams and columns, the structure is subdivided into bents running each way through the frame. The entire slab width, that is, one-half panel width on each side of the column, is considered in establishing the slab load and slab stiffness. Stiffness is based on gross concrete area, but considers a variable moment of inertia because the stiffness at column or drop panel becomes part of the problem.

If the service live load is not more than 75% of the dead load, the frame analysis is made with full design (factored) load on all spans instead of a pattern loading. For higher live load ratios pattern loads must be used (Sec. 16.7) but, as in frame analysis for beam design, certain simplifications are permissible:

a. Columns fixed at their far end (at floor above and below).
b. Two adjacent spans loaded with live load for maximum negative moment.
c. Alternate spans loaded with live load for maximum positive moment.
d. Slab-beams may be considered fixed two panels away from the point where moment is being computed, provided the slab continues further.

In addition the pattern loading may be applied with only 75% of the design live load.

Although spans to centerline of columns are used in analysis, the critical negative moments are considered those at the face of any rectangular column or, if not rectangular, at the face of the equivalent (equal area) square column.

The chief complication is the use of equivalent column stiffness K_{ec}, recognizing the part that the slab or transverse beam torsional stiffness plays in the system. Equivalent stiffness has already been discussed in Sec. 15.9 and calculated for an exterior column in Sec. 15.13b (but there using a simplified version of K_c for the column).

16.5 FRAME CONSTANTS

Since no member of the equivalent frame has a uniform I, the stiffness is not $4\,EI/\ell$ but something more than 4 for the constant and something more than 0.5 for the carry-over factor, C_f* or C.O.F. Tables can tabulate k for use in kEI/ℓ and C_f, although C_f is directional, not the same from each end except with symmetry of cross section. Likewise the fixed-end moment is normally greater than for uniform I.

Table 16.5, slightly modified from Reference 1, lists for flat plates the slab stiffness factors, moment factors, carry-over factors, and at the very bottom the useful† quantity $(1 - c_2/\ell_2)^3$. The EI variation used in preparing this table is sketched alongside. With a drop panel the EI for the drop length is that of a very flat wide T-shape of width ℓ_2; this requires a

---

*The symbol C_f (or other subscript) is used since the Code has preempted C for a torsional constant.
†Useful in the torque stiffness calculation.

TABLE 16.5 Moment Distribution Factors for Slab-Beam Elements (Flat Plate with or without Column Capital)

c_1/ℓ_1 \ c_2/ℓ_2		0.00	0.05	0.10	0.15	0.20	0.25	0.30
0.00	M	0.083	0.083	0.083	0.083	0.083	0.083	0.083
	k	4.000	4.000	4.000	4.000	4.000	4.000	4.000
	C_f	0.500	0.500	0.500	0.500	0.500	0.500	0.500
0.05	M	0.083	0.084	0.084	0.084	0.085	0.085	0.085
	k	4.000	4.047	4.093	4.138	4.181	4.222	4.261
	C_f	0.500	0.503	0.507	0.510	0.513	0.516	0.518
0.10	M	0.083	0.084	0.085	0.085	0.086	0.087	0.087
	k	4.000	4.091	4.182	4.272	4.362	4.449	4.535
	C_f	0.500	0.506	0.513	0.519	0.524	0.530	0.535
0.15	M	0.083	0.084	0.085	0.086	0.087	0.088	0.089
	k	4.000	4.132	4.276	4.403	4.541	4.680	4.818
	C_f	0.500	0.509	0.517	0.526	0.534	0.543	0.550
0.20	M	0.083	0.085	0.086	0.087	0.088	0.089	0.090
	k	4.000	4.170	4.346	4.529	4.717	4.910	5.108
	C_f	0.500	0.511	0.522	0.532	0.543	0.554	0.564
0.25	M	0.083	0.085	0.086	0.087	0.089	0.090	0.091
	k	4.000	4.204	4.420	4.648	4.887	5.138	5.401
	C_f	0.500	0.512	0.525	0.538	0.550	0.563	0.576
0.30	M	0.083	0.085	0.086	0.088	0.089	0.091	0.092
	k	4.000	4.235	4.488	4.760	5.050	5.361	5.692
	C_f	0.500	0.514	0.527	0.542	0.556	0.571	0.585
$X = (1 - c_2/\ell_2)^3$		1.000	0.856	0.729	0.613	0.512	0.421	0.343

$FEM = Mw\ell_2\ell_1^2$
$K = kEI_s/\ell_1$ with
$I_s = \ell_2 h^3/12$
C.O.F. $= C_F$

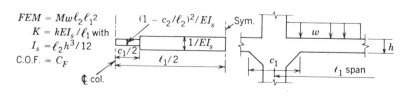

separate table. Reference 1 has similar tables for drop panel thicknesses of 1.25 h_s and 1.50 h_s in symmetrical panels. The 1977 ACI Code Commentary now includes moment distribution constants (FEM, stiffness, carry-over factor) with equal or unequal columns, not only for slabs with and without **drop panels as in 1971, but now also for 'slab-beam' members with column** capitals, both with and without drop panels (of 1.25 h_s thickness).

Values of column K_{ec} cannot be readily tabulated, but K_c values for the column are easily obtained from the coefficients in Table 16.6, also slightly modified from Reference 1. The *EI* variations used are indicated on the sketch with the table, $1/EI$ as zero for the thickness of the slab (and of the drop if any) and for the column capital height. Since such a stiffness is different at the two ends (if there is either drop or capital or both), care must be taken to use a/ℓ_c for the *near end* and b/ℓ_c for the *far end* in every case. In this way the one table works equally well for the column above or below a joint.

Alternatively, for K_c the curve data of Fig. 16.3a should give the same values plus the carry-over factors, *provided one used the column length* ℓ_c

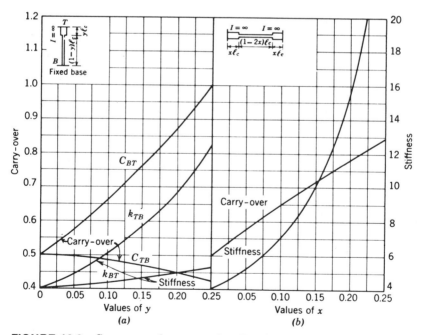

FIGURE 16.3 Carry-over factors and stiffness factors for member with infinite *I* over part of length. (Adapted from Reference 3, Portland Cement Assn.)

TABLE 16.6 Column Stiffness Coefficients, k_c*

a/ℓ_c \ b/ℓ_c	0.00	0.02	0.04	0.06	0.08	0.10	0.12	0.14	0.16	0.18	0.20	0.22	0.24
0.00	4.000	4.082	4.167	4.255	4.348	4.444	4.545	4.651	4.762	4.878	5.000	5.128	5.263
0.02	4.337	4.433	4.533	4.638	4.747	4.862	4.983	5.110	5.244	5.384	5.533	5.690	5.856
0.04	4.709	4.882	4.940	5.063	5.193	5.330	5.475	5.627	5.787	5.958	6.138	6.329	6.533
0.06	5.122	5.252	5.393	5.539	5.693	5.855	6.027	6.209	6.403	6.608	6.827	7.060	7.310
0.08	5.581	5.735	5.898	6.070	6.252	6.445	6.650	6.868	7.100	7.348	7.613	7.897	8.203
0.10	6.091	6.271	6.462	6.665	6.880	7.109	7.353	7.614	7.893	8.192	8.513	8.859	9.233
0.12	6.659	6.870	7.094	7.333	7.587	7.859	8.150	8.461	8.796	9.157	9.546	9.967	10.430
0.14	7.292	7.540	7.803	8.084	8.385	8.708	9.054	9.426	9.829	10.260	10.740	11.250	11.810
0.16	8.001	8.291	8.600	8.931	9.287	9.670	10.080	10.530	11.010	11.540	12.110	12.740	13.420
0.18	8.796	9.134	9.498	9.888	10.310	10.760	11.260	11.790	12.370	13.010	13.700	14.470	15.310
0.20	9.687	10.080	10.510	10.970	11.470	12.010	12.600	13.240	13.940	14.710	15.560	16.490	17.530
0.22	10.690	11.160	11.660	12.200	12.800	13.440	14.140	14.910	15.760	16.690	17.210	18.870	20.150
0.24	11.820	12.370	12.960	13.610	14.310	15.080	15.920	16.840	17.870	19.000	20.260	21.650	23.260

*$K_c = \dfrac{k_c E_c I_c}{\ell_c}$

449

TABLE 16.7 Values of Torsion Constant, C* (in 10^6 mm^4)

y mm \ x mm	100	125	150	175	200	225	250	300	350	400
300	79	144	231	339	464	601	742	999	—	—
350	96	177	288	428	597	791	1 002	1 449	1 851	—
400	113	209	344	518	731	980	1 263	1 899	2 565	3 157
450	129	242	400	607	864	1 170	1 523	2 349	3 280	4 224
500	146	274	456	696	997	1 360	1 784	2 799	3 995	5 291
550	163	307	513	786	1 130	1 550	2 044	3 249	4 709	6 357
600	179	339	569	875	1 264	1 740	2 305	3 699	5 424	7 424
750	229	437	738	1 143	1 664	2 309	3 086	5 049	7 568	10 623
900	279	535	906	1 411	2 064	2 879	3 867	6 399	9 711	13 824
1 050	329	632	1 075	1 679	2 464	3 448	4 648	7 749	11 855	17 024
1 200	379	730	1 244	1 912	2 864	4 018	5 430	9 099	13 999	20 224
1 350	429	827	1 413	2 215	3 264	4 588	6 211	10 449	16 143	23 424
1 500	479	925	1 581	2 483	3 664	5 157	6 992	11 799	18 286	26 623

$*C = (1 - 0.63\, x/y)\, \dfrac{x^3 y}{3}$

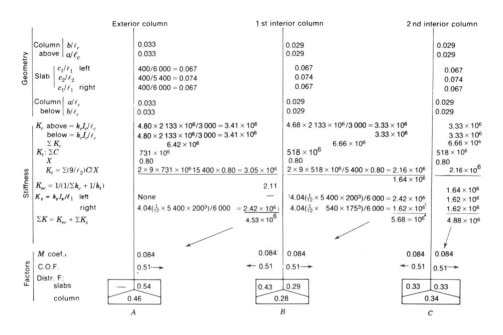

FIGURE 16.4   Factors needed for moment distribution, flat plate.

reduced to the distance to the far fixed end created by the face of the slab above or below (or the start of the capital above where a capital is used).‡ Since both stiffness and carry-over factor are directional, the first subscript designates the joint which is rotating. Thus k_{TB} applies for a joint rotation at the top and C_{TB} gives the C.O.F. from top to bottom of the column; k_{BT} and C_{BT} are in the opposite direction.

The torsional stiffness of the transverse beam or slab (see Sec. 15.13b) is $K_t = 9\,EC/[\ell_2(1 - c_2/\ell_2)^3]$ for the transverse span *on one side* of the column. Values of $(1 - c_2/\ell_2)^3$ have already been tabulated at the bottom of Table 16.5. Values of $C = \Sigma(1 - 0.63\,x/y)x^3y/3$ have been tabulated for *individual* rectangles in Table 16.7, also adapted slightly from Reference 1. The values for individual rectangles must be summed to care for an L- or T-section.

The use of these tables and the calculation of K_{ec} and the distribution factors make up a rather involved process that needs some formalizing to maintain clarity of presentation. An example is developed in the next

---

‡The moment then carried over acts at the face of slab (the assumed far fixed end).

section using a columnar form (Fig. 16.4) suggested by Reference 1, but paralleling this with the detailed number calculations.

16.6 EQUIVALENT FRAME METHOD—DESIGN EXAMPLE

Although neither the spans nor the loading require it, use the data of Sec. 15.7 (interior panel) and 15.13 (exterior panel) to work out corresponding design moments for the 6 m span by the equivalent frame method and compare the results.

Panels are 5.4 m by 6.0 m (center to center of columns) with the 4.8 m length parallel to the exterior edge. Columns are 400 mm square, both interior and exterior, with a 3.0 m story height, and the slab runs to the exterior face of the outside column without a spandrel beam. For simplicity, any wall load has been excluded from the example. Live load is given as 3 kPa and total dead weight 5 kPa (including slab weight). Slab thickness for interior panel is 175 mm with 200 mm for the exterior panel. The difference in slab weight has been ignored (5 kPa used as the total in both panels), making the design dead load 7.0 kPa and the design live load 5.1 kPa. Here the break in slab stiffness will be assumed on the centerline of the first interior column row. $f'_c = 25$ MPa for all columns and slabs.

Solution

The sequence of tabulation in Fig. 16.4 is used as a guide for discussion, followed by items e and f.

a. Geometry
b. Stiffnesses
c. Factors (FEM, COF, distribution factors)
d. Distribution (moment distribution for each loading)
e. Design moments from each loading
f. Possible moment reductions because the slab qualifies for the direct design method.

(a) Geometry

Geometry is simplified by the uniform spans and equal column sizes. For the exterior column the half-depth of slab gives $a/\ell_c = b/\ell_c = \frac{1}{2} \times 200/3\,000 = 0.033$. For the typical interior columns $a/\ell_c = b/\ell_c = \frac{1}{2} \times 175/3\,000 = 0.029$. Where the break in slab thickness occurs, one might use average thickness of slab to fix a and b, but that much accuracy is not really inherent in the process; use value for typical interior column, 0.029.

For the slab $c_1/\ell_1 = 400/6\,000 = 0.067$ at each end and c_2/ℓ_2 for width is $400/5\,400 = 0.074$.

(b) Stiffnesses

For K_c the a/ℓ_c and b/ℓ_c are available above. For the main part of the column above and below the joint, I_c is $\frac{1}{12} \times 400^4 = 2\,133 \times 10^6$ mm^4, E_c may be omitted (that is, used as 1 since all concrete is alike in the frame and only relative stiffnesses are finally involved). From Table 16.6 the value of k_c for exterior column ($a/\ell_c = b/\ell_c = 0.033$) is 4.80, based on interpolation.

$$K_c = 4.80 \times 2\,133 \times 10^6/3\,000 = 3.413 \times 10^6$$
$$\Sigma K_c = 2K_c = 6.826 \times 10^6$$

For the transverse beam at the column line the Code assumes a constant cross section, using the largest of:

1. That part of the slab having a width equal to the width of the column, bracket, or capital, in this case at the exterior column a member 400 mm by 200 mm.
2. The slab just described plus that part of any transverse beam above and below the slab, in this case none.
3. The beam of Fig. 15.10b, in this case none.

From Table 16.7, C for the 400 mm by 200 mm "beam" at the exterior column line is 731×10^6 mm^4. (Note that interpolations in the y direction are nearly linear, but not in the x direction; any reasonable interpolation is probably as accurate as the method justifies.) Since there is a beam of this type on each side of the column,

$$K_t = 2[9\ EC/\ell_2(1 - c_2/\ell_2)^3] = 18\ C/(\ell_2 X)$$

where X is the bottom line of Table 16.5 and shows as 0.80 for c_2/ℓ_2 of 0.074

$$K_t = 18 \times 731 \times 10^6/5\,400 \times 0.80 = 3.046 \times 10^6\ \text{mm}^4$$

For the exterior column,

$$K_{ec} = 1/(1/\Sigma K_c + 1/K_t) = 10^6/(1/6.826 + 1/3.046) = 2.106 \times 10^6\ \text{mm}^4, \text{ with } E \text{ of } 1.$$

For the slab in the exterior panel, $c_2/\ell_2 = 0.074$ and $c_1/\ell_1 = 0.067$, which in Table 16.5 leads to

$$M \text{ coef.} = 0.084,\ k = 4.04,\ C_f = 0.51$$
$$K_s = 4.04(\tfrac{1}{12} \times 5\,400 \times 200^3)/6\,000 = 2.424 \times 10^6\ \text{mm}^4, \text{ with } E \text{ of } 1.$$

For the exterior joint, $\Sigma K = (2.106 + 2.424)\,10^6 = 4.530 \times 10^6\ \text{mm}^4$

For the first interior column line with a/ℓ_c and b/ℓ_c reflecting a 175 mm slab, K_c is changed to 4.212×10^6 mm^4 and K_t is changed by the transition between the 200 mm and 175 mm slab shown in Fig. 16.5.

For C the beam shown can be divided into whatever rectangles give a maximum total which appears to be the 400 mm by 175 mm area plus a 200 mm $\times$ 25 mm area. From Table 16.7.

$$C_1 = 518 \times 10^6\ \text{mm}^4$$
$$C_2 = (1 - 0.63 \times 25/200)\tfrac{1}{3} \times 25^3 \times 200 = 1 \times 10^6\ \text{mm}^4 \text{ (negligible)}$$
$$\text{Use } C = 518 \times 10^6\ \text{mm}^4$$

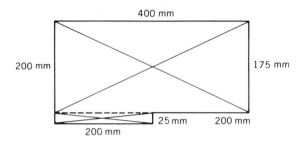

FIGURE 16.5 "Beam" cross section at first interior column.

Since all other factors aside from C remain unchanged, $K_t = 3.046 \times 10^6 \times 518/731 = 2.158 \times 10^6$ mm^4. For the interior column, equivalent

$$K_{ec} = 10^6/(1/6.826 + 1/2.158) = 1.641 \times 10^6 \text{ mm}^4$$

With the slab changing only in thickness from 200 mm to 175 mm,

$$K_s = 2.424 \times 10^6 \, (175/200)^3 = 1.624 \times 10^6 \text{ mm}^4 \text{ for an interior slab.}$$

The ΣK for the joint at the first interior column $= (1.624 + 2.424 + 1.641) \times 10^6 = 5.689 \times 10^6$ mm^4.

The ΣK for the joint at the second interior column $= (1.624 + 1.624 + 1.641) \times 10^6$
$$= 4.889 \times 10^6 \text{ mm}^4$$

The moment distribution factors are:

$$\text{Slab at } A: \frac{2.424 \times 10^6}{4.530 \times 10^6} = 0.54$$

$$\text{Column at } A: 1 - 0.54 = 0.46$$

$$\text{Slab at } B, \text{ exterior span:} \frac{2.424 \times 10^6}{5.689 \times 10^6} = 0.43$$

$$\text{Slab at } B, \text{ interior span:} \frac{1.624 \times 10^6}{5.689 \times 10^6} = 0.29$$

$$\text{Column at } B: 1 - 0.43 - 0.29 = 0.28$$

$$\text{Slab at } C, \text{ both spans:} \frac{1.624 \times 10^6}{4.889 \times 10^6} = 0.33$$

$$\text{Column at } C: 1 - 0.33 - 0.33 = 0.34$$

These data are shown diagrammatically in Fig. 16.4.

(c) Fixed-end Moments

Since $w_l/w_d = 3/5 = 0.60 < 0.75$, the Code accepts the use of *full* design load on all the frame with no pattern loadings required. For the end span, Table 16.5 shows the moment factor as 0.084 for fixed-end moment, just a trifle more than the nominal 1/12.

Load per linear meter $= (7 + 5.1) \times 5.4 = 65.34$ kN/m

End span fixed-end $M = M^F = 0.084 \times 65.34 \times 6.0^2$

$$= 198 \text{ kN} \cdot \text{m}$$

Interior span fixed-end $M = M^F = 198$ kN · m

At the exterior column, there is a 200 mm width of slab which is beyond the column center line and adds a negative $M^F = -\frac{1}{2} \times 12.1(5.4 - 0.4) \times 0.2^2 - 1.2$ kN · m. This is almost negligible but is mentioned here since there would usually be some wall weight and the slab often cantilevers significantly beyond the column itself.

(d) Moment Distribution

Consider, closely enough, that the frame system, Fig. 16.6, is fixed (roughly symmetrical) about the fourth column line. It is fully loaded, has uniform spans,

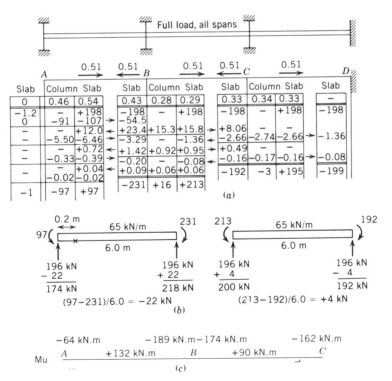

FIGURE 16.6 Moment distribution and steps toward design moments. (a) M distribution. (b) Shears. (c) Design moments.

and should have very little rotation several joints away from the exterior column. The alternate joint sequence is used for the moment distribution.

(e) Design Moments

The moments thus found by moment distribution are at the center of column lines and should be corrected to the face of support for rectangular (or equivalent square supports), but in no case is the correction to be farther than $0.175\,\ell_1$ from the column center line. In case of an exterior column with a bracket or capital, the correction must be reduced to that for only half the actual projection used at the column. In this example all corrections are to face of column, 0.2 m from the column center line.

On the basis of the beam free bodies sketched in Fig. 16.6b the simple beam shear as half of the beam load, and the continuity shear as the difference between the end moments divided by the span have been calculated. Summation of moments about the face of column then leads to the following design moments.

$$\text{Ext. } M_{AB} = -97 + 174 \times 0.2 - \tfrac{1}{2} \times 65 \times 0.2^2 = -64 \text{ kN} \cdot \text{m}$$

$$\text{1st Int. } M_{BA} = -231 + 218 \times 0.2 - \tfrac{1}{2} \times 65 \times 0.2^2 = -189 \text{ kN} \cdot \text{m}$$

$$M \text{ at midspan} = +\tfrac{1}{8} \times 65 \times 6.0^2 - \tfrac{1}{2}(97 + 231) = +129 \text{ kN} \cdot \text{m}$$

$$V \text{ at midspan} = 174 - 65 \times 3.0 = -21.0 \text{ kN}$$

$$\text{For max. } M, e \text{ from midspan} = -21/65 = 0.32 \text{ m (to left)}$$

$$M = \text{area of } V \text{ diagram over length } e = +\tfrac{1}{2} \times 21 \times 0.32 = +3.4 \text{ kN} \cdot \text{m}$$

$$\text{Max. pos. } M = +129 + 3 = +132 \text{ kN} \cdot \text{m}$$

In similar fashion for span BC:

$$M_{BC} = -213 + 200 \times 0.2 - \tfrac{1}{2} \times 65 \times 0.2^2 = -174 \text{ kN} \cdot \text{m}$$

$$M_{CB} = -192 + 192 \times 0.2 - \tfrac{1}{2} \times 65 \times 0.2^2 = -155 \text{ kN} \cdot \text{m}$$

$$\text{Pos. midspan } M = +\tfrac{1}{8} \times 65 \times 6.0^2 - \tfrac{1}{2}(213 + 192) = +90 \text{ kN} \cdot \text{m}$$

The slight shift in the positive moment maximum point is negligible in this case with the end moments differing only 11%.

(f) Check Against M_o For Possible Economy

Since this slab is eligible for the direct design method, check the total moment indicated above against the M_o total that would be required by the direct design (Sec. 15.13). $M_o = \tfrac{1}{8} \times 65.34 \times (6 - 0.4^2) = 256 \text{ kN} \cdot \text{m}$. The total that is comparable is the sum of the positive moment plus the average end moment, which is the simple beam moment.

$$\text{Simple beam } M, \text{ ext. span} = 132 + \tfrac{1}{2}(64 + 189) = 258 \text{ kN} \cdot \text{m}$$

is a trifle more than M_o.

On the first interior span the M_o value is also 256 kN · m, from Sec. 15.7. For the design moments,

$$\text{Simple beam } M = +90 + \tfrac{1}{2}(174 + 155) = 254.5 \text{ kN} \cdot \text{m}$$

This is a trifle *less* than M_o; no change is warranted and the apparent difference is probably not a significant number anyway.

It is noted that the negative moments at the joints on the third span are higher than M_{CB} (at the joint) by 3 kN · m and 7 kN · m and corrections to face of column will be similar. The author feels it would be good judgment to design at all interior column lines (except the first) for a moment of something like 7 kN · m more than recorded for M_{CB}, say, a total of 162 kN · m. This change is indicated on Fig. 16.6c.

The distribution in Fig. 16.6 has not been extended as many bays as the Code seems to require (two spans beyond the critical moment computed), but inspection indicates that under full load on all spans another span would make no noticeable difference.

The moments for design are indicated in Fig. 16.6c. The distribution to middle and column strips is to follow the same procedure used in earlier examples.

16.7 PROCEDURE WHERE LIVE LOAD IS LARGER

If live load exceeds 75% of dead load (at service load stage), more loadings in pattern form are necessary to obtain design moments. In each such case the pattern live load is at the lower level of 75% of the design live load. The cases to be analyzed are sketched in Fig. 16.7, where all frame constants and all correction procedures to the critical section would be handled exactly as in the example just given. The pattern loading moments in Fig. 16.7 may not be taken smaller than those of full live load on all spans. (Where the loading qualifies for the direct design method there is no corresponding provision; the M_o calculation takes its place.)

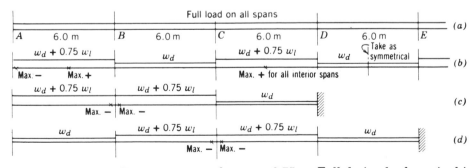

FIGURE 16.7 Loadings necessary when $w_l > 0.75w_d$. Full design load required in (a).

16.8 SLAB MOMENT PERCENTAGES IN DIRECT DESIGN METHOD—STIFF BEAMS

For *interior* panels on stiff beams this section gives the equations defining the percentage of M_o assigned to column and middle strip and tabulates some of these in Table 16.8.

For an interior panel, as in Sec. 16.2, M_o is divided into total negative moment, 65%, and total positive moment, 35%. For stiff beams (such that $\alpha_1 \ell_2/\ell_1 \geqslant 1.0$), Fig. 16.2b,c shows that the amount of this negative or positive moment going to the column strip varies from 90% for $\ell_2/\ell_1 = 0.5$ to 45% for $\ell_2/\ell_1 = 2$; or this percentage can be stated algebraically as $105 - 30\ \ell_2/\ell_1$ percent which goes to the column strip. This is further subdivided into 85% to the beam and 15% to the column strip slab. All moment not thus assigned to the column strip becomes middle strip moment. Below, the various percentages are expressed algebraically.

Col. strip *slab* neg. $M_u = -0.15 \times 0.65(105 - 30\ \ell_2/\ell_1)M_o$

$\qquad = -(10.25 - 2.93\ \ell_2/\ell_1)M_o$, as a percentage.

Col. strip *slab* pos. $M_u = 0.15 \times 0.35(105 - 30\ \ell_2/\ell_1)M_o$

$\qquad = +(5.53 - 1.58\ \ell_2/\ell_1)M_o$, as a percentage.

Mid. strip neg. $M_u = -0.65(100 - 105 + 30\ \ell_2/\ell_1)M_o$

$\qquad = -(19.5\ \ell_2/\ell_1 - 3.25)M_o$, as a percentage.

Mid. strip pos. $M_u = +0.35(100 - 105 + 30\ \ell_2/\ell_1)M_o$

$\qquad = +(10.5\ \ell_2/\ell_1 - 1.65)M_o$, as a percentage.

The beam moments (not shown) make up the missing 0.85 of the total column strip moments.

TABLE 16.8 Slab Moments for Interior Spans with Stiff* Beams on All Sides and at Least Three Continuous Spans

		M_o for long span			M_o for short span	
ℓ_2/ℓ_1:		0.50	0.75	1.00	1.33	2.00
Middle	Neg.	$-6.50\% M_o$	$-11.40\% M_o$	$-16.25\% M_o$	$-22.75\% M_o$	$-35.75\% M_o$
Strip	Pos.	$+3.60\% M_o$	$+6.21\% M_o$	$+8.85\% M_o$	$+12.35\% M_o$	$+19.35\% M_o$
	Width	$0.5\ \ell_2$	$0.5\ \ell_2$	$0.5\ \ell_2 = 0.5\ \ell_1$	$0.83\ \ell_1$	$1.50\ \ell_1$
Column	Neg.	$-8.79\% M_o$	$-8.05\% M_o$	$-7.33\% M_o$	$-6.33\% M_o$	$-4.39\% M_o$
Strip	Pos.	$+4.74\% M_o$	$+4.34\% M_o$	$+3.95\% M_o$	$+3.43\% M_o$	$+2.37\% M_o$
	Width	$0.5\ \ell_2$	$0.5\ \ell_2$	$0.5\ \ell_2 = 0.5\ \ell_1$	$0.5\ \ell_1$	$0.5\ \ell_1$

*Technically, with $\alpha_1 \ell_2/\ell_1 \geqslant 1.0$.

For several cases of ℓ_2/ℓ_1 slab moments are tabulated in Table 16.8. The widths of strips are added below since otherwise the size of moment per unit width would be difficult to visualize. This table is really divided vertically at the middle between coefficients on the left to be used with M_o for the long span and its strips and on the right with M_o for the short span and its strips. The picture is improved when one pairs data from two columns to make one complete set:

ℓ_2/ℓ_1 of 0.5 and 2.0 for 2:1 slab proportions

ℓ_2/ℓ_1 of 0.75 and 1.33 for 4:3 slab proportions

ℓ_2/ℓ_1 of 1.00 alone for a square slab

This table can give some idea of how moments change with panel proportions. But these percentages are rather minimum values for interior panels only. They should obviously change, somewhat upward, as boundary restraints are relaxed. A corresponding table for an exterior panel is much more difficult since these moments involve K_{ec} and α_{ec}, values neither simple to obtain nor easy to tabulate.

Thus Table 16.8 still leaves designers with inadequate guidance when they come to the numerous small slabs that do not qualify for the direct design method. The conservative approach used in Chapter 12 is recommended as giving detailed slab moment coefficients for a wide range of restraints, applicable wherever beam deflections do not greatly increase the slab deflections.

SELECTED REFERENCES

1. Sidney H. Simmonds and Janko Music, "Design Factors for the Equivalent Frame Method," *Jour. ACI*, 68, No. 11, Nov. 1971, p. 825.
2. W. G. Corley and J. O. Jirsa, "Equivalent Frame Analysis for Slab Design," *Jour. ACI*, 67, No. 11, Nov. 1970, p. 875.
3. "Frame Analysis Applied to Flat Slab Bridges," Portland Cement Assn., Skokie, Ill.

PROBLEMS

PROB. 16.1.

(a) Pick a slab thickness based on Code deflection requirements for a two-way slab, interior panel 5.7 m by 6.6 m (to centers of supporting beams and columns), assuming beam stems are 300 mm wide, columns 400 mm square, $f'_c = 20$ MPa, Grade 400 bars, dead load (including slab weight) = 5 kPa, live load = 6 kPa. Assume $\alpha_m = 2$.

(b) Check the direct moment requirements against the strength of this thickness.

(c) Design the slab reinforcement.

PROB. 16.2 A 6.0 m square interior panel is supported on beams 450 mm deep overall (and assumed 250 mm width) which are carried by columns 350 mm square and 3.0 m high story-to-story. $f'_c = 30$ MPa, Grade 400 steel, $w_d = 4.3$ kPa (including slab weight), $w_\ell = 3.8$ kPa. Find slab thickness for deflection; check slab and beam* sizes for moment; and check beam size for shear. Then design the reinforcement for the slab.

PROB. 16.3. Assume the same data as in Prob. 16.2 to be used for an exterior panel (not a corner panel).

(a) Design the slab.

(b) Design the beam running perpendicular to the wall line for flexure and check it for shear.

(c) Check the spandrel beam for shear and torsion, assuming a wall load of 6 kN/m of beam, applied on the axis of the spandrel beam and column center line. What stirrups are required at d from the face of column?

PROB. 16.4. Assume the flat plate designed in Sec. 15.7 has equal columns above and below with a 3.0 m story height. Use the equivalent frame method of design to establish the following for the *short* strips:

(a) Maximum negative design moment and its distribution to column and middle strips.

(b) Maximum positive moment for design and its distribution to design strips.

(c) Maximum column design moment, about this axis, at face of slab.

(d) Compare these moments with those from the direct design method used in Sec. 15.7.

*The 1977 Commentary tables can often assist in beam calculations.

17

DISTRIBUTION OF
CONCENTRATED LOADS
AND OTHER SPECIAL
PROBLEMS

17.1 CONCRETE STRUCTURES DISTRIBUTE CONCENTRATED LOADS

The ordinary reinforced concrete structure is either monolithic or is tied together to act as a unit. Although parallel members of the structure may be analyzed somewhat independently of each other under uniform live loads, actually the entire structure is a three-dimensional frame. When moving concentrated loads are considered, their spacing and their number suggest that all parallel slab strips and all neighboring beams will not be equally loaded. The interaction of the several slab strips and beams is usually such as to make the effective slab loading less severe than if each set of loads acted separately on the individual members.

When a heavy wheel rolls over a plank floor, each plank in turn must support the total load. In constrast, when a wheel moves over a concrete slab the wheel deflects the slab locally into a saucerlike pattern and this depression moves with the wheel across or along the slab. Thus a slab strip is deflected (and must be loaded) without a wheel actually resting on it. As the wheel passes over a particular strip the deflection increases, but the single 1 m strip of slab never carries the entire wheel load unassisted. The designer describes this by saying the wheel load in Fig. 17.1a is distributed over an effective width E (Fig. 17.1b), meaning that the moment on the most heavily loaded 1 m strip is that which would be produced by $1/E$ parts

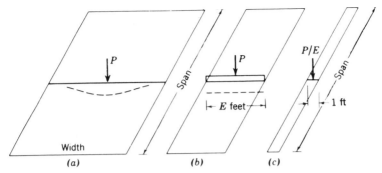

FIGURE 17.1 Effective width under a single concentrated load. (*a*) Cross section through midspan showing unequal deflections. (*b*) Design assumption of a width *E* with uniform deflections. (*c*) Equivalent 1-ft (0.3 m) strip.

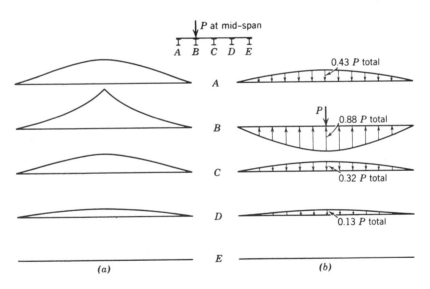

FIGURE 17.2 The distribution of a single concentrated load to the girders of a bridge span, girder spacing of 0.1 span. Reproduced from Reference 1, Highway Research Board. (*a*) Moment diagrams. (*b*) Approximate load distributions and the resulting total loads from the slab.

462

of the total load, as in Fig. 17.1c. Likewise, closely spaced beams share in carrying concentrated loads when the beams are connected by stiff floor slabs or stiff diaphragms.

The result of a theoretical study as to how a single wheel load is carried by a simple girder highway span is shown in Fig. 17.2. The load is applied to midspan, directly over beam B, and the girder stiffness assumed is five times that of the slab for a width equal to the girder span. Girder B then deflects more than its neighbors A and C. The slab (attached to the beams) is pulled down by beam B, but it resists this movement and exerts upward forces on the beam, as shown by the shaded ordinates on the right of the figure. These upward forces on beam B total $0.88\,P$, leaving the net downward load only $0.12\,P$. The resulting moment diagram on the beam is shown to the left. The neighboring beams deflect less and carry less moment than beam B. Actually, the slab imposes heavier net loads on these beams than on beam B, but the load of $0.43\,P$ on A and $0.32\,P$ on C is better distributed and produces less moment. The load on beam D is only $0.13\,P$ and the load on beam E is negligible.

Although such interaction of members can be approximately evaluated on a theoretical basis, design rules depend equally on field tests. This chapter will not attempt to demonstrate how load distribution factors are established nor to tabulate them for the many possible conditions. Rather its objective is to call attention to the problem of load distribution and illustrate how it can be handled in a few typical cases.

17.2 LOAD DISTRIBUTION IN A CONCRETE SLAB

Load distribution in a slab can be approached on two different bases that may be described as (1) the service load or deflection basis and (2) the ultimate strength or yield-line basis. The distribution at service loads is the one more commonly considered. For a wide slab having a 3 m simple span, the effective width thus determined for a simple load is between 1.8 m and 2.4 m, depending somewhat on the size of the load contact area and the particular algebraic formula[2] used. For comparison, Johansen has shown[3] that at ultimate load, the effective width would be twice the span multiplied by $\sqrt{\mu}$, where μ is the ratio of the perpendicular top steel to the longitudinal bottom steel. If μ is about one-third, this would make the effective width $2 \times 3\sqrt{0.33} = 3.45$ m.

One complication that may make calculations at ultimate strength uncertain is the shear capacity of the slab around the load. Richart and Kluge[4] found shear failures occurring from diagonal tension, with a truncated cone of concrete punched out below the load. When those shear stresses were calculated on a surface at a distance d beyond the load, the unit shear stress was low, in one series from $0.044\,f'_c$ to $0.057\,f'_c$. Since the shear failures came at loads 50% greater than those producing local yielding of the steel, these low shear stresses were not considered serious. For a yield-line analysis shear stresses around the load might be more significant.

It appears that the distribution based on elastic conditions, as commonly used, is on the safe side. Its use also tends to reduce crack size at working loads. For elastic conditions, Westergaard[5] established an extreme value of maximum positive moment on a slab as $0.315\,P$ for any simple span when P is distributed over a circular area with the diameter equal to one-tenth of the span, the slab thickness is one-twelfth of the span, and Poisson's ratio is 0.15. (This local moment is quite sensitive to the size of the bearing area.) The corresponding transverse moment is $0.248\,P$. Jensen[6] extended these results to show the effect of a rigid beam support at right angles, that is, an effect like that in a two-way slab. At this crossbeam the maximum negative moment is $-P/2\pi = -0.159\,P$ and it occurs with the wheel quite close to the beam.

When closely spaced multiple wheels occur, an extra slab width acts, but the effective width per wheel is reduced. The AASHTO *Standard Specifications for Highway Bridges*[7] specify such an effective width E for a slab carrying a single wheel (traveling in the direction of the span) that the resultant design will be safe for multiple wheels without further calculations. Special transverse distribution steel is also specified as a percentage of the positive moment steel, in the amount $100/\sqrt{S}$ but not over 50%, where S is the span in feet ($55/\sqrt{S}$, where S is the span in meters).* When the wheels travel perpendicular to the span, the distribution steel is a larger percentage, $220/\sqrt{S}$, where S is the span in feet ($120/\sqrt{S}$, where S is the span in meters), but not over 67%.

17.3 CALCULATION FOR CONCENTRATED LOAD ON SLAB

A specification calls for a live load of 3 kPa or a moving concentrated load of 9 kN. Determine which loading controls for a continuous 3.0 m span where the moment coefficients for uniform load are $+\frac{1}{16}$ and $-\frac{1}{12}$, and for a concentrated load $+\frac{1}{8}$ and $-\frac{1}{8}$. Consider $E = 0.68\,S + 2\,c$, where S is the span and c the diameter of the loaded area.[2] Use ACI load factors.

Solution

For uniform live load, $M_{u\ell} = -\frac{1}{12} \times 1.7 \times 3 \times 3.0^2 = -3.83$ kN·m/m.
For concentrated load, with $c = 0$ in the absence of better data,

$E = 0.68 \times 3.0\text{ m} + 0 = 2.04$ m for effective width

*The AASHTO notations S and E are retained in this chapter.

Effective load, $P_e = 9/2.04 = $ 4.41 kN/m strip

$M_{u\ell} = -\frac{1}{8} \times 1.7 \times P_e S = -\frac{1}{8} \times 1.7 \times 4.41 \times 3.0 = -2.81$ kN·m/m < -3.83 kN·m/m

Therefore the uniform load moment governs the slab design.

It might be noted that the effective width for shear would call for the concentrated load near the support that would give less slab deflection and a much reduced effective width. The AASHTO specification says that slabs designed for moment will be considered safe in bond (development length) and shear. As an extreme assumption, consider the entire load resisted by a 0.3 m (1 foot) strip and the design shear existing at a distance d from the support. Assume also a slab having $h = 125$ mm, $d = 100$ mm, and a weight of 3 kPa.

$V_u = 1.7 \times 9 \times (3.0 - 0.1)13.0 + 1.4 \times 3 \times 0.3(1.5 - 0.1) = 16.5$ kN

$v = V/bd = 16.5 \times 10^{-3}/0.3 \times 0.1 = 0.55$ MPa $< \frac{1}{6}\sqrt{20} = 0.75$ MPa

Although there are limited data as to reaction distribution, it is difficult to imagine a diagonal tension failure that would involve less than a width of four to five times the slab thickness, in the assumed case at least 0.5 to 0.6 m.

17.4 HIGHWAY BRIDGE LOADINGS

The basic units of loading for highway bridges are the H truck, a two axle loading, and the H-S truck with trailer, a three-axle loading. The 20-ton (178 kN) truck is designated as H20-44, the last number denoting the year 1944 when this loading was established. Figure 17.3 shows the distribution of the load between the various wheels. The corresponding truck and trailer combination HS 20-44 represents the standard 20-ton (178 kN) truck plus a second 16-ton (143 kN) axle, as shown in Fig. 17.4. For either type of truck an alternate uniform load of 640 plf (9.34 kN/m) for each lane plus a concentrated load of 18 000 lb (80 kN) for moment or 26 000 lb (116 kN) for shear is to be used wherever it gives larger values, which will be the case on longer spans. The lane load occupies a 3.0 m width and is to be placed in the worst position when actual lanes are wider. The reader is referred to the AASHTO specification[7] for such details as the number of loaded lanes and proper reduction factors, loads for continuous spans, impact allowance, and distribution of loads. In the following examples, only the governing portion of the specification is mentioned. The AASHTO notation has been retained where no equivalent has yet been set in the new ACI notation.

17.5 DESIGN OF A HIGHWAY SLAB SPAN

Design an interior panel of a 2.5 m span continuous slab for H20-44 loading, $f'_c = 20$ MPa, Grade 300 steel, AASHTO specification. Consider that this slab spans

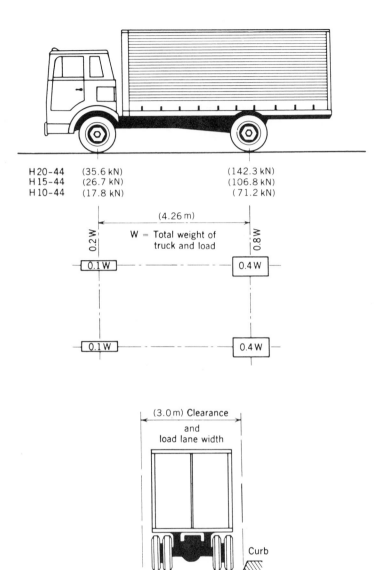

H 20–44	(35.6 kN)	(142.3 kN)
H 15–44	(26.7 kN)	(106.8 kN)
H 10–44	(17.8 kN)	(71.2 kN)

(4.26 m)

0.2 W

W = Total weight of truck and load

0.8 W

0.1 W   0.4 W

0.1 W   0.4 W

(3.0 m) Clearance
and
load lane width

Curb

(0.6 m)   (1.8 m)   (0.6 m) *for beam design

(0.3 m) – for slab design

Standard H trucks

FIGURE 17.3 The H truck of the AASHTO *Standard Specifications for Highway Bridges*. For slab design the center line of wheels shall be assumed to be 1 ft (0.3 m) from face of curb.

466

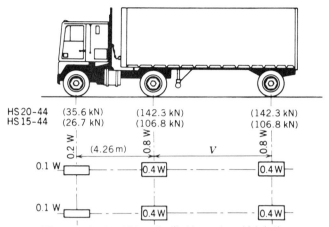

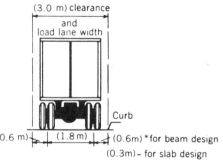

W = combined weight on the first two axles which is the same
as for the corresponding H truck.

V = variable spacing–(4.26 to 9.14 m) inclusive. Spacing to
be used is that which produces maximum stresses.

Standard HS trucks

FIGURE 17.4 The H-S truck of the AAHSTO *Standard
Specifications for Highway Bridges.* For slab design the
center line of wheels shall be assumed to be 1 ft (0.3 m)
from face of curb.

longitudinally between transverse girders.* Use service load design method of
Appendix A (Sec. A.9). AASHTO (1977) now permits either load factor or service
load designs.

*More typically, highway slabs span transversely between longitudinal girders. Un-
fortunately for the purpose of this text (showing generally how to handle concentrated loads)
AASHTO for that case gives resultant moment formulas that do not show the effective width
E used. That **E** appears to lie in the range of 6.3 to 7 ft (1.9 m to 2.1 m) for the two wheels
(full-axle load) on the strip.

Solution

The AASHTO specification allowable stresses* are $f_c = 0.40\, f'_c = 8$ MPa, $n = 10$, $f_s = 140$ MPa.

$$c_b = \frac{8}{8 + 140/10}\, d = 0.364\, d, \qquad z_b = d - c_b/3 = 0.879\, d$$

$$k_b = (0.5\, f_c bc)z/bd^2 = 0.5 \times 8 \times 0.364\, d \times 0.879\, d/d^2 = 1.28 \text{ MPa}$$

Since the reinforcement is parallel to traffic, this slab falls under Case B in the specification with an effective width $E = 1.2 + 0.06\, S$ where S is the span in meters.

$$E = 1.2 + 0.06\, S = 1.2 + 0.06 \times 2.5 = 1.35 \text{ m} < 2.1 \text{ m max.}$$
$$P = 0.4 \times 178 = 71 \text{ kN}$$
$$P_e = P/E = 71/1.35 = 53 \text{ kN/m}$$

If the moment for the concentrated load is taken as 80% of the simple span moment,

$$M_L = 0.8\, P_e S/4 = -0.8 \times 53 \times 2.5/4 = 26.5 \text{ kN} \cdot \text{m/m}$$

The impact fraction is

$$I = \frac{50}{3.28\, S + 125} = \frac{50}{3.28 \times 2.5 + 125} = 0.375$$

but not to exceed 0.30, which governs here.

$$\text{Impact } M_I = -0.30 \times 26.5 = 8.0 \text{ kN} \cdot \text{m/m}$$

Assume dead load $w_d = 5.4$ kPa(for 225 mm slab) and moment coefficients of $-1/12$ and $+1/16$.

$$M_D = -5.4 \times 2.5^2/12 = -2.8 \text{ kN} \cdot \text{m/m}$$
$$M_T = -26.5 - 8.0 - 2.8 = -37.3 \text{ kN} \cdot \text{m/m}$$
$$d = \sqrt{M/k_b b} = \sqrt{37.3 \times 10^{-3}/1.28 \times 1} = 0.171 \text{ m} = 171 \text{ mm}$$

Since AASHTO specifies a minimum of 25 mm of cover

$$h = 171 + \tfrac{1}{2}\, d_b(\text{say, } 13 \text{ mm}) + 2.5 = 209 \text{ mm}$$
$$\text{USE } h = 225 \text{ mm}, \; d = 225 - 25 - 13 = 187 \text{ mm}$$

Neg. $M_T = -37.3$ kN $\cdot$ m/m

Pos. $M_T = +37.3 \times 12/16 = +28.0$ kN $\cdot$ m/m

Neg. $A_s = \dfrac{37.3 \times 10^6}{140 \times 0.879 \times 187} = 1620$ mm²/m (#25 at 300 mm or #20 at 175 mm)

Pos. $A_s = 1620 \times 28.0/37.3 = 1220$ mm²/m (#20 at 200 mm)

USE #20 at 175 mm for both positive and negative M.

This design ignored the possibility of using compression steel that would be available at the supports but possibly not at midspan. AASHTO also says slabs designed for moment "shall be considered satisfactory in bond and shear."

*The equations for service load analysis are denied in Appendix A.

The concentrated load requires distribution steel perpendicular to the moment steel. The amount is specified as a percentage of the positive moment steel given by $55/\sqrt{S} = 55/\sqrt{2.5} = 34.8\% < 50\%$ maximum. Transverse steel must be at least $0.348 \times 1220 = 424$ mm^2/m (#10 at 200 mm)

17.6 DESIGN MOMENTS AND SHEARS FOR HIGHWAY GIRDER

Calculate the unfactored live load and impact load moments and shears for a 12 m simple span girder built integrally with a slab. Consider both H20-44 and H20-S16-44 loadings on the center girder of the cross section shown in Fig. 17.5a.

Solution

For shear calculations the AASHTO says there shall be no lateral or longitudinal distribution of the wheel load at the end of the span. For other wheels the distribution shall be that applying for moment. The provision for moment when two or more lanes of traffic are involved calls for the wheel loads to each stringer to be $S/1.8 = 2.9/1.8 = 1.61$ of the full load.*

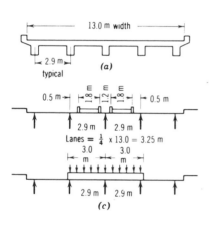

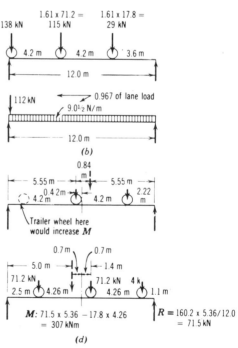

FIGURE 17.5 Highway girder bridge. (*a*) Bridge cross section. (*b*) Location of wheels and lane load for maximum shear. (*c*) Position of loads laterally for maximum stresses in center girder.

*For $S > 3$ m, the load to the beam is the reaction computed from "simple span" slabs.

The longitudinal arrangement of wheels for the truck and uniform lane loads are shown in Fig. 17.5b, ignoring any complication that the end diaphragm may cause. Figure 17.5c shows the lateral arrangement of the loads on the cross section, with two trucks assumed to be passing. For the truck the simple beam slab reactions would give for the end wheels:

$$(2 \times 2 \times 1.4/2.9) \times 71.7 = 138.5 \text{ kN}$$

For the lane load of Sec. 17.4, the end concentrated load for shear becomes $2 \times 0.5 \times 115 \times 2.9/3.0 = 111$ kN and the uniform load $9.34 \times 2.9/3.0 = 9.0$ kN/m. For the trucks, maximum $V = 138 + 115 \times 6.8/12 + 29 \times 3.6/12 = 212$ kN. For the lane load, maximum $V = 111 + 9.0 \times 12/2 = 165$ kN.

$$\text{Max } V_L = 212 \text{ kN*}$$

By specification, L is from the shear point to the far reaction, that is, the full 12.0 m span.

$$I = \frac{50}{3.28 \times 12.0 + 125} = 0.304$$

Since $I = 0.30$ governs, $V_I = 0.30 \times 212 = 64$ kN

For moment three loadings must be investigated, as worked out in Fig. 17.5d. For simplicity these loadings are compared on the basis of one line of wheels or one-half lane load. The HS 20-44 loading governs maximum moment. Since the number of lines of wheels is $S/b = 1.61$,

$$M_L = 1.61 \times 307 = 495 \text{ kN} \cdot \text{m}$$

The loaded length for moment, by definition, is the full length, $L = 12.0$ m

$$I = \frac{50}{3.28 \times 12 + 125} = 0.304$$

Since $I = 0.30$ governs, $M_I = 0.30 \times 495 = 148$ kN $\cdot$ m

The above illustrates the manner specified by AASHTO for handling load distribution on stringers or longitudinal girders. The remainder of the design of the T-beams would be very similar to that of a building T-beam with these minor differences, which are usually more restrictive:

1. Different allowable unit stresses for f_c and v, compared to ACI Code.
2. Option of service load or factored load design requires two different sets of allowable stresses.
3. Different maximum moment diagram, due to type of loading.
4. Different maximum shear diagram, due to type of loading.
5. Webs of T-beams must carry stirrups full length, at a maximum spacing not to exceed 0.75 d where not required for stress.
6. Slightly different bar extensions.

---

*AASHTO designs for shear *at* the face of support rather than d from the support.

All these are the small differences that can be expected in changing from one specification to another.

17.7 OPENINGS IN SLABS

When openings in slabs are large, as for stairs or elevators, beams must be used around the openings. Good practice usually requires that these beams be framed into columns sufficiently to provide a stable unit without the slab.

Small openings such as pipe sleeves, if not too numerous, can be made almost anywhere in a slab, except adjacent to the columns in flat slab construction. What can be done about larger openings may require at least some rough calculations. Obviously, openings are least dangerous where shear stresses are small and bending moments are below maximum.

Electric conduits, unless closely spaced or crossing at small angles, can be included without considering any loss in moment strength. The detailed requirements of Code 6.3 might be noted in this connection.

The effect on shear of openings around columns in flat plate construction was discussed briefly in Sec. 6.3 and illustrated in Fig. 6.3. These shear provisions recognize the coexistence of large bending moments.

An equally common and troublesome problem is that of relatively large openings that interrupt the normal flexural action. Openings of any size are permitted by Code 13.5.1 if analysis shows that both strength and deflection are still acceptable. For strength the methods of Chapter 14 on the strip method seem appropriate even if they might appear rather arbitrary.

Without analysis considerable leeway is still given by Code 13.5.2:

13.5.2—In lieu of special analysis as required by Sec. 13.5.1, openings may be provided in slab systems without beams only in accordance with the following.

13.5.2.1—Openings of any size may be located in the area common to intersecting middle strips, provided total amount of reinforcement required for the panel without openings is maintained.

13.5.2.2—In area common to intersecting column strips, not more than 1/8 the width of column strip in either span shall be interrupted by openings. An amount of reinforcement equivalent to that interrupted by an opening shall be added on the sides of the opening.

13.5.2.3—In the area common to one column strip and one middle strip, not more than 1/4 the reinforcement in either strip shall be interrupted by openings. An amount of reinforcement equivalent to that interrupted by an opening shall be added on the sides of the opening.

13.5.2.4—Shear requirements of Sec. 11.11.5 shall be satisfied.

In two-way slabs supported on all four sides, openings in the corners of the slab are least damaging. Since for architectural reasons openings for ducts frequently need to be near the columns, this is a definite advantage of this type of slab. The negative moment zone of the short middle strips (near midspan of the longer beams) is the least favorable zone for openings.

Often rough checks can be made on the strength of the construction after the openings are located. If openings should reduce a critical design section for moment, the required bd^2 for moment must be maintained by providing extra depth to offset the reduced width. The steel may be more closely spaced on each side of the opening to maintain the necessary A_s. Often it will be possible to locate openings where moment is well below the compression capacity of the slab, thereby leaving the arrangement of reinforcement as the only problem. Of course, shear strength must be maintained, but this is rarely a problem except near the columns in flat slab types, as noted above.

The arrangement to bars around any but minor openings can constitute a real problem. Bars running perpendicular to the face of an opening are

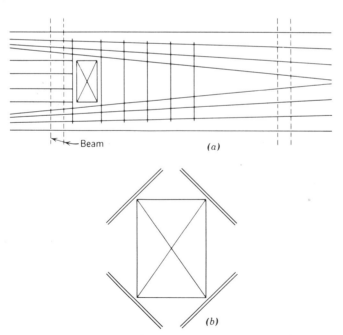

FIGURE 17.6   Bars at openings in slabs. (*a*) Bars fanned out to miss opening. (*b*) Corner reinforcement to reduce shrinkage cracking.

not fully effective when simply cut off at the opening. This would be all right if there were a beam at the opening to act as a reaction for the slab. If there is no beam it is better to fan the bars out or splay them to go around the opening, as shown in Fig. 17.6a. If this leaves too wide an area without steel, extra bars can be placed parallel to the side of the opening, as indicated.

If minor cracking at the corners of an opening is objectionable, it is always well to add one or two diagonal bars at each corner, especially at large openings (Fig. 17.6b). This is always desirable around window and door openings in concrete wall slabs. Such reinforcement helps to take care of shrinkage stresses.

17.8 OPENINGS IN BEAMS

Openings through slabs that encroach on the flange width of T-beams require a check on the bending strength remaining.

Openings through a beam web are increasingly important for installation of building services. In a region of small shear, as near the middle of a beam span, a horizontal pipe sleeve should not be serious. Elsewhere, shear strength must be closely watched and in many places bending strength as well. Large openings in beams are particularly weakening. They destroy beam action and force this reduced section to act much as a Vierendeel truss (a truss without diagonals). In such a truss the average bending moment over the length of the opening is resisted by axial compression in one chord and tension in the other, with these two forming a couple in the case of pure flexure. Where shear is present the change in the moment over the length of the opening superimposes a reversed bending resistance in each chord, the total of the four end moments on the chords equaling the external shear times the length of the opening. How the shear and these reversed moments are shared by the two chords depends on the relative chord stiffness. The designer in such a case should note that for members with significant axial tension Code 11.3.1.3 requires all shear to be resisted by stirrups (none assigned to the concrete).

SELECTED REFERENCES

1. C. P. Siess and A. S. Veletos, "Distribution of Loads to Girders in Slab-and-Girder Bridges: Theoretical Analyses and Their Relation to Field Tests," Highway Research Board *Report 14-B*, Washington, 1953, p. 58.
2. Clyde T. Morris, "Concentrated Loads on Slabs," Ohio State Univ. Eng. Exp. Sta. *Bull. No. 80*, 1933.

3. K. W. Johansen, "Bruchomente Der Kreuzweise Bewehrten Platten" (Moments of Rupture in Cross-Reinforced Slabs), International Association for Bridge and Structural Engineering, Liége, Vol. 1, 1932, p. 277.

4. Frank E. Richart and Ralph W. Kluge, "Tests of Reinforced Concrete Slabs Subjected to Concentrated Loads," Univ. of Ill. Eng. Exp. Sta. *Bull. No. 314*, 1949.

5. H. M. Westergaard, "Computation of Stresses in Bridge Slabs Due to Wheel Loads." *Public Roads*, 11, No. 1, Mar, 1930, p. 1.

6. Vernon P. Jensen, "Solutions for Certain Rectangular Slabs Continuous Over Flexible Supports," Univ. of Ill. Eng. Exp. Sta. *Bull. No. 303*, 1938.

7. *Standard Specifications for Highway Bridges*, AASHTO, Washington, 12th ed., 1977.

PROBLEMS

Note The service load method of Appendix A is to be used for these highway bridge type problems, unless otherwise instructed.

PROB. 17.1. Design a simple span slab to carry its own weight and a single 90 kN concentrated load over a 6.0 m span. Assume the impact as 25% of live load and an effective width $E = 1.7$ m. Allowable $f_c = 8$ MPa, $f_s = 140$ MPa, $n = 10$, cover of 40 mm to center of steel. Assume the load is distributed over a sufficient bearing area to avoid a punching shear failure.

PROB. 17.2. An interior span of a continuous slab with a clear span of 3.9 m is to be designed for a live load of 5 kPa or a single concentrated live load of 22.5 kN, whichever is worse. If the effective width assumed for the concentrated load is 1.5 m when the load is at its worst position for moment, design the slab using the moment coefficients of Sec. 10.5 (Fig. 10.8) for uniform load and $\pm 0.20 \, P\ell$ for the concentrated load. Assume zero impact. $f'_c = 30$ MPa, $f_s = 140$ MPa, $n = 8$, allowable $f_c = 7$ MPa.

18
AXIAL LOAD PLUS BENDING—
SHORT COLUMNS

18.1 THE PRACTICAL COLUMN PROBLEM

All practical columns are members subject not only to axial load but also to moment from direct loading or end rotations. This chapter covers *short* columns, those where lateral deflections are not significant. Chapter 19 covers *long* columns where deflections have an important effect on member strength.

From a 1970 survey the ACI-ASCE Column Committee estimated that 90 percent of all braced columns and 40 percent of all unbraced columns can be designed as short columns. Braced columns are those where shear walls, diagonal bracing, and so on, largely prevent relative lateral movement of the two column ends (or joints).

Creep and shrinkage are important in column behavior and the corresponding stresses under service conditions can only be estimated. However, as Sec. 18.3 indicates, the strength of a given cross section at failure is reasonably uninfluenced. This chapter covers column strengths ranging from where moments are small to the other limiting condition of flexure alone (axial load zero).

Because columns are structurally more important than beams (they carry more floor area) and are subject to moments less accurately known,* the strength reduction factor ϕ is lowered to 0.70 or 0.75, depending on the type of column involved.

---

*Other reasons for a lower ϕ lie in (1) the larger approximation involved when the rectangular stress block covers a greater portion of the cross section, and (2) the heavy dependence on the concrete for the larger portion of the strength.

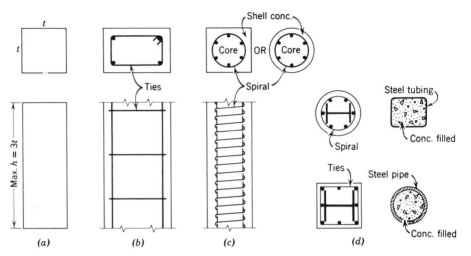

FIGURE 18.1 Types of columns. (*a*) Plain concrete pedestal. (*b*) Tied column. (*c*) Spiral column. (*d*) Composite columns, four types.

18.2 TYPES OF COLUMNS

Plain concrete is not used for columns, but may be used for pedestals in which the height does not exceed three times the least lateral dimension (Fig. 18.1*a*).

Reinforced concrete columns normally contain longitudinal steel bars and are designated by the type of lateral bracing provided for these bars. *Tied columns* (Fig. 18.1*b*) have the bars braced or tied at intervals by closed loops called ties. *Spiral columns* have the bars (and the core concrete) wrapped with a closely spaced helix or spiral of small diameter wire or rod (Fig. 18.1*c*).

Composite columns may contain a structural steel shape surrounded by longitudinal bars with ties or spirals or may consist of high strength steel tubing filled with concrete or a steel pipe so filled.*

Tied and spiral columns are the most common forms. Either may be made circular, octagonal, square, or rectangular in cross section, as desired. Tied columns may also be of L shape.

18.3 COLUMN TESTS

For nearly 50 years it has been evident that in a reinforced concrete column under sustained axial load one could not calculate f_c, the actual

---

*The 1977 Code no longer assigns different names to the several kinds of composite columns.

unit stress in the concrete, nor f_s, the actual unit stress on the steel. If the materials were really elastic it would be possible to use the transformed area (Chapter 8) to establish these stresses. However, actual observations show that the steel stress is much larger than this calculation would indicate, because of both shrinkage and creep of the concrete under load.

Starting about 1930, a very large research project on columns was carried out at the University of Illinois[1] and at Lehigh University.[2] These tests indicated clearly that even under axial load alone there was no fixed ratio of steel stress to concrete stress in the ordinary column. The ratio of these stresses depended on the amount of shrinkage, which in turn depended on the age of the concrete and the method of curing. It also depended on the amount of creep in the concrete. Creep is greater when the load is applied at an early stage of the hardening or curing process. The amount of creep is influenced by any of the factors that determine the quality of concrete, such as cement content, water content, curing, and even type of aggregate used.

A load applied for only a short time, such as the ordinary live load, causes very little creep, especially after the concrete is well cured. The usual live load thus produces an increment or increase of stress in steel and concrete that can be calculated reasonably well by elastic analysis (Sec. 8.4). However, stresses produced by dead load or any permanent or semipermanent load depend on the entire history of the column. It is even possible to have a loaded column with tension in the concrete and compression in the steel under very special circumstances (such as a large percentage of steel and a heavy initial loading later greatly reduced in amount).

Historically, in the United States, these tests initiated the slow switch in emphasis from service load to ultimate strength for columns. They showed that *ultimate* column strength did *not* vary appreciably with the history of loading. If, as the loading was increased, the steel reached its elastic limit first, the increased deformation then occurring built up stress in the concrete until its ultimate strength was reached. If the concrete approached its ultimate strength before the steel reached its elastic limit, the increased deformation of the concrete near its maximum stress forced the steel stress to build up more rapidly. Thus, regardless of loading history, a column reached what might be called its yield point only when the load became equal to approximately 85% of the ultimate strength of the concrete (as measured by standard cyclinder tests) plus the yield-point strength of the longitudinal steel. The 85% factor for the concrete is probably due to less ideal compaction of concrete in columns (around the steel) than in cylinders, together with the reduction in apparent strength caused by the slower application of load and the longer specimen.

Up to the column yield point, tied columns and spiral columns act

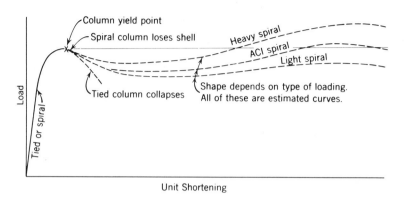

FIGURE 18.2   Comparison of strains in tied and spiral columns.

almost identically and the spiral adds nothing measurable to the yield-point strength. The stress-strain curves for the tied columns and the spiral column up to this point are essentially identical, similar to Fig. 18.2.

After the yield-point load is reached, a tied column immediately fails with a shearing diagonal failure of the concrete (as in a test cylinder) and a buckling failure of the column steel between ties as shown in Fig. 18.3. The yield point and ultimate strength of a tied column are thus the same thing. In a spiral column, the yield-point load results in cracking or complete destruction of the shell of concrete outside the spiral (Fig. 18.4). The spiral comes into effective action only with the large increased deformation that follows yielding of the column and loss of the shell concrete. At the stage shown in the Fig. 18.4 columns, the spiral provides radial compressive forces on the concrete within the core of the column and these confining stresses add significantly to the load the core concrete can carry. Pound for pound, steel in the spiral has been found to be from 2.0 to 2.4 times as effective as longitudinal steel in contributing to the ultimate strength of the column. The spiral steel never becomes significantly effective until after the destruction of the shell concrete that covers it. Moreover, excessive longitudinal column shortening is then involved (Fig. 18.2) that makes this spiral steel of questionable value *except* as a factor of safety against complete collapse.

A heavy spiral can add more strength to the column than that lost in the spalling or failure of the shell, in which case the column will carry an ultimate load greater than the yield-point load, but with unsuitable shortening. If too light a spiral is used, the column will continue to carry some load beyond the column yield point, but not as much as that which caused the spalling of the shell. The ACI Building Code specifies that

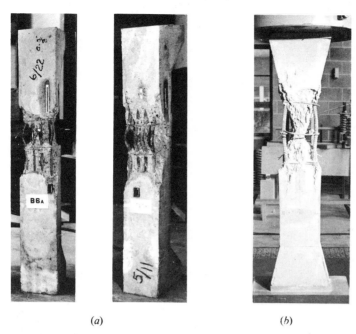

(a)   (b)

FIGURE 18.3   Failure of tied columns. (a) Note the bars buckled
between ties. Column height-thickness ratio $l_u/h = 7.5$. Special
ends were cast to permit comparative tests with eccentric loads.
(From Reference 3, Univ. of Ill.) (b) Column in which a tie seems
to have failed after yield point of column was reached. (Courtesy
Portland Cement Assn.)

amount of spiral steel that will just replace* the strength lost when the
shell concrete spalls. The initial cracking of the shell gives some warning
of overload prior to failure. The spiral also adds a considerable element of
toughness to the column. Toughness is valuable in resisting explosion or
blast, since it measures the energy that can be absorbed. Two columns
from the same story of a building severely damaged by a strong earthquake (Fig. 18.5) show that only the heavily damaged spiral columns
prevented a total collapse of this story.

More recently emphasis has been given to the fact that to some extent
column ties also confine the concrete, although their shape makes them
much less efficient than spirals. Heavy ties on columns or around compression steel in beams can establish a considerable degree of toughness in

---

*Actually, an estimated 10% in excess of the shell strength is used just to be sure the
strength after spalling is not less than before. See Sec. 18.13 for the design of a spiral.

(a)

(b)

(c)

FIGURE 18.4 Spiral column tests under concentric loads. (From Univ. of Illinois tests, References 3, 4, and 5.) (a) Failure of 32-in. (800 mm) diameter column; $\ell_u/h = 6.6$. (b) Failure of 12-in. (300 mm) diameter column; $\ell_u/h = 7.3$. Shell has completely spalled off: (The special ends were cast to permit comparative tests with eccentric loads.) (c) Failure of column with thin cover or shell; $\ell_u/h = 10.0$.

FIGURE 18.5 Columns almost destroyed in severe earthquake. Notice that spiral column still has its core acting. (Bars outside a spiral or tie become totally ineffective.)

these members. Hence ties are an important requirement in earthquake resistance and limit design where members are expected to maintain their peak resistance while forming the so-called plastic hinges (Chapter 11). Code Appendix A on seismic resistance calls for heavy ties in columns where they go through the beams, which would have helped the joint above the spiral column of Fig. 18.5.

INTERACTION OF AXIAL LOAD AND MOMENT

18.4 DEFINITIONS AND SOME CODE BASIC REQUIREMENTS

A short column is one where the length effect or deflection response under load is very small and may be considered negligible. This limiting length is more fully developed in Sec. 18.14. For preliminary discussion here, the limiting length-to-thickness ratio ℓ_u/h for a short column may be thought

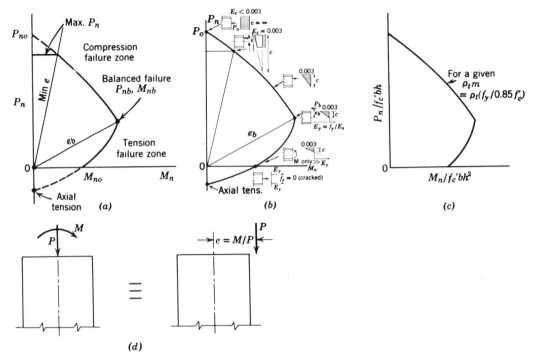

FIGURE 18.6 Column interaction diagram. (*a*) For a given column. (*b*) Strains involved. (*c*) Dimensionless form. (*d*) Equivalent eccentricity.

of as ranging roughly from 7 to 12 in a braced frame and several units lower in an unbraced frame. The short column can be analyzed or designed from the strength of its cross section alone.

A plot of the column axial load capacity against the moment it can simultaneously carry is called a column interaction diagram. The axial load capacity decreases as moment is increased. Schematically, Fig. 18.6*a* shows such a diagram, with key points and areas noted, for later discussion. Any loading that plots within this area is a possible loading; any combination outside the area represents a failure combination. A radial line from point O represents a constant ratio of moment to load, that is, a constant eccentricity of load. Figure 18.6*b* shows how strains and the neutral axis distance c shift along such a diagram.

The interaction diagram can be put into dimensionless form as indicated in Fig. 18.6*c* for a given steel ratio and arrangement of steel, which facilitates assembly of graphs into groups for design charts.*

---

*The designer must be careful to check whether such a chart is plotted in terms of P_n, M_n or P_u, M_u (a difference of ϕ in the numbers shown).

Since all concrete columns are subject to some moment, past codes set minimum eccentricities of loading to be used, $0.10\,h$ for tied columns and $0.05\,h$ for spiral columns. The 1977 Code requires something similar by setting an upper limit on the maximum axial load P_n, as shown by the horizontal line below P_{n0} in Fig. 18.6a. This limit is given in detail in the next section. The actual maximum $e = M/P$ (Fig. 18.6d) may exceed these minima and then control the design. There remain some significant nonlinear problems when establishing the design M in frames that the Code does not recognize, discussed in Sec. 18.22.

All columns are required (Code 10.9.1) to contain longitudinal bars sufficient to make the steel ratio, $\rho_t = A_s/hb$, at least 0.01, because of the shrinkage and creep stresses on smaller areas, and ρ_t must not exceed 0.08. At 0.08 crowding in the member is very severe. At least 6 bars must be used when in a circular arrangement or 4 bars in a rectangular arrangement.

Ties (Code 7.10.5) must be at least #10. Their spacing shall not exceed 16 bar diameters, 48 tie diameters, or the least column dimension. Every corner bar and every alternate bar must be braced by the tie and no bar shall be more than 150 mm from such a laterally supported bar.

Spirals must fully replace the strength lost when the shell of concrete outside the spiral spalls off. The minimum spiral is #10 with its clear vertical) spacing between 25 and 80 mm. Spiral design is covered in Sec. 18.13.

18.5 NOMINAL AXIAL LOAD CAPACITY, P_{n0}*, AND CODE MAXIMUM, $\phi P_{n(max)}$

Although in design, axial load without moment is not a practical case, P_{n0} is a convenient theoretical limit and one well documented experimentally.

The tests discussed in Sec. 18.3 established the ultimate strength of either a tied or a spiral column, axially loaded, as:

$$P_{n0} = 0.85\,f_c'A_n + f_yA_s$$
$$= 0.85\,f_c'(A_g - A_{st}) + f_yA_s \qquad (14.1)$$

where P_{n0}† = ultimate load capacity (yield-point strength) of tied or spiral column when eccentricity is zero (for ideal materials and dimensions)

A_n = net area of concrete = $A_g - A_s$

---

*The Code omits the subscript n where nominal is obvious.

†The subscript n will sometimes be dropped wherever it appears obvious that the nominal value is intended.

A_g = gross area of concrete, mm²
A_{st} = area of vertical column steel, mm²
f_c' = standard cylinder strength of concrete, MPa
f_y = yield point stress for steel, MPa

This is the ideal or nominal strength and the Code would consider the design strength ϕP_{n0}, where ϕ would be 0.70 for a tied column and 0.75 for a spiral column.

Another extreme limit, generally only of theoretical interest, is the axial tension limit, which is $A_s f_y$ in magnitude, with the concrete fully cracked at such a limit.

The Code does not permit a nominal load $P_{n0}*$ on a column. For a tied column the design strength is limited (Code 10.3.5.2) to

$$\phi P_{n(max)} = \phi\{0.80[0.85\,f_c'(A_g - A_{st}) + f_y A_{st}]\}$$

which in this book will usually be used as the nominal maximum:

$$P_{n(max)} = 0.80[0.85\,f_c'(A_g - A_{st}) + f_y A_{st}] = 0.80\,P_{n0}$$

For a spiral column the design strength limit (Code 10.3.5.1) is

$$\phi P_{n(max)} = \phi\{0.85[0.85\,f_c'(A_g - A_{st}) + f_y A_{st}]\}$$

which will usually be used here as the nominal maximum:

$$P_{n(max)} = 0.85[0.85\,f_c'(A_g - A_{st}) + f_y A_{st}] = 0.85\,P_{n0}$$

Where used, ϕ is 0.70 for tied columns or 0.75 for spiral columns.

These maximum load limits govern wherever the moment is small enough to keep the eccentricity for a tied column under $0.10\,h$ or for a spiral column under $0.05\,h$. These are approximate rather than exact limits, but entire short column designs can be based on satisfying these maximum load limits when e is noticeably under these limits. The following shows this simple process.

Example

Design a short tied column for a total dead load of 1 300 kN and live load of 1 100 kN, a moment of 56 kN · m (from live load), f_c' = 30 MPa, f_y = 400 MPa. Try for ρ_t of about 0.03.

Solution

$$P_u = 1.4 \times 1\,300 + 1.7 \times 1\,100 = 3\,690 \text{ kN} \qquad M_u = 1.7 \times 56 = 95.2 \text{ kN · m}$$
$$e = M_u/P_u = 95.2/3\,690 = 0.0258 \text{ m} = 25.8 \text{ mm}$$

The loading appears to require a column h much greater than 250 mm, which means

$e/h < 0.10$. Try the equation above in the short form:

$$P_{n(\text{max})} = 0.80\, P_{n0} = P_u/\phi = 3\,690/0.70 = 5\,271\,\text{kN}$$

Reqd. $P_{n0} = 5\,271/0.80 = 6\,589\,\text{kN} = 6.59\,\text{MN}$

$$P_{n0} = 0.85\, f'_c A_g + (f_y - 0.85\, f'_c) A_{st}$$
$$= 0.85 \times 30\, A_g + (400 - 0.85 \times 30)0.03\, A_g = 36.7\, A_g = 6.59$$
$$A_g = 0.180\,\text{m}^2 \quad h = \sqrt{0.18} = 0.424\,\text{m} = 424\,\text{mm}$$
$$e/h = 25.8/424 = 0.061 < 0.10 \quad \text{The equation used does control.}$$

USE 450 mm $\times$ 450 mm column $A_g = 202.5 \times 10^3\,\text{mm}^2 = 0.2025\,\text{m}^2$

Return to P_{n0} equation.

$0.85 \times 30 \times 0.2025 + (400 - 0.85 \times 30)\, A_{st} = 6.59$, $A_{st} = 3.808 \times 10^{-3}\,\text{m}^2 = 3\,808\,\text{mm}^2$

USE 6-#30 $= 4\,200\,\text{mm}^2$

USE 3 bars on opposite faces.

For #10 ties, clear bar spacing $= \frac{1}{2}(450 - 2 \times 40 - 2 \times 10 - 3 \times 30)$
$$= 130\,\text{mm} < 150\,\text{mm max.)}$$

A single tie around all bars is adequate.

Spacing: $h = 450\,\text{mm}$ 48 tie $d_b = 450\,\text{mm}$ 16 bar $d_b = 480\,\text{mm}$

USE #10 ties at 450 mm on centers.

These calculations are noticeably shorter than the general solution in Sec. 18.10, even although interaction charts were used there.

18.6 BALANCED LOADING, P_{nb}, M_{nb}*

A balanced section for a beam was defined in Chapter 3 as one developing simultaneously a concrete compression strain of 0.003 and a steel tension strain of f_y/E_s. For this condition a unique area of tension reinforcement was required in the beam.

For *any* column the same definition of balanced strains holds (Fig. 18.6b). *Any* column, regardless of its reinforcement, will reach its balanced ultimate load when the load is so placed as to maintain the eccentricity $e_b = M_b/P_b$.* Balance in a column is a matter of *loading*, and it is more descriptive to speak of *balanced loading* rather than of a balanced column. Furthermore, although it is possible to avoid balanced beams in order to avoid compression failures and thus obtain ductility, it is not possible to avoid either compression failures or balanced failures in columns; these are primarily compression members.

For a given column it is very easy to establish the nominal balanced load P_{nb} for ideal conditions, and the accompanying e_b, after the fashion of

*The n subscript is omitted where confusion appears improbable.

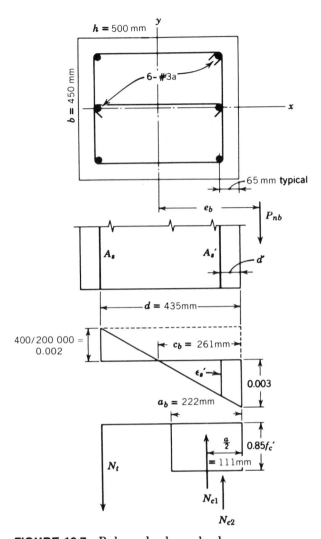

FIGURE 18.7   Balanced column load.

Fig. 18.7. The tension steel A_s and the compression steel A_s' are each 3-#$30 = 2\,100$ mm², $f_c' = 20$ MPa, $f_y = 400$ MPa. The maximum strain of 0.003 in compression and f_y/E_s give c from similar triangles, most simply by thinking of the large dotted triangle.

$$c_b = \frac{0.003}{0.003 + 0.002} \times 435^* = 261 \text{ mm}$$

*Or this equation could be used in the form $c_b = \dfrac{600}{600 + f_y}\, d.$

As for beams, for $f'_c \lesssim 30$ MPa, $a_b = 0.85\,c_b = 222$ mm

$$N_{c1} = 0.85 \times 20 \times 0.222 \times 0.450 = 1.70 \text{ MN}$$

$$\epsilon'_s = \frac{261 - 65}{261} \times 0.003 = 0.00225 > \epsilon_y$$

$$N_{c2} = 2\,100 \times 10^{-6}(400 - 0.85 \times 20) = 0.80 \text{ MN}$$

This value takes account of concrete displaced by steel. This correction is often omitted, but there seems to be no reason to do so when numbers are being used; algebraically, the correction is more awkward.

$$N_t = 2\,100 \times 10^{-6} \times 400 = 0.84 \text{ MN}$$

These three forces are in equilibrium with P_b.

$$\Sigma F_y = 0 = P_b + N_t - N_{c1} - N_{c2} = P_b + 0.84 - 1.70 - 0.80$$

$$P_{nb} = 1.66 \text{ MN}, \quad P_{ub} = \phi P_{nb} = 0.7 \times 1.66 = 1.162 \text{ MN}$$

$$\Sigma M \text{ about center of column} = 0$$

$$N_t \times (250 - 65) + N_{c1}(250 - 111) + N_{c2} \times (250 - 65) - P_b e_b = 0$$

$$1.66\,e_b = 0.84 \times 0.185 + 1.70 \times 0.139 + 0.80 \times 0.185 = 0.540$$

$$e_b = 0.325 \text{ m}$$

$$M_{nb} = P_{nb}e_b = 1.66 \times 0.325 = 0.540 \text{ MN} \cdot \text{m}$$

$$M_{ub} = \phi M_{nb} = 0.7 \times 0.540 \text{ MN} \cdot \text{m} = 0.378 \text{ MN} \cdot \text{m}$$

When a column is unsymmetrical as shown in Fig. 18.8, with A'_s of 5-#30 = 3500 mm^2, the reference line for eccentricity is taken from the *plastic centroid*, which is simply the location of the resultant load that would give a uniform strain all across the column. The plastic centroid falls a distance x_0 off the center of column:

Force	x	M abt. center line
$N_{c1} = 3.83$ MN	0	0
$N_{c3} = 1.34$	$+0.185$ mm	$+0.248$ MN $\cdot$ m
$N_{c2} = 0.80$	-0.185	-0.148
$\Sigma N_c = \overline{5.97}$		$\overline{+0.100}$ $x_0 = +0.10/5.97 = +0.017$ m $= +17$ mm

The method for finding P_b and e_b is then not basically changed, but from a nomenclature point of view it is convenient to define by d'' the distance from this plastic centroid to the centroid of the tension steel.

   If column steel were distributed along four column faces, the numbers used in finding P_b would be increased by an additional term for each group of bars falling at different distances from the neutral axis, but no other complication would exist.

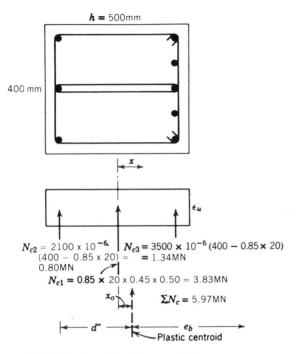

$h = 500$mm

400 mm

$N_{c2} = 2100 \times 10^{-6}$ $(400 - 0.85 \times 20) =$ 0.80MN

$N_{c3} = 3500 \times 10^{-6}(400 - 0.85 \times 20)$ $= 1.34$MN

$N_{c1} = 0.85 \times 20 \times 0.45 \times 0.50 = 3.83$MN

$\Sigma N_c = 5.97$MN

Plastic centroid

FIGURE 18.8 Plastic centroid.

18.7 OTHER INTERACTION POINTS AND SPECIAL ϕ VALUES

For flexure without axial load, the interaction point at the moment axis of the diagram, the procedure used for the balanced load can be used unchanged except (1) that $\epsilon_s > \epsilon_y$, and (2) the neutral axis is unknown and must be located by trial-and-error such that the total tension equals the total compression. For curve plotting, this M_0 point need not be exactly established numerically because a solution with a very small tension resultant and another with a very small compression resultant establishes the curve across the moment axis. Since this is pure flexure, the ϕ here is 0.90.

Other points below the balanced loading also represent primary failures in tension with the tension steel past the yield strain and ϵ_c still 0.003. Plotting points are easily found with assumed c values $< c_b$.*

Curve points above P_b can be established by using $c > c_b$ with $\epsilon_c = 0.003$. As c increases the tensile steel stress must drop and the failure is in

---

*Analysis of such a rectangular column is shown in Sec. 18.16 and for a circular column in Sec. 18.17.

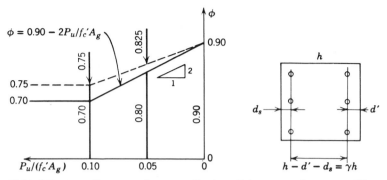

FIGURE 18.9 Increase in ϕ as P_u ($= \phi P_n$) approaches zero. Note limitations where $f_y < 400\,\text{MPa}$ or $\gamma h < 0.70\,h$.

primary compression. The rectangular stress block is valid for this column up to 400 mm. For larger values it becomes approximate because it is based on a full triangle of strain.

As eccentricity beyond P_b, M_b increases, the member becomes more like a beam, justifying $\phi = 0.90$ when P is zero. When longer columns in this region are considered, it is found that buckling problems can exist much below P_b, indicating that the value of 0.70 should not be discarded or increased too soon. Code 9.3.2 now includes two paragraphs:

> For members in which f_y does not exceed 400 MPa, with symmetrical reinforcement, and with $(h - d' - d_s)/h$ not less than 0.70, ϕ may be increased linearly to 0.90 as ϕP_n decreases from $0.10\, f'_c A_g$ to zero. For other reinforced members, ϕ may be increased linearly to 0.90 as ϕP_n decreases from $0.10\, f'_c A_g$ or ϕP_b, whichever is smaller, to zero.

In this, h is the overall column thickness and d' and d_s are the covers to the centroid of compression and tension steel, respectively. The relationship is plotted in Fig. 18.9. For higher f_y or a smaller distance between A_s and A'_s, the balanced load condition moves closer to the moment axis and the second Code paragraph is then necessary to assure a normal ϕ at the balanced load level.

18.8 ACCOUNTING FOR BARS NEAR NEUTRAL AXIS

Bars very near the neutral axis will not be effective in carrying stress; for that combination of M and P any bars near the axis will have stresses lower than the yield stress. For any given neutral axis one should sketch the unit deformations to establish the status of nearby bars, starting with

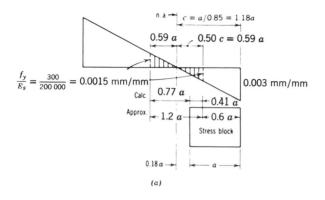

(a)

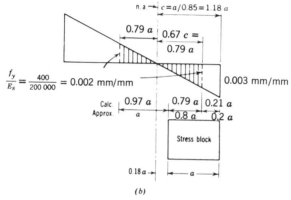

(b)

FIGURE 18.10 Deformation studies for $f_c' \lesssim$ 30 MPa. (a) For $f_y = 300$ MPa. (b) For $f_y = 400$ MPa.

$\epsilon_c = 0.003$ as in Fig. 18.10a. To make $f_s = f_y = 300$ MPa, the steel deformation must be at least 0.0015. With the stress block depth as $a = 0.85\,c$ for f_c' not over 30 MPa, bars falling in the shaded zone will have f_s less than f_y. A rough rule would be to say that f_s is less than an f_y of 300 MPa whenever the bar falls at a depth between 0.6 a and 1.8 a. For $f_y = 400$ MPa the corresponding zones (Fig. 18.10b) would be roughly between depths of 0.4 a and 2.0 a.

The deformation sketch is quite simple to use whenever c is known or assumed. It also facilitates the inclusion in an analysis of bars with f_s or f_s' values less than f_y.

The placement of bars near the neutral axis modifies the shape of the interaction diagram considerably, rounding it upward or outward and making the balanced loading point less conspicuous because extreme bars

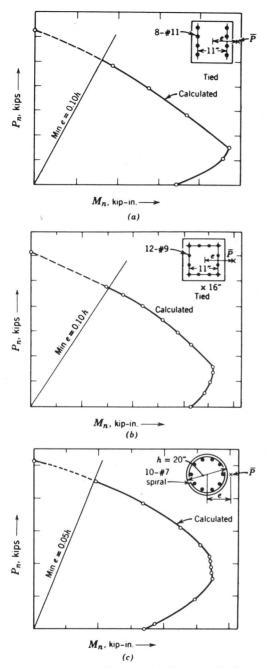

FIGURE 18.11 The varied shapes of column interaction diagrams. Note that minimum *e* values shown are not valid for the 1977 Code.

491

yield first* and other bars later. Typical interaction diagrams are developed in Fig. 18.11, and their different shapes should be noted.

18.9 COLUMN INTERACTION DIAGRAMS AS DESIGN AIDS

Although the designer should understand analytical methods of design (Secs. 18.6 and 18.7), especially for unusual shapes or occasional usage, most design for loads greater than P_b will be from interaction diagrams or tables. Hence their use for checking and for design is introduced here, even though loading patterns, maximum lengths for short columns, and other relevant discussion must follow. These designs for given loads and moments will be extended through the choice of ties. The equivalent details involved in spiral columns are also part of this introduction into short column design.

A single set of four charts for the usual range of steel placement ratios, described by four values of the ratio $\gamma = (h - 2\,d')/h$, for $f'_c \leqslant 30$ MPa and $f_y = 400$ MPa are given in Fig. 18.12a to d. Since each plot contains ratios for axial load, moment, eccentricity, and steel, any two of these, in the analysis of a given column, establish the other two. For example, P and M in a given short column fix e/t and the necessary $\rho_t m$ † $= \rho_t(f_y/0.85\,f'_c)$.

A warning about some variations in charts used must be noted. The charts here (Fig. 18.12a–d) are in terms of P_u and M_u, *not* the P_n ($= P_u/0.70$) and M_n used in Fig. 18.6; either system is acceptable. In the charts here the concrete displaced by the compression bars is ignored in evaluating the resistance provided by these bars. The curves are labeled with values of $\rho_t m$, where $m = f_y/0.85\,f'_c$. These particular charts also fail to reflect the increases in ϕ below $P_u = 0.10\,f'_c bh$, as given in Sec. 18.7. This increase would cause the curves to change slope at that point, "swelling out" farther to the right if projected along the e/h line and returning to a value $(0.90/0.70)$ times that now shown at P_u (or α) of zero. This break in slope is pronounced; a casual glance will tell whether this higher ϕ value near the bottom has been introduced.

18.10 DESIGN OF SHORT TIED COLUMNS USING INTERACTION DIAGRAMS

Design a square tied column, assuming it qualifies as a short column,‡ for a total dead load of 900 kN, a live load of 1 100 kN and a moment of 100 kN · m (all from live load). Use $f'_c = 30$ MPa, $f_y = 400$ MPa, and try for ρ_t of about 0.02.

---

*The Code (10.3.2) is not specific in defining how many bars must yield at the balanced condition. The author uses the yielding of the outermost bar as his criterion.

†$m = f_y/0.85\,f'_c$ is 1963 notation; such a ratio should now be a Greek letter.

‡Sec. 18.14 and Table 18.1 relate to the criterion.

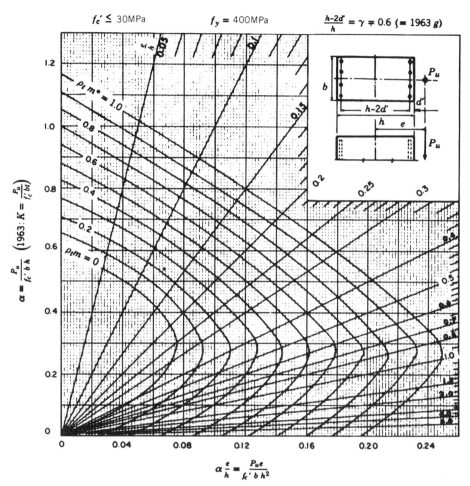

FIGURE 18.12a Column Interaction Chart. $\gamma = 0.6$. $m = f_y/0.85\, f'_c$ is 1963 notation. m now should become a lower-case Greek letter. Chart readings include the effect of ϕ, but not correctly below ordinate of 0.10.

Solution

$$P_u = 1.4 \times 900 + 1.7 \times 1\,100 = 3\,130 \text{ kN} = 3.13 \text{ MN}$$

$$M_u = 100 \times 1.7 = 170 \text{ kN} \cdot \text{m} = 0.17 \text{ MN} \cdot \text{m}$$

$$e = M_u/P_u = 0.054 \text{ m} = 54 \text{ mm} \quad m = f_y/0.85\,f'_c = 400/0.85 \times 30 = 16$$

Desired $\rho_t m = 0.02 \times 16 = 0.32$

Try 500 mm square column. $e/h = 54/500 = 0.109$. Assume $\gamma = 0.70$ (Fig. 18.12b)

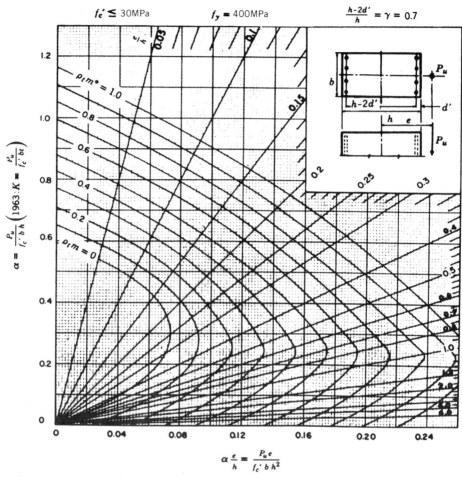

$f_c' \leq 30\text{MPa}$ $f_y = 400\text{MPa}$ $\dfrac{h-2d'}{h} = \gamma = 0.7$

FIGURE 18.12b Column Interaction Chart. $\gamma = 0.7$. Chart readings include the effect of ϕ, but not correctly below ordinate of 0.10.

Enter diagram at intersection of $e/h = 0.109$ and $\rho_t m = 0.32$, reading $\alpha = 0.61$ on the left scale. These charts automatically introduce the ϕ.

$$0.61 = P_u/f_c'h^2 = 3.13/30\ h^2;\ \ h = 0.414\ \text{m} = 414\ \text{mm}$$

Try 425 mm by 425 mm column.

$$\alpha = P_u/f_c'h^2 = 3.13/30 \times 0.425^2 = 0.58;\quad e/h = 54/425 = 0.127$$

For #25 bars and #10 ties, $\gamma = (425 - 2 \times 40 - 2 \times 10 - 25)/425 = 0.706$

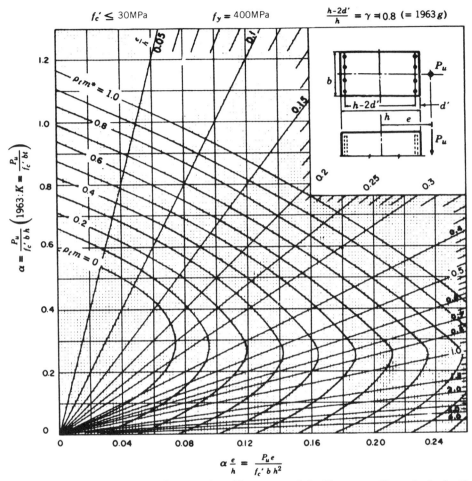

FIGURE 18.12c Column Interaction Chart. $\gamma = 0.8$. Chart readings include the effect of ϕ, but not correctly below ordinate of 0.10.

$\gamma = 0.706$ is sufficiently close to 0.7 to make interpolation between the charts for $\gamma = 0.7$ and $\gamma = 0.8$ unnecessary.

Enter $\gamma = 0.7$ chart with $\alpha = 0.58$ and $e/h = 0.127$, and read $\rho_t m = 0.31$.

$$\rho_t = 0.31 \times 0.85 \times 30/400 = 0.0198$$

The required bar area $A_{st} = 0.0198 \times 425^2 = 3\,580 \text{ mm}^2$

Check $P_{u(\max)} = \phi(0.8)\,P_0)$

$$= 0.7 \times 0.8\,[0.85 \times 30 \times 0.425^2 + (400 - 0.85 \times 30) \times 3\,580 \times 10^{-6}]$$

$$= 3.330 \text{ MN} > 3.130 \text{ MN}$$

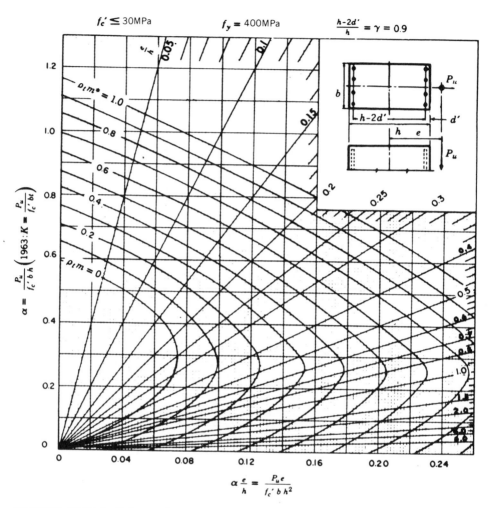

FIGURE 18.12d Column Interaction Chart. $\gamma = 0.9$. Chart readings include the effect of ϕ, but not correctly below ordinate of 0.10.

USE 425 mm by 425 mm column, 8-#25 bars ($A_{st} = 4\,000$ mm²)

12-#20 bars ($A_{st} = 3\,600$ mm²) are more economical, but it would be difficult to place 12 bars in a column of this size.

The minimum ties are #10.

USE #10 ties as shown in Fig. 18.13e or f. An alternate would be as in Fig. 18.13a with a single interior crosstie as in Fig. 18.13d.

A complete design also includes analysis for M about the other axis.

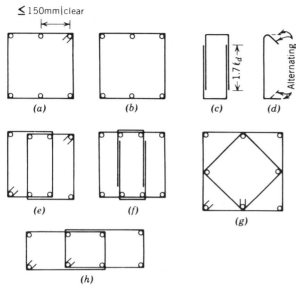

FIGURE 18.13 Simple tie arrangements.

18.11 COLUMN TIES

Column ties must be made from at least #10 bars. The details of Code 7.10.5 are rather simple, in effect requiring at least alternate bars to be braced and no bar more than 150 mm from a braced bar. The maximum tie spacing is the smallest of:

1. $16\,d_b$ for the column bars
2. $48\,d_b$ for the tie itself
3. The minimum column thickness.

A few typical tie layouts are shown in Fig. 18.13. The crosstie with a 135° hook on one end and a 90° hook on the other has been developed to facilitate placement when the vertical bars are already in place. Obviously, the 90° hook end must be securely tied to the vertical bar to avoid displacement and alternate ties should be reversed such that each 90° hook falls vertically between two 135° hooks. The 90° hook is less conservative as to strength but placement is much easier.

The 90° hook shown in Fig. 18.13*b* is a great convenience for placement, but it should not be used where the member is subjected to heavy reversing moment, as in earthquake resistant design; it tends to develop spalling over the tail of the hook in such places, which cancels its benefit as a tie. Likewise, it is sometimes convenient to use the lapped ties of Fig. 18.13c to supplement square overall ties as in Fig. 18.13f. When lapped thus in the interior with a full splice length this is more dependable than when lapped as an outside tie leg; under earthquake or other reversing loading the exterior face splice is *not* recommended.

If in Fig. 18.13e the column bars are not separated by more than 150 mm clear, a single outside tie plus a crosstie over one middle bar on opposite faces, as in Fig. 18.13d is a possible solution. If the same column had two more bars on the sides, the ties as in Fig. 18.13d connecting them would be adequate.

The spacing of ties for the design problem of Sec. 18.10 must not exceed any of the three criteria listed at the start of this section:

$16 d_b = 16 \times 25 = 400$ mm for the main bar diameter.

$48 d_b = 48 \times 10 = 480$ mm for the tie diameter

Column thickness $= 425$ mm. Max. tie spacing is 400 mm

18.12 DESIGN OF SHORT SPIRAL COLUMNS USING INTERACTION DIAGRAMS

Spiral columns may be either round or square, although the spiral itself must be round, as shown in Fig. 18.1c. Under axial load alone a spiral column has the same ultimate strength as a tied column having the same A_g and A_{st}. It could be argued that spiral columns should be designed for the same ϕ as tied columns. However, their greater ductility before failure (loss of some stiffness but not strength) justifies the Code use of ϕ of 0.75, compared to 0.70 for tied columns. The author would like to see a larger differential because, even in Fig. 18.5, the spiral resistance picks up the strength lost by the spalling cover concrete.

The limitation on the maximum axial load on the spiral column, at the middle of Sec. 18.5, shows a reduction factor of 0.85 (compared to 0.80 for tied column). The 1971 Code accomplished nearly the same thing by using a minimum eccentricity of 0.05 h for the spiral column (compared to 0.10 h for the tied column). The design of short spiral columns where e/h is significantly under 0.05 h may follow the same design sequence as shown for tied columns in Sec. 18.5, with three coefficient changes: ϕ for spiral column 0.75; e less than 0.05 h; $P_{n(\max)}$ coefficient 0.85. The number of bars used must be at least 6 but need not be an even number. The spiral design is shown in Sec. 18.13.

The algebra of handling the circular pattern of bars is more complex than the single face layers covered by the charts of Fig. 18.13, but the final corresponding charts for spiral columns differ chiefly in a more rounded shape (Fig. 18.11c). The method of design using them is identical with that for the tied column in Sec. 18.10.

Biaxial bending is not a complication when the spiral column is circular, since resistance is equal in all directions. However, spirals are often used in square columns and occasionally dual spirals (overlapping) are used in narrow rectangular columns. Interaction diagrams for the latter are not readily available.

18.13 SPIRAL DESIGN

The spiral must replace the strength of the concrete shell that can be expected to spall off when the column yields.

The spiral reinforcement is specified by ρ_s in Code 10.9.3:

$$\rho_s = 0.45(A_g/A_c - 1)f'_c/f_y \qquad (10\text{-}5)$$

where ρ_s = ratio of volume of spiral reinforcement to volume of concrete core (out to out of spiral)
A_c = area of core of column
f_y = yield strength of spiral steel but no greater than 400 MPa

This equation reflects the strengthening effect of the radial compressive forces from the spiral on the core of the column as it shortens enough to make the unsupported shell spall off. This relation can be arranged as follows:

$$\rho_s A_c f_y = 0.45(A_g - A_c)f'_c$$
$$2\,\rho_s A_c f_y = 0.90\,f'_c(A_g - A_c)$$

Since $A_g - A_c$ is the shell area, which spalls off, the right-hand side represents the assumed ultimate strength of the shell with an increase in the coefficient from 0.85 to 0.90. The quantity $\rho_s A_c$ represents the spiral steel reduced to equivalent volume of longitudinal steel. This steel can develop a stress f_y and in spiral form is roughly twice as effective as longitudinal steel. Thus the spiral is designed to replace the strength of the shell with a very slight extra allowance for safety. For a 525 mm diameter column with Gr 400 bars and $f'_c = 25$ MPa:

$$\rho_s = 0.45\left(\frac{A_g}{A_c} - 1\right)\frac{f'_c}{f_y} = 0.45\left(\frac{525^2}{445^2} - 1\right)\frac{25}{400} = 0.0110$$

By definition, ρ_s = (vol. of spiral in one round) $\div$ (vol. of core in height s_s), where s_s is the pitch as shown in Fig. 18.14. The pitch s_s is small enough for the volume of one round or turn of the spiral to be taken as $A_{sp}\pi(d_c - d_b)$. The volume of the core in height s_s is $s_s\pi d_c^2/4$.

$$\rho_s = A_{sp}\pi(d_c - d_b) \div (s_s\pi d_c^2/4)$$
$$= 4\,A_{sp}(d_c - d_b) \div (s_s d_c^2)$$

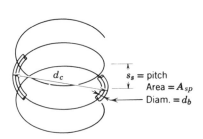

FIGURE 18.14 Spiral notation.

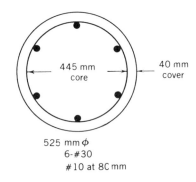

FIGURE 18.15 Spiral column design of Sec. 18.13.

(Many designers ignore the difference between $d_c - d_b$ and d_c, and use $\rho_s = 4\,A_{sp} \div s_s d_c$.) It is recommended that the size of spiral wire be assumed and s_s then be calculated, since the wire will usually be 10 mm* or 15 mm. The spacing s_s can be specified as closely as desired.

Try 10 mm spiral rod, $A_{sp} = 100$ mm^2

$$\rho_s = 0.0110 = \frac{4 \times 100(445 - 10)}{s_s \times 445^2}$$

$$s_s = 79.9 \text{ mm (69.9 mm clear)}$$

Specification maximum: 80 mm clear spacing

Specification minimums: 25 mm clear

                        1.33 max aggregate size, clear

USE #10 at 80 mm.

The design would be as shown in Fig. 18.15, with the 6-#30 assumed in order to show a total column description in a form frequently used.

18.14 LENGTH LIMITATIONS FOR SHORT COLUMNS

In a chapter devoted to short columns the maximum height or length qualifying as short for design has thus far been avoided. The particular effective column length in a braced frame that demands the special long column technique depends on the deflection the column moment creates.

*10 mm is the minimum for cast-in-place construction (Code 7.10.4.2).

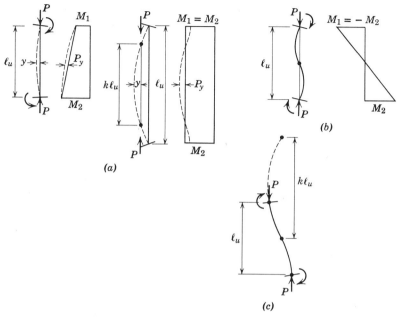

FIGURE 18.16 Effective lengths $k\ell_u$ of columns. (a) Single curvature in braced frame, short and long. (b) Reversed curvature in braced frame. (c) Reversed curvature in unbraced frame.

This in turn depends on the column curvature and its joint restraints at the ends.

Three cases must be separately recognized in setting proper limits on short column length. The first two are for braced frames, with bracing implying negligible lateral movement of the end joints. The third is the general case of unbraced frames, which can deflect laterally.

The single curvature case of Fig. 18.16a is the more flexible of the two braced frame cases and includes $M_1 < M_2$ as well as $M_1 = M_2$, either producing a simple deflection curve at service loads. P times the deflection y of this curve measures an added moment sensitive to length in a nonlinear fashion. In the long columns, as the load increases the curvature changes, because the end joints do not rotate freely and the center deflection increases rapidly. Points of inflection appear and it becomes the reduced effective length $k\ell_u$ between inflection points, which becomes the critical length. The usual range for k is from 0.65 to 0.90, dependent on the relative stiffness ψ of columns-to-beams at the end joints. The use of $k = 1$ is required by the Code if there is essentially no moment at the joints. (See Code 10.11.5.4 for more detail.) The limiting short column

TABLE 18.1 Typical Maximum Short Column Lengths in Terms of Thickness, ℓ_u/h

Values of $\psi_1 = \psi_2 = \psi^* =$		Rectangular Column			Circular Column		
		0.5	1.0	2.0	0.5	1.0	2.0
Braced Frame k =		0.68	0.77	0.86	0.68	0.77	0.86
Single curvature	$M_1 = M_2$	9.7	8.6	7.7	8.1	7.2	6.4
	$M_1 = 0$	15.0	13.3	11.9	12.5	11.1	9.9
Reversed curvature	$M_1 = -0.4\,M_2$	17.1	15.1	13.6	14.2	12.6	11.3
	$M_1 = -M_2$	20.3	17.9	16.1	17.0	14.9	13.4
Unbraced Frame k =		1.17	1.31	1.59	1.17	1.31	1.59
All *M* values		5.6	5.0	4.1	4.7	4.2	3.4

*See Fig. 19.6 for definition of ψ as relative column stiffness.

length is defined as

$$k\ell_u/r = 34 - 12\,M_1/M_2$$

The radius of gyration r may be used as $0.30\,h$ for rectangular columns or $0.25\,h$ for circular columns (Code 10.11.3). These r values have been used by Dr. Furlong in developing Table 18.1 in terms of ℓ_u/h. The numbers tabulated, when multiplied by h, give the maximum lengths for the short column analysis for several listed ψ values.* In design a majority of single curvature columns qualify as short. Section 19.8 gives more detail, including in Fig. 19.6 a nomograph to assist in evaluating k.

The second curvature case for the braced frame is shown in Fig. 18.16b. Because the reversed curvature case, typical in an exterior column, creates less column deflection, larger $k\ell_u/r$ or ℓ_u/h values are permissible. The limiting equation above is also used here, with M_1 negative and M_2 positive, thus increasing $k\ell_u/r$ above 34. In braced frames most reversed curvature columns are short columns, as the ℓ_u/h ratios in Table 18.1 would indicate.

The third curvature case (Fig. 18.16c) applies to *all* columns in an unbraced frame, because under wind or other lateral forces the frame deflects laterally. The top of each column moves relative to the joint at its lower end and this sets up a reversed curvature case with joints at top and bottom rotating in such fashion as to increase the lateral deflection and the Py moment. Table 18.1 shows all k values greater than unity; only *very* stiff beams could reduce k to 1.0. Note the same limitation applies to

*See Fig. 19.6 for definition of ψ and relationship to k.

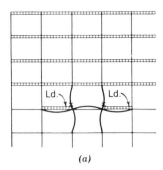
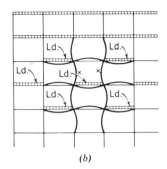

(a)   (b)

FIGURE 18.17   Column moments without sidesway. (a) Maximum load and moment loading but minimum long column effect. (b) Single curvature loading.

the limiting short column length even if only vertical loads are involved; as long columns these also fail by becoming unstable and swaying laterally. The Code limit for short columns in the unbraced frame is a $k\ell_u/r$ of 22.

The Code Commentary gives an alternate design method for long columns under a limited range of conditions. This somewhat simpler method is called the R method. It builds from the same short column design but uses a different technique for long columns and slightly different limiting short column lengths are appropriate. Details are given in Sec. 19.15. One case that often controls design is significantly different, shown in Fig. 18.17a. Under the R method such a column is short up to ℓ_u/r of 54, that is, ℓ_u/h of 16.2 for a rectangular column.

For other moment loadings, in both braced and unbraced frames, the R method in Fig. B-2 in Appendix B shows at the top of the graph the start of sloping lines marking the reduced R of long columns. Where these intersect, the top ($R = 1.0$) marks the maximum ℓ_u/r appropriate for short column design.

18.15   FRAME LOADINGS FOR MAXIMUM COLUMN MOMENTS

The designer must distinguish sharply between columns where sidesway is prevented (diagonal bracing, shear walls, etc.) and those free to sway.

Three different frame loadings are involved in the three different curvatures of Sec. 18.14. Two of these cases on interior columns occur in a braced frame as shown in Fig. 18.17. For critical stresses in a relatively uniform frame at the point marked x, the loading of Fig. 18.17a gives both a maximum axial load and the maximum moment consistent therewith. The column behavior is essentially that of Fig. 18.16b. There is a minimum of long-column effect. This case will be referred to as the reversed moment case, whether the reversal is small, as in the loading diagram, or a complete reverse, as it might be on an exterior column. The case shown in the loading diagram might also be called a column restrained at the far end (by the moment resisting the upper joint rotation).

A checkerboard loading adjacent to a column (Fig. 18.17b) produces single curvature in the columns, as in Fig. 18.16a. Fortunately, this curvature requires that some potential load be omitted from the floor above and analysis also shows smaller end moments in this case than in the one previously discussed. Single curvature can accompany maximum axial load only when dealing with unequal spans, and then usually with lower moments than for the case of far end restrained. Nevertheless, the long column effect is large at x and could make this govern on very long columns or some unequal span situations.

In the absence of shear walls, sidesway may permit a whole story height in Fig. 18.17a or b to deflect laterally and shift the moment conditions of Fig. 18.16a,b to one very nearly like Fig. 18.16c. Symmetry of loading or frame delays this, but elastic analysis shows sidesway to be the expected collapse mode for the entire story. Length effects that might be small without sidesway would then become large when sidesway develops.

The sidesway case under shear load is worst when the shear is maximum. The distribution of these shears to the individual columns requires analysis of the entire story rather than individual columns and the Code specifies use of the entire story in computing the long column effects (Sec. 19.9). Although the sidesway moment gives the most severe long-column effect, in a low structure the basic moments may still be small enough to make the vertical loading control the design.

18.16 TENSION FAILURES—ANALYSIS

(a) Analysis Procedures

Although compression failures have to be handled on the basis of interaction diagrams or strain analyses, tension failures of short columns of ideal materials are as easily analyzed as beams, from basic principles, when e is given.

When a tension failure occurs, the yielding of the tension steel causes the neutral axis to move toward the compression face, reduces the compression area, and finally brings about a secondary compression failure. Although this shift of the neutral axis lowers the strain on A'_s and makes it necessary finally to check whether $f'_s = f_y$, it is convenient to start by assuming f_y and neglecting the concrete displaced by A'_s. With these simplifications equilibrium of nominal forces is very simply established

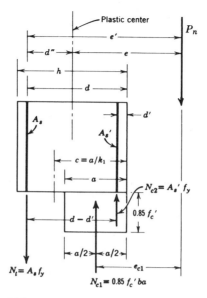

FIGURE 18.18 Equilibrium of nominal forces with tension failure, ultimate strength method.

from Fig. 18.18. For $A_{s1} = A_s/2 = A'_s$, $N_t = N_{c2}$, and it follows from $\Sigma F_y = 0$ that $N_{c1} = P_n$. For $\Sigma M = 0$, the forces divide into two couples, one composed of N_t and N_{c2}, the other of N_{c1} and P_n.

$$N_{c2}(d - d') = P_n e_{c1}$$

where $e_{c1} = e + d'' - d + 0.5\,a$ which for symmetrical sections becomes $e_{c1} = e - 0.5\,h + 0.5\,a$. These are powerful and simple equations and are adequate for many problems.

If displaced concrete is considered, the effective N_{c1} is a little greater than P_n, and the first equation is replaced by ΣM about some convenient center.* Unsymmetrical columns can be treated similarly.

If a is small, the strain on compression steel should be checked using $c = a/k_1$ and $\epsilon_c = 0.003$. It may be necessary to substitute $\epsilon'_s E_s$ in place of f_y on A'_s. Also, if a is large, it could be that the tension f_s is less than f_y; if this occurs it is proof that P_n is really greater than P_{nb} and the compression failure solution of Sec. 18.7 is needed.

If bars are not grouped in single layers near opposite faces, but are at varying distances from the column centerline, the same two equations

---

*A center on A'_s is often convenient.

$\Sigma F_y = 0$ and $\Sigma M = 0$, must be satisfied, probably by trial and error methods starting with an assumed neutral axis location.

(b) Example—Tension Failure

The short 450 mm by 500 mm column of Fig. 18.7 will be analyzed to establish the load it may be assigned if placed on the x-axis at an eccentricity of 375 mm about the y-axis. $f_c' = 20$ MPa, $f_y = 400$ MPa. $A_s = 6$-#30 = 4 200 mm^2.

Solution

Cover of 65 mm to center of bar provides 40 mm cover with #10 ties **O.K.** First proceed (approximately and simply) by assuming $f_s' = f_y$ and neglecting displaced concrete. Find the nominal short column strength as an ideal material.

$N_t = N_{c2} = 2\,100 \times 10^{-3} \times 400 = 840$ kN

$M_1 = N_{c2}(d - d') = 840 \times 0.375 = 315$ kN $\cdot$ m

Try $a = 250$ mm, $e_{c1} = 375 - 250 + \frac{1}{2} \times 250 = 250$ mm (compare Fig. 18.19)

$M_1 = e_{c1}N_{c1};\ N_{c1} = 315/0.250 = 1\,260$ kN

$a = N_{c1}/(0.85\,f_c'b) = 1\,260 \times 10^{-3}/0.85 \times 20 \times 0.45 = 0.165$ m $= 165$ mm

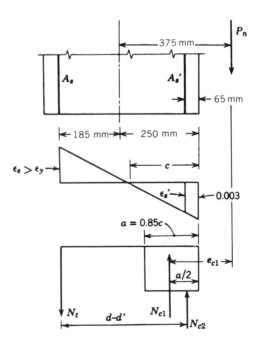

FIGURE 18.19   Column of Sec. 18.16b.

Try $a = 190$ mm; $e_{c1} = 375 - 250 + \frac{1}{2} \times 190 = 220$ mm (Fig. 18.19)

$0.22\, N_{c1} = 315$, $N_{c1} = 1\,432$ kN $= P_n$

$a = 1.432/0.85 \times 20 \times 0.45 = 0.187$ m $= 187$ mm $\qquad$ say **O.K.**

Check f_s': $c = a/0.85 = 190/0.85 = 224$ mm

$$\epsilon_s' = \frac{(224 - 65)0.003}{224} = 0.002\,13$$

$$f_s' = \epsilon_s' E_s = 0.002\,13 \times 200\,000$$
$$= 426\,\text{MPa} > f_y \quad \therefore f_s' = f_y$$
$$P_n = 1\,432\,\text{kN, subject to reduction for } \phi$$

The more exact solution would consider the displaced concrete. In effect, the 2 100 mm^2 of A_s' steel displaces concrete counted at $0.85\, f_c' = 17$ MPa, or a total of about 36 kN, which must come from N_{c1} (and P_n since P_n is used here as N_{c1}). The second trial value of $a = 190$ mm above is a good starting point, giving

$$c = 224\,\text{mm as above}$$
$$\epsilon_s = 0.003(435 - 224)/224 = 0.0028 \gg \epsilon_y$$

This proves the failure is in tension. (Alternate: Compare P_n with P_{nb}.) We can now calculate the forces and moments more accurately:

$$N_t = 2\,100 \times 10^{-3} \times 400 = 840\,\text{kN}$$
$$N_{c2} = 2\,100 \times 10^{-3}(400 - 0.85 \times 20) = 804\,\text{kN}$$
$$N_{c1} = (190 \times 450) \times 10^{-3} \times 0.85 \times 20 = 1454\,\text{kN}$$
$$P_n = N_{c1} + N_{c2} - N_t = 1\,454 + 804 - 840 = 1\,418\,\text{kN}$$
$$\Sigma M = 0 = N_t \times 185 + N_{c1}(250 - \tfrac{1}{2} \times 190) + N_{c2} \times 185 - P_n \times 375$$
$$840 \times 185 + 1\,454 \times 155 + 804 \times 185 = P_n \times 375$$
$$P_n = 1\,412\,\text{kN}$$

This is almost identical with the result obtained from vertical equilibrium, say $P_n = 1\,414$ kN

$$\text{Short column } P_u = P_n\, \phi = 1\,415 \times 0.7 = 991\,\text{kN}$$
$$M_u = P_u e = 991 \times 0.375 = 371\,\text{kN} \cdot \text{m}$$

18.17 CIRCULAR COLUMNS—ANALYSIS

The column of Fig. 18.20 will be analyzed to establish the ultimate load capacity for a load at a 300 mm eccentricity, assuming $f_c' = 20$ MPa, Grade 300 steel, and a short column length.

Solution

The geometrical coefficients shown in Fig. B.3 in Appendix B make the analysis of the curved shape feasible. A trial-and-error solution starting from assumed c is practical.

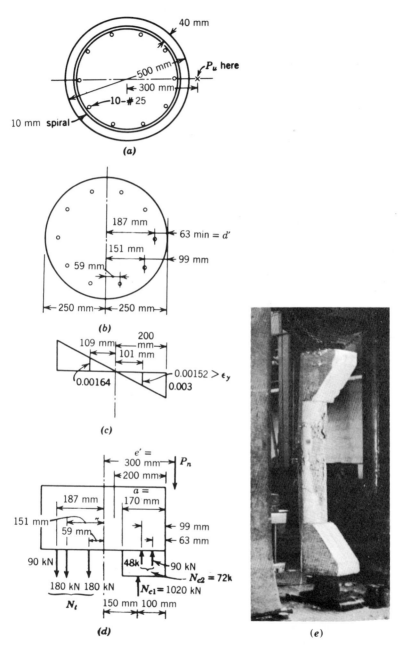

FIGURE 18.20 Analysis of strength of circular column. (*a*) Column cross section. (*b*) Bar locations. (*c*) Strain triangles. (*d*) Forces in equilibrium. (*e*) The failure of an eccentrically loaded spiral column. (Courtesy University of Illinois.)

Try $c = 200$ mm, $a = 0.85\,c = 170$ mm, $a/h = 170/500 = 0.340 = x/h$ of Fig. B.3 and corresponds to $B = 0.24$, $A = 0.60$.* Hence, $\bar{x}$ of the segment from the center of the column $= Ar = 0.60 \times 250 = 150$ mm, and the area $= Bh^2 = 0.24 \times 500^2 = 60\,000$ mm^2.

Permissible $N_{c1} = 0.85 \times 20 \times 60\,000 \times 10^{-3} = 1\,020$ kN

The assumed bar locations are shown in Fig. 18.20b. One pair of bars is so close to the assumed neutral axis as to be only slightly stressed. This pair is totally neglected in this calculation; the problem has been discussed in Sec. 18.8. All other bars carry the yield stress, as shown by the strain triangles of Fig. 18.20c which makes it a tension failure case and results in the forces of Fig. 18.20d (which here neglect displaced concrete). For one #20 bar the force is $300 \times 300 \times 10^{-3} = 90$ kN. Moments of these forces about the center of the column must balance the moment Pe:

$$2 \times 90 \times 0.187 = \quad 33.7$$
$$2 \times 180 \times 0.151 = \quad 54.4$$
$$180 \times 0.059 = \quad 10.6$$
$$N_{c1}\!: 1\,020 \times 0.150 = 153.0$$
$$\overline{251.7} = 0.300\,P_n$$
$$P_n = 839.0 \text{ kN}$$
$$N_t - N_{c2} = (90 + 180 + 180) - (180 + 90) = 180 \text{ kN}$$
$$\text{Reqd. } N_{c1} = \overline{839\,\text{k}} + 180 = 1\,019 \text{ kN}$$

This is sufficiently close to the permissible N_{c1} of $1\,020$ kN for the assumed c.   **O.K.** Hence the ultimate load $P_u = \phi P_n = 0.75 \times 1\,019 = 764$ kN.

The failure of such a column is shown in Fig. 18.20c.

This type of analysis is suitable for either a tension or compression failure, with some question remaining when the neutral axis is moved just outside of the cross section or (practically) when the stress block just covers the entire depth of member h. This questionable point would arise close to the case of $P_{n(\text{max})}$ set by the Code (*near e* of $0.10\,h$ for a rectangular column or $0.05\,h$ for a spiral column). For the extreme cases the author would assume N_{c1} on the concrete uniform over the total area at $0.85\,f_c'$ and the steel stress varying as the strains indicated. The author would also recognize that the rectangular stress block is inherently less accurate when it covers the entire member as here than in a beam where the detailed shape of the stress-strain curve makes little numerical difference. Hence, some over-design rather than skimping is appropriate here.

18.18 BIAXIAL MOMENTS ON COLUMNS

Many columns are subject simultaneously to moments about both major axes, especially corner columns. The mathematics of such cases is quite involved, although for any given neutral axis the analysis can follow very

---

*The ACI standard notation would use lower-case Greek letters for ratios A, B, and C, but for this limited use here a change from Professor Shank's notation seems unnecessary.

simple ideas like those of Sec. 18.6 for P_b and e_b. For example, in Fig. 18.21a, let the neutral axis be arbitrarily chosen as shown. One may construct, perpendicular to this axis, strain triangles precisely as done in Fig. 18.20c for the circular column. The maximum concrete strain of 0.003 is probably too small and the equivalent rectangular stress block may not be exactly $a = \beta_1 c$, but a thorough investigation[13] has indicated that in combination the two are satisfactory. Then N_{c1} is simply 0.85 f_c' times the shaded area and acts at its centroid (or the area can be broken up into triangles and rectangles if preferred). The strains lead easily to N_t, N_{c2}, N_{c3} and N_{c4}. $\Sigma F_y = 0$ leads to the value of P_n, $\Sigma M_x = 0$ leads to x_P, (equal to e_x along the x-axis), and $\Sigma M_y = 0$ leads to y_P (equal to e_y along the y-axis). This completely establishes the load that would create this neutral axis. The method is awkward chiefly because of the dimensions, which might be easily obtained graphically from a scale layout. The real problem is that one needs to start with given loads and eccentricities and not with a location of the neutral axis; and this neutral axis is not usually perpendicular to the resultant eccentricity.

For a circular column, the interaction diagram for moments about any axis is the same and no real problem exists. It is easy to visualize a three-dimensional interaction diagram for this case. It would be a surface of revolution obtained by rotating the interaction diagram about the P_n-axis, a little like Fig. 18.21b except with the same diagram on each axis. With a square column having equal steel on each face, the interaction diagram on the x-axis is the same as on the y-axis; but when these are rotated they give values on the unsafe side for intermediate angles. Similarly, if a rectangular column is considered, it is easy to visualize a varying radius of rotation, as in Fig. 18.21b, creating an ellipse on any horizontal plane which connects the x- and y-axis diagrams. This would also indicate values on the unsafe side. In other words, both the circle for the square column and the ellipse for the rectangular column must be considered as upper bounds on proper values, the real three dimensional surfaces being a little flattened on the diagonals. This difference is quite small for the upper part of the diagrams and is maximum near the P_b level.

The author's earlier suggestion was that these surfaces be used as a starting point and then the moment capacity at any given P_n level be reduced, by a maximum (for ordinary amounts of reinforcing) of about 15%* when the resultant e is on the 45° line between the x- and y-axes. A study of Furlong's[14] and Pannell's[9] work would permit some refinements on this approximation.

A more rigorous study by Gouwens[10] indicates more specific methods

---

*Pannell shows up to 32% for high-steel ratios.

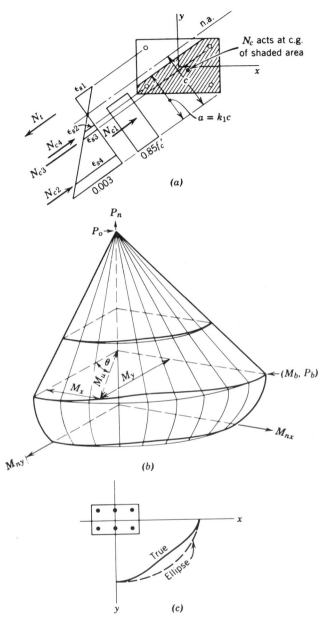

FIGURE 18.21 Biaxial bending on column. (a) Forces acting on cross section. (b) Typical interaction surface (from ACI Ref. 14). (c) Contour at level P and P_b.

that are usable as very reasonable approximations. The horizontal section through the interaction diagram at a constant P_n in Fig. 18.21b is a function of the height P_n/C_c where $C_c = f'_c bh$. Gouwens follows Meek's approximation of using this curve as two straight lines (when moments are expressed as ratios) fixed by the β_b at their intersection, as shown in Fig. 18.22a. Note that $\beta_b = M_y/M_{yo} = M_x/M_{xo}$ when axes in this figure are plotted as dimensionless ratios. Approximate β_b values for various combinations of P_n/C_c values and ratios of C_s/C_c (i.e., $A_s f_y/f'_c bh = \rho_t f_y/f'_c$) are shown in Fig. 18.22b. Gowens notes a desirable correction for β_b values for columns having *only* corner bars: Lower the plotted β_b values by 0.02.

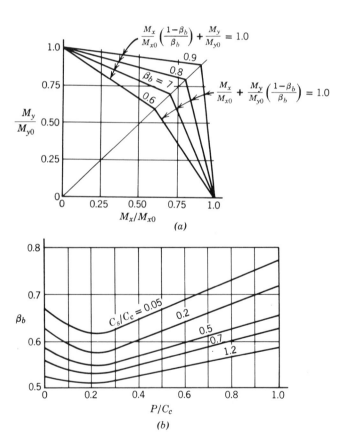

FIGURE 18.22 Charts by Gowens (see Reference 10) for biaxial column moments. (a) Charts and equations for various β_b. (b) Chart establishing β_b to use in (a).

A method* published by Bresler[15] in 1960 relates the desired P_u (or now ϕP_a) under biaxial loading (e_x and e_y) to three other P_u values:

$P_{uy} = \phi P_{ny}$ = design strength for same column under the same e_y (Fig. 18.23b)

$P_{ux} = \phi P_{nx}$ = design strength for same column under the same e_x (Fig. 18.23c)

$P_{uo} = \phi P_{no}$ = theoretical axial load design strength for same column when $e_x = e_y = 0$ (as if the Code permitted such design)

This equation is

$$\frac{1}{P_u} = \frac{1}{P_{ux}} + \frac{1}{P_{uy}} - \frac{1}{P_{uo}} \quad \text{or} \quad \frac{1}{\phi P_n} = \frac{1}{\phi P_{nx}} + \frac{1}{\phi P_{ny}} - \frac{1}{\phi P_{no}}$$

With charts (or tables) for P_u plus uniaxial moment (similar to Fig. 18.12) about the x axis and similar ones for P_u plus moment about the y axis, values of P_{ux}, P_{uy}, and the theoretical P_{uo} are easy to establish. Substitution in the equation then gives the desired biaxial P_u. (Usually the ordinates of charts give $P_u/f'_c bh$, with ϕ normally included so that in 1977 notation this would lead to ϕP_n.) As an approximate method it is one of the best when one has the necessary charts or tables, *provided* the resulting P_u or ϕP_n is above $0.10 f'_c bh$. If P_u is lower than the balanced design level (or say the Code $0.10 f'_c bh$ level) the errors by this method can increase. In typical cases it is then on the safe side to design for biaxial moment alone, since tension failure then controls.

For *square* columns loaded on the diagonal, interaction diagrams are now available in the ACI handbook[11]. Reference 12 will contain curves that appear to be very helpful in the general problem of biaxially loaded columns (not yet available in metric units).

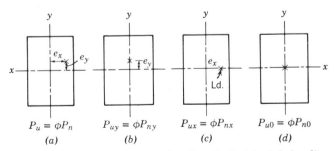

$$P_u = \phi P_n \qquad P_{uy} = \phi P_{ny} \qquad P_{ux} = \phi P_{nx} \qquad P_{uo} = \phi P_{no}$$

$$(a) \qquad\qquad (b) \qquad\qquad (c) \qquad\qquad (d)$$

FIGURE 18.23 Notations for Bresler's biaxial loading equation. (a) Biaxial moments. (b)(c) Eccentricities about x and y axes. (d) Axial load alone.

*Reportedly from the Russian Code.

For long columns (Chapter 19) the magnifier is to be separately calculated and applied to the moment about each axis independently. The relative values will then determine the plane of the principal moment. The minimum e/h applies about each axis, but not simultaneously.

18.19 COLUMN SPLICES

The general subject of splice design is adequately covered in Chapter 7 but some further general comments are appropriate for columns.

Location of column splices in the past has typically been just above the floor level, in effect starting new column bars at the lowest possible point. Certain complications can result. If the dowel projections and the new column bars are placed side by side so that they are equidistant from the axis of the column, $2\,d_b$ must be deducted from the center-to-center spacing to find the clear bar spacing. When 5% or more of longitudinal steel is used, this spacing can become critical and limit the permissible column steel below what might be desired.

Several factors now tend on occasion to make other splice points look attractive. Code 12.18.3 requires at least 25% of the vertical steel on each face to be developed to f_y in tension (or more bars at a lower stress to give the same total tension capacity). If butt splices are not used, the compression lap may provide the required tension. Another way to achieve this is to run bars in two-story heights and stagger the splices. Although such bars are awkward to support on the job, this relieves the congestion in a single splicing section. Since tension splices tend to be long, elimination of some of these is also attractive. The awkwardness in construction lies not only in bracing the loose bars but also, after columns ties are in place, in interweaving the beam and girder steel through the column steel.

Appendix A of the Code, covering seismic design, limits the minimum splice to $30\,d_b \geq 400$ mm and the Commentary points out that splices at mid-height are preferable because there the moment is lowest. The maximum design moments for wind, earthquake, and even vertical loads (in typical multistory buildings) typically occur at the *end* (or ends) of the column; the splice at mid-height escapes some of this moment.

For compression splices, bars may be butted and held in place by pipe or commercial mechanical splices.

For tension splices, bars can be welded together, usually with butt welds, but it should be noted that a number of welding passes are required for large high strength bars. American Welding Society preheat requirements can also be formidable.[16] A Cadweld splice, using molten metal to hold the bar to a sleeve (deformed on its inside surface), in effect splices the bar to the sleeve and the sleeve to the next bar. The sleeve in this case is notably larger than the bar and uses up some of the clear bar spacing and may encroach on the specified clear cover unless this is specially considered in design.

18.20 BRIEF COMMENTS ON ECONOMY

Economy in column design favors the use of higher strength concrete on well-controlled jobs, since extra concrete strength costs only slightly more than low strength. Concrete will generally be cheaper than intermediate grade steel. Steel of Grade 400 is cheaper than Grade 300 because of its higher f_y value. The higher strengths will often be as cheap as concrete in load-carrying capacity. Tied columns will generally be cheaper than spiral columns, particularly if square columns are needed rather than round. However, the cost of the column can rarely be considered alone. Spiral columns and heavy steel save floor space which has an annual rental value. Form costs are also a major item and beam forms can be reused from floor to floor most simply if column sizes are kept constant. Hence it is quite common practice to keep the column size constant over several stories and take care of increasing load with increasing steel, stronger concrete and steel, or the use of spiral steel.

18.21 REDUCTION IN COLUMN LIVE LOADS

Building codes specify uniform live loads large enough to represent ordinary local concentrations of loading. Over larger areas the probability that this same load intensity will exist everywhere is reduced. Codes usually permit some reduction in the assumed live load on floor areas in excess of 10 to 15 m². Many codes permit such reduction only when the live load is 5 kPa or less; heavier loads imply storage or machinery loads that are easily concentrated over large areas.

The live load on columns may also be reduced under the same reasoning. The reductions permitted vary considerably in the different city codes and it will be necessary for the designer to consult the particular code that governs his or her design. The magnitude of the reduction depends both on the occupancy, or size of the unit live load, and on the number of floors carried by the column; it may vary from none to as much as 60% of the total live load.

18.22 CORRECTIONS NEEDED FOR FRAME MOMENTS ON ALL COLUMNS

Discrepancies between nominal and actual column moments cannot all be ascribed to column-length effects. A computer study[6,7,8] of braced frames, using fully cracked beams (with $\rho = \rho'$ over their full length) and columns programmed to their P-M-ϕ (curvature) response indicated that ordinary elastic analyses based on gross moments of inertia from the concrete alone were not totally adequate for design purposes. The analysis of these braced frames loaded for maximum moments as in Fig. 18.24 on the basis of gross I of concrete alone, led to what is termed the nominal eccentricity.

For a ratio of column-to-beam stiffness of $\psi = 1$ (based on I_g values) and nominal e/h of 0.10, the load-moment responses of two frames,[6] identical except for the beam steel, are plotted in Fig. 18.24. The lower curve represents $\rho = 0.005$ of Grade 300 beam steel and the upper curve $\rho = 0.023$ of the same steel. The resulting differences are:

Ultimate loads:	1 370 kN	and	1 690 kN versus nominal 1 550 kN
Final e/h:	0.147	and	0.068 versus nominal 0.10
Initial e/h:	0.167	and	0.120 versus nominal 0.10

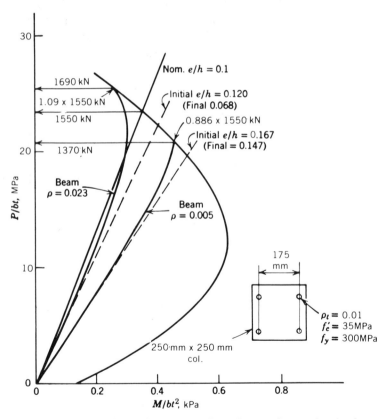

FIGURE 18.24   Effect of beam steel on short columns in single curvature.

With the heavily reinforced beams the column became relatively less stiff as its compression steel yielded on one face (about 1 300 kN) and its resisting moment actually *decreased* somewhat as it went on to increased loading.

In the 121 frames of this type studied, the ratio of failure e/h to the nominal, which would also be the ratio of ultimate column moment to the nominal, varied from 0.53 to 2.12, with more cases increasing than decreasing. Column design moments thus need some correction,[7] although the matter is not yet codified.

The best "elastic-type" of analysis found for this loading considered beams cracked with a transformed steel area based on 4 n while the columns were considered uncracked with 2.5 n for the steel.

Alternatively, if the gross concrete I is used for both beam and column, for $\psi = K_{col}/K_{bm} = 1$, the resulting e/h or moment on the column can be multiplied by the empirical factor:

$$1.38 + 5.5\,\rho_t - 30\,\rho$$

where $\rho_t = A_{st}/bh$ for the column

$\qquad \rho = A_s/bd$ for the beam $= \rho'$

This factor plots as a plane surface in Fig. 18.25 where less steep planes

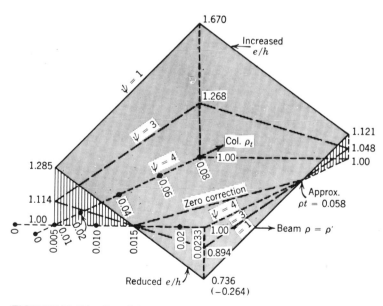

FIGURE 18.25   Graphical presentation of empirical relationships for e/h multiplier, single curvature cases.

are shown for larger ψ values. The horizontal axes are ρ_t for the column (toward the rear) and ρ for the beam (toward the right) and the vertical ordinates are the correction multiplier. For $\psi = 0.5$ the factor is much larger and erratic, suggesting that the combination is not a good one or the data are inadequate. The larger ψ becomes, the less the influence of the beam steel and the smaller the empirical correction.

The figure and numbering on it imply an accuracy that is not attainable. Multipliers shown are average values, not an upper bound, not always on the safe side. The actual increments added or subtracted (by the multiplier) will scatter in a typical ± 15 to $\pm 20\%$ range.

Columns loaded by reversing moment diagrams (Fig. 18.16b type) are effectively stiffer and the above multiplier is smaller, approximated by using ψ equal to two times the actual ψ.

SELECTED REFERENCES

1. F. E. Richart and G. C. Staehle, "Column Tests at University of Illinois," *Jour. ACI*, 2, Feb., Mar. 1931; *Proc.*, 27, pp. 731, 761; *Jour.*, 3, Nov. 1931, Jan. 1932; *Proc.*, *28*, pp. 167, 279.

2. W. A. Slater and Lyse, "Column Tests at Lehigh University," *Jour. ACI*, 2, Feb., Mar. 1931; *Proc.*, 27, pp. 677, 791; *Jour.*, 3, Nov. 1931, Jan. 1932; *Proc.*, 28, pp. 159, 317.

3. Eivind Hognestad, "A Study of Combined Bending and Axial Load in Reinforced Concrete Members," Univ. of Ill. Eng. Exp. Sta. *Bull. No. 399*, 1951.

4. Frank E. Richart and Rex L. Brown, "An Investigation of Reinforced Concrete Columns," Univ. of Ill. Eng. Exp. Sta. *Bull. No. 267*, 1934.

5. Frank E. Richart, Jasper O. Draffin, Tilford A. Olson, and Richard H. Heitman, "The Effect of Eccentric Loading, Protective Shells, Slenderness Ratios, and Other Variables in Reinforced Concrete Columns," Univ. of Ill. Eng. Exp. Sta. *Bull. No. 368*, 1947.

6. S. N. Pagay, P. M. Ferguson, and J. E. Breen, "Importance of Beam Properties on Concrete Column Behavior," *Jour. ACI*, 67, No. 10, Oct. 1970, p. 808.

7. H. Okamura, S. N. Pagay, J. E. Breen, and P. M. Ferguson, "Elastic Frame Analysis—Corrections Necessary for Design of Short Concrete Columns in Braced Frames," *Jour. ACI*, 67, No. 11, Nov. 1970, p. 894.

8. Phil M. Ferguson, Hajime Okamura, and S. N. Pagay, "Computer Study of Long Columns in Frames," *Jour. ACI*, 67, No. 12, Dec. 1970, p. 955.

9. F. N. Pannell, "Failure Surfaces for Members in Compression and Biaxial Bending," *Jour. ACI*, No. 1, Jan. 1963, p. 129.

10. Albert J. Gouwens, "Biaxial Bending Simplified," *Reinforced Concrete Columns*, SP-50, American Concrete Institute, Detroit.

11. *Ultimate Strength Design Handbook, Vol. 2, Columns*, SP-17A, American Concrete Institute, Detroit, 1970.

12. *Design Handbook, Vol. 2 Columns*, SP-17A(78), American Concrete Institute, Detroit, 1978.

13. Alan H. Mattock and Lidislay B. Kriz, "Ultimate Strength of Non-rectangular Structural Concrete Members," *Jour. ACI*, 32, No. 7, Jan. 1961; *Proc.* 57, p. 737.

14. Richard W. Furlong, "Ultimate Strength of Square Columns Under Biaxially Eccentric Loads," *Jour. ACI*, 32, No. 9, Mar. 1961; *Proc.* 57, p. 1129.

15. Boris Bresler, "Design Criteria for Reinforced Concrete Columns under Axial Load and Biaxial Bending," *Jour. ACI*, V. 57, 1960, p. 481.

16. "AWS Standard D12.1, Reinforcing Steel Welding Code," Amer. Welding Soc., Miami, Florida.

PROBLEMS

Note If any column in these problems does not qualify as short, so indicate; then calculate capacity as a short column and state permissible *must* be lower.

PROB. 18.1. Given $f'_c = 20$ MPa and bars Grade 400 with cover of 68 mm from center of bars. If column qualifies as short under the Code, find the permissible ultimate column load, using graphs of Fig. 18.12 if desired.

(a) For the column of Fig. 18.26a under a load on the x axis 150 mm to the right of center line at each end, $\ell_u = 2.7$ m, single curvature $M_1 = M_2$, $k = 0.77$ in a braced frame, $\psi = 1.0$.*

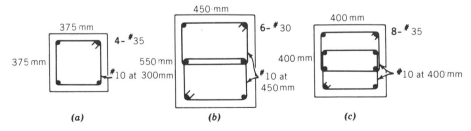

FIGURE 18.26 Column cross sections for problems.

*The average relative column stiffness ψ at the two ends fixes the k values. Both are given to avoid the necessity of using Fig. 19.6 here.

(b) For the column of Fig. 18.26b under a load on the x axis 250 mm to the right of the center line, $\ell_u = 2.7$ m, single curvature $M_1 = M_2$, $k = 0.77$ in a braced frame, $\psi = 1.0$.*

(c) For the column of Fig. 18.26c under a load on the x axis 200 mm off the center of column, $\ell_u = 3.0$ m, single curvature, $M_1 = M_2$, $k = 0.77$ in a braced frame, $\psi = 1.0$.

PROB. 18.2.   Conditions and requirements same as in Prob. 18.1 except columns are in reversed curvature (e at top and bottom on opposite sides of the axis) with $M_1 = -M_2$, $\psi = 2$, changing k to 0.85.

(a) Same as Prob. 18.1a.

(b) Same as Prob. 18.1b.

(c) Same as Prob. 18.1c.

PROB. 18.3.   Conditions, including values of e producing M_2, same as in Prob. 18.1 except columns are in unbraced frame with $\psi = 1.0$, $k = 1.3$, and $M_1 = -0.6 M_2$. Do any columns in Prob. 18.1a,b,c still qualify as short columns? If so, identify.

PROB. 18.4.   If $f_c' = 30$ MPa and Grade 400 steel is used with cover of 68 mm to center of steel, find the ultimate load permitted by the Code on the column of:

(a) Prob. 18.1a with e increased to 500 mm

(b) Prob. 18.1b with e increased to 500 mm

(c) Prob. 18.1c with e increased to 300 mm

PROB. 18.5.†

(a) Design a square tied column for service loads of 530 kN dead plus 450 kN live and for single curvature moments of 30 kN·m dead and 80 kN·m live, using $f_c' = 30$ MPa, about 2% of Grade 400 steel with 68 mm cover to center of bars, $\psi = 1.2$, $k = 0.79$, $\ell_u = 3.0$ m, braced frame, $M_1 = M_2$. Use a column size large enough to qualify as a short column, even if this lowers the percentage of A_s required.

(b) Same as (a) except *reversed* curvature with M_2 as in (a) and $M_1 = -0.5 M_2$.

(c) Same as in (a) except reversed curvature moments in an *unbraced* frame with $M_1 = -M_2$, $k = 1.37$. What minimum column size is required to classify this as a short column?

---

*The average relative column stiffness ψ at the two ends fixes the k values. Both are given to avoid the necessity of using Fig. 19.6 here.

†*Note:* If column interaction diagrams similar to Fig. 18.12 are available for spiral columns, Prob. 18.5 is also suitable for assignment as a spiral column design.

PROB. 18.6.

(*a*) Evaluate and locate the ultimate load that gives a horizontal neutral axis 150 mm below the top edge of the column of Fig. 18.26*c*. Assume cover 63 mm to center of steel, $f_c' = 30$ MPa, Grade 400 steel, and consider deformations to establish the steel stresses.

(*b*) Repeat for a horizontal axis 150 mm below the top edge of Fig. 18.26*b*.

PROB. 18.7.   Design a column spiral of Grade 300 steel that will be adequate for a 500 mm square column with longitudinal steel in a circular pattern that permits the spiral to have an outside diameter of 425 mm $f_c' = 30$ MPa.

19
LONG COLUMNS

19.1 THE NATURE OF THE LONG COLUMN EFFECT— BRACED FRAME

Columns always carry moment in the practical case, both by virtue of slight initial crookedness and by the nature of the loading from beams and slabs that rotates frame joints and thus imposes curvatures. Moments in columns also produce curvatures that lead to deflections between joints and sometimes sidesway deflections of the joints (similar to that occurring under wind load), even when no external horizontal loads exist.

Long columns have already been discussed briefly in Sec. 18.14 in rationalizing the upper limits for short column analyses. In those limits the Code in effect ignores any loss of strength up to 5%. Here a clearer understanding of behavior is attempted by approaching the long column deflection shown in Fig. 18.16a in two stages.

Figure 19.1 shows a statically determined column loaded to give single curvature.* If y_0 is large as shown, a large moment Py_0 is added at midspan and the ends at A and B rotate, not only because of the original Pe moment but also because of Py moments; and the two rotations are additive. This extra movement could be called the long column effect. In a frame, however, the usual column length ends in joints where beams frame in. The column end is not free to rotate except as the column pulls the beams around with it—that is, the beams tend to resist extra joint rotation. Thus the column deflection caused by Py is not like that of a simple beam but partially like that of a restrained or fixed-end member, as shown in Fig. 18.16a. This means the end moment on the column is

---

*This term also covers unequal end moments.

522

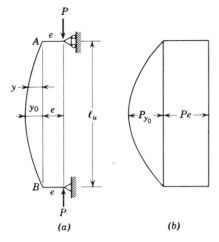

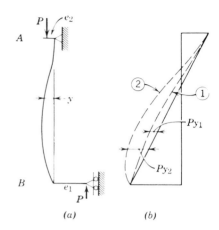

FIGURE 19.1 Single curvature column. (*a*) Curvature. (*b*) Moment.

FIGURE 19.2 Column with end moments of opposite kind. (*a*) Curvature. (*b*) Moment.

slightly reduced and the column develops points of inflection, making the reduced effective length $k\ell_u$ the most accurate to use in design.

The column in reversed curvature in Fig. 19.2 is so loaded that a point of inflection must occur. If its deflection is small, the added Py_1 moment does not increase any critical moment. If the column is long enough to deflect more, the moment Py_2 may develop an increased moment away from the joint. The reversed curvature case includes any value of e_2 from small to equality with e_1. The ratio of height to radius of gyration has to be very large before there is an ill effect (long column effect) from this loading.

Both the above cases apply in a braced frame where the upper joint cannot deflect laterally. The reversed curvature fits outside columns and the critical loading shown in Fig. 18.17*a* fits interior columns. The long column effect is less severe in the braced frame case discussed above than in the unbraced frame where lateral movement of the joints is involved.

19.2 INCREASED COLUMN DEFLECTION UNDER SHEAR LOADING

A major case is that of a column in an unbraced frame subjected to a shear load, as in resisting wind or earthquake forces. Consider first the situation where the floor system is very heavy, that is, for the purpose of this initial discussion, infinitely stiff. The column deflects as in Fig. 19.3*a* and can be analyzed by considering the half-length shown in Fig. 19.3*b* that shows

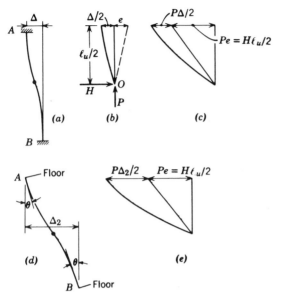

FIGURE 19.3   Columns with sidesway carrying horizontal shears. (*a*) Curvature. (*b*) Statics of a half height. (*c*) Moment. (*d*) With joints which rotate. (*e*) Moment for (*d*).

the moment increased at A, much like it is at mid-height of the column of Fig. 19.1*a*. Although the magnitude of the Py term compared to the Pe or $H\ell_u/2$ moment is different, the two cases are closely related to each other. However, infinitely stiff floors with zero joint rotations at A and B are not common.

If the joints at A and B rotate, Δ increases to Δ_2. This follows because the $H\ell_u/2$ moment is unchanged and the curvature from A down to mid-height is changed only by the Py term that increases with Δ. Thus the deflection of the cantilever in Fig. 19.3*b* changes from that of a member fixed at the end to one where deflection is augmented by joint rotation as in Fig. 19.3*d*. The moment diagram of Fig. 19.3*e* then involves a greater end moment. If the joint rotation is large, the end moment is multiplied several times, that is, $P\Delta_2$ is several times as much as the static force moment $H\ell_u/2$. Hence the long column action can be very important in this case. The degree of importance is determined by the angle θ at each end. While this angle depends on the total moment that the column carries, for any given value of this moment the angle is controlled almost linearly by the stiffness of the beams; a beam half as stiff will double the angle.

19.3 INFLUENCE OF LENGTH ON COLUMN BEHAVIOR

The behavior response of even the very short column to an increasing load, as already shown in Fig. 18.22 is not linear, possibly raising a question about the most appropriate design e/h to use. The discussion in that section selected the final value, which is usually a sound policy when one cannot extrapolate other values linearly. The Code committee has not yet undertaken the codifying of the design e/h, but in the area of length effects it has set up a specific procedure* that defines the increasing value of critical moment with increasing axial load (at a given monimal eccentricity).

Consider a column, say in single curvature as in Fig. 19.1. As one doubles the load, y_0 more than doubles, developing possibly a critical moment as indicated by curve A in Fig. 19.4a.

The amount of curvature in the curve for a particular case depends on the length of the member. The moment for a very short column might plot

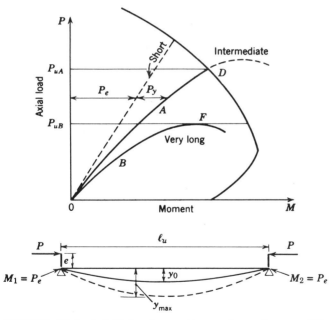

FIGURE 19.4 (a) Behavior patterns for columns. (b) When the end moments are caused by an eccentric axial load the deflection resulting from M_1 and M_2 is increased from y_0 to y_{max}

*But see also the alternate R Method in the Commentary (Sec. 19.15).

essentially straight (y_0 negligible), following the dotted line to failure at the interaction diagram. The medium-length column of curve A would fail at point D. These failures would be in compression, and the author calls this failure at the interaction diagram a materials failure.

A longer column might give the curve B which turns horizontal before reaching the interaction diagram. This column fails from instability or buckling in the classical sense, the inability to resist more load, although no material has reached the failure point. Mathematically, some like to treat point D also as a special case of instability*, but the author believes the separate names for the two quite different phenomena help to clarify one's understanding of columns.

The method of the following sections permits one to compute the increased or "magnified" moment[5] and to plot this type of curve if one wishes. For design or analysis, only the portion around D or F would be important; and at F the method only indicates a horizontal line, representing increasing moment with no increase in axial load.

19.4 CODE METHODS FOR SLENDERNESS EFFECTS

(a) Frame Analysis Approach

The Code (10.10) encourages the designer to evaluate the design forces and moments from an analysis of the structure, stating:

> Such analysis shall take into account influence of axial loads and variable moment of inertia on member stiffness and fixed-end moments, effect of deflections on moments and forces, and the effects of duration of loads.

Beyond these general requirements, the Code Commentary speaks of a second order analysis, including the effects of sway deflections on the axial loads and moments. It also recommends realistic moment-curvature or moment-end-rotation relationships, finally stating five controls, including:

> In lieu of more precise values, it is satisfactory to take EI as $E_c I_g (0.2 + 1.2\, \rho_t E_s/E_c)$ in computing the column stiffnesses and $0.5\, E_c I_g$ when computing the beam stiffnesses.
>
> It is necessary to consider the effect of axial loads on the stiffness and carry-over factors for very slender columns ($\ell/r > 45$).

Analyses such as these are basically computer problems and beyond the scope of this book. The use of such programs is gradually increasing because such use does avoid the many detailed requirements of the moment magnifier method covered in Code 10.11.

---

*With the curve turning sharply downward at point D.

(b) Moment Magnifier as an Approximate Evaluation

In the mechanics of elastic members a member that is bent in a specific shape, either from initial crookedness or some external loading, will have its deflection increased by the addition of an axial load P acting along the reference chord, from an initial y_0 to y_{max}. A reasonable approximation is

$$y_{max} = y_0/(1 - P/P_c)$$

where P_c is the critical buckling load for that member (Euler load). In the hinged column of Fig. 19.4b, with $M_1 = M_2 = Pe$,

$$
\begin{aligned}
M_{max} = M_1 + Py_{max} &= M_1 + Py_0/(1 - P/P_c) \\
&= [M_1(1 - P/P_c) + Py_0(M_1/Pe)]/(1 - P/P_c) \\
&= M_1(1 - P/P_c + y_0/e)/(1 - P/P_c)
\end{aligned}
$$

Since $y_0 = M_1\ell_u^2/8\ EI = Pe\ell_u^2/8\ EI$, and $P_c = \pi^2 EI/\ell_u^2$,

$$
\begin{aligned}
M_{max} &= M_1(1 - P/P_c + P\ell_u^2/8\ EI)/(1 - P/P_c) \\
&= M_1(1 - P/P_c + \pi^2 P/8\ P_c)/(1 - P/P_c) \\
&= M_1[1 + (P/P_c)(\pi^2/8 - 1)]/(1 - P/P_c) \\
&= M_1(1 + 0.23\ P/P_c)/(1 - P/P_c) = M_1(1 - P/P_c).
\end{aligned}
$$

For the single curvature case the error in omitting $0.23\ P/P_c$ varies from 2.3% when $P/P_c = 0.1$ to 11.5% when $P/P_c = 0.5$; for other initial curvatures it can be either more or less.

If $M_1 \neq M_2$, the maximum moment is not at mid-height initially, and the largest increase is near mid-height, resulting in a smaller total at the new maximum point. The Code cares for this by including the factor C_m which is to take care of different shapes of moment diagrams and shifts in points of maximum moment. Based on the larger end moment M_2,

$$M_{max} = M_2[C_m/(1 - P/\phi P_c)] = M_2\delta$$

where δ is the quantity in brackets and is called the moment magnifier or amplifier.

$$\delta = \frac{C_m}{1 - P_u/\phi P_c} \geqslant 1.0$$

In braced frames for members without transverse loads between supports, C_m may be taken (Code 10.11.6.3)

$$C_m = 0.6 + 0.4\ M_1/M_2 \geqslant 0.4$$

Here M_1 is positive for a single curvature case, negative for a double curvature case, and the moment ratio itself is always in the range between $+1$ and -1.

With this procedure the long column must then be designed* for P_u and $M_2\delta = M_c$, where δ is always greater than unity but may be used as unity for "short columns," a term for which exact boundaries were set in Sec. 18.14.

One important consideration in adopting this moment magnifier method was its present use in structural steel design. The designer must be alert, however, for detailed differences in values recommended for C_m and $k\ell_u$, and the addition here of a new term β_d (Sec. 19.7) for creep.

19.5 ADAPTING THE MOMENT MAGNIFIER TO CONCRETE COLUMNS IN FRAMES—GENERAL

The moment magnifier is based on an analysis of an elastic curve that increases amplitude but does not change in shape as axial load is applied at the column ends or at the joints of the frame. Although for concrete the deflected shape will change, this normally does not involve serious error.

In a general frame five important problems arise, the last three not being limited to reinforced concrete.

1. The effective EI of reinforced concrete is dependent on the magnitude and type of loading as well as the materials, and varies along the column length (Sec. 19.6).
2. The creep of concrete occurs when load is sustained and creep modifies the effective EI needed in item 1 (Sec. 19.7).
3. The effective length of column for the calculation of P_c may be either more or less than ℓ_u (Sec. 19.8).
4. The column sway is limited to the story deflection (Sec. 19.9).
5. The maximum moment does not always occur at mid-height (Sec. 19.10).

The answers provided for these problems by any approximate method cannot give an exact result; there are too many basic variables. For example, the amount of beam steel or the degree of beam cracking alone can create a scatter of at least 10% in column strength.

19.6 EFFECTIVE *EI* FOR MAGNIFIER CALCULATION

Stiffness against curvature is measured by EI. In a reinforced concrete column, stiffness decreases (1) as concrete in compression approaches the flatter part of the stress-strain curve, (2) as creep develops under sustained

---

*But note the limited alternate in Sec. 19.15.

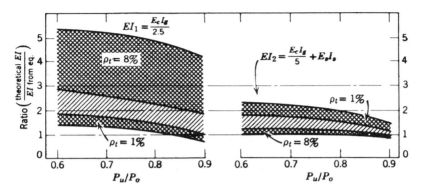

FIGURE 19.5 The accuracy of the formula values of the *EI* equations. The horizontal line at unity would be a perfect fit. (Modified from Reference, 5, *ACI Jour.*)

load, (3) as compression steel yields, or (4) as the concrete cracks (for larger eccentricities). The varying degree of stress distribution over the column length means no single value of *EI* can be a true one for use under all types of loading.

Especially where ρ_t of the column is small (0.01, or up to 0.02 for small columns), and especially also for columns that nearly qualify as short columns (where the moment magnifier will be small anyway), Code 10.11.5.2 suggests

$$EI_1{}^* = (E_cI_g/2.5)/(1 + \beta_d)$$

where β_d is an allowance for creep discussed in Sec. 19.7. This equation would greatly underestimate *EI* where the value of ρ_t is large, leading to over-design of the column. For cases having a large ρ_t, unless the column length is still so short as to make the calculation unimportant, economy dictates the use of the other Code equation for *EI*:

$$EI_2{}^* = (E_cI_g/5 + E_sI_s)/(1 + \beta_d)$$

With either equation the scatter is broad, as shown in Fig. 19.5, especially so for EI_1. The scatter is essentially all on the safe side. Reference 6 suggests a reduction in these *EI* values to account for sustained load effects. Since the larger *EI*, either EI_1 or EI_2, gives the smaller δ multiplier, and since both are safe values, it is sometimes desirable to know the particular ρ_t that makes EI_2 the larger. The breakpoint will be established[1] by setting the ratio EI_2/EI_1 at unity, considering

*The EI_1 and EI_2 subscripts are not in the Code but are convenient for reference.

TABLE 19.1 Minimum Values of ρ_t Making EI_2 Govern over EI_1

f_c'	20 MPa	25 MPa	33 MPa
n	9	8	7
ρ_t	$1/(135\ \gamma^2)$	$1/(120\ \gamma^2)$	$1/(105\ \gamma^2)$
$\gamma = 0.6$	0.0205	0.0231	0.0264
0.7	0.0151	0.0170	0.0194
0.8	0.0115	0.0130	0.0149
0.9	0.0091	0.0103	0.0117

steel on two faces of a rectangular column:

$$1 = \left(\frac{E_cI_g/5 + E_sI_s}{1+\beta_d}\right) \div \left(\frac{E_cI_g/2.5}{1+\beta_d}\right) = \frac{bh^3/60 + (E_s/E_c)\rho_t bh(\gamma h)^2/4}{bh^3/30}$$

where γ (gamma) $= (h - 2\,d')/h =$ relative distance between the steel on the two faces. Multiplying top and bottom by $30/bh^3$:

$$0.5 + 7.5(E_s/E_c)\rho_t\gamma^2 = 1$$

Let $E_s/E_c = n$, as in transformed area calculations. The resulting equation

$$7.5\ n\rho_t\gamma^2 = 0.5$$

based on $EI_1 = EI_2$ indicates that EI_2 is the larger when ρ_t exceeds $1/(15\ n\gamma^2)$, that is, when Table 19.1 values of ρ_t are exceeded.

For design it is often more convenient to use the ratio of EI_2/EI_1 times an EI_1 value than to start fresh on an EI_2 calculation, that is, to use: $EI_2 = EI_1\,(0.5 + 7.5\ n\rho_t\gamma^2)$ or a similar equation for other type columns.

19.7 THE CREEP PROBLEM

At $\ell_u/h = 10$ creep effects tend to offset each other, the tendency toward increase of column curvature from creep being offset by the decreasing distribution factor to the column because of its reducing stiffness. Around ℓ_u/h of 16 or 20 it becomes more important, except for columns with reversed moments in braced frames, which exhibit long-column behavior only in extreme cases.

The Code provides for creep by reducing the approximate EI by the $(1 + \beta_d)$ factor in the denominator, where β_d is the ratio of the maximum factored dead load *moment* on the column to the maximum factored total load moment on the column. (It would appear that the ratio of service load moments might be more appropriate, since creep is a function of sustained loading.) The author feels that this β_d correction may be unimportant in the assessment of an EI value that probably involves a typical error of at least 30%. Fortunately, the effect of β_d (and even of EI) is far removed from the end result of a design. Experience may show that it is only on very unusual long columns that β_d is really of significance.

19.8 EFFECTIVE COLUMN LENGTH, $k\ell_u$

In simple mechanics the concept of effective column length is well established; a fixed ended column has an effective length of half its overall height between fixed ends. If it were only a matter of the frame reaction to a statically determined moment applied directly as a load on the column, the effective length would be a relatively simple matter of frame or joint stiffness.

Jackson and Moreland[2] solved the $k\ell_u$ problem in terms of relative

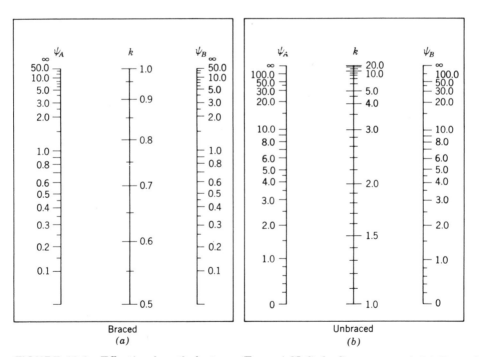

FIGURE 19.6 Effective length factors. (From ACI Code Commentary.) (a) Braced frames. (b) Unbraced frames. ψ = ratio of $\Sigma EI/\ell_c$ of compression members to $\Sigma EI/\ell$ of flexural members in a plane at one end of a compression member; k = effective length factor.

member stiffness and published the nomographs of Fig. 19.6. One enters with the values of the relative column stiffness ψ at each end of the column; a straight line between the two ψ values reads k on the center scale. A separate nomograph is required for the braced and unbraced frames. The greater k values for unbraced frames follow from the knowledge that final failure will tend to be a sidesway mode for the entire story. Table 19.2 tabulates some of the unbraced frame values.

One of the basic problems here is that in reinforced concrete, the true ψ changes as beams crack, as column deflection builds up, as column concrete in compression enters the flatter part of the f_c stress-strain curve, and as compression steel yields. These complications combine with the similar uncertainties in the value of EI just discussed.

The effect of *cracked* beams on the effective length is important in long columns, possibly in the sidesway case more important than its effect on end moments generally, already discussed in Sec. 18.22. The code (10.11.2.2) specifies:

> For compression members not braced against sidesway, effective length factor k shall be determined with due consideration of cracking and reinforcement on relative stiffness, and shall be greater than 1.0.

The best thinking for this sidesway case at present* is that relative stiffness should be based on the I for the cracked beam (a simple transformed area) and for the column on EI_2 as in Sec. 19.6, with $\beta_d = 0$, and all divided by E_c, that is,

$$I_{col} = EI_2/E_c = (E_c I_g/5 + E_s I_s)/E_c = I_g/5 + nI_s$$

TABLE 19.2   Tabulation of k Factors for *Unbraced* Frames (by Dr. R. W. Furlong)

ψ	0.0	0.2	0.4	0.6	0.8	1.0	1.2	1.4	1.6	1.8	2.0
k	1.00	1.07	1.16	1.23	1.29	1.34	1.39	1.44	1.48	1.52	1.56

ψ	2.0	2.5	3.0	3.5	4.0	5.0	6.0	8.0	10.0	15.0	20.0
k	1.56	1.68	1.80	1.91	2.01	2.20	2.38	2.70	2.99	3.60	4.12

$k = \dfrac{20 - \psi}{20} \sqrt{1 + \psi}$ for $\psi < 2$

$k = 0.9\sqrt{1 + \psi}$ for $\psi > 2$

*Notice that use of beam I_g reduces safety (Sec. 19.8 and Fig. 19.10c). Appendix B, Fig. B.8, shows some ratios I_{cr}/I_g.

With ψ calculated on this basis, the value of k for kl_u in the sidesway case may be substantially higher. Since P_c varies with $1/(k\ell_u)^2$, the difference can be important.

At column footings in an unbraced frame, the AISC (for steel columns) recognizes the possibility of fixed connections justifying a ψ of 1.0; but at the same time it recognizes the usual lack of frictionless pin connections by limiting the maximum ψ to 10. The designer of reinforced concrete should also consider the need for some limit not representing complete fixity at footings.

For the braced frame the Code is less specific: "...k shall be taken as 1.0, unless analysis shows that a lower value may be used." This seems to be encouraging the use of $k = 1.0$. Analysis might be simply the use of the Jackson-Moreland charts of Fig. 19.6 or might go so far as a ψ calculated similarly to that for the sway cases. The author feels that $k = 1.0$ may be the best answer, but the Sec. 19.12 example uses the chart k value.

19.9 COLUMN SWAY LIMITED TO STORY DEFLECTION

Except in isolated cantilever columns, when a single column sways, an entire floor or level must move relative to another. This story sidesway is also a typical mode of failure in an unbraced frame under vertical load alone. A floor system is typically stiff enough to make all columns deflect alike, unless a torsional loading adds a rotation to the structure. Thus each column is not free to deflect independently as required to carry a fixed shear. Instead, the shear carried is a function of its stiffness relative to all the columns in the story. This discussion will omit the more complex effect of wind on unsymmetrical structures that can result in a torsional rotation; even this case marshals the unequal sway of the columns into an ordered, interrelated group movement that depends on the group stiffness.

Thus there are few cases where a column free to sway can be designed by itself; a single column supporting a hyperbolic paraboloid may qualify. Possibly also a bent or a series of single bents carrying a roof with overhanging ends and not much eccentricity of loading on the columns would qualify in that parts of the system might all tend to react alike. Typically, there will be different column sizes, different beam restraints or, if nothing more, exterior columns braced by one beam combined with interior columns braced by two beams at each floor level.

In a story height the stiffest column tends to pick up horizontal load, possibly more than the designer tends to assign to it, while the flexible column is braced by the limited deflection of the stiffer column. Any design method for lateral loading involves an initial distribution of the horizontal shear to the individual columns, that is, some type of frame analysis. Code 10.11.6.2 requires that a common moment magnifier δ be computed for the story and used for all columns in that story. This δ is to be based on $\Sigma P_u / \Sigma P_c$, with the Σ representing a summation over all the columns in that story, including any short columns.* In determining k for use in this P_c calculation, the relative column stiffness ψ should be based on the cracked section of the beam and the reinforcement ratios of beam and column (Code 10.11.2.2), as discussed in the previous subsection.

The actual data required for the P_c calculations are available only *after* the beams and columns have been designed, which means the design method must start more approximately. In a sense this is true of nearly all design situations, but here the approximations may initially have to be rougher than usual until design experience builds up in this area.

To start the column design in this sidesway case one could (1) make a rough estimate of all columns and beams sizes involved, (2) arbitrarily modify their relative stiffnesses at joints by some typical factor to represent beam cracking† and column reinforcement percentages expected

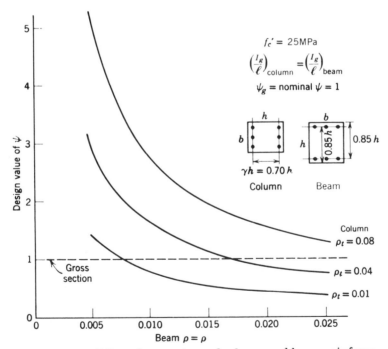

FIGURE 19.7 Effect of percentage of column and beam reinforcement and beam cracking on the design ψ for $\gamma = 0.70$ and $f'_c = 25$ MPa. (Modified from chart by Dr. J. E. Breen.)

*The ϕ in the denominator does stay in the equation for δ.
†Figure B.4, Appendix B, might help here.

to be used, and on this basis (3) make an initial trial for $\Sigma P_u / \Sigma P_c$. A reference chart such as Fig. 19.7 would be helpful, although limited here to a particular f'_c, γ, and column steel arrangement. The ψ design value of this chart is based on equal gross I/ℓ of column and beam. Similar charts can be made for other gross ψ values, and this chart itself can be used as a multiplier of any nominal ψ in roughly estimating the second step above, since here it is in effect a ratio term.

Alternatively,* designers might compare this story with some other they have computed and simply estimate a trial δ for initial use. Tentative column designs may then be made with the trial δ until it is possible to get a better feel for the real $\Sigma P_u / \Sigma P_c$ for the story; then the better δ value will determine whether some revisions are essential.

This story system is more complex than individual column designs, but also more realistic and possibly in the long run more economical. It is also probable that the story δ will act as a form of stabilizer in design calculation. A change in a single-column size will modify the story δ less than it would the individual member δ, and changes in all the column sizes should not usually be in the same direction.

The story procedure thus far described does not guarantee an individual slender column against overload. Such a column design might be too weak to handle the maximum moment and vertical load in a *braced* frame. The Code (10.11.6.3) requires this individual load check to be made (as if in a braced frame); of course, the larger of the two requirements controls that column. This check should only be necessary on the most slender column as a rule.

19.10 MAXIMUM MOMENT AWAY FROM MID-HEIGHT

The use of C_m has already been noted in Sec. 19.4 along with the governing equation, which is to be used only where sidesway is prevented and where no lateral load acts on the column. This equation is a straight-line approximation to a curve originally developed by Massonnet from steel column tests.

In sidesway under a shear loading, the worst moment stays at the end of the column and the maximum deflection is also there, making $C_m = 1$ appropriate.

In the reversed curvature case without sidesway (Fig. 19.2), which looks

---

*The author also suggests in the middle of Sec. 19.14 that in the absence of column tables[3] or charts[4] the R method may be a wise choice for preliminary sizing of unbraced columns. Although rather crude, the following has also been discussed as a starting point:

For preliminary design, using $0.5\,I_g$ for flexural members and I_g for compression members will usually result in reasonable estimates of member sizes.

in the moment diagram slightly similar to the sidesway case, the deflection adds no *end* moment. Hence behavior is that of a short column until deflection can build up a larger moment away from the joint. Most such practical columns would actually be short columns.

19.11 REVISED FORMAT FOR *EI* AND C_m EQUATIONS

A design problem starts with known, or easily established, data for f'_c, f_y, P_u, e (or M_u), β_d (the ratio of dead load moment to total design moment), ℓ_u and k. The major unknowns are h (or b and h) and ρ_t, although there may be a desired ρ_t which can be used as a target.

For long column design it is helpful to restate the *EI* and P_c equations in terms of the unknown column thickness h as Furlong has done[1]; this eliminates some awkward number repetitions. For square tied columns with steel on two faces:

$$EI_1 = \frac{E_c I_g/2.5}{1+\beta_d} = \frac{4\,700\sqrt{f'_c}\,h^4/(2.5\times 12)}{1+\beta_d}$$

$$= \frac{160\sqrt{f'_c}\,h^4}{1+\beta_d} = (\text{constant})\,h^4$$

$$\phi P_c = \phi\pi^2 EI/(k\ell_u)^2 = 0.7\times 9.86\,EI_1/(k\ell_u)^2 = (\text{another constant})\,h^4$$

If ρ_t exceeds the values of Table 19.1 (Sec. 19.6), the other equation for

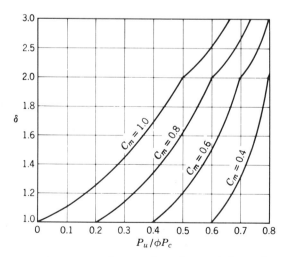

FIGURE 19.8 Chart for determining the moment magnifier δ. (Adapted from Furlong, Reference 1, ACI.)

EI controls:

$$EI_2 = (E_c I_g/5 + E_s I_s)/(1 + \beta_d)$$

The ratio of this equation to the EI_1 value above was established in Sec. 19.6 as:

$$EI_2/EI_1 = 0.5 + 7.5\ n\rho_t\gamma^2$$

Since ρ_t is not known in advance, it is frequently simplest to start with EI_1 and, when ρ_t has been roughly fixed, to find EI_2 and the new P_c with the above ratio applied to the old EI_1 values.

The curves of Fig. 19.8 are sometimes helpful instead of solving the equation for δ directly, particularly when $C_m \neq 1.0$.

19.12 DESIGN OF LONG COLUMN IN BRACED FRAME USING INTERACTION DIAGRAMS

Design a square tied column for a total dead load of 900 kN, a live load of 1 100 kN, and a moment of 100 kN · m (all from live load) at each end creating single curvature. $\ell_u = 6.0$ m, $\psi = 1$ in braced frame,* $f'_c = 30$ MPa, Grade 400 steel. Try for ρ_t of about 0.02.

Solution

$P_u = 1.4 \times 900 + 1.7 \times 1\ 100 = 3\ 130$ kN $M_u = 1.7 \times 100 = 170$ kN · m

$e = M_u/P_u = 0.054$ m $= 54$ mm $m = f_y/(0.85\ f'_c) = 400/0.85 \times 30 = 16$

For ρ_t of 0.02, $\rho_t m = 0.32$

From Fig. 19.6, k for ψ of 1 is 0.77, making $k\ell_u = 0.77 \times 6.0 = 4.62$ m.

This appears to be a long column and the moment and e will be magnified. Assume $\delta = 1.3$, making the design $e = 1.3 \times 54 = 70$ mm. If h is assumed as 450 mm, $e/h = 70/450 = 0.156$. For assumed #10 stirrups and #25 bars, the trial γ is:

$$\gamma = (h - 2\ d')/h = (450 - 2 \times 40 - 2 \times 10 - 25)/450 = 0.72$$

Try chart of Fig. 18.12*b* with $\gamma = 0.70$, entering with $e/h = 0.156$ and $\rho_t m = 0.32$ and reading $\alpha = 0.55$ as $P_u/(f'_c h^2)$. These charts automatically introduce the ϕ.

$$h^2 = 3\ 130 \times 10^{-3}/30 \times 0.55,\ \ h = 0.436\ \text{m} = 436\ \text{mm using the assumed }\delta.$$

Continue with the 450 mm column and calculate the magnifier needed. For $\rho_t = 0.02$ Table 19.1 in Sec. 19.6 indicates that EI_2 should control unless γ is less than 0.65. Although one is tempted to use the simpler EI_1 for the first trial EI_2 will be used, with $\beta_d = M_{uD}/M_{u(D+L)} = 0$ since all the moment is from live load in this

---

*A given ψ in an example is always an oversimplification. It assumes a specific relative column stiffness that the design may not provide.

case. Code 8.5.1 sets E_c.

$$E_c = 5\,000\sqrt{f_c'} = 5\,000\sqrt{30} = 27\,400 \text{ MPa}$$
$$E_s/E_c = n = 200\,000/27\,400 = 7.3$$
$$\text{Trial } A_s = 0.02 \times 450^2 = 4\,050 \text{ mm}^2$$
$$EI_2 = (E_c I_g/5 + E_s I_s)/(1 + \beta_d) = E_c(I_g/5 + nI_s)$$
$$I_g = \tfrac{1}{12} \times 450^4 = 3.417 \times 10^9 \text{ mm}^4 = 3.417 \times 10^{-3} \text{ m}^4$$
$$I_s = A_s\,(\tfrac{1}{2}\gamma\,h)^2 = 4\,050(\tfrac{1}{2} \times 0.72 \times 450)^2 = 106.3 \times 10^6 \text{ mm}^4 = 0.1063 \times 10^{-3} \text{ m}^4$$
$$EI_2 = E_c(\tfrac{1}{5}I_g + nI_s) = 27\,400\ (\tfrac{1}{5} \times 3.417 + 7.3 \times 0.106\,3) \times 10^{-3} = 40.0 \text{ MN} \cdot \text{m}^2$$
$$\phi P_c = \phi\pi^2\,EI_2/(k\ell_u)^2 = 0.70 \times \pi^2 \times 40.0/4.62^2 = 13.0 \text{ MN*}$$

With $M_1 = M_2$, $C_m = 0.6 + 0.4\,M_1/M_2 = 1.0$

$$\delta = C_m/(1 - P_u/\phi P_c) = 1/(1 - 3.13/13.0) = 1.32$$

Enter Fig. 18.12b for $\gamma = 0.70$ using the new values of $\delta e/h = 1.32 \times 54/450 = 0.158$ and $\rho_t m = 0.32$ to read $\alpha = 0.555 = 3.13/30\,h^2$, giving $h = 0.434 \text{ m} = 434 \text{ mm}$.

USE 450 mm by 450 mm, subject to further evaluation of ρ. From the same chart, for $\alpha = 3.13/30 \times 0.45^2 = 0.515$ and $\delta e/h = 0.158$, read $\rho_t m = 0.25$. This leads to $\rho_t = 0.25/16 = 0.0156$.

Revise EI_2 for this ρ_t.

$$EI_2 = 27\,400\ (\tfrac{1}{5} \times 3.417 + 7.3 \times 0.1063 \times 0.0156/0.02) \times 10^{-3} = 35.3 \text{ MN} \cdot \text{m}^2$$
$$\phi P_c = 0.7\pi^2\,EI_2/(k\ell_u)^2 = 0.7 \times \pi^2 \times 35.3/4.62^2 = 11.4 \text{ MN}$$
$$\delta = 1/(1 - 3.13/11.4) = 1.38 \quad \delta e/h = 1.38 \times 54/450 = 0.165$$

This change from 0.158 should not change the column size, but it changes A_s. Enter charts with $\delta e/h$ and $\alpha = 0.515$ as earlier, to read $\rho_t m$.

$$\gamma = 0.70 \qquad \rho_t m = 0.27$$
$$\gamma = 0.80 \qquad \rho_t m = 0.23$$

For $\gamma = 0.72$, $\rho_t m = 0.26$,[†] and $\rho_t = 0.016$

The required $A_s = 0.016 \times 450^2 = 3\,240 \text{ mm}^2$

This new ρ_t is 80% of the 0.02 originally tried. The error is on the safe side because the larger A_s will give a larger P_c that lowers δ. For the purpose here this A_s is close enough.

It is desirable to use a symmetrical bar layout, and we will therefore use 8-#25 bars, although only 7 bars are required.

USE 8-#25 bars with #10 ties. Tie spacing: 48 d_b tie = 480 mm; h = 450 mm

(A_s = 4 000 mm²)   16 d_b for A_s = 400 mm

USE #10 ties at 400 mm, in pairs as in Fig. 18.13e or f.

---

*Note that if k were used here as 1.0 (as suggested by Code quoted at close of Sec. 19.8), ℓ_u would be 240 in. and δ would sharply increase, probably calling for a larger column.
†Larger charts are desirable to permit greater accuracy.

Alternatively, one could use one overall tie plus one single crosstie (Fig. 18.13d) between *one* pair of center bars, since clear bar spacing is under 150 mm.

A complete design would also involve at least a separate check on moment about the other axis.

19.13 DESIGN OF LONG COLUMNS IN UNBRACED FRAMES USING INTERACTION DIAGRAMS

(a) Typical Case

To indicate the more serious length problem in an unbraced frame, redesign the column in Sec. 19.12 considering the frame unbraced, $P_u = 3\,130$ kN, M_u (all LL) = 170 kN · m, $\ell_u = 6.0$ m, $f'_c = 30$ MPa, $f_y = 400$ MPa, $\beta_d = 0$, and desired ρ_t about 0.02. Assume the story δ is either known or assumed at $\delta = 1.90$, $\psi = 1$.

Solution

In actual unbraced frames one must start with a known or assumed δ and later verify this value *for the entire story.* With many columns in a story, the selected individual column size will have only a minor effect on δ, but the cumulative effect could be important. In this example the verification, or lack of verification, and its potential influence on the design cannot be resolved in terms of the single column. This design is thus oversimplified. (The design in b is at the other extreme and appears overcomplex.)

From Fig. 19.6, $k = 1.31$, $k\ell_u = 1.31 \times 6.0 = 7.86$ m

$e = M_u/P_u = 170/3\,130 = 0.054$ m $= 54$ mm

Try $h = 500$ mm, $\delta e/h = 1.90 \times 54/500 = 0.205$

$m = f_y/0.85 \times 30 = 16$ Desired $\rho_t m = 0.02 \times 16 = 0.32$

$h - 2\,d' = 500 - 125 = 375$ mm $\gamma = 0.75$

For $\gamma = 0.70$ (Fig. 18.12b)* enter with e/h and $\rho_t m$ to read $\alpha = 0.49$

   $\gamma = 0.80$ (Fig. 18.12c) similarly $\alpha = 0.51$

   $\gamma = 0.75$ $\alpha = 0.50 = P_u/f'_c h^2 = 3.13/30\,h^2$ $h = 0.457$ m $= 457$ mm

Try $h = 475$ mm   Desired $\rho_t m = 0.32$ $\delta e/h = 1.9 \times 54/475 = 0.216$

$h - 2\,d' = 475 - 125 = 350$ mm $\gamma = 350/475 = 0.737$

For $\gamma = 0.70$ $\alpha = 0.48$ For $\gamma = 0.737$, $\alpha = 0.487 = P_u/f'_c h^2 = 3.13/30\,h^2$

   $\gamma = 0.80$ $\alpha = 0.50$ $h = 0.463$ m $= 463$ mm

USE 475 mm by 475 mm column

$\alpha = 3.13/30 \times 0.475^2 = 0.462$ $\gamma = 0.70$ $\rho_t m = 0.27$

   $\gamma = 0.80$ $\rho_t m = 0.25$

   $\gamma = 0.737$ $\rho_t m = 0.263$, $\rho_t = 0.0164$

*These charts automatically introduce the ϕ.

Reqd. $A_s = 0.0164 \times 475^2 = 3\ 700$ mm^2 $8\text{-}\#25 = 4\ 000$ mm^2
USE 8-#25 with #10 ties in pairs as in Fig. 18-13e or f. Tie spacing governed by 16 bar diameters, which is 400 mm.

Although the δ for the entire story (as used) is little influenced by one column, the column must also be checked individually as though it were in a braced frame. Usually this proves critical only where the column is smaller than the average for the story. In this case the braced frame column already designed for the same loading in Sec. 19.12 shows that δ is much smaller (as expected) and that a 450 mm column would be adequate; no more exact check is needed.

(b) Extreme Case

Assume now that the column initially described in (a) is one of only four columns in a single structure, so situated that all four columns (and their connecting beams) appear to be identical cases that would lead to identical δ values whether separate or in a group. Based on a known stiffness of the beams (cracked section) and an estimated column size of 450 mm square, with ρ_t about 0.02, the ψ is estimated on the basis of EI_1 to be 1.0. P_u is 3130 kN, $M_u = 170$ kN $\cdot$ m, $\ell_u = 6.0$ m, $f'_c = 30$ MPa, $f_y = 400$ MPa, assumed unchanged (for example simplicity) as column size varies in design. Redesign for these conditions.

Solution

When one has a new situation where estimated column size is apt not to be good, it is often advantageous to check the first guess with the much simpler EI_1 even though the design intent is to use a larger ρ_t that seems to demand EI_2 (Table 19.1 in Sec. 19.6). As the planned ρ_t gets over 3% this advantage is probably lost.
For the estimated 450 mm column:

$E = 5\ 000\sqrt{f'_c} = 5\ 000\sqrt{30} = 27\ 400$ MPa

$EI_1 = EI_g/2.5 = 27\ 500 \times 0.45^4/12 \times 2.5 = 37.6$ MN $\cdot$ m^2

For $\psi = 1$, Fig. 19.6 shows $k = 1.31$, $k\ell_u = 1.31 \times 6.0 = 7.86$ m

$\phi P_c = 0.70\pi^2\ EI/(k\ell_u)^2 = 0.70 \times \pi^2 \times 37.6/7.86^2 = 4.206$ MN

$C_m = 1.0$ for unbraced frames, $\delta = 1/(1 - P_u/\phi P_c) = 1/(1 - 3.130/4.206) = 3.91$

This δ is too large to be practical. Somewhere around 2, or maybe 3, is a top practical limit.
Try a 500 mm by 500 mm column.

$EI_1 = (500/450)^4 \times 37.6 = 57.3$ MN $\cdot$ m^2

Revised $\psi = 1.0 \times 57.3/37.6 = 1.52$ $k = 1.46$ $k\ell_u = 8.76$ m

$r = 500/\sqrt{12} = 144$ mm

$k\ell_u/r = 8.76/0.144 = 60.8 < 100$ (Code 10.11.4.3). Method is O.K.

$\delta = 1/(1 - 3.130/5.160) = 2.54$ $\gamma = (500 - 125/500) = 0.75$

For $\rho_t m = 0.02 \times 16 = 0.32$, Fig. 18.12* shows

For $\gamma = 0.70$ $\alpha = 0.45$

    $\gamma = 0.80$ $\alpha = 0.43$

Interpolating for $\gamma = 0.75$, $\alpha = 0.44$

$h = \sqrt{P/f'_c \alpha} = \sqrt{3.13/30 \times 0.44} = 0.487$ m $= 487$ mm   **O.K.**

Check for EI_2, with $\rho_t = 0.02$:

$$EI_2 = E_c(\tfrac{1}{5}I_g + nI_s) = E_c[\tfrac{1}{5} \times \tfrac{1}{12} h^4 + n\,\rho_t\,h^2(\tfrac{1}{2}\gamma\,h)^2]$$

$$= 27\,400 \times 0.500^4(\tfrac{1}{5} \times \tfrac{1}{12} + 7.3 \times 0.02 \times 0.375^2) = 63.7 \text{ MN} \cdot \text{m}^2$$

$$\phi P_c = 0.70\pi^2 \times 63.7/8.76^2 = 5.736 \text{ MN}$$

$$\delta = 1/(1 - 3.130/5.736) = 2.20$$

$$\delta e/h = 2.20 \times 0.1086 = 0.239$$

$$\rho_t m = 0.02 \times 16 = 0.32$$

From Fig. 8.12

For $\gamma = 0.70$ $\alpha = 0.49$

    $\gamma = 0.80$ $\alpha = 0.47$

    $\gamma = 0.75$ $\alpha = 0.48$

    $h = \sqrt{3.13/30 \times 0.48} = 0.466$ m $= 466$ mm   **O.K.**

To determine A_s:

$$\delta e/h = 0.239 \qquad \alpha = P/f'_c\,h^2 = 3.13/30 \times 0.5^2 = 0.417$$

From Fig. 8.12

    $\gamma = 0.70$ $\rho_t m = 0.25$

    $\gamma = 0.80$ $\rho_t m = 0.20$

    $\gamma = 0.75$ $\rho_t m = 0.225$ $\rho_t = 0.225/16 = 0.014$ **O.K.**

Required $A_s = 0.014 \times 500^2 = 3\,500$ mm^2

USE 500 mm $\times$ 500 mm column, 8-#25(4000 mm^2)

Use #10 ties, at 400 mm spacing.

This design was simplified by assuming all columns in the story the same. This will rarely be feasible and the final δ for the story cannot be known until at least tentative designs have been made. As noted earlier, this means that the story δ should be much less influenced by a single column δ.

If, as stated initially, the structure is the same about both axes, A_{st} spread on all four faces would fit better, but for this arrangement a different set of column design charts is needed (available in Reference 3). The effective γ would be reduced by this less effective steel arrangement (less effective for the single axis), and a little more total steel would be necessary.

*These charts automatically introduce the ϕ.

(c) Beams in Unbraced Frames

Emphasis needs to be added as to the role of the beam in the unbraced frame. *All* the increased column moment must also be resisted by the beams. These increases are secondary beam moments to be added to the beam in addition to the first-order beam moments given by frame analysis. With such a large δ as in this example, there must be large resisting beam moments. The designer might consider whether a column larger than 500 mm square with a ρ_t lower than 0.02 and a δ lower than 2.20 might not be a stiffer and better design.

Although, in the particular case above, the 500 mm column is barely inside the limit for the R method, the reader should compare the alternate design by the R method in Sec. 19.17. Within the boundaries suggested in the Commentary (and in Sec. 19.15), the author prefers the simpler approach of that method. Very slender columns *must* be designed by the moment magnifier method or a computer analysis; the R method is not adequate beyond its stated limits.

19.14 ALTERNATE *R* METHOD

The Code Commentary (just prior to its Sec. 10.13 discussion) introduces the "modified R method" as an alternate design method for columns within the limits noted. The R method provides a reduction or discount factor R (less than unity), linear with regard to column length ℓ_u/r. Beyond the R limits stated in the Commentary the method can be unsafe. Within these limits accuracy is comparable with that of the Code magnifier method. This does not mean the two methods give the same answer, but each is, on the average, providing a similar margin of safety. The "modified R" reflects several changes in equations and length limitations from those used in the 1963 ACI Code.

In usage the R method in effect assumes the usable portions of both P_u and M_u for a given short column are reduced by the *same* multiplier R, as sketched in Fig. 19.9, with no change in nominal eccentricity.

In design the process for the long column needing P_u and M_u is the reverse, that is, to design a *short* column capable of caring for P_u/R and M_u/R.

The allowable R values, given in detail in Sec. 19.15, were designed to define a lower-bound measure of the column strength. Computer analyses and tests at the University of Texas at Austin verified that, although generally safe, the 1963 R method was in some areas overconservative, was somewhat erratic for stability failures, and could be unsafe in some cases of instability. Accordingly, the method was revised to that in the Commentary that now excludes the unsafe cases and liberalizes some

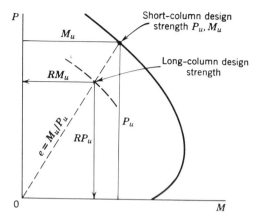

FIGURE 19.9   The R method concept.

formerly overconservative cases. It leaves many of the *unbraced frame* cases for the Code moment magnifier method.

For columns with sidesway under shear loading the R method is limited to columns of about the length used in the design of Secs. 19.12 and 19.13, but within this limit it is as accurate as the magnifier method, as can be seen by comparing the R method in Fig. 19.10a and the Code magnifier method in Fig. 19.10b. The latter used P_c in the δ equation rather than the present ϕP_c. The inclusion of ϕ would shift points to the left, especially in the medium to low P range, and would thus eliminate most of the slightly unsafe points. In Fig. 19.10a the special test value (beam $\rho = \rho' = 0.0074$) and all the points noticeably low are eliminated by the Commentary restrictions, namely, limiting effective length ℓ'_u to ℓ'_u/r of 40 and requiring beam negative moment steel of at least $\rho = 0.01$.

For the R method the ψ value is simply that given by the ratio of *gross* I values; this value will here be designated as $\psi_g{}^*$ for identification as such. Accurate values of I_g and ψ_g can enter into the design process much earlier than can the δ for the magnifier method. Apparently it is only on the longer columns that the physical difference is significant.

For the Code magnifier method the relative column stiffness ψ in an *unbraced* frame must be based on the cracked beam section and a column I consistent with the EI_2 equation (Sec. 19.6). Such values lead to the excellent results plotted in Fig. 19.10b. The need for *this* value of ψ for the

---

*The Code Commentary uses α_m and ψ_g. The subscript emphasizes the use of gross I values in this evaluation.

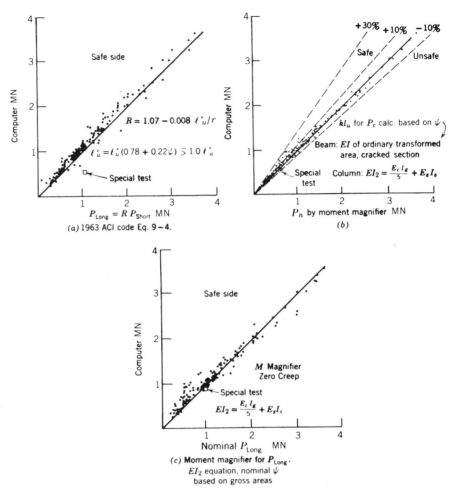

FIGURE 19.10 Columns with *sidesway*. Comparative safety of different methods. All data are for 250 mm × 250 mm columns *without allowance for creep*.

magnifier method is shown by the poorer results of Fig. 19.10c where ψ_g values were used in the calculations.

In the sidesway case the Code calls for consideration of the entire story in fixing the uniform magnifier to be used for all the columns in that story. This seems broad enough to require that columns that are officially "short columns" must go into the determination of the story δ and must have their design moments likewise increased by the factor. Although initiating this process into a design appears awkward, it probably provides overall economy in that more flexible columns will be designed for lower δ values than they individually would justify, and the increased δ for the heavier columns will mean only a small increase. Since tests at the University of Texas indicate that, after the most heavily stressed column reaches its maximum capacity, that column can maintain most of this capacity over a considerable added drift, it appears that either the Code approach or the single-column approach may prove adequate. The *R* method might be a wise choice (where columns are in its range) to use in sizing columns before a story δ can be established for the magnifier method. Familiarity with tables[3] or charts[4] for allowable column loads will likewise lead to a feel for such sizing.

For braced columns a similar comparison of the two methods, using in both cases ψ_g, is given in Fig. 19.11, where the magnifier method in *b* tends to predict loads a little high for the shorter columns, the upper half of the load range. The *R* method in *a* shows slightly better for this case, in the author's opinion. If the ψ_g used in establishing *k* for Fig. 19.11*b* were made ψ based on the moments of inertia (cracked beam and EI_2/E_c) used in developing the data for Fig. 19.10*b*, it is probable that this excess would be reduced or eliminated. However, there is a Code difference here. For the sidesway case such a technique is required; for the braced frame this technique (by implication) is *not required*, but may be used. If *k* = 1.0 is used with the magnifier method in these braced frames, no study results are available. The greater *k* and the inclusion of ϕP_u mentioned earlier in discussing Fig. 19.10*b* would each tend to lower permissible P_n, thereby reducing the number of unsafe values.

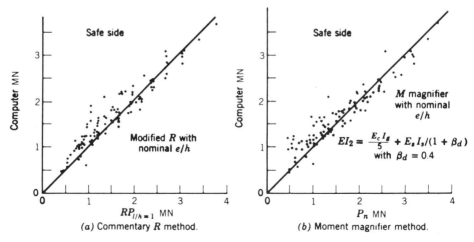

FIGURE 19.11   Comparative safety of two methods for single curvature columns in *braced* frames, nominal 250 mm × 250 mm columns. Both use ψ_g.

Since the R method for braced frames, as now limited, still includes most practical columns not loaded at large eccentricities and is simpler and faster than the magnifier method, the designer's choice must depend on whether he or she wishes to keep familar with *two* methods: the R method for the usual case, and the magnifier method for the very long column where moment is a major loading. The moment magnifier deals with more realistic behavior and properly emphasizes that strength is lower *because moment is higher*; it gives moment values easily available for possible revision of the beam strength in the sidesway case. Within the Commentary limits for R design, the moment magnifier is *not safer* and, on the average for the University of Texas investigation, *not more economical*. This economy statement could also be discounted for very large columns (larger than 0.6 m) and very high ρ values that have not been widely sampled.

As design tables[3] and charts[4] for long columns evaluated on the moment magnifier basis become more familiar and therefore easier to use, the need for the shorter R method should decrease correspondingly.

19.15 THE ALTERNATE R METHOD IN DETAIL

The R method uses a reduction factor R, linear with respect to the ℓ_u/r ratio, which is applied to the short-column strength, P_n and M_n, to define the nominal permissible long-column capacity. This operation is similar to, and accumulative with, the use of ϕ applied to the ideal column to give the permissible short column load. Thus one can sketch three interaction diagrams in Fig. 19.12a. The middle one is the one used in the ACI column interaction charts, except below $P_u = 0.10 \, f'_c \, bt$ where ϕ has here been changed to agree with the ϕ factor changes in Sec. 18.7.

For a design loading $P_{u\ell}$ and $M_{u\ell}$ on a long column one can move radially from this point on the inner interaction diagram to $P_u = P_{u\ell}/R$ and $M_u = M_{u\ell}/R$ and, if he wishes, on to $P_n = P_u/\phi$ and $M_n = M_u/\phi$. With the ACI interaction charts, only the first shift is required to give P_u and M_u. (The moment magnifier method is a different concept and is not presented in terms of such an inner interaction diagram.)

The required value of R is an empirically derived bounding straight line for each of three cases, two of which are subdivided.

1. In a braced frame for a column loaded by moment at one end but restrained at the far end such that a point of contraflexure occurs:

$$R = 1.32 - 0.006 \, \ell_u/r \gtrless 1.0 \quad \text{for} \quad 54 < \ell_u/r < 100$$

Up to ℓ_u/r of 54, or ℓ_u/h of 16 for a rectangular column, this is a short column with $R = 1$.

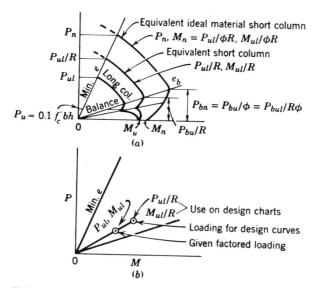

FIGURE 19.12 Interaction diagram showing relation of long to short and ideal columns. (*a*) Note $\phi = 0.70$ or 0.75 only above $P_u = 0.10 f_c'bh$; below that see Fig. 18.9. (*b*) Shift from long column loading to design curve. Minimum e to be used as $0.10\ h$ for tied columns, $0.05\ h$ for spiral columns.

2. In a braced frame for a column bent in single curvature:
    a. With the nominal $e \gtrless 0.1\ h$

    $$R = 1.23 - 0.008\ \ell_u/r \gtrless 1.00 \quad \text{for} \quad 29 < \ell_u/r < 100$$

    Up to ℓ_u/r of 29, or ℓ_u/h of 8.6 for a rectangular column, this is a short column with $R = 1$.
    b. With the nominal $e > 0.10\ h$

    $$R = 1.07 - 0.008\ \ell_u/r \gtrless 1.0 \quad \text{for} \quad \ell_u/r < 100$$

    Few practical columns are short columns in this case.
3. If an unbraced frame having $\ell_u'/r \gtrless 40$ *and* having beams with $\rho \gtrless 0.01$:
    a. For loads of short duration

    $$R = 1.07 - 0.008\ \ell_u'/r \gtrless 1.0$$
    $$\text{where } \ell_u' = \ell_u(0.78 + 0.22\ \psi_g) \gtrless 1.0\ \ell_u{}^*$$

---

*The Code Commentary uses α_m for ψ_g. The subscript emphasizes the use of gross I values in this evaluation.

b. For loads of longer duration

$$R = 0.97 - 0.008 \, \ell'_u/r \lessgtr 1.0$$

In the equation for ℓ'_u, the ratio ψ_g of column-to-beam stiffness is the average value at the two ends based on gross I of members. When $\ell'_u/r > 40$, the moment magnifier should be used.

The above R equations are diagramed in Fig. B.6 in Appendix B.

Of these equations, the one in 2a is considerably liberalized compared to the 1963 Code, but it is for a limited e/h. Also the one in 3a is more restrictive in that it formerly applied only to sustained loads. The one in 3b is new.

In any case where $P_u \lessgtr 0.10 \, f'_c bh$, the formula value of R, say R_f, should be increased to unity at P_u of zero in the fashion used for ϕ in Fig. 18.9. The result is: Eff. $R = 1 - 10(P_u/f'_c bh)\,(1 - R_f)$.

In Fig. 19.3 the column moment is shown much increased by the sideway. This increased column moment must be resisted by the bracing beams, a fact that designers have not always fully realized. The increase shown is largely from lateral load. However, much the same increase occurs at ultimate whether a lateral load is applied outside the structure or whether instability, the common mode of failure, develops internally in the frame. The moment can be estimated from Fig. 19.13 by using similar

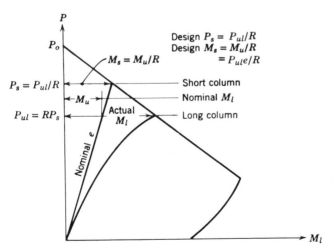

FIGURE 19.13 Approximation of actual M_ℓ for beam design. The Code Commentary notation has been altered here to keep $P_{u\ell}$ for the factored long column load (to match Code 9.3.2c).

triangles that lead to

$$M_{bm} = \text{Col. } M_{\ell} = \text{Nom. } P_{u\ell}e(1 - P_{u\ell}/P_0)/(R - P_{u\ell}/P_0)$$

The actual interaction diagram is curved upward above the straight line and the straight-line simplification is on the safe side for this calculation. P_0 is defined in Sec. 18.5 and $P_0 = \phi P_{n0}$.

19.16 DESIGN OF LONG COLUMN IN BRACED FRAME BY *R* METHOD

The most often governing case in design for braced frames is that of columns in reversed curvature, but these nearly always turn out to fall in the short column classification with an $\ell_u/h \gtrsim 16$. Hence the following design example is taken in single curvature, almost matching the example using the magnifier method in Sec. 19.12.*

Example

Design a square tied column with $\ell_u = 6.0$ m in single curvature in a braced frame, using $f_c' = 30$ MPa, Grade 400 reinforcement. Total dead weight (including its own) is 900 kN, live load is 1 100 kN, and live load moment is 80 kN · m. Try for ρ_t of about 0.02.

Solution

Assume $\ell_u/r < 100$, which is not seriously restrictive.

$$P_u = 900 \times 1.4 + 1\,100 \times 1.7 = 3\,130 \text{ kN} \qquad M_u = 80 \times 1.7 = 136 \text{ kN} \cdot \text{m}$$

$$e = 136/3\,130 = 0.044 \text{ m} = 44 \text{ mm} \qquad m = f_y/0.85\,f_c' = 16 \qquad \text{Desired } \rho_t m = 0.32$$

At start one must assume R or column size. Try $h = 500$ mm column.

For $d = 500 - 40 - 10 - 13 = 437$ mm, the usable R equation is:

$$R = 1.23 - 0.008\,\ell_u/r \gtrsim 1.0 \text{ for } 29 < \ell_u/r < 100$$

$$\ell_u/r = r = 500/\sqrt{12} = 144 \text{ mm} \qquad \ell_u/r = 6\,000/144 = 42 \qquad R = 1.23 - 0.008 \times 42 = 0.90$$

Equivalent short column $P_s = P_u/R = 3\,130/0.90 = 3\,478$ kN

$$d' = 40 + 10 + 13\text{(for half bar)} = 63 \text{ mm}, \qquad \gamma = (500 - 2 \times 63)/500 = 0.75$$

From Fig. 8.12c† for $\gamma = 0.70$, $e/h = 44/500 = 0.088$, $\rho_t m = 0.32$

read $\alpha = 0.66 = P/f_c'h^2 = 3.13/30\,h^2$ $h = 0.397$ m/397 mm

---

*Except for M of 80 kN · m instead of 100 kN · m changed because when $e > 0.10\,h$, the R equation decreases R to the old 1963 value (such cases are not included in the University of Texas study).

†These charts automatically introduce the ϕ.

USE 450 mm by 450 mm

$$r = 450/\sqrt{12} = 130$$
$$\ell_u/r = 6\,000/130 = 46$$
$$R = 1.23 - 0.008 \times 46 = 0.86$$
$$P_s = 3.13/0.86 = 3.640 \text{ MN}$$
$$\gamma = (450 - 2 \times 63)/450 = 0.72$$
$$e/h = 44/450 = 0.098$$
$$\alpha = 3.64/30 \times 0.45^2 = 0.60$$

From Fig. 8.12b, $\gamma = 0.70$, $\rho_t m = 0.27$

This gives $\rho_t = 0.27/16 = 0.017$

Required $A_s = 0.017 \times 450^2 = 3\,443 \text{ mm}^2$ USE 8-#25 bars (4 000 mm²)

USE #10 ties at 400 mm spacing.

Note that the magnifier method solution of Sec. 19.12 gave essentially the same required design.

19.17 DESIGN OF LONG COLUMN IN UNBRACED FRAME BY R METHOD

Redesign the unbraced square columns of Sec. 19.13 by the R method for loads of short duration. $P_u = 3\,130 \text{ kN}$, $M_u = 170 \text{ kN} \cdot \text{m}$, $f'_c = 30 \text{ MPa}$, $f_y = 400 \text{ MPa}$, $\ell_u = 6.0 \text{ m}$, $\psi = 1$. The story δ is not used in the R method (a probable weakness). Although Fig. 19.7 shows beam reinforcement greatly influences ψ, the magnifier method does not require any specific ρ for beams nor differentiate because of short duration of loads, as for wind load. Assume beam $\rho \geq 0.01$,* the minimum for the R method in the unbraced case. Try for column ρ_t near 0.02.

Solution

The effective unsupported lengths of the column

$$\ell'_u = \ell_u(0.78 + 0.22\,\psi) = 6.0 \text{ m since } \psi = 1.$$

For the R method ℓ'_u must not exceed $40\,r = 40\,h/\sqrt{12} = 12\,h$. Thus the method cannot be used unless $h = 6\,000/12 = 500 \text{ mm}$. Try 500 mm column,† $r = h\sqrt{12} = 144 \text{ mm}$.

$$R = 1.07 - 0.008 \times 6\,000/144 = 0.74$$

Design $P_u = 3\,130/R = 4\,230 \text{ kN}$ $M_u = 170/R = 230 \text{ kN} \cdot \text{m}$

$$e = 230/4\,230 = 0.054 \text{ m} = 54 \text{ mm}$$

---

*The author believes this minimum negative moment steel ratio ρ of 0.01 should be mandatory for beams in all unbraced frames.

†The first trial normally would be simply a guess, with the chart used to find the required α and through this the required h.

$\gamma = (500 - 2 \times 40 - 2 \times 10 - 25)/500 = 0.75$

$\alpha = P_u/f_c'h^2 = 4.23/30 \times 0.5^2 = 0.564$ $e/h = 54/500 = 0.108$

From Fig. 8.12* entering with $\alpha = 0.564$, and $e/h = 0.108$:

for $\gamma = 0.80$ $\rho_t m = 0.22$

$\gamma = 0.70$ $\rho_t m = 0.24$

$\gamma = 0.75$ $\rho_t m = 0.23$ $\rho_t = 0.23/16 = 0.014$

For this method a smaller column cannot be used and here $\rho_t = 0.02$ is not needed.

USE 500 mm $\times$ 500 mm column, reqd. $A_s = 0.014 \times 500^2 = 3\,500$ mm^2

USE 8-#25(4 000 mm^2) with #10 ties at 400 mm spacing.

Note that the R design is nearly the same as that found in Sec. 19.13b by the magnifier method.

Code 10.11.2.2, already quoted in Sec. 19.8, emphasizes the consideration of cracking in establishing ψ as a step toward fixing k for a braced frame when using the magnifier method. This would also be appropriate when calculating ψ for use in the ℓ_u equation, although not specified in the Commentary statement.

19.18 DESIGNS AGAINST TENSION FAILURES WITHOUT USE OF INTERACTION DIAGRAMS

(a) Advantages

For tension failures the internal force methods of Sec. 18.16 are more accurate than the usual interaction diagrams and give the designer a clearer feel for the problem. Especially for these cases having axial load less than $0.10\,f_c'bt$ (with larger ϕ factors and larger R factors where the R-method is used), the direct approach is not distinctly longer than the interaction-diagram approach. The interaction diagrams are not so easily read for large eccentricities,† and the required ρ_t is sensitive to γ values, often calling for interpolation between two charts.

(b) Design for Tension Failure—R Method

Design a square tied column 4.2 m high with a service load of 90 kN dead load plus 100 kN live load and an eccentricity of 600 mm. Use $f_c' = 30$ MPa, $f_y = 400$ MPa, $\psi = 1$. The column is not restrained against sidesway and the R method for length is to be used. (The beams have the requisite 1% of longitudinal steel.) Try for ρ_t of about 0.02.

---

*These charts automatically introduce the ϕ.

†Unless one reads $\alpha e/h$ from the abscissa instead of α from the ordinate.

Solution

A 600 mm eccentricity and light axial loads indicate a probable tension failure where the internal stress distribution at ultimate is easy to handle by the method of Sec. 18.16. When the compression steel yields, the two couples of Fig. 18.18 balance each other. When the stress block depth a is so small that the compression steel does not yield, the designer can substitute $\Sigma M = 0$ about the compression steel; N_{c1} then will be so close to the moment center that an approximate value of N_{c1} and an approximate value of its arm will have only a small influence on the calculations. This is equivalent to noting that, when one cannot deal with simple couples, the problem almost resolves into a beam type of calculation for the steel, with a minor correction term included to account for the axial load. Note that a specification for a low ρ_t is equivalent to requiring a large arm for the tensile steel in the moment equation, that is, a relatively deep column.

$$P_u = 1.4 \times 90 + 1.7 \times 100 = 296 \text{ kN} = 0.296 \text{ MN}$$

Since $\psi = 1$, the effective long-column length

$$\ell_u' = \ell_u(0.78 + 0.22 \; \psi) = \ell_u = 4.2 \text{ m}$$

The R method may be used only if $\ell_u'/r \geq 40$, which means for this length $r = h/\sqrt{12} = 0.29 \; h \geq \ell_u'/40 = 4\,200/40 = 105$ mm. The minimum column for this method is thus 362 mm. (A column size should not be selected on this basis, but the designer should be alert to the need for a change to the moment magnifier method if a smaller column seems to fit the loading.)

Try a 375 mm column.

$$R = 1.07 - 0.008 \; \ell_u/r = 1.07 - 0.008 \times 4\,200/0.29 \times 375 = 0.76$$

Check whether ϕ and R can be increased because $P_u < 0.10 \; f_c'h^2$:

$$P_u/f_c'h^2 = 0.296/30 \times 0.375^2 = 0.070 < 0.10$$

$$\phi = 0.90 - 2 \times 0.070 = 0.76 \text{ (Fig. 18.9)}$$

Effective $R = 1 - 0.070 \times 10(1 - 0.76) = 0.83$ (Sec. 19.15)

$$P_n = P_u/R\phi = 296/0.83 \times 0.76 = 469 \text{ kN} = 0.469 \text{ MN}$$

Assume $f_s' = f_y$. The two couples in Fig. 19.14, which neglects displaced concrete, show that N_{c1} must equal P_n, leading to

$$a = 0.469/0.85 \times 30 \times 0.375 = 0.049 \text{ m} = 49 \text{ mm}$$

This places the compression steel very close to the neutral axis (low f_s'); but continue with the couples and the false assumption until a trial ρ_t is indicated. The method is not overly sensitive to shifting between N_{c1} and N_{c2} at this stage.

$$N_t = \tfrac{1}{2}A_sf_y, \; h - 2 \; d' = 375 - 2 \times 40 - 2 \times 10 - 25 = 200 \text{ mm}$$

$$P_n(e - \tfrac{1}{2} h + \tfrac{1}{2} a) \;\; = N_t(h - 2 \; d') = \tfrac{1}{2} A_sf_y\gamma \; h$$

$$A_s = 2 \times 0.469 \times (600 - \tfrac{1}{2} \times 375 + \tfrac{1}{2} \times 49)/400 \times 200 = 5\,124 \times 10^{-6} \text{ m}^2 = 5\,124 \text{ mm}^2$$

$$\rho_t = 0.0364 > 0.002$$

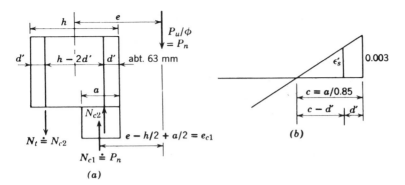

FIGURE 19.14 Large eccentricity of load. (a) Equilibrium forces.
(b) Compressive strains.

Try 425 mm column and recycle, first without even revising the depth of stress block.

$R = 1.07 - 0.008 \times 4\,200/0.29 \times 425 = 0.80$

$P_u/f'_c h^2 = 0.296/30 \times 0.425^2 = 0.0546$

$\phi = 0.90 - 2 \times 0.0546 = 0.791$

Effective $R = 1 - 0.0546 \times 10(1 - 0.80) = 0.891$

$P_n = 296/0.891 \times 0.791 = 420$ kN $= 0.420$ MN

$A_s = 2 \times 0.420(600 - \frac{1}{2} \times 425 + \frac{1}{2} \times 49)/400 \times 250 = -3\,470 \times 10^{-6}$ m^2 $= 3\,470$ mm^2,

$\rho_t = 0.0192$

USE 425 mm by 425 mm column

For A'_s at f_y this gives $a = 0.42/0.85 \times 30 \times 0.425 = 0.039$ m $= 39$ mm. With this small value of a, the N_{c2} steel could be in the tension zone, but, if N_{c2} is missing, N_{c1} must increase to maintain the total, and this in turn increases a. Obtaining the exact N_{c1}, a, and f'_s is almost a marginal need, since N_{c1} will act not far from A'_s (which will be used as the moment center) and will account for only a small part of the total moment. The designer might prefer to use a little more steel and not refine the A_s requirement further.

More to demonstrate the stability of the solution than actually to refine it investigate N_{c1}, a, and f'_s further. As an extreme, if c falls at N_{c2}, $f'_s = 0$, leaving $\Sigma F_y = 0$ or $N_{c1} - P_n - 0.5\ A_s f_y = 0$.

$$N_{c1} = 0.42 + 0.5\ A_s f_y = 0.42 + \tfrac{1}{2} \times 3.470 \times 10^{-3} \times 400 = 1.114 \text{ MN}$$

$$a = 1.114/0.85 \times 30 \times 0.425 = 0.103 \text{ m} = 103 \text{ mm}$$

The depth of the stress block must lie between 39 mm and 103 mm. Try $a = 70$ mm (about the average), $c = 70/0.85 = 82$ mm.

$$\epsilon'_s = 0.003(82 - 63)/82 = 0.000\ 69$$

$$f'_s = 200\ 000 \times 0.000\ 69 = 138 \text{ MPa}$$

$$N_{c1} = P_n + N_t - N_{c2} + \text{correction for displaced concrete}$$

$$N_{c1} = 0.42 + \tfrac{1}{2} \times 3.47 \times 10^{-3}(400 - 138 + 0.85 \times 30) = 0.919 \text{ MN}$$

The second term allows for the concrete displaced by the reinforcement.

Take moments about N_{c2}:

$$\Sigma M = 0 = \tfrac{1}{2}A_s \times 400(425 - 2 \times 63) - 0.42(600 - \tfrac{1}{2} \times 425 + \tfrac{1}{2} \times 70) + 0.919(63 - \tfrac{1}{2} \times 70)$$

$$A_s = 203.2/59\,800 = 3\,400 \times 10^{-6} \text{ m}^2 = 3\,400 \text{ mm}^2$$

$$a = 0.919/0.85 \times 30 \times 0.425$$

$$= 0.085 \text{ m versus } 0.070 \text{ m averaged earlier}$$

USE 8-#25 bars (4 000 mm²) with 10 ties at 400 mm

(c) Design for Tension Failure—Moment Magnifier Method

Design the square tied column for the conditions of (b) except use the moment magnifier for the length effect. Note the special basis on which ψ must be computed (Sec. 19.9).

Solution

The comments which opened the solution in (b) are all appropriate here, but will not be repeated.

$$P_u = 1.4 \times 90 + 1.7 \times 100 = 296 \text{ kN} = 0.296 \text{ MN}$$

Since $\psi = 1$, Fig. 19.6 for the unbraced frames shows $k = 1.31$

$$k\ell_u = 1.31 \times 4.2 = 5.5 \text{ m}$$

Assume $\delta = 1.1$ as a start, noting that this column load is very small and the moment large. This gives $\delta e = 600 \times 1.1 = 660$ mm. Try a 350 mm column. Check whether ϕ may exceed 0.70, that is, whether $P_u/f'_c h^2 < 0.10$.

$$P_u/f'_c h^2 = 0.296/30 \times 0.35^2 = 0.081 < 0.10$$

$$\phi = 0.90 - 0.081 \times 2 = 0.738 \text{ (Fig. 18.9)}$$

$$P_n = P_u/\phi = 0.296/0.738 = 0.400 \text{ MN}$$

Assume $f'_s = f_y$. The two couples of Fig. 19.14, which neglect displaced concrete, show that N_{c1} must equal P_n. For $N_{c1} = P_n$,

$$a = 0.40/0.85 \times 30 \times 0.35 = 0.045 \text{ m} = 45 \text{ mm}$$

Although this indicates that f'_s is slightly negative and N_{c1} must be increased to make up the deficit, continue with the false assumption that $f'_s = f_y$ until a trial ρ_t is indicated. The method is not overly sensitive at this stage to shifting resistance between N_{c1} and N_{c2}.

$$N_t = \tfrac{1}{2}A_s f_y$$

$$N_{c1} = P_n(\text{approx.})$$

$$d' = 40 + 10 + \tfrac{1}{2} \times 25 = 63 \text{ mm}$$

$$\tfrac{1}{2}A_s f_y(h - 2\,d') = P_n\,\delta e - P_n(\tfrac{1}{2}\,h - \tfrac{1}{2}\,a)$$

$$\tfrac{1}{2}A_s \times 400\,(350 - 126) = 0.400 \times 660 - 0.400(\tfrac{1}{2} \times 350 - \tfrac{1}{2} \times 45)$$

$$A_s = 4\,530 \times 10^{-6}\,\mathrm{m}^2 = 4\,530\,\mathrm{mm}^2$$

$$\rho_t = 0.037$$

Since ρ_t is so high, a larger column is indicated, although there are several assumed values in the calculations so far. The most critical is δ, and it cannot be significantly smaller. Try a 400 mm square column, and check δ.

$$P_u/f_c' h^2 = 0.296/30 \times 0.4^2 = 0.0617$$

$$\phi = 0.90 - 2 \times 0.0617 = 0.777$$

$$P_n = 0.296/0.777 = 0.381\,\mathrm{MN}$$

$$E_c = 5\,000\,\sqrt{30} = 27\,400\,\mathrm{MPa}$$

For $\beta_d = 0$,
$$EI_1 = E_c I_g/2.5 = 27\,400 \times \tfrac{1}{12} \times 0.4^4/2.5 = 23.38\,\mathrm{MN \cdot m^2}$$

$$k\ell_u = 5.5\,\mathrm{m}$$

$$\phi P_c = \phi \pi^2\,EI/(k\ell_u)^2 = 0.777\,\pi^2 \times 23.38/5.5^2 = 5.929\,\mathrm{MN}$$

$$\delta = 1/(1 - 0.381/5.929) = 1.07$$

$$\delta e = 1.07 \times 600 = 642\,\mathrm{mm}$$

$$\tfrac{1}{2}A_s \times 400(400 - 126) = 0.381(642 - \tfrac{1}{2} \times 400 + \tfrac{1}{2} \times 45)$$

$$A_s = 3\,230 \times 10^{-6}\,\mathrm{m}^2 = 3\,230\,\mathrm{mm}^2; \qquad \rho_t = 0.020$$

The compression steel is so close to the neutral axis that it is of doubtful value. Assume N_{c1} has to pick up all of $N_{c2} = \tfrac{1}{2} \times 3\,230 \times 10^{-6} \times 400 = 0.646\,\mathrm{MN}$.

$$N_{c1} = 0.381 + 0.646 = 1.027\,\mathrm{MN}$$

$$a = 1.027/0.85 \times 30 \times 0.4 = 0.100\,\mathrm{m} = 100\,\mathrm{mm}$$

With this a, the steel becomes partially effective, which decreases N_{c1} and a. Assume that $a = 75$ mm, $c = 75/0.85 = 88$ mm.

$$\epsilon_s = 0.003(88 - 63)/63 = 0.0012$$

$$Ff_s' = 200\,000 \times 0.0012 = 240\,\mathrm{MPa}$$

$$N_{c1} = 0.381 + \tfrac{1}{2} \times 3\,230 \times 10^{-6}\,(400 - 240) + \tfrac{1}{2} \times 3\,230 \times 10^{-6} \times 0.85 \times 30*$$
$$= 0.681\,\mathrm{MN}$$

$$a = 0.681/0.85 \times 30 \times 0.4 = 0.067\,\mathrm{m} = 67\,\mathrm{mm}$$

Since this operation is not sensitive to N_{c1} and a, use $N_{c1} = 0.681$ MN and $a = 67$ mm.

$$\Sigma M = 0 \text{ about the compression steel as center:}$$

$$\tfrac{1}{2}A_s \times 400(400 - 2 \times 63) - 0.381(642 - \tfrac{1}{2} \times 400 + 63) + 0.681(63 - \tfrac{1}{2} \times 67) = 0$$

*To establish a, N_{c1} is increased (in lieu of reducing its area) to offset the concrete displaced by $\tfrac{1}{2}A_s$.

$$A_s = 3\,144 \times 10^{-6}\,\text{m}^2 = 3\,144\,\text{mm}^2$$
$$\rho_t = 0.0197$$

Another cycle with this A_s would reduce N_{c1} and a, a pair of changes that have opposite effects on the moment. Say O.K. as calculated.

USE 400 mm × 400 mm column

8-#25 (4 000 mm²)

SELECTED REFERENCES

1. Richard W. Furlong, "Column Slenderness and Charts for Design," *Jour. ACI*, 68, No. 1, Jan. 1971, p. 9.
2. B. C. Johnston (ed.), *The Column Research Council Guide to Design Criteria for Metal Compression Members*, 2nd ed., John Wiley and Sons, Inc., New York, 1966.
3. *Design Handbook, V. 1, SP-17(73)*, American Concrete Institute, Detroit, 1973.
4. *Design Handbook, V. 2, SP-17A(78)*, American Concrete Institute, Detroit, 1978.
5. J. G. MacGregor, J. E. Breen, and E. O. Pfrang, "Design of Slender Columns," *Jour. ACI*, 67, No. 1, Jan. 1970, p. 6.
6. J. G. MacGregor, V. H. Oelhafen, and S. E. Hage, "A Re-examination of the EI Value for Slender Columns," *Reinforced Concrete Columns, SP-50*, American Concrete Institute, Detroit, 1975, p. 1.
7. *Reinforced Concrete Columns, SP-50*, American Concrete Institute, Detroit, 1975.

PROBLEMS

PROB. 19.1. Analysis of a braced frame shows an interior column in single curvature with $M_1 = M_2$ consisting of service load $M_d = 0$, $M_\ell = 183$ kN · m, $P_d = 450$ kN, $P_\ell = 700$ kN, $\psi = 1.0$, $\ell_u = 6.0$ m, $C_m = 1.0$, $\beta_d = 0$, $k = 0.77$ (Table 18.1), $f_c' = 30$ MPa, Gr 400 bars. Design a square tied column with ρ about 0.02.

PROB. 19.2. In the braced frame of Prob. 19.1 an interior column has M_2 as given there but the member is in reversed curvature with $M_1 = -0.4\,M_2$, likewise all from live load. Here $P_d = 450$ kN, $P_\ell = 800$ kN, $\psi = 1.0$, $\ell_u = 6.0$ m, $C_m = 0.6 + 0.4\,M_1/M_2 \gtrless 0.4$, $\beta_d = 0$, $k = 0.77$ (Table 18.1), $f_c' = 30$ MPa, Gr 400 bars. Design a square tied column with ρ about 0.02.

PROB. 19.3.   Same as Prob. 19.1 except frame *unbraced* making reverse curvature critical with $M_1 = -M_2$ where ψ leads to $k = 1.31$, $\delta = 1.45$ for entire story. (Designer must satisfy the average δ for the entire story in unbraced frame.)

PROB. 19.4.   Calculate the permissible load by the R method for the following columns. (Note Fig. B-6 in Appendix B.)
(*a*) Column of Prob. 18.2*b*.
(*b*) Same as (*a*) except triple the column length.

PROB. 19.5.   Calculate the permissible load by the R method for the following columns:
(*a*) Column of Prob. 18.1*a*.
(*b*) Same as (*a*) except with the length of column doubled.

PROB. 19.6.   Calculate the permissible load by the R method for:
(*a*) Column of Prob. 18.3*c* considering load as a short time load.
(*b*) Same as (*a*) except column length doubled; here 6.0 m.

PROB. 19.7.   For $f_c' = 30$ MPa, Gr 400 bars, $P_d = 450$ kN, $P_\ell = 900$ kN, $e = 125$ mm, $\ell_u = 6.0$ m, single curvature in braced frame, design a square tied column by the R method with ρ about 0.03. Consider this loading primarily short time.

PROB. 19.8.   For the data of Prob. 19.7 except the given loading now to be considered as *factored* loads and the frame *unbraced* with $\ell_u = 4.8$ m, $\psi_g = 0.90$, design the column by the R method.

20
FOOTINGS

20.1 TYPES OF FOOTINGS

Although underground conditions call for many variations in foundation design, the majority of building footings can be classified as one of the following types.

Bearing walls may be supported on a continuous strip of concrete called a wall footing, as shown in Fig. 20.10.

Isolated column footings under individual columns may be square (as shown in Fig. 20.1b), rectangular, or round in plan. Alternatively, the column foundation may be a drilled pier, that is, a shaft drilled into the ground, often flared at the bottom for greater bearing area, and finally filled with concrete.

When a single footing supports two or more column loads, as in Figs. 20.13 and 20.14, it is called a combined footing.

A cantilever footing, as in Fig. 20.16 also supports two or more columns. It is characterized by the fact that it is really two footings joined by a beam instead of by a bearing portion of the footing.

Under poor soil conditions it is sometimes desirable to support the entire structure on a single mat or slab. Such a foundation is often called a floating or raft foundation, more commonly a mat foundation.

Any of these footings may be directly supported on the soil or they may rest on piling.

20.2 TESTS ON FOOTINGS

Two noteworthy series of footing studies have influenced American practice, Talbot's tests[1] in 1907 and Richart's tests[2] in 1946. Moe's tests at the

558

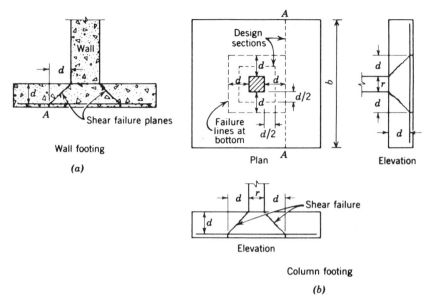

Wall footing

(a)

Design sections

Failure lines at bottom

Shear failure planes

Plan

Elevation

Shear failure

Elevation

Column footing

(b)

FIGURE 20.1 Diagonal tension failure in footings.

Portland Cement Association Development Laboratory have added more detail on the effect of openings near the column.

Shear failures never occur on vertical planes along the wall or around the column. For wall footings that fail in shear, a diagonal tension crack develops on an approximate 45° plane parallel to the wall, as shown at A in Fig. 20.1a. The shear causing this crack is that produced by the upward load to the left of A, that is, the load beyond a plane a distance d from the face of the wall. In an isolated square column footing a similar failure occurs, the column pushing ahead of it a truncated pyramid with an approximate 45° slope on all faces. This pyramid has a base width equal to the column width plus approximately twice the effective depth of the footing (Fig. 20.1b). The probable reason for this type of failure lies in the heavy compressive stresses between the diagonal cracks as the column load spreads out into the surrounding footing, with this compression accentuated by the upward soil reaction. In this zone the diagonal tension normally caused by shear stresses is somewhat counteracted or reduced by the vertical compressive stress. (Compare Fig. 5.3a and the related discussion in Sec. 5.4c). The 45° failure planes mark the approximate boundaries between which substantial vertical compression is developed.

In a column footing the initial diagonal cracks described above occur much before ultimate load and are not generally visible because they are in the interior of the concrete. The 1962 report[3] of the Joint ACI-ASCE

Committee on Shear and Diagonal Tension recognized that, *after* these shear failure planes first developed, the critical shear was largely carried by the flexural compression area above the cracks in what can best be described as the old idea of punching shear of this limited compression depth around the perimeter of the column. Failure might then be described as a two-step process, initial diagonal cracking extending a distance d from the wall or column and finally a shearing failure at the face of wall or column. Usually, a considerable portion of the total load resistance develops after the diagonal crack is formed. Rather than introduce two checks for the two kinds of behavior the Committee recommended that the shear be calculated on a pseudo-critical plane between the two, that is, at a distance $d/2$ from the column. Then as a safety check, which probably controls only on long, narrow footings, the shear strength as a one-way slab should be checked on sections all across the footing at a distance d from the face of column.

With both walls and columns of concrete, the critical section for moment and bar development is found at the face of the wall or column. In the wall footing the one-way cantilever moment due to forces beyond the critical section is uniformly distributed along the wall (Fig. 20.10); any 1 m wide strip can be used in this part of the design. In the column footing Richart found the distribution of this cantilever moment (in each direction) to be very nonuniform at working loads, being largest for strips passing under the column and least for strips near an edge. However, near ultimate loads the yielding of the steel in the central strips causes more moment to shift to the edge strips and moment failure does not occur until essentially all the steel has reached its yield point. Bond stress was found to be less critical than expected. In this connection, the 1977 Code requires only that the bar length beyond the face of column be a full development length (Chapter 7).

20.3 SQUARE COLUMN FOOTINGS—ANALYSIS

(a) Critical Stresses

Column footing must be checked or designed for six strength conditions:

1. Bearing (compression) from column on top of footing.
2. Dowels into the footing.
3. Strength of soil beneath footing, soil pressure q_s.
4. Shear strength.
5. Reinforcement provided.
6. Development length of bars.

(b) Bearing under Column

The bearing from the column on the footing is permitted to be larger than the value of the same concrete in the column, that is, lower strength concrete is permissible in the footing without lowering the column capacity or necessarily increasing the dowels provided. The usual permitted bearing strength (Code 10.16.1) of ($\phi \times 0.85\ f'_c A_1$) may be multiplied by $A_2/A_1 \gtrless 2$, where A_1 is the bearing area and A_2 is the lower area sketched in Fig. 20.2. Area A_2 is concentric to and geometrically similar to A_1 and is established by going down and out from the loaded area at a slope of 1 vertically and 2 horizontally* until one intersects the boundary closest to the column. This boundary fixes one side of A_2 and it is simple to construct the complete area on the plan view similar to A_1. From similar areas

$$A_2/A_1 = (x_2/x_1)^2 \qquad \sqrt{A_2/A_1} = x_2/x_1 \gtrless 2$$

The total bearing value of the area A_1 of a steel bearing plate or a concrete column becomes:

$$N_{nc} = 0.85\ f'_c A_1 \sqrt{A_2/A_1} = 0.85\ f'_c A_1 (x_2/x_1)$$

where f'_c is the footing concrete strength and A_1 the bearing area. For the concrete column, this part of the load is also limited to the capacity of the concrete in the column, that is, $N_{nc} = 0.85\ f'_c A_1$ where f'_c is the column concrete strength.

(c) Dowels into the Footing

Since the Code permits dowels into the footing designed for strength instead of requiring them to match the column steel, some savings in compression dowels are occasionally possible. However, any column bar that can be subject to tension must either be anchored into the footing or matched with dowels into the footing.

A small saving can occur if a column requires almost no steel to carry its load, but is reinforced with $\rho = 0.01$ to satisfy the Code minimum. The dowel minimum is now lowered to $\rho = 0.005$ of the column area with a minimum of four bars. Although not stated, the four bars should obviously be on four sides and probably at the column corners.

If the column concrete strength exceeds the footing concrete strength, the increased bearing strength could still mean that no *extra* dowels were required to transfer the difference in concrete strengths into the footing.

In designing compression dowels, one must consider both the footing below and the column above:

---

*This slope is not intended to represent the direction of the pressure, but rather to insure some confining concrete around the actual pressure distribution.

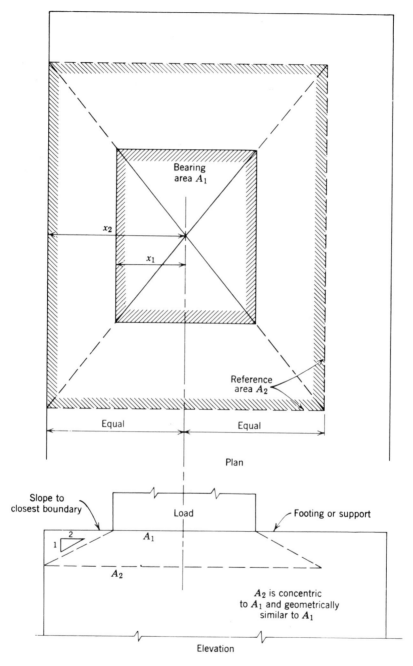

FIGURE 20.2 Bearing pressure limit is based on $\sqrt{A_2/A_1} = x_2/x_1 \gtrless 2$.

562

1. For the footing, dowels must carry the total load less the bearing value of the footing concrete. More exactly stated, the dowels should fit the demands of the column interaction chart for the design eccentricity and A_1 when f'_c is used as the footing value times the ratio $\sqrt{A_2/A_1} \lessgtr 2$.
2. For the column, the load, including moment from eccentricity at the base, must be carried by the concrete in the column assisted by *only* those bars which are extended or doweled into the footing.

The second rule is necessary because any bars neither extended nor doweled into the footing have zero load capacity at the point where they are stopped and very little capacity nearby; strength is always limited to the lower available development length on the two sides of any design section.

It is quite important to note that the second condition says nothing done in the footing itself lowers the demand for effective column reinforcement, that is *developed* column steel.

(d) Soil Pressure

The discussion of actual soil pressure under a footing is beyond the scope of this book. (Reference 6 is recommended.) This pressure commonly will be higher near the center of a column footing, but occasionally higher at the edges, depending on whether the soil is sandy or clayey.

The concentration of pressure directly under the load when a footing is on rock can be reduced only by a very stiff footing as discussed in Sec. 20.10. Although not necessary in simple cases, elastic analysis would be appropriate for foundations on rock and would be helpful in special cases of loading or footing shape. The same approach is usable in other stiff and well-consolidated foundation materials if there is enough uniformity to permit a meaningful determination of the foundation modulus.

Elastic analyses, however, are not a general answer to the pressure distribution problem. If measurable foundation settlement is expected, the soil is only momentarily elastic and the initial response to the load will be only a transient response. Time (and settlement) will smooth out this initial response, lowering peak pressures and building up low pressures. In such cases, the author considers elastic analyses of little value. More uniform (oversimplified) pressure distributions appear more appropriate as a design approach where significant settlement is expected.

The soil pressure is usually considered as uniform for a centered loading and trapezoidal or triangular for eccentric loadings. The assumed uniform pressure is usually on the safe side in the calculation of internal moments and shears in isolated footings, but a different situation in the middle of a combined footing is discussed in the last paragraph of Sec. 20.12.

The allowable soil pressure is also a matter of soil mechanics and cannot be discussed here. However, this again brings up the problem mentioned in connection with retaining wall design (Sec. 9.5f). Allowable soil pressure is usually given in terms of the service load permissible while footing design is by strength (at ultimate). The designer has at least two equivalent) ways to keep the relationships proper in the case of isolated column footings.

The Code (15.2.2) specifies, for the purpose of establishing the footing size, that the (unfactored) external forces and moments be used to match the allowable soil pressure; but also that soil pressures based on design (factored) loads be used to calculate design moment and shear.

The author suggests an alternate giving the same answers for concentric loads, namely, increasing the allowable soil pressure in the ratio of the ultimate U values to the service loads involved, that is, multiplying the allowable soil pressure by

$$\frac{1.4D + 1.7L}{D + L}, \qquad \frac{1.4D + 1.7L + 1.7W}{D + L + W}, \text{ etc.}$$

This method will be used here because the soil value thus computed is available for direct use in computing the strength demands for shears and moments in the footings.

Because the ratio of dead to live load varies from column to column, there is no unique relationship between soil pressure at service loading and soil pressure under ultimate loading. It appears to the author that design would be better served by the general use of an allowable or dependable *ultimate* soil pressure. It is noted, however, in settlement situations that a permissible pressure at *usual* loads might also be significant and possibly governing, just as deflection can govern over strength in a flexural member.

The footing base area must be adequate to care for the column load, footing weight, and any overburden weight, all within the permissible soil pressure, often assumed uniformly distributed* under the footing. Sometimes this over-all pressure is called the gross soil pressure to distinguish it from net soil pressure, which is a convenient *design* concept. Since the weight of footing and overburden is usually nearly uniform over the footing area, the design moment or shear on any section will be the result of the gross soil pressure upward less the design value of the footing and other overburden weight downward. It is convenient to think in terms of the resultant load or "net soil pressure" which is the difference between these upward and downward unit pressures. For vertical column load

---

*But variation caused by any moment loading must also be considered.

without moment net pressure is usually found most simply by dividing the design column load alone by the footing area.

(e) Shear Strength

Shear quite frequently controls the footing thickness.

The unit shear as a measure of the diagonal tension is first calculated for the shear caused by loads outside the inner dashed square in Fig. 20.1b and is resisted by a width equal to the perimeter of that square, $b_o = 4(r + d)$. The use of d_v as the average depth to the two layers of steel appears justified. There is two-way bending here with permissible unit shear of $\frac{1}{3}\sqrt{f'_c}$, or $V_{nc} = \frac{1}{3}\sqrt{f'_c}b_od$.*

The diagonal tension on section AA at a distance d must also be checked, using permissible one-way slab shear of $\frac{1}{6}\sqrt{f'_c}$, or $V_{nc} = \frac{1}{6}\sqrt{f'_c}bd$. This usually governs only on long, narrow footings.

(f) Reinforcement Provided

A proper reinforcement design always gives an underreinforced member, automatically avoiding compressive failure.

The moment is critical on the section at each face of column and calls for two-way bottom bars. The moment in each direction is a matter of statics and can be varied only by a change in the distribution of soil pressure. The separate steel in each direction should be adequate for the moment in that direction; excess steel in the y-direction cannot make up for a steel shortage in the x-direction. Hence, it appears desirable that the smaller depth to the upper steel layer in the bottom of the footing be used for calculation of the moment resistance in a square footing. In a square footing the use of equal steel areas in the two directions helps to avoid field errors in placement. Economy will be achieved by using a z value corresponding to the actual percentage of steel used, this value usually being larger than $0.9d$.

The minimum reinforcement requirement of $1.4/f_y$, unless A_s is at least 4/3 that required by analysis, should be applied to footings. Although one could argue legally that a footing is a slab, and thereby exempt from this requirement, the combination of high shear and low ρ is not good.

(g) Development Length of Bars

In a single column footing, the available development length is from the face of column to the edge of footing (less the end cover). No bar requiring

---

*Note that Code 11.11.2, for rectangular areas, uses $(2 + 4/\beta_c) \gtrless 4$, where β_c is ratio of long to short side of concentrated load or reaction area. In Canadian units, $0.17(1 + 2\beta_c)$ or 0.33, whichever is smaller, using β_c as the ratio of short to long, the *inverse* of ACI usage.

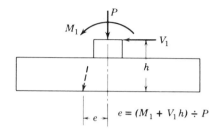

FIGURE 20.3 Equivalent eccentricity of load.

a longer development length can be used, unless it is understressed at the maximum moment section or unless an end hook makes up the deficiency.

20.4 ANALYSIS FOR MOMENT LOADING

Footings must often carry moment from a column or wall. Combined with direct load this gives the equivalent of an eccentric load, as in Fig. 20.3. If the moment were constant it would be desirable to put the center of the footing under this eccentric load. For the retaining wall (Sec. 9.5g) the adjustment of the base to make the resultant load fall between the center and third point was discussed. Usually the varying nature of the moment makes it impossible to avoid all eccentricity on the footing and a trapezoidal soil pressure results. This modifies the magnitude of the design moment and shears, but not the general design procedure. There might be doubts as to how to handle the two-way slab shear. For important eccentricities the method indicated in Sec. 6.4 is available. For minor eccentricities the author would simply apply the shear developed as in Fig. 20.5a from the heavier loaded half of the footing onto *one* half of the perimeter around the column at the distance $d/2$.

Although office practice in footing design has often neglected nominal column eccentricities of loading, considering them a minor matter, other eccentricities are important and should more commonly be included.* Columns designed for significant moment *at their base* need the resisting moment or eccentricity developed by the footing reactions.

20.5 SQUARE FOOTING—ANALYSIS EXAMPLE

Check the footing of Fig. 20.4, assuming $f'_c = 20$ MPa for both column and footing, Grade 300 steel, and an allowable soil pressure of 150 kPa. The column dead load is 0.85 MN and the live load 1.3 MN. Assume no overburden of soil. The weight of reinforced concrete is 24 kN/m³.

---

*But see Code 15.2.4 and its footnote.

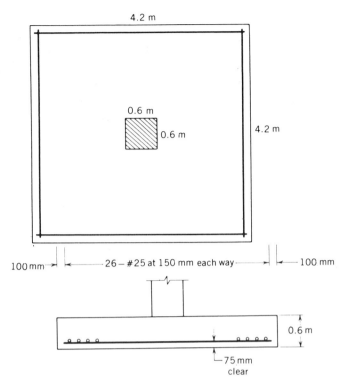

FIGURE 20.4   Square column footing.

Solution

(a) Since the column is of the same strength concrete as the footing, the bearing on top of the footing will be adequate. Assuming the column fully stressed, dowels should be provided to match the column steel, extended vertically into the footing a compression development length l_d.

No moment is specified on the footing. Assume concentric loading.

$$\text{Factored column load} = 1.4 \times 0.85 + 1.7 \times 1.3 = 3.4 \text{ MN} = 3\,400 \text{ kN}$$

$$\text{Average factored soil pressure} = q_{net} = 3\,400/4.2^2 = 193 \text{ kPa}$$

Allowable soil pressure	=	150 kPa
Subtract weight of footing $= 0.6 \times 24 =$		15
Allowable $q_{net} =$		135 kPa (at service load)
Equivalent allowable at ultimate =		135 (factored load)/(dead + live load)
	=	$135 \times 3.4/(0.85 + 1.3)$
	=	213 kPa > 193 kPa actual **O.K.**

In a square footing the use of average d is recommended only for shear calculations around the column, with the depth to the upper steel layer for moment

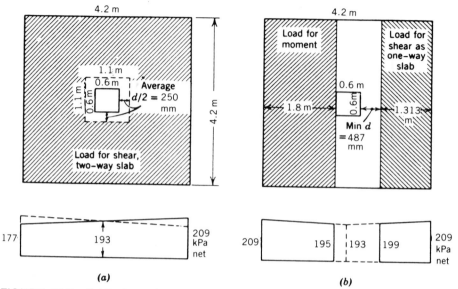

FIGURE 20.5 Critical sections on square footing and, for solution (*b*) only, net factored load pressures under eccentric loading. (*a*) Shear as two-way slab. (*b*) Moment and shear as one-way slab.

calculations and for shear all across the footing.

$$\text{Average } d_v = 600 - 75 - 25 \text{ (bar diam.)} = 500 \text{ mm}$$

Shear will first be checked for two-way action all round the column at a distance $d/2$ outside the column as indicated in Fig. 20.5*a*, with the net soil pressure used as the applied load. The net soil pressure omits that part of the reaction created by the footing weight because everywhere in moment and shear evaluation the downward weight of the footing balances this portion of the reaction.

$$V_u = 193(4.2^2 - 1.1^2) = 3\ 170 \text{ kN} = 3.17 \text{ MN}$$
$$\text{Reqd. } V_n = V_u/\phi = 3.17/0.85 = 3.73 \text{ MN}$$
$$v = V_n/b_o d = 3.73/(4 \times 1.1 \times 0.5) = 1.69 \text{ MPa}$$
$$\text{Allowable } v = \tfrac{1}{3}\sqrt{f_c'} = 1.5 \text{ MPa} < 1.69 \text{ MPa} \qquad \textbf{N.G.} \text{ (}v \text{ not O.K.)}$$

Although it will normally govern only on long narrow footings, shear will also be checked on the full width as a one-way slab (Fig. 20.5*b*).

$$\text{Min. } d_v = 600 - 75 - 1.5 \times 25 = 487 \text{ mm}$$
$$V_u = 193 \times 1.313 \times 4.2 = 1\ 064 \text{ kN} = 1.064 \text{ MN} \quad V_n = V_u/0.85 = 1.252 \text{ MN}$$
$$v = \frac{V_n}{bd} = \frac{1.252}{4.2 \times 0.487} = 0.61 \text{ MPa}$$
$$\text{Allowable } v \text{ on one-way slab} = \tfrac{1}{6}\sqrt{20} = 0.75 \text{ MPa} > 0.61 \text{ MPa} \quad \textbf{O.K.}$$

Moment must be calculated at the face of the column (**Fig. 20.5b**).

$M_u = 4.2 \times 1.8 \times 193 \times \frac{1}{2} \times 1.8 = 1\,313$ kN $\cdot$ m $= 1.313$ MN, $M_n = M_u/0.90 = 1.459$ MN

$d_M = 600 - 75 - 1.5 \times 25 = 487$ mm

$A_s = 26 \times 500 = 13\,000$ mm²

$\rho = \dfrac{13\,000}{4\,200 \times 487} = 0.0635 \quad \begin{array}{l} < \text{allowable Table 3.1}\quad\textbf{O.K.} \\ > \text{min. of } 1.4/f_y \\ = 0.005 \quad \textbf{O.K.} \end{array}$

$a = A_s f_y/(0.85\, f_c' b) = 13\,000 \times 10^{-3} \times 300/0.85 \times 20 \times 4.2 = 0.054$ m $= 54$ mm

$z = d - a/2 = 487 - 27 = 460$ mm

Reqd. $A_s = 1.459/300 \times 0.460 = 10\,570 \times 10^{-6}$ m² versus $13\,000 \times 10^{-6}$ m²

The critical section for bar development is at the maximum moment section, where there is available for $\ell_d = 1.8$ m less end cover, say, 1.7 m.

$$\text{Reqd. } \ell_d = 0.019\, A_b f_y/\sqrt{f_c'} = 0.018 \times 500 \times 300/\sqrt{20} = 604 \text{ mm}$$

A much larger bar could be used, but this spacing is quite reasonable.
Conclusions: The footing must be deepened for shear. The A_s is about 15% more than needed for the present depth. A larger bar with wider spacing could be used. (These conclusions should be compared with those of the following solution.)

(b) *Analysis with moment loading*. Assume the design moment from the column is equivalent to 200 kN $\cdot$ m at the soil bearing level.

$$e = 200/3\,400 = 0.059 \text{ m} = 59 \text{ mm}$$

Dowel requirements are assumed simply to match column bars. Average design net soil pressure remains unchanged at 193 kPa and e/b is $59/4\,200 = 0.0140$.

$$q_{net} = 193(1 \pm 6 \times e/b)^* = 193(1 \pm 6 \times 0.014) = 209 \text{ or } 177 \text{ kPa}$$

This is in addition to the footing weight of 15 kPa that causes neither moment nor shear.

At service load assume maximum net pressure of 209 kPa decreases in ratio of service load to factored load $= 2.15/3.40 = 0.63$, to 132 kPa. With footing weight of 15 kPa the gross pressure is 147 kPa, or essentially the allowable 147 kPa.

For the one-way slab shear (**Fig. 20.5b**) the average net pressure is 150 kPa.

$$\tfrac{1}{2}(209 + 199) = 204 \text{ kPa}$$

$V_n = V_u/\phi = 204 \times 4.2 \times 1.313/0.85 = 1\,323$ kN $= 1.323$ MN

$v_n = V_n/bd = 1.323/4.2 \times 0.487 = 0.65$ MPa < 0.75 MPa allowable   **O.K.**

For the two-way slab shear, since the eccentricity of loading is small, compute V_n on half the loaded area (heavy side) using the trapezoidal loading of **Fig. 20.5.a**.

$V_n = [\tfrac{1}{2} \times (209 + 193) \times 4.2 \times 2.1 - (\text{estimated})\ 197 \times 1.1 = 0.55]/0.85$

$ = (1\,773 - 119)/0.85 = 1\,945$ kN $= 1.945$ MN

$v_n = V_n/bd = 1.945/2 \times 1.1 \times 0.500 = 1.768$ MPa > 1.5 MPa   **N.G.**

---

*e/h would be better except that h has so many other meanings nearby.

The moment is calculated using two triangles for the trapezoid.

$$M_n = M_u/\phi = (\tfrac{1}{2} \times 209 \times 1.8 \times 1.2 \times 4.2 + \tfrac{1}{2} \times 195 \times 1.8 \times 0.6 \times 4.2)/0.90$$

$$= 1\,545 \text{ kN} \cdot \text{m} = 1.545 \text{ MN/m}$$

Reqd. $A_s = 1.545/300 \times 0.460 = 11\,196 \times 10^{-6} \text{ m}^2 = 11\,196 \text{ mm}^2 < 13\,000 \text{ mm}^2$ **O.K.**

Bar development length is unchanged from (a).
Conclusions: Soil pressure is increased about 5%, shear by 2%, and moment by 6%. Except for the original low soil pressure, a larger footing would be needed for soil pressure; a deeper footing is required for shear.

20.6 SQUARE COLUMN FOOTING—DESIGN

Design a square column footing using 20 MPa concrete, Grade 400 reinforcement, and allowable soil pressure of 170 kPa (at service load) to carry a 460 mm square spiral column made of 35 MPa concrete and carrying a dead load of 890 kN and live load of 1560 kN. The weight of reinforced concrete is 24 kN/m³.

Solution

The nominal design load assigned to the concrete in the spiral column is based on 0.85 f'_c, or an ultimate based on $\phi(0.85\,f'_c) = 0.75 \times 0.85 \times 35 = 22.3$ MPa. Since the footing will be large, $\sqrt{A_2/A_1}$ will have (in Fig. 20.2) an $x_2/x_1 \geqslant 2^*$ and the allowable ultimate from the column on this concrete will be $2\phi \times 0.85 \times 20 = 2 \times 0.70 \times 17 = 23.8$ MPa. This is more than the ultimate from the column. (**O.K.**)

Dowels that match the column bars will be adequate. Required embedment length for compression bars (Sec. 7.16) = $\ell_d = 0.24\,f_y d_b/\sqrt{f'_c} = 0.24\,d_b \times 400/\sqrt{20} = 21.5\,d_b$, where ℓ_d and d_b are in mm. This ℓ_d, as a vertical embedment into the footing, fixes a minimum d in mm for the footing expressed as a multiple of d_b, the column bar diameter. (An alternate sometimes available is more dowels of smaller diameter and smaller ℓ_d).

$$\text{Column } P_u = 1.4 \times 890 + 1.7 \times 1\,560 = 3\,900 \text{ kN} = 3.900 \text{ MN}$$

Assume footing 750 mm thick exerting a gravity force of $0.75 \times 24 = 18$ kPa.† Then in terms of ultimate loads:

---

*$x_2/x_1 = 2$ where d of footing is at least half the width of the column (because of the 1:2 slope involved in Fig. 20.2).

†It might appear at first that a 0.9 load factor should be used for the footing weight in establishing q_{net}, but the opposite conclusion is reached here. Whatever net soil pressure the column load produces is increased by the local footing weight. The weight of the footing and its reaction are two forces in equilibrium locally as well as overall; they therefore produce no moments or shears. If the designer estimates the footing weight incorrectly, this results in too much, or too little, *gross* reaction pressure computed; but, for the area of footing used, the moments and shears are unchanged by a bad estimate. It is thus appropriate to use the ratio of factored *column* load to service *column* load to establish the ratio of factored q_{net} to service load q_{net} for use in moment and shear calculations.

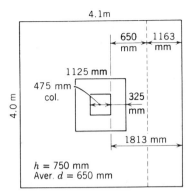

Figure 20.6 Sections for shear in square footing design.

Allowable ultimate $q_{net} = (170 - 18) \times 3\,900/(890 + 1\,560) = 242$ kPa

Reqd. $A = P_u/q_{net} = 3\,900/242 = 16.12$ m²

Try 4.1 m square footing ($A = 16.81$ m²). Actual $q_{net} = 3\,900/16.81 = 232$ kPa

Since it is not typical that one can fit the required A exactly, it is better to use the *actual* q_{net} in M and V calculations. In this example the close fitting of the area makes this precaution appear useless.

For the assumed h of 750 mm and 75 mm clear cover (required for casting directly on the ground), the average $d_v = 750 - 75 - d_b =$ say, 650 mm for #25 bars. The critical shear for the two-way slab action is then at $\frac{1}{2}d = \frac{1}{2} \times 650 = 325$ mm from the face of columns, as shown in Fig. 20.6, with a face width of $475 + 2 \times 325 = 1\,125$ mm.

$$V_u = 232\,(4.1^2 - 1.125^2) = 3\,600 \text{ kN} = 3.6 \text{ MN}$$

Reqd. $V_n = V_u/\phi = 3.60/0.85 = 4.24$ MN

Allowable $v_c = \frac{1}{3}\sqrt{f_c'} = \frac{1}{3}\sqrt{20} = 1.5$ MPa, $b_o = 4 \times 1.125$ m

Reqd. $d_v = V_n/b_o v_c = 4.24/4 \times 1.125 \times 1.5 = 0.628$ m

Although shear all across the footing at distance d from the column rarely, if ever, governs, check the section shown in Fig. 20.6 against the allowable v_c for one-way action, limited to $\frac{1}{6}\sqrt{f_c} = \frac{1}{6}\sqrt{20} = 0.75$ MPa.

$V_u = 242 \times 1.163 = 281$ kN/m width   Reqd. $V_n = V_u/\phi = 281/0.85 = 331$ kN/m

$v_c = V_n/bd = 331 \times 10^{-3}/1 \times 0.65 = 0.509$ MPa < 0.75 **O.K.**

USE footing 4.1 m square, $h = 750$ mm

For moment at face of column the *lesser* depth controls, not the average.

$$d_m = 750 - 75 - 1.5\,d_b = 637 \text{ mm for } d_b = 25 \text{ mm}$$

On full width, $M_u = \frac{1}{2} \times 242 \times 10^{-3} \times 4.1 \times 1.813^2 = 1.631$ MN · m

$$M_n = M_u/0.90 = 1.812 \text{ MN} \cdot \text{m}$$

Assume $z = 0.9\,d = 0.9 \times 637 = 573$ mm $= 0.573$ m

Reqd. $A_s = M_n/f_y z = 1.812/400 \times 0.573 = 7\,905 \times 10^{-6}$ m$^2 = 7\,905$ mm^2

$a = A_s f_y/0.85\,f_c'\,b = 7\,905 \times 10^{-6} \times 400/0.85 \times 20 \times 4.1 = 0.45$ m $= 45$ mm

$z = 637 - 45 = 592$ mm

Reqd. $A_s = 7\,905$ (above) $\times 573/592 = 7\,651$ mm^2

Check on minimum $\rho = 1.4/f_y$ (Code 10.5.1)

$$= 1.4/400 = 0.0035$$

Minimum $A_s = 0.0035 \times 4.1 \times 0.637$

$$= 9\,141 \times 10^{-6} \text{ mm}^2 = 9\,141 \text{ mm}^2 > 7\,651 \text{ mm}^2$$

USE $19 - 25$ (9 500 mm^2) *each way*, spacing about 200 mm centers.

The 19 bars control the spacing, assumed uniform, although precision is not needed.

Bar development length, $\ell_d = 0.019\,A_b f_y/\sqrt{f_c'} = 0.019 \times 500 \times 400 \times \sqrt{20} = 850$ mm

Available $\ell_d = 1\,813$ mm minus end cover.   **O.K.**

20.7 RECTANGULAR COLUMN FOOTINGS—ANALYSIS

Analyze the rectangular footing of Fig. 20.7 assuming $f_c' = 20$ MPa, Grade 400 steel, and an allowable soil pressure at service load of 150 kPa. The column service load is 0.9 MN dead and 1.4 MN live load without any moment and there is no overburden.

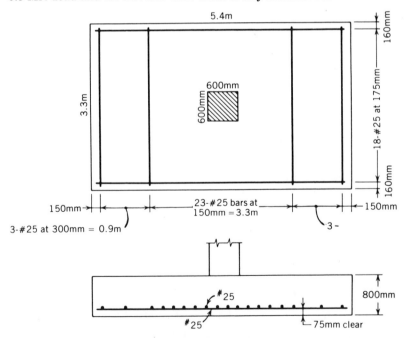

FIGURE 20.7   Rectangular column footing.

Solution

Design $P_u = 1.4 \times 0.9 + 1.7 \times 1.4 = 3.64$ MN $= 3\,640$ kN

Net soil pressure at ultimate $= \dfrac{3\,640}{5.4 \times 3.3} = 204$ kPa

p_{net} at service load $= 204 \times 2.3/3.64 = 129$ kPa

Footing weight $= 24 \times 0.80 \qquad\qquad = \underline{19}$

$\qquad\qquad\qquad$ Service load $p_{gross} = \overline{148}$ kPa < 150 kPa   **O.K.**

No information on dowels is given, but these are obviously necessary. For the long projection, $d = 800 - 75 - 13 = 712$ mm. For the short projection, $d = 800 - 75 - 25 - 13 = 687$ mm. Diagonal tension around the column will be considered critical on the average depth of 700 mm at a distance $d/2$ beyond the column (Fig. 20.8) considering two-way slab action.

$b_o = 4(0.6 + 0.7) = 5.2$ m

$V_u = 204(5.4 \times 3.3 - 1.3^2) = 3\,290$ kN $= 3.29$ MN

$V_n = V_u/\phi = 3.29/0.85 = 3.87$ MN

Permissible $v = \frac{1}{3}\sqrt{f_c'} = 1.50$ MPa

Reqd. $d_v = V_n/vb = 3.87/1.50 \times 5.2 = 0.496$ m $= 496$ mm $\ll 700$ mm   **O.K.**

Check one-way slab shear on long projection at distance $d = 0.712$ m from column, as shown in Fig. 20.8.

$V_u = 204 \times 3.3 \times 1.688 = 1\,136$ kN $= 1.136$ MN

$V_n = 1.138/0.85 = 1.337$ MN

Permissible $v = \frac{1}{6}\sqrt{f_c'} = 0.75$ MPa

Reqd. $d_v = 1.337/3.3 \times 0.75 = 0.540$ m $\ll 0.712$ m   **O.K.**

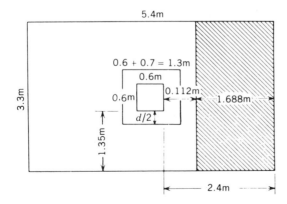

FIGURE 20.8 Diagonal tension analysis for rectangular footing.

The footing could be thinner insofar as shear is concerned. In the long direction
$M_{uL} = 204 \times 3.3 \times 2.4 \times \frac{1}{2} \times 2.4 = 1\,939$ kN · m $= 1.939$ MN · m

$M_{nL} = M_{uL}/\phi = 1.939/0.9 = 2.154$ MN · m

Assume $z = 0.92 \times 0.712 = 0.655$ m

Reqd. $A_s = \dfrac{2.154}{400 \times 0.655} = 8\,222 \times 10^{-6}$ m$^2 = 8\,222$ mm^2 vs. $18 \times 500^2 = 9\,000$ mm^2

This is O.K. without a check on z.

Minimum A_s for flexure* $= (1.4/f_y)bd$
$$= 3.3 \times 0.712 \times 1.4/400$$
$$= 8\,224 \times 10^{-6} \text{ m}^2 = 8\,224 \text{ mm}^2 < 9\,000 \text{ mm}^2 \quad \textbf{O.K.}$$

In the short direction, $M_{ns} = 204 \times 5.4 \times 1.35 \times \frac{1}{2} \times 1.35/0.9 = 1\,115$ kN · m $= 1.115$ MN · m

Assume $z = 0.94 \times 687 = 646$ mm

Reqd. $A_s = \dfrac{1.115 \times 10^6}{400 \times 0.646} = 4\,315$ mm$^2 < 29 \times 500 = 14\,500$ mm^2

Min. $A_s = (1.4/f_y)\,bd = 5.4 \times 0.687 \times 1.4 \times 10^6/400 = 12\,984$ mm^2 **O.K.**

For temperature and shrinkage, min. $A_s = 0.0018\,bh = 0.0018 \times 5.4 \times 0.8 = 7\,776$ mm^2 **O.K.**

Code 15.4.4 requires much of the transverse reinforcement to be concentrated within the central width equal to the short side of the footing, the specified proportion being $2/(1 + \beta)$ where β is the ratio of the sides of the footing, $5.4/3.3 = 1.64$ in this case:

$2/(1 + 1.64) = 0.76$ vs. 23 bars/29 bars $= 0.79$ in the 3.3 m width. This proportion is satisfactory.

For bar development the short projection appears the more critical, but probably adequate. $\quad \ell_d = 0.019\,A_b f_y/\sqrt{f'_c} = 0.019 \times 500 \times 400/\sqrt{20} = 850$ mm

The available is $\frac{1}{2}(3\,300 - 600) - 50$(end cover) $= 1\,300$ mm.

20.8 WALL FOOTINGS

(a) Basic Concepts

The uniform nature of a wall loading leads to the use of a typical 1-m width as a design strip, as in Fig. 20.9. For heavy walls, the design is quite similar to that of column footings, but simpler because the bending is in only one direction. For light walls, the design is frequently based on rather arbitrary minimum sizes. Here plain concrete is often the appropriate material.

When the wall is of reinforced concrete, the critical moment section is taken at the face of the wall (Code 15.4.2). With masonry walls a different critical section is specified, as illustrated in the following examples.

*Minimum A_s will always be much under $0.75\,A_{sb}$ (balanced).

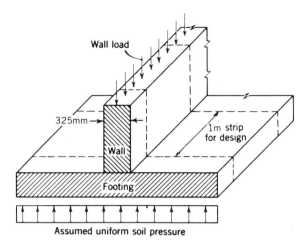

FIGURE 20.9 Part of wall footing, with heavy loading.

(b) Examples of Design

(1) Design a wall footing for a 325 mm brick wall carrying a service load per running meter of 80 kN dead load plus 220 kN live load using $f'_c = 20$ MPa, Grade 400 steel, and an allowable soil pressure at service load of 210 kPa, negligible overburden and moment.

Solution

Assume a 300-mm-thick footing with a footing weight of $0.3 \times 24 = 8$ kPa. Service $q_{net} = 210 - 8 = 202$ kPa

Ultimate applied load = $1.4 \times 80 + 1.7 \times 220 = 486$ kN/m of wall

At ultimate, $q_{net} = 202 \times 486/(80 + 220) = 327$ kPa allowable

Footing width = $486/327 = 1.486$ m

USE 1.525 m, giving a 600 mm cantilever on each side of the wall.

$q_{net} = 486/1.525 = 319$ kPa **O.K.**

If d is assumed 200 mm, the critical section for diagonal tension (Fig. 20.10a) lies $600 - 200 = 400$ mm from the edge.

$$V_u = 319 \times 0.4 = 127.6 \text{ kN/m}$$

$$V = 127.6/0.85 = 150.0 \text{ kN/m} = 0.15 \text{ MN/m}$$

Reqd. $d_v = V/bv = 0.15/1 \times \frac{1}{6}\sqrt{20} = 0.201$ m = 201 mm

Required $h = 201 + 75$ cover + $d_b/2 = 286$ mm for #20 bars.

USE $h = 300$ mm, $d = 215$ if #20 bars are used

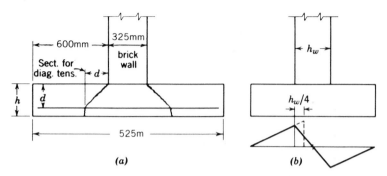

FIGURE 20.10   Reinforced concrete wall footing, design. (*a*) Critical shear section. (*b*) Shear diagram and equivalent cantilever length.

The footing weight changes so little it will be neglected. For a brick wall the critical section for moment lies at the quarter-point of the wall thickness, that is, in this case $\frac{1}{4} \times 325 = 81$ mm inside the face of the wall. The use of this equivalent cantilever length is explained by Fig. 20.10*b*. There the shear diagram indicates that the maximum moment is under the middle of the wall, but if the cantilever projection were extended back to the quarter point of the wall thickness, as shown dotted, the area under the shear diagram would be nearly the same.

$$M_u = 319 \times \tfrac{1}{2}(0.600 + 0.081)^2 = 74.0 \text{ kN} \cdot \text{m/m strip}$$
$$M_n = M_u/\phi = 74.0/0.90 = 82.2 \text{ kN/m}$$

Assume $z = 0.94d = 0.94 \times 215 = 202$ mm

$$A_s = M_n/(f_y z) = 82.2 \times 10^3/400 \times 0.202$$
$$= 1\,017 \text{ mm}^2/\text{m}$$
$$a = 1\,017 \times 400/0.85 \times 20 \times 1\,000 = 24 \text{ mm}$$
$$z = 215 - \tfrac{1}{2} \times 24 = 203 \text{ mm} \ldots \text{estimate was good}$$

Min. $A_s = 1\,000 \times 202 \times 1.4/400 = 707$ mm²/m < 698 mm²/m

#15 at 175 mm = 1 143 mm²

For #15, $\ell_d = 0.019 \times 200 \times 400/\sqrt{20} = 340$ mm

The projection of 600 mm beyond the wall is ample for ℓ_d plus end cover. It should be noted that the full ℓ_d is needed only from the point of maximum moment, but since the moment increases more slowly under the wall, the really critical ℓ_d might be for the smaller f_s at face of wall. Here it was obviously on the safe side to compare the full ℓ_d with the available projection in the footing.

USE #15 at 175 mm (with 4-#15 longitudinal bars as spacers)

Minimum for temperature and shrinkage reinforcement

$A_s = 0.0018 \, bh = 0.0018 \times 1\,525 \times 300 = 823$ mm²/m   **O.K.**

(2) Redesign the wall footing of (1) as a plain concrete footing in accordance with Code 15.11.*

Solution

Ultimate applied load = 486 kN/m, as before

Assume the footing is 0.7 m thick, so that its weight is $0.7 \times 24 = 16.8$ kPa

Service load $q_{net} = 210 - 16.8 = 193$ kPa

At ultimate $q_{net} = 193 \times 486/300 = 313$ kPa allowable

Footing width = 486/313 = 1.553 m, Say 1.575 m

$q_{net} = 486/1.575 = 309$ kPa

Moment is taken at the quarter-point of the wall, Fig. 20.11.

$M_u = \frac{1}{2} \times 309 \times 0.706^2 = 77$ kN · m/m

The allowable tension in such footings is given in Code 15.11.1 as $0.4\phi \sqrt{f'_c}$ with $\phi = 0.65$, or $0.26\sqrt{20} = 1.16$ MPa, to be used with M_u (not M_n).

$$77 \times 10^{-3} = f_t bh^2/6 = 1.16 \times 1 \times h^2/6$$
$$h = 0.631 \text{ m} = 631 \text{ mm}$$

The bottom 25 mm of concrete placed against the ground should usually not be considered for strength. Hence a 700 mm overall depth is indicated.

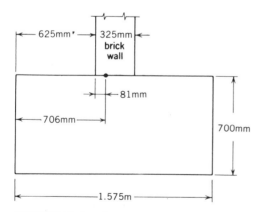

FIGURE 20.11 Plain concrete wall footing, design.

---

*It is usually not feasible to *combine* plain concrete concepts with reinforced concrete concepts. For example, a designer may see in a tower leg footing the possibility of uplift. If tempted to compute flexure from uplift on a plain concrete footing (having cared for positive load flexure with reinforcement), the designer should note that he or she has probably assumed tension cracks in the lower 80% of the footing and the gross area is *not* available for the reversed bending.

USE $h = 700$ mm overall, 675 mm effective

Diagonal tension is rarely significant in plain concrete footings. If d is considered 675 mm, the critical section for shear will fall 675 mm from the face of the wall or outside the footing. Unit shear in this footing is obviously lower than for the reinforced footing of Fig. 20.10. This more than offsets the fact that shear for the plain concrete must be calculated from $1.5\,V/bh$, as a homogeneous section. Shear at the *face* of wall cannot be critical.

(3) Design a wall footing for a 325 mm brick wall carrying an ultimate load of 70 kN/m using $f'_c = 20$ MPa and an allowable soil pressure at *ultimate* load of 240 kPa.

Solution

By inspection, soil pressure would call for a footing width less than the thickness of the wall. A minimum projection of 25 to 100 mm is generally used on walls.

USE footing width $= 375$ mm $h = 200$ mm

Code 15.11.4 calls for the minimum thickness of 200 mm for plain concrete, which will be used. The moment will be calculated (at the quarter-point of wall thickness), although one senses that it will be negligible.

Actual $q_{net} = 70/0.375 = 187$ kPa

$M_u = \frac{1}{2} \times 187 \times 0.106^2 = 1.05$ kN $\cdot$ m/m

This is of no significance, even on 175 mm of plain concrete, giving a calculated stress of around 0.2 MPa.

20.9 FOOTINGS ON ROCK

Very often only a nominal footing is required on rock since good rock is usually at least as strong as concrete and, moreover, is usually loaded on only a small part of its surface. Where a significant footing is required, as under a heavy steel column, deformation and relative stiffness become more significant than calculated unit stresses. A reinforced concrete footing cannot function normally unless the steel elongates and the cantilever elements deflect. Since the underlying rock is stiffer against compressive deformation than the reinforced concrete footing is against bending deflection, such a footing cannot effectively distribute the load laterally to any significant extent. A reinforced concrete footing on rock is thus of little value. The designer must depend more on the transfer of stress through shear with its smaller deformations. This calls for a deep footing and suggests that plain concrete may be just as effective as reinforced. It should be noted that, on the usual *rough* rock surface, friction (and interlock) forces are ample to prevent any significant stretching of the bottom of the footing. A footing on rock requires a large depth to give it shear stiffness, usually more depth than the length of its projection. The tension developed on the bottom face of the concrete is largely a function of the rock quality below, its surface roughness (interlock), and its lack of volume change. Under ideal rock conditions, tension from flexure can be neglected; on a poor rock, the concrete should be designed to resist the entire moment as for a footing on soil.

20.10 BALANCING FOOTING PRESSURES

Many years ago it was commonly accepted that footings would undergo essentially uniform settlements if they developed the same soil loadings per unit area under the usual loads on the structure. Such footings were said to have balanced footing areas.

Modern soil mechanics has greatly restricted the field of usefulness of this concept. It has shown in most practical cases that uniform soil pressure does not lead to uniform settlement. Proportioning for balanced footing pressures usually results in an increased size for outside wall footings. In a heavy structure on a thick stratum of material that consolidates slowly under additional load, such an increase in exterior footing size will actually *increase* the final differential settlement. In many areas, such as in the Southwest, foundation movement on ordinary residence and commercial structures is as likely to be upward due to expansive soils as downward due to applied loads. The engineer who uses the balanced footing areas should be certain that his special foundation situation justifies this approach. Frequently, it increases rather than reduces differential settlements.

For a balanced footing design, the term "usual load" will be applied to the average load or the load most commonly on the structure. This might vary from dead load alone for a church or a school, up to nearly full live load plus dead load for certain types of warehouses.

If all footings were proportioned for the usual loads, at full allowable soil pressure, the footing that cared for the greatest percentage of live load (compared to dead load) would most overstress the soil when full live load was acting. This gives the clue for balanced footing area design. That footing having the largest ratio of live load to dead load is designed first by the usual procedures, using maximum load and the full allowable soil pressure. The usual soil pressure under this footing is next established for the usual load, most frequently something between dead load alone and dead load plus half live load. The bearing areas of all other footings are then chosen to give this same usual soil pressure under the usual footing loads. (This will automatically give them less than the maximum permissible soil pressure under the maximum live load.) The individual footings are then designed for thickness and steel using these balanced areas and the full design live load.

Example

Balance the areas for footings given in Table 20.1 for dead load plus one-fourth live load if the allowable net soil pressure is 150 kPa at service load.

TABLE 20.1

Footing	Dead Load	Live Load
A	0.5 MN	1.0 MN
B	0.4	1.0
C	0.2	0.8
D	0.5	0.8

TABLE 20.2

Footing	Dead Load	Live Load	Usual Load	Usual q	Area	Max q
A	0.5 MN	1.0 MN	0.75 MN	60 kPa	12.50 m²	120 kPa
B	0.4	1.0	0.65	60	10.83	129
C	0.2	0.8	0.4	60	6.67	150
D	0.5	0.8	0.7	60	11.67	111

Solution

Footing C has the largest ratio of live load to dead load, that is, $0.8/0.2 = 4.0$. Choose the area of this footing for full dead load plus live load and the full allowable soil pressure.

Area $C = (0.2 + 0.8) \times 10^3/150 = 6.67$ m²
Usual $q = (0.2 + \frac{1}{4} \times 0.8) \times 10^3/6.67 = 60$ kPa

Choose the other footing areas for this usual soil pressure and the specified usual load; see Table 20.2. The maximum pressure column is based on full dead load and live load and is shown only to demonstrate that under this procedure one need not consider maximum soil pressures after the usual pressure has been properly established and used as shown.

20.11 COMBINED FOOTINGS

When two or more columns are carried on a single footing, it is called a combined footing. Two circumstances, especially, call for a combined footing: (1) two columns so closely spaced that separate footings would overlap or be of uneconomic proportions, Fig. 20.12a; (2) an exterior column footing which cannot be made symmetrical because of property

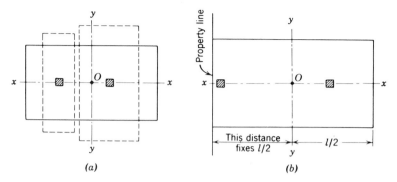

FIGURE 20.12 Combined footings.

line limitations or other restrictions, Fig. 20.12b. In each case the centroid of the combined footing must coincide with the resultant of the two column loads, noted as O.

The point O would shift slightly as one considers the design (factored) loads,* the service loads, or the usual loads of Sec. 20.10. Designers may use as the centroid of the footing the one they feel most important in their design situation.

In Fig. 20.12b, the length of a rectangular footing is fixed by the fact that if rectangular it must be symmetrical about axis y-y through this resultant load point O. The width in turn is fixed by the required soil-bearing area to carry the combined column loads.

In Fig. 20.12a, the designer has some choice in the length and width chosen, but the centroid of the area chosen must fall at O.

Design of combined footings, like the design of many other reinforced concrete structures, has not been standardized. One approach, sketched in Fig. 20.13, assumes structural behavior as consisting of transverse slab strips under the column which act to distribute load laterally and longitudinal slab strips that act like slabs supported on these "crossbeams." This type of analysis seems appropriate to long narrow footings. Another approach is to think of the projecting ends of the footing as similar to half of an isolated column footing and the slab between columns acting as a longitudinal beam. This method seems appropriate to footings having projections of such proportions as might be used for isolated columns, that is, projections somewhat of the same order of magnitude in the two

---

*Varying ratios between service dead and live loads on the two columns also invalidate the mathematical interchangeability of the two permissible soil pressures (at service and at ultimate conditions) discussed in Sec. 20.3d. The designer may take a choice and be consistent with it. The author considers the usual allowable soil pressure not so precisely established as to make the choice a critical one.

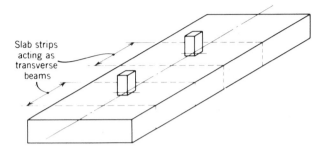

FIGURE 20.13 Transverse "beams" in combined footing.

directions. This type of design calls for a wider distribution of transverse steel than the transverse "beam" idea.

Shears should be checked as for other footings, which means a double check, around the column at a distance $d/2$ outside the column for two-way slab action, and at a distance d from the column all the way across the footing as a one-way slab.

Reasonably typical moment and shear diagrams for the longitudinal strips of a combined footing are shown in Fig. 20.14a. Design is concerned only with the portions outside the column widths since moments are considered critical at the face of column. The moment diagram can be visualized as a simple beam moment between columns (negative due to upward loads), superimposed on the positive moment diagram produced by the upward loads on the two end cantilevers. These two diagrams going into the total moment diagram of Fig. 20.14a are shown separately in Fig. 20.14b. These diagrams present no special design problem except one relating to the negative moment between columns. If the columns are

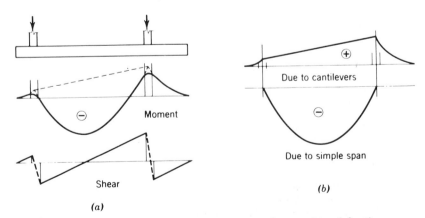

FIGURE 20.14 Moment and shear diagrams for combined footing.

closely spaced, no top steel may be required, because the simple span moment may be small compared to the cantilever moment.

Attention should be called to the fact that it is *not* on the safe side to assume a uniform soil pressure in the design of this center negative steel. If the pressure is nonuniform, it is probably larger near the center than it is toward the ends. A uniform load assumption gives too much positive cantilever moment and too little negative moment from the center span. These errors add to indicate a negative moment smaller than the actual value. The designer should be liberal in providing this negative moment steel, even introducing some steel when the uniform pressure assumption indicates a small remnant of positive moment.

20.12 CANTILEVER FOOTINGS

When an exterior column footing, because of property line restrictions, calls for some kind of combined footing but when the nearest interior footing is some distance away, the ordinary combined footing becomes long and narrow and subject to excessive bending moment. The cantilever footing is then more economical.

The cantilever footing is really two separate column footings joined by a beam that preferably should not contribute to the bearing area. In Fig. 20.15a, footing A under the exterior column is eccentric by a distance e from the column. By itself it would produce a very unsatisfactory soil pressure distribution. The stiff beam C is attached to footing A to balance the overturning moment P_1e and is carried to the concentric footing B for its necessary balancing reaction. This beam must be stiff in order to function well. It functions most simply if it is relieved of soil pressure from below, except possibly enough to carry the beam's weight. This condition is approached in several ways, most commonly by loosening or spading the soil under the beam, occasionally by forming the beam free from the soil, and so forth. Each of these measures leaves some question as to its effectiveness, which causes some engineers to shun this type of footing.

Under the simplest assumption, that the beam weight is carried by the soil, the analysis of the cantilever footing is quite simple. The beam is then effectively weightless. The reaction from the exterior column alone is R_1 on footing A (Fig. 20.15b) and ΔR downward on the beam at the interior column, which means ΔR is an uplift on this footing at B (Fig. 20.15c). Hence

$$\Delta R = P_1e(\ell - e)$$
$$R_1 = P_1 + \Delta R$$
$$R_2 = P_2 - \Delta R$$

Hence footing A must be slightly larger than needed for P_1 alone and

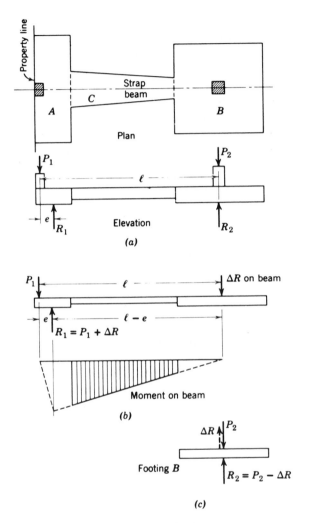

FIGURE 20.15 Cantilever footing.

footing B can be slightly reduced in area below that needed for P_2 alone. The beam C must carry a constant shear ΔR and a moment as shown in Fig. 20.15b.

In sizing footings A and B, either the service loads with the usual allowable soil pressure or the design loads (fractored) with the increased soil pressure (Sec. 20.3d) may be used, with R computed on a comparable basis. At ultimate, the increased soil pressure will not be equal on A and B unless their ratios of dead to live load happen to be equal.

20.13 RAFT FOUNDATIONS

It is sometimes desirable to utilize the soil pressure under the entire building area by covering the site with a thick slab or mat foundation. Such a foundation is often called a raft foundation, particularly when enough soil is excavated to offset the increased load due to the structure. Such a foundation may consist of a uniform slab a meter or more thick, or a slab stiffened by beams (either above or below the slab), or an inverted flat slab floor (Chapter 15).

No special design problems are involved in the concrete for such foundations other than those of a heavy floor design to resist upward pressures, with the columns furnishing the downward reactions. The distribution of the design upward soil pressures is the major problem here; it is dependent on the variability of the soil reaction properties and the resulting settlement response.

20.14 SLABS-ON-GROUND

A slab supported directly on a stable soil develops very little moment or shear in carrying a uniform load over its surface. When loads are unequally distributed or concentrated, the distributed reaction produced on the soil does create some moment and shear. (This is the principle used in pavement design.) On very stable soils, light loads such as residence loads can be carried on a simple 100 mm slab with a wire fabric reinforcement, providing no partition load is in excess of 7 kN/m. Chimney and other heavy loads, such as columns, must be carried to separate footings and the slab should not rest on or be tied to these separate elements or footings, which often will not move exactly the same as the slab.

Many soil types are subject to shrinkage in drying and expansion in wetting. Slabs alone are not adequate on such soils and experience shows that one portion of the slab may be pushed upward or settle relative to another. If this is not to crack walls and partitions, the slab must be stiffened by monolithically cast beams in both directions. In effect, a residence may then be supported in the center and the stiffened slab will have to cantilever out in all directions to prevent excessive floor warpage; or the structure may be supported on two corner areas and the slab will have to span as a simple span between these. Thus, the stiffened slab must ride like a boat over differential ground movements. Then the floor must be designed for the above loadings and a stiffening grid similar to that shown in Fig. 20.16 will be necessary. If such a system appears excessively costly, the final alternative is for spot footings to be carried deeper to a better soil and then to design a fully framed floor not in contact with the soil. Alternatively, soil stabilization measures are now being used on some bad soils.

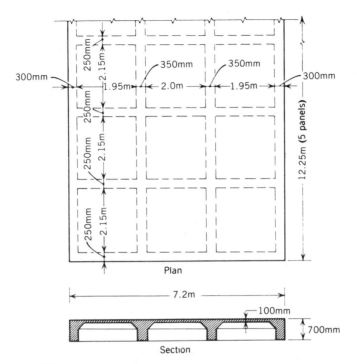

Plan

7.2m

100mm

700mm

Section

FIGURE 20.16 A slab-on-grade stiffened as required for a poor soil condition.

20.15 FOOTINGS ON PILES

Many footings must be built on piles. The piles act as a series of relatively concentrated reactions for the footing and in design are usually considered as concentrated loads. Since it is usually difficult to drive piles exactly where they are desired, the designer must expect that piles will be from 100 mm to 300 mm from the spot assigned to them. Accordingly, the Code (15.5.3) provides specific rules for counting piles that are near the critical section and sets somewhat stricter minimum thickness in Code 15.7, at least 300 mm above bottom reinforcement.

In Reference 5 a much more detailed discussion of special shear problems that occur at points closer than d from the column is given. Usually this is a case with only a small number of piles.

20.16 STEPPED AND SLOPED FOOTINGS

Specifications now permit thinner footings than those formerly used. There is thus less reason now for stepped or sloped footings and fewer are used. Such footings must definitely be cast as a monolith to avoid horizontal shearing weakness. Stresses must usually be checked at more than one section. For instance, steel stress at a step or development length beyond a step may be more critical than from the face of the column. Both these and shear must be checked from each step, the shear at a distance d or $d/2$ as in other cases.

Stepped footings are more appropriate for massive footings. A step can sometimes be used to advantage with a combined footing.

20.17 UPLIFT ON FOOTINGS

If uplift can exist on a footing, several special points must be considered. First, uplift probably results from lateral loading on the structure and Code Eq. 9.3 probably governs with its lower factor on dead load because it opposes the resultant uplift.

$$U = 0.9D + 1.3W$$

Second, the axial tension (plus the horizontal shear) may require tension splices and tension anchorage into the footing. Third, special design considerations are necessary. The footing must either be heavy enough in itself to balance the uplift or it must engage a mass of earth adequate for this balance. This reverses the type of bending on the footing and normally calls for some measure of top reinforcement. The footing will normally have some flexural cracks from service load positive moment. The assumption of an uncracked plain concrete section for resisting the negative moment from uplift would not be realistic.

SELECTED REFERENCES

1. A. N. Talbot, "Reinforced Concrete Wall and Column Footings," Univ. of Ill. Eng. Exp. Sta. *Bull. No. 67*, 1913.
2. F. E. Richart, "Reinforced Concrete Wall and Column Footings," *Jour. ACI*, 20, Oct. and Nov. 1948; *Proc.*, 48, pp. 97, 237.
3. ACI-ASCE Committee 326, "Shear and Diagonal Tension," *Jour. ACI., Proc.* 59, Jan. 1962, p. 30; Feb. 1962, p. 277; Mar. 1962, p. 1962.
4. ACI Committee 436, "Suggested Design Procedure for Combined Footings and Mats," *Jour. ACI, Proc.* 63, Oct. 1966, p. 1041.
5. *CRSI Handbook*, Concrete Reinforcing Steel Institute, Chicago, 2nd Ed., 1975.
6. R. B. Peck, W. E. Hanson, and T. H. Thornburn, *Foundation Engineering*, John Wiley & Sons, New York, 2nd Ed., 1974.

PROBLEMS

PROB. 20.1.   Redesign the rectangular footing of Sec. 20.7 and Fig. 20.7 as a square footing.

PROB. 20.2.   Redesign the square column footing of Sec. 20.6 as a rectangular footing having a width of 1 m.

PROB. 20.3.   Design a square column footing for a 400 mm by 400 mm tied column (of 30 MPa concrete) to carry a service column load of 670 kN dead load and 620 kN live load using permissible *ultimate* soil pressure of 400 kPa, $f'_c = 20$ MPa, Grade 400 steel.

PROB. 20.4.   Redesign the footing of Prob. 20.3 as a plain concrete footing of uniform thickness.

PROB. 20.5.   Design a reinforced concrete wall footing to carry service loads of 90 kN/m dead and 60 kN/m live load from a concrete wall 300 mm thick. Use an allowable soil service load pressure of 135 kPa, $f'_c = 20$ MPa, Grade 400 steel.

PROB. 20.6.   Redesign the footing of Prob. 20.5 using plain concrete.

PROB. 20.7.   Redesign the wall footing of Prob. 20.5 if the allowable service load soil pressure is only 100 kPa.

PROB. 20.8.

(*a*) Lay out the plan view of a combined rectangular footing for the columns of Fig. 20.17 where the service loads are:
   A: 670 kN dead plus 450 kN live.
   B: 900 kN dead plus 1 100 kN live.

Based on an allowable *net* soil pressure of 150 kPa and a centroid based on full design loads, try to achieve equal projections around the heavier column.

(*b*) Lay out again on the assumption that there is a property line 100 mm beyond the outside face of column *A*.

(*c*) Lay out again for conditions of (*a*) on the assumption that lateral

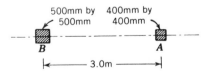

FIGURE 20.17   Column layout for Prob. 20.8

FIGURE 20.18. Column layout for Prob. 20.9.

projections beyond column B are to be about half the longitudinal projection.

(*d*) Repeat (*a*) with the centroid based on full service loads.

PROB. 20.9. If the columns of Fig. 20.18 carry the service loads of Prob. 20.8(*a*), lay out the plan view of a cantilever footing (without dimensions on the beam) based on an allowable net soil pressure of 150 kPa. Draw the M and V diagrams for the strap beam which connects the two footing elements:

(*a*) If the footing under column A extends 1.2 m from the property line.
(*b*) If the dimension in (*a*) is changed to 1.5 m.
(*c*) If the dimension in (*a*) is changed to 1.8 m.

PROB. 20.10. Design a square column footing using $f'_c = 25$ MPa, Grade 300 reinforcement, and allowable soil pressure 150 kPa at service load to carry a 600 mm square tied column (same f'_c) delivering a dead load of 1 100 kN and live load of 2 000 kN. The weight of concrete $= 24$ kN/m³.

PROB. 20.11. Design a rectangular footing having a width of 3 m to carry a centered column 500 mm square delivering a service load of 1 000 kN dead load and 1 800 kN live load, with zero moment and overburden. $f'_c = 30$ MPa, reinforcement Grade 400, allowable net ultimate soil pressure 250 kPa (above footing weight), weight of concrete $= 24$ kN/m³.

21
PRESTRESSED CONCRETE ANALYSIS

21.1 SCOPE OF PRESENTATION

This discussion of prestressed concrete covers some of the fundamental concepts and presents a general idea of how such concrete responds to loads. No attempt is made to develop design procedures. For the ease of presentation, simple rectangular sections are used in the examples rather than the more practical I or T shapes.

Originally there was a wide gap between prestressed concrete and ordinary reinforced concrete because it was recognized that a small prestress would be lost due to shrinkage and creep. Now, however, partial prestressing is also finding a place and the gap is disappearing. In this summary treatment only full prestressing is discussed.

The term tendon is often used to refer to the prestressing element, without differentiation between wires, strands, or bars. Bars are large, up to 36 mm diameter, and have a breaking stress of 1 030 to 1 100 MPa. Wires are small with diameter up to 7 mm, and have a strength of 1 550 to 1 760 MPa. Strands normally have 7 wires, diameters from 9.5 mm to 15.7 mm and have strengths of 1 760 to 1 860 MPa. Only strand is used in pretensioned concrete, at least in the United States.

21.2 THE NATURE OF PRESTRESS

Reinforced concrete becomes prestressed concrete when permanently loaded in such a way as to build up initial stresses opposite to those which later will be developed by service loads.

A circular tank, designed to resist circumferential or ring tensile stres-

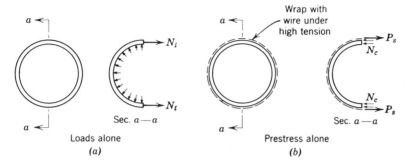

FIGURE 21.1   Prestress largely eliminates ring tension in tanks.

ses, as in Fig. 21.1a, is prestressed by being placed in initial circumferential compression. This compression is produced by wrapping the tank with wire under high tensile stress as indicated in Fig. 21.1b.

A simple span beam is prestressed (Fig. 21.2) by introducing (1) a negative moment to offset the expected positive moment and at the same time (2) a longitudinal compression to offset the tensile stresses from bending moment. Both effects are obtained by embedding highly stressed tension members eccentrically (below the middle of a simple beam) giving both a P/A effect from the load P_s and a negative bending moment effect measured by $P_s e$. The tension prestressing element, called a tendon, may be wire, twisted wire strands, or rods.

Continuous beams are prestressed in similar fashion but for best results require an effective eccentricity above mid-depth in negative moment zones. Their analysis is more involved because the prestress may modify the external reactions, shears, and moments.

For piles, plank, columns, and other members subject to moments both positive and negative, or subject to loading on either face, longitudinal prestressing wires without any resultant eccentricity are used solely for

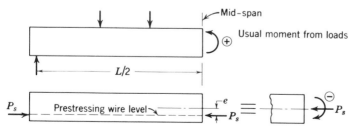

FIGURE 21.2   Beam prestress produces axial compression and a moment opposite in sign to dead load moment.

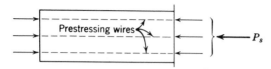

FIGURE 21.3   Beam with uniform prestress.

their P/A effect, as in Fig. 21.3. Since eccentric prestressing wires or tendons in a simple span build up negative moment to offset the ordinary positive moment, such prestressing would actually weaken the member if it were inverted or put under negative moment.

21.3 OBJECTIVES OF PRESTRESSING

Proper prestressing ordinarily prevents the formation of tension cracks under working loads. This alone is adequate reason for prestressing tanks to avoid leakage, or for prestressing structural members subject to severe corrosion conditions.

With this lack of cracking, the entire cross section remains effective for stress; a smaller required section normally results. With lighter sections, precasting of longer members becomes possible. Precasting has been a most important economic factor in promoting prestressed concrete in the United States.

Since prestress puts camber into a member under dead loading, deflections at working loads are much reduced.

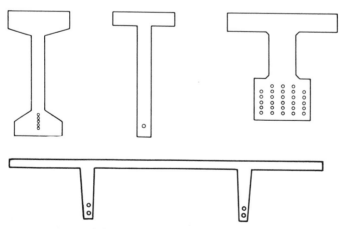

FIGURE 21.4   A few of the cross sections used for prestressed members.

Because tension cracking of prestressed concrete at working loads is avoided, very high strength steel (and smaller steel areas) can be used to provide the necessary tension. In some countries prestressed concrete is thus a device to make small supplies of steel serve more efficiently. In this country prestress is a device to permit very high strength steel to be used economically and to increase permissible span lengths while keeping the concrete uncracked and hence more effective.

Prestress reduces diagonal tension stresses at working loads. This has led to the general use of modified I and T shapes with less web area than conventional beams, as in Fig. 21.4.

Most of the advantages of prestressed concrete are construction or service load advantages. Under ultimate load conditions the strength of the usual (underreinforced) prestressed beam is about the same as without prestress; its deflection increases sharply after cracking, and stirrups are then necessary. The Code requires at least minimum stirrups in all prestressed beams.

21.4 NOTATION AND SIGNS

The key to prestressed concrete analysis lies in a good bookkeeping system. All the usual calculations are individually simple, but they consist of a number of small pieces that must be assembled into various combinations. Good bookkeeping requires a consistent set of signs for stresses and an adequate notation. For the beginner numerous sketches, such as those in Figs. 21.7, 21.8, and 21.9, are helpful.

In this chapter a tensile stress will carry a plus sign and a compressive stress a negative sign. Although this is contrary to the signs used by many writers for prestressed concrete, it is in accord with general structural practice and is recommended also in Professor Lin's book.[1] No sign convention is needed for ordinary reinforced concrete since the kind of stress in such calculations is never in doubt; in prestressed concrete the concrete stress often changes from tension to compression (or the reverse) as the loading changes.

Much of the checking of a prestressed member is at service loads (unfactored loads) and thus requires the calculation of stresses; but ultimate strength must also be checked.

21.5 STRESSES BY SUPERPOSITION OF STRESS BLOCKS

In the following sections, detailed calculations show "exact" methods for computation of both bonded and unbonded tendons. A simpler, slightly approximate approach in terms of gross area of concrete and the final

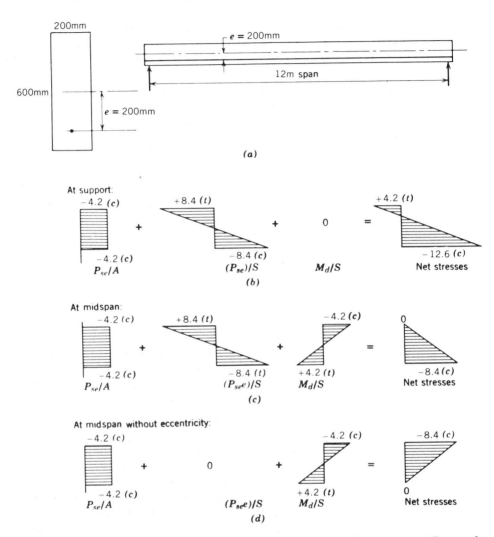

FIGURE 21.5 Influence of tendon eccentricity on service load stresses. (a) Beam of Sec. 21.5. (b) Stresses at support. (c) Stresses at midspan. (d) Stresses at midspan without tendon eccentricity.

prestress (effective prestress) is often useful to the designer; it will also probably be easier to visualize here for the first numerical example.

Consider the simple span rectangular beam of Fig. 21.5a with effective prestress P_{se} of 500 kN from a straight tendon 200 mm below middepth. Establish the service load stresses existing prior to the application of live load.

Solution

The effect of the prestress is an axial load from the 500 kN = 0.5 MN prestress plus a negative moment from the eccentricity = $P_{se}e = -0.5 \times 0.2 = -0.1$ MN $\cdot$ m. The axial load produces the uniform stress block of Fig. 21.5b with $f_c = -0.5/0.2 \times 0.6 = 4.2$ MPa compression. The moment gives the antisymmetric triangles of that same figure.

Section modulus $S = bh^2/6^* = 0.2 \times 0.6^2/6 = 12 \times 10^{-3}$ m^3

$f_c = -0.1/12 \times 10^{-3} = -8.4$ MPa compression on bottom

 or $+8.4$ MPa tension on top

  At the end of the beam the dead load moment is zero, leading to the combined stress block shown at the right of Fig. 21.5b. The 4.2 MPa tension on the top is more than would be permissible because it exceeds the probable cracking stress. If the tendon were draped upward at the end, reducing the end eccentricity and thus $P_{se}e$, this stress could be made acceptable.

  At midspan the dead load moment of the beam weight of $0.2 \times 0.6 \times 24 = 2.8$ kN/m is $M_d = \frac{1}{8} \times 2.8 \times 12^2 = 50.4$ kN $\cdot$ m, leading to stresses:

$$f_c = +50.4 \times 10^{-3}/12 \times 10^{-3} = +4.2 \text{ MPa tension at bottom}$$

$$\text{or } -4.2 \text{ MPa compression at top.}$$

The superimposed stress blocks then show in Fig. 21.5c. It should be noted that this beam now has bottom compression all the way across the span and thus deflects upward under dead load and prestress. Also, this substantial compression stress at midspan indicates no tension would develop there until at least 5.6 kN/m of live load (twice the dead load) is added.

  Had the prestress been added on the axis instead of at 200 mm eccentricity, the stresses in Fig. 21.5b at the support would have been only the rectangular stress block of -4.2 MPa compression. (The cable could also be draped so as to obtain the same effect.) At the mid-span the stress blocks would then drop to those of Fig. 21.5d, with a zero stress at the bottom. Hence no live load could be added if one designed to the very strict requirement occasionally used of *no* tension under service load. Lowering the tendon at midspan to the 200 mm eccentricity as above has increased the load capacity by 5.6 kN/m in this case, and at the same time increased the ultimate moment capacity because of the greater effective depth of the steel.

  The solution is not as complete as in those examples which follow, but it is often used as being reasonably close. M_u must still be checked.

21.6 PRE-TENSION METHOD OF PRESTRESSING

Since prestressing calculations follow closely the physical process of prestressing, both the pre-tensioning and post-tensioning procedures must be clearly understood. Sketches of the beam and stresses at each stage will be helpful to the beginner in making summations. The individual calculations are quite simple.

---

*With symmetry S is more convenient than the use of I/c.

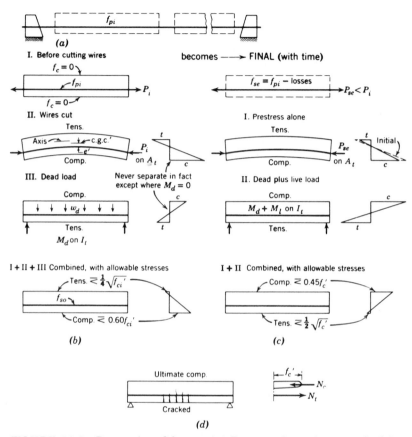

FIGURE 21.6 Pretensioned beam. (a) Prestressing wire stretched between casting yard abutments. (b) Initial stresses at midspan, separated for calculation purposes; also allowable concrete stresses. (c) Stresses at mid-span under full service load, separated for calculation purposes; also allowable concrete stresses. (d) Ultimate strength conditions.

In the pre-tension process the first step is to stretch high strength wires between abutments or end piers of a (long) prestressing bed, as indicated schematically in Fig. 21.6a. This total initial wire tension is here designated P_i, the unit stress f_{pi}. The member forms are placed around the wire, with a number of such units along the total length, and in these forms the members are cast and cured (usually by accelerated curing). The wires are then cut between members (Fig. 21.6b).

The resulting release of wire tension P_i is equivalent to P_i applied as an external compressive force on the entire transformed area (Sec. 8.4) of the member A_t. The wires are usually eccentric, below the centroid of the transformed area of the member, and thus the external force P_i introduces a negative moment P_ie, which causes the beam to curve upward and pick up its dead load moment M_d. Because of P_i alone, rather large calculated tensile stresses develop on the top beam fibers, but this need not be dangerous because these never exist without the counteracting compression from dead load moment.* This combined top tensile stress is often a critical design stress with a recommended limit of $\frac{1}{4}\sqrt{f'_{ci}}$ unless top steel is used, f'_{ci} being the cylinder strength at this early age. To turn such a beam on its side, or invert it, or lift it by a sling at the middle will be unsafe and may cause failure. The bottom stress in compression may also be a controlling stress, limited to $0.60 f'_{ci}$. Shrinkage is usually omitted from this calculation. Although some shrinkage will have occurred by the time the wires are cut, and these calculations apply, the shrinkage reduces both these critical stresses.

With time, shrinkage of concrete and the creep of concrete and steel under stress reduce the length of the concrete holding the wires and thereby reduce P_i to a lower effective prestress P_{se}. Formerly the reduction was often taken as 240 MPa but now it is commonly more carefully appraised.[6,7] The loss tends to increase with small ratios of volume to surface area (thin members) and with exposure to lower humidities, but it varies considerably for different members in different regions and under different exposures. It increases if high initial stresses are used on the concrete. As a result of the lowered prestress the top tension and bottom compression are reduced below the initial values. It is upon these lowered stresses that live load stresses are superimposed, resulting in the bottom sketch of Fig. 21.6c. The final result is compression critical at the top and tension at the bottom, with the allowable stresses for buildings noted in the combined load sketch in Fig. 21.6c.

At ultimate load, the beam will be cracked as indicated in Fig. 21.6d and conditions will not differ greatly from those of an ordinary reinforced concrete beam, except for the very high f_{ps}.

21.7 POST-TENSION METHOD OF PRESTRESSING

In the post-tension method the plain concrete member is first cast with a tube or slot for future introduction of steel as indicated at I in Fig. 21.7a. A variation in procedure is to embed the steel in the member but separate it from the concrete by an enclosing metal or plastic tube or mastic coating—

---

*With straight wires and straight members, stresses would be critical near the ends where M_d is small. Present practice often reduces the eccentricity near the ends to keep the center stresses the critical ones (Figs. 21.7 and 21.12a–d).

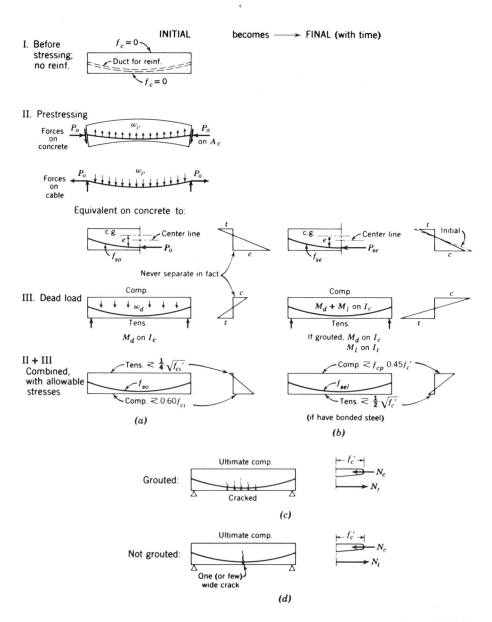

FIGURE 21.7 Post-tensioned beam. (*a*) Initial stresses separated for calculation purposes; also allowable concrete stresses. (*b*) Stresses under full service load separated for calculation purposes; also allowable concrete stresses. (*c*) Ultimate strength when grouted. (*d*) Ultimate strength, when ungrouted, is lower; hence Code 18.9 generally requires added bonded reinforcement.

this prevents grouting and leaves the reinforcing unbonded. (See last paragraph of this section.) After the concrete has cured, the steel tendons are stressed by jacking against the concrete and are then locked under stress by appropriate end anchorage or clamps. Since the concrete simultaneously shortens under compression, the jacking must account for both steel stretch and concrete shortening; and this is complicated by friction losses and by the fact that beam curvature under this stressing causes the dead load moment also to become active (Fig. 21.7a).

The figure shows the resultant steel tension as P_o acting on the plain concrete section A_c* (compared to P_i and A_t for pre-tension). In the pretensioned case P_o could also have been calculated from the known value of P_i;† in the post-tensioned case there exists nothing comparable to P_i. The forces w_p between cable and concrete are discussed in Sec. 21.14. The allowable stresses are also marked at the bottom of Fig. 21.7a.

The calculations for full service load are diagrammed in Fig. 21.7b. These calculations should start with a loss of prestress slightly less than was used for the pre-tension case; some shrinkage will have occurred prior to prestressing and all the elastic deformation of the concrete occurs during jacking to the f_{so} stress. This loss, at one time taken as 240 MPa, is now calculated more carefully,[6,7] considering the volume-to-area ratio and exposure actually expected.

If the tendons are grouted effectively, as is often specified, the live load stresses can be based on the entire transformed section area. The change in steel stress, although larger than when ungrouted, is still not important. Ultimate strength of a grouted beam (Fig. 21.7c) is about the same as for a pretensioned beam.

Unless tendons are grouted into place, Code 18.9 now requires minimum bonded reinforcement in essentially all cases to avoid failures from a single flexural crack, as sketched in Fig. 21.7d. (Such failure results from the live load being resisted primarily by plain concrete.) The bonded reinforcement then becomes part of the transformed area resisting the live load.

21.8 OTHER PROCEDURES IN PRESTRESSED CONCRETE

The original sharp distinction between pre-tensioned and post-tensioned members is now less sharp than might be implied in the above discussion. Many variations on the original patterns are now common practice.

Precast systems may include either pre-tensioning or post-tensioning on

---

*The Code now says loss of area must be *considered*; it does not say it must be used.
†In the pre-tension case P_o on the net concrete area would give the same result as P_i on the transformed area, but P_i is the known starting force there.

a partial basis adequate for handling stresses with later additional post-tensioning used after members are in place to bring them to design capacity. This includes the possibility of adding such post-tensioning as is needed to assure continuity of adjacent members, the addition of inverted U strands over supports, and the like.

Both grouted and ungrouted systems are in use with strong differences of opinion over the merits of ungrouted tendons. The wide crack formation at failure when unbonded, indicated in Fig. 21.7d, is avoided by casting in ordinary reinforcing bars of small diameter for crack control. Likewise the ultimate strength of either a pre-tensioned or post-tensioned member can be raised where necessary by casting in ordinary reinforcement as supplemental steel. Supplemental steel may run only a limited distance, wherever reserve moment strength or crack control is needed.

The possibility of aligning the post-tensioned tendons in any desired shape gives this construction an advantage in taking care of shear and varying moments, especially in cast-in-place construction. For precast simple spans, however, it is now feasible with pre-tensioning to depress the strands in a precise fashion to accomplish much the same advantages.

21.9 CRACKING LOAD AND ULTIMATE STRENGTH

European specifications have placed considerable emphasis on the cracking load, requiring that no cracks be formed for a given (usually small) overload. American practice has placed greater emphasis on ultimate strength.

Each of these methods is an attempt to provide an adequate factor of safety, the first indirect, the second direct. The need for both working stress and ultimate strength calculations lies in the radical change in beam behavior when cracks form. Prior to cracking the gross area of the beam is effective. As a crack develops, all the tension from the concrete must be picked up by the steel. If the percentage of steel is small, there may be very little added capacity between cracking and failure. Cracking may be assumed to take place when the calculated tensile stress reaches $0.6\sqrt{f'_c}$.

21.10 EXAMPLE OF PRESTRESS WITHOUT ECCENTRICITY

The hollow member of Fig. 21.8a is reinforced with the equivalent of four wires of 60 mm^2 each, pre-tensioned to $f_{si} = 1\,000$ MPa. If $f'_c = f'_{ci} = 35$ MPa, $n = 7$, determine the stresses when the wires are cut between members; also determine the moment that can be carried at a maximum tension of $\frac{1}{2}\sqrt{f'_c}$ and a maximum f_c of 0.45 f'_c. If 240 MPa of the prestress is lost (in addition to the elastic load deformation) determine this limiting moment.

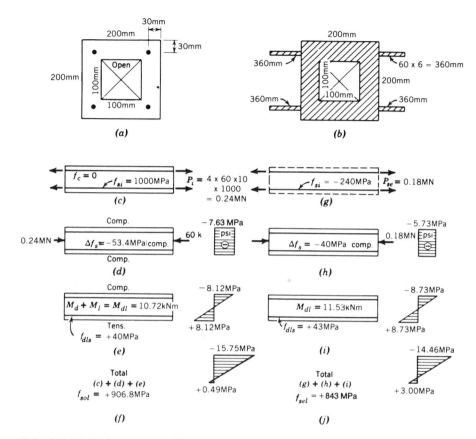

FIGURE 21.8 Prestress without eccentricity, pretensioned. (*a*) Beam cross section. (*b*) Effective transformed area. (*c*) Tendons under initial tension. (*d*) Effect of cutting tendons. (*e*) Permissible moment for initial loading. (*f*) Total stresses for initial loading. (*g*) Equivalent initial tension after losses. (*h*) Equivalent reduction due to compression of concrete. (*i*) Permissible moment for later loading. (*j*) Total stresses for later loading.

Solution

Concrete area $= 200^2 - 100^2 = 30\,000\text{ mm}^2 = 30 \times 10^{-3}\text{ m}^2$

Transformed area (Fig. 21.8*b*) $= A_t = 200^2 - 100^2 + 4 \times 60(7 - 1)$

$$= 31\,444\text{ mm}^2 = 31.44 \times 10^{-3}\text{ m}^2$$

$I_t = \frac{1}{12} \times 200 \times 200^3 - \frac{1}{12} \times 100 \times 100^3 + 4 \times 360 \times 70^2 = 132.1 \times 10^6\text{ mm}^4$

$$= 132.1 \times 10^{-6}\text{ m}^4$$

The starting stresses of Fig. 21.8*c* are $f_c = 0$ and $f_{si} = 1\,000$ MPa. When the tendons are cut, the 0.24 MN force acts on the transformed area (Fig. 21.8d) to add these stresses:

$$f_c = -P_i/A_t = -0.24/31.44 \times 10^{-3} = -7.63 \text{ MPa compression}$$
$$\Delta f_s = n f_c = -7 \times 7.63 = -53.4 \text{ MPa compression}$$

The total stresses before loading are thus:

$$f_c = 0 - 7.63 = -7.63 \text{ MPa compression}$$
$$f_{so} = 1\,000 - 53.4 = 946.6 \text{ MPa}$$

Before any prestress losses take place, $M_d + M_\ell$ can be enough to bring f_c at the top to $0.45 f_c' = 15.75$ MPa compression or to bring f_c at the bottom to $\frac{1}{2}\sqrt{f_c'} = 3.0$ MPa tension.

This leaves for dead and live load moment, $M_d + M_\ell = M_{d\ell}$:

$$\text{On top, } f_c = -15.75 + 7.63 = -8.12 \text{ MPa}$$
$$\text{On bottom, } f_c = +3.0 + 7.63 = 10.63 \text{ MPa}$$

The smaller value limits $M_{d\ell}$, as shown in Fig. 21.8e.

$$M_{d\ell} = 8.12 \times I_t/y_t = 8.12 \times 132.1 \times 10^{-6}/0.1 = 10.72 \times 10^{-3} \text{ MN} \cdot \text{m} = 10.72 \text{ kN} \cdot \text{m}$$

Under this loading:

Total concrete stress on top $= -7.63 - 8.12 = -15.75$ MPa compression

Total concrete stress on bottom $= -7.63 + 8.12 = +0.49$ MPa tension

Total steel stress $= +1\,000 - 53.4 - 7 \times 8.12 \times 70/100$
$$= 906.8 \text{ MPa tension}$$

After 240 MPa of prestress is lost, P_i is, in effect, reduced to $(1\,000 - 240) \times 4 \times 60 \times 10^{-6} = 0.18$ MN in Fig. 21.8g,h, giving uniform stresses:

$$f_c = -0.18/31.44 \times 10^{-3} = -5.73 \text{ MPa}$$
$$f_{se} = +1\,000 - 240 + 7 \times 5.73 = +800.1 \text{ MPa}$$

$M_{d\ell}$ may safely change the top stress by $-15.75 + 5.73 = -10.02$ MPa and the bottom stress by $+3.0 + 5.73 = +8.73$ MPa. The smaller stress increment controls.

$$M_{d\ell} = 8.73 \times 132.1 \times 10^{-6}/0.1 = 11.53 \times 10^{-3} \text{ MN} \cdot \text{m} = 11.53 \text{ kN} \cdot \text{m}$$

The lower $M_{d\ell}$ value of 10.72 kN $\cdot$ m controls (Fig. 21.8e,f).

The final stresses, if the higher $M_{d\ell}$ could be used, would have been:

$$\text{Concrete on top} = -5.73 - 8.73 = -14.46 \text{ MPa} < 0.45 f_c'$$
$$\text{Concrete on bottom} = -5.73 + 8.73 = +3.00 \text{ MPa} = \tfrac{1}{2}\sqrt{f_c'}$$
$$\text{Steel stress} = +800.1 + 7 \times 8.73 \times 70/100$$
$$= 842.9 \text{ MPa tension}$$

Ordinarily there is no need to calculate the steel stresses. The values given indicate steel stresses cannot become higher than initial values.

The capacity cannot be accepted as above until the value of M_u is also established as shown in Sec. 21.13.

21.11 EXAMPLE OF PRE-TENSIONED BEAM WITH ECCENTRICITY

The pre-tensioned beam of Fig. 21.9 will be investigated for $f'_c = 35$ MPa, $n = 7$, $f_{si} = 930$ MPa, and a creep and shrinkage loss of 240 MPa. $M_d = 90$ kN $\cdot$ m, $M_\ell = 100$ kN $\cdot$ m. The concrete strength when tendons are released is assumed to be $f'_{ci} = 30$ MPa, but for simplicity the same n will be used. Assume that the beam axis (or tendon) is curved so that stresses under dead load and prestress do not require a check other than at midspan. Allowable concrete stresses are shown in Fig. 21.6b,c.

Solution

$$A_t = 250 \times 500 + (7-1) \times 1\,000 = 131\,000 \text{ mm}^2 = 0.131 \text{ m}^2$$
$$\bar{y} \text{ from mid-depth} = -6\,000 \times 175/131\,000 = 8 \text{ mm (below)}$$
$$I_t = \tfrac{1}{12} \times 250 \times 500^3 + 250 \times 500 \times 8^2 + 6\,000 \times (175-8)^2$$
$$2\,780 \times 10^6 \text{ mm}^4 = 2.78 \times 10^{-3} \text{ m}^4$$

Figure 21.9c shows forces before cutting tendons and Fig. 21.9d and e shows the stress changes occurring after cutting tendons. Actually, these two sets of changes occur together and cannot be separated, except in the calculations.

From cutting the tendons:

$$f_c \text{ at top} = -P_i/A_t + P_i e y_t/I_t = -\frac{0.93}{0.131} + \frac{0.93 \times 0.167 \times 0.258}{2.78 \times 10^{-3}}$$
$$= +7.31 \text{ MPa tension}$$
$$f_c \text{ at bottom} = -\frac{0.93}{0.131} - \frac{0.93 \times 0.167 \times 0.242}{2.78 \times 10^{-3}}$$
$$= -20.62 \text{ MPa compression}$$
$$\Delta f_s = n f_c \text{ at same level} = -\frac{7 \times 0.93}{0.131} - \frac{7 \times 0.93 \times 0.167 \times 0.167}{2.78 \times 10^{-3}}$$
$$= -115.00 \text{ MPa compression}$$

From dead load moment:

$$f_c \text{ at top} = M_d y_t/I_t = -90 \times 10^{-3} \times 0.258/2.78 \times 10^{-3} = -8.35 \text{ MPa}$$
$$f_c \text{ at bottom} = +90 \times 10^{-3} \times 0.242/2.78 \times 10^{-3} = +7.83 \text{ MPa tension}$$
$$f_s = 7 \times 7.83 \times 0.167/0.242 = +37.85 \text{ MPa tension}$$

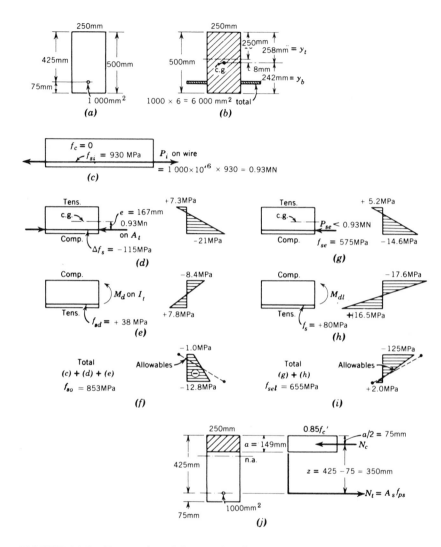

FIGURE 21.9 Pretensioned beam. (*a*) Cross section. (*b*) Transformed area, A_t. (*c*) Tendons under initial tension. (*d*) Effect of cutting tendons, eccentric load on A_t. (*e*) Dead load stresses. (*f*) Total initial stresses. (*g*) Effective pretension stresses. (*h*) Dead load plus live load stresses. (*i*) Total stresses under full service load. (*j*) Ultimate strength calculation.

604

Total stresses without live load:

$$f_c \text{ at top} = 0 + 7.31 - 8.35 = -1.04 \text{ MPa compression} < \tfrac{1}{4}\sqrt{f_{ci}} \text{ tension}$$
$$f_c \text{ at bottom} = 0 - 20.62 + 7.83 = -12.79 \text{ compression} < 0.6\, f'_{ci} = 18.0 \text{ MPa}$$
$$f_s = +930.00 - 115.00 + 37.85 = +852.85$$

A slightly larger prestress would be possible since both stresses in Fig. 21.9f are low. (One objection to the rectangular beam shape is the large prestress required for full effectiveness.)

When f_s drops 240 MPa (due to losses) from $930 - 115 = 815$ MPa to 575 MPa tension, as in Fig. 21.9g, the prestress effect, without M_d, drops to:

$$f_c \text{ at top} = +7.31 \times 575/815 = +5.15 \text{ MPa tension}$$
$$f_c \text{ at bottom} = -20.62 \times 575/815 = -14.55 \text{ MPa compression}$$

Total moment $= M_d + M_\ell = 90 + 100 = 190 \text{ kN} \cdot \text{m}$, giving stresses:

$$f_c \text{ at top} = -190 \times 10^{-3} \times 0.258/2.78 \times 10^{-3} = -17.63 \text{ MPa compression}$$
$$f_c \text{ at bottom} = +190 \times 10^{-3} \times 0.242/2.78 \times 10^{-3} = +16.54 \text{ MPa tension}$$
$$\text{Total } f_c \text{ at top} = +5.15 - 17.63 = -12.48 \text{ MPa compression} < 0.45\, f'_c$$
$$= 15.75 \text{ MPa}$$
$$\text{Total } f_c \text{ at bottom} = -14.55 + 16.54 = +1.99 \text{ MPa tension} < \tfrac{1}{2}\sqrt{f'_c} = 3.0 \text{ MPa}$$

Total $f_s = +575.0 + 7 \times 16.54 \times 0.167/0.242 = 575.0 + 79.9 = 654.9$ MPa tension. The total moment could be increased by 6% or the prestress could be reduced by 6% and still keep the bottom tension within the allowable.

The check cannot be considered complete without an investigation of the ultimate moment capacity (Sec. 21.13).

21.12 EXAMPLE OF POST-TENSIONED BEAM WITH ECCENTRICITY

The post-tensioned beam of Fig. 21.10a will be investigated for $f'_c = f'_{ci} = 35$ MPa, $n = 7$, $f_{so} = 930$ MPa, 170 MPa loss in prestress with time, $M_d = 80$ kN $\cdot$ m, $M_\ell = 90$ kN $\cdot$ m. It is assumed that the cables are positioned along the span so as to make the mid-span section the critical one. The cables will first be considered as ungrouted; then the effect of grouting will be investigated. Allowable concrete stresses are shown in Fig. 21.7a,b.

Solution

$$A_c = 250 \times 500 - 100 \times 60 = 119 \times 10^3 \text{ mm}^2 = 0.119 \text{ m}^2 \text{ (Fig. 21.10b)}$$
$$\bar{y} \text{ from mid-depth} = 100 \times 60 \times (250 - 50)/119 \times 10^3 = 10.0 \text{ mm (above)}$$
$$I_c = \tfrac{1}{12} \times 250 \times 500^3 + 250 \times 500 \times 10^2 - \tfrac{1}{12} \times 100 \times 60^3 - 100 \times 60(250 - 50 + 10)^2$$
$$= 2\,350 \times 10^6 \text{ mm}^4 = 2.35 \times 10^{-3} \text{ m}^4$$
$$e = 420 - 250 + 10 = 180 \text{ mm}$$

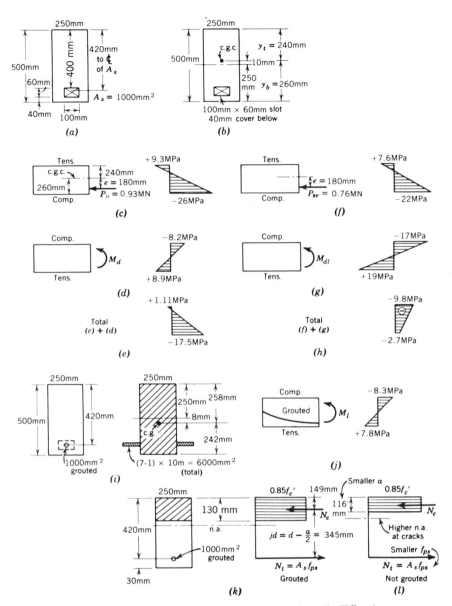

FIGURE 21.10 Post-tensioned beam. (*a*) Cross section. (*b*) Effective area, considered here as plain concrete only, A_c. (*c*) Initial prestressing. (*d*) Dead load stresses. (*e*) Total initial stresses. (*f*) Prestress effects after losses. (*g*) Dead plus live load service stresses (ungrouted). (*h*) Total service load stresses (ungrouted). (*i*) Effective section for live load when effectively grouted. (*j*) Service live load stresses, effectively grouted. (*k*) Ultimate strength calculation, grouted. (*l*) Effect of lack of grouting on ultimate strength conditions.

Figure 21.10c and d shows prestress effects artifically separated from M_d stresses. For prestress alone:

$$P_o = 1\,000 \times 10^{-3} \times 930 = 0.93 \text{ MN}$$

At top, $\quad f_c = -P_o/A_c + P_o e y_t/I_c = -\dfrac{0.93}{0.119} + \dfrac{0.93 \times 0.18 \times 0.24}{2.35 \times 10^{-3}} = +9.28 \text{ MPa tension}$

At bottom, $\quad f_c = \dfrac{0.93}{0.119} - \dfrac{0.93 \times 0.18 \times 0.26}{2.35 \times 10^{-3}} = -26.34 \text{ MPa compression}$

For M_d alone:

At top, $\qquad f_c = -80 \times 10^{-3} \times 0.24/2.35 \times 10^{-3} = -8.17 \text{ MPa compression}$

At bottom, $\qquad f_c = +80 \times 10^{-3} \times 0.26/2.35 \times 10^{-3} = +8.85 \text{ MPa tension}$

Combined prestress and dead load (Fig. 21.10e):

At top, $\qquad f_c = +9.28 - 8.17 = +1.11 \text{ MPa} < \frac{1}{4}\sqrt{f'_{ci}} = 1.5 \text{ MPa}$ **O.K.***

At bottom, $\qquad f_c = -26.34 + 8.85 = -17.49 \text{ MPa} < 0.60\, f_{ci} = 21 \text{ MPa}$ **O.K.**

With time, f_{so} drops to $f_{se} = 930 - 170 = 760 \text{ MPa}$, and the effect of prestress drops in like ratio, as indicated in Fig. 21.10f.

At top, $\qquad f_c = +9.28 \times 760/930 = +7.58 \text{ MPa tension}$

At bottom, $\qquad f_c = -26.34 \times 760/930 = -21.52 \text{ MPa compression}$

For dead and live load, $M_{d\ell} = 80 + 90 = 170 \text{ kN} \cdot \text{m} = 0.17 \text{ MN} \cdot \text{m}$ these stresees are:

At top, $\qquad f_c = -0.17 \times 0.24/2.35 \times 10^{-3} = -17.36 \text{ MPa compression}$

At bottom, $\qquad f_c = +0.17 \times 0.26/2.35 \times 10^{-3} = +18.80 \text{ MPa tension}$

Total at top, $\quad f_c = +7.58 - 17.36 = -9.78 \text{ MPa compression}$

$$< 0.45\, f'_c = 15.75 \text{ MPa}$$

Total at bottom, $\quad f_c = -21.52 + 18.80 = -2.72 \text{ MPa compression vs.}$

$$\text{allowable tension} = \tfrac{1}{2}\sqrt{f'_c}. \qquad \textbf{O.K.}$$

For an ungrouted beam the changes in f_s due to live load are small (significantly less than n times the concrete live load stress at that level). These changes in f_s are not usually calculated.

Neglecting the complications the grouted section may involve in the matter of loss of prestress, the only change in working stress calculations resulting from grouting is that the live load stresses will be based on the transformed area.

If the beam were effectively grouted, and again neglecting the variation in n, the transformed area would become that of Fig. 21.10i.

$$A_t = 250 \times 500 + (7-1) \times 1\,000 = 131 \times 10^3 \text{ mm}^2 = 0.131 \text{ m}^2$$

$$\bar{y} \text{ from mid-depth} = -6\,000 \times (420 - 250)/131\,000 = -8 \text{ mm}$$

---

*Note that if left ungrouted Code 18.9 requires supplementary reinforcing bars to control cracking, wherever calculated tension exceeds $\frac{1}{6}\sqrt{f'_c} = 1.0 \text{ MPa}$. Such bars would modify the member properties used here.

$$I_t = \tfrac{1}{12} \times 250 \times 500^3 + 250 \times 500 \times 8^2 + 6\,000 \times (420 - 242)^2 = 2\,802 \times 10^6 \text{ mm}^4$$
$$= 2.8 \times 10^{-3} \text{ m}^4\,|$$

At top, $f_\ell = -90 \times 10^{-3} \times 0.258/2.8 \times 10^{-3} = -8.29$ MPa comp.

At bottom, $f_\ell = +90 \times 10^{-3} \times 0.242/2.8 \times 10^{-3} = +7.78$ MPa tension

Total on steel, $f_s = 930 - 170 + 7 \times 90 \times 10^{-3} \times 0.178/2.8 \times 10^{-3}$
                         $= +800.0$ MPa

Total at top, $f_c = +7.58 - 8.17 - 8.29 = -8.88$ MPa
                       compression $< 0.45\,f'_c$ **O.K.**

Total at bottom, $f_c = -21.52 + 8.85 + 7.78 = -4.89$ MPa
                        compression $< \tfrac{1}{2}\sqrt{f'_c}$ tension

These totals put together stresses from Fig. 21.10f,d, and j.

The differences as a result of grouting are small at working stress, about 1 MPa less compression on the top and 2 MPa more compression on the bottom. The ultimate moment is more noticeably improved by grouting, as noted in Sec. 21.13.

21.13 ULTIMATE STRENGTH OF PRESTRESSED BEAMS

Since the high strength steels lack a sharp and distinct yield point (Fig. 21.11) and since prestress may modify the neutral axis location at failure, it is not possible to use ultimate strength relations developed for ordinary beams without some concern or caution. Theoretical considerations seem to call for a calculation process correlating strains on steel and concrete at failure.

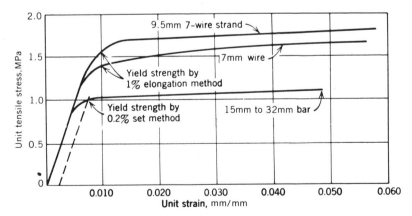

FIGURE 21.11 Typical stress-strain curve for prestressing steels. (From Lin.[1])

Nevertheless, for beams with bonded tendons and a percentage of steel low enough to ensure a tension failure, the strength method of analysis (Sec. 3.9) gives approximate values that seem to agree fairly well with tests. Since there is no sharp yield point, the Code, following a report by the ACI-ASCE Joint Committee on prestressed concrete, uses steel stress f_{ps} at nominal strength in this calculation rather than f_y. Fundamentally, the determination of f_{ps} depends on stress-strain diagrams and strain relationships, which means that f_{ps} is not a quantity easily established without detailed data. This situation is eased by two approximate formulas for f_{ps} for use in typical cases:

For bonded members, $f_{ps} = f_{pu}(1 - 0.5\, \rho_p f_{pu}/f'_c)$

For unbonded members, $f_{ps} = f_{se} + 70 + f'_c/(100\, \rho_p)$

but for unbonded members not more than the specified yield strength f_{py} or $f_{se} + 400$ MPa. The ultimate moment for rectangular beams, or for flanged beams with the neutral axis within the flange, is expressed by the relation

$$M_n = A_{ps}f_{ps}d(1 - 0.59\, \rho_p f_{ps}/f'_c)$$

which is the relation given in Sec. 3.8 for ordinary reinforced concrete except that f_{ps}, A_{ps}, and ρ_p take the place of f_y, A_s, and ρ.

To avoid approaching the overreinforced beam condition, steel for rectangular sections must not exceed that giving $\rho_p f_{ps}/f'_c = 0.30$. For larger steel ratios the value of M_n is limited more by compression conditions; the Commentary suggests $M_n = \phi(0.25\, f'_c bd^2)$ is reasonable.

As an example of approximate ultimate strength analysis, the pretensioned beam of Sec. 21.11 and Fig. 21.9 will be investigated assuming $f_{pu} = 1\,350$ MPa*. The failure conditions of Fig. 21.9j give:

$$\rho_p = 1\,000/250 \times 425 = 0.009\,41$$
$$f_{ps} = f_{pu}(1 - 0.5\, \rho_p f_{pu}/f'_c)$$
$$= 1\,350(1 - \tfrac{1}{2} \times 0.009\,41 \times 1\,350/35) = 1\,105 \text{ MPa}$$
$$\rho_p f_{ps}/f'_c = 0.009\,41 \times 1\,105/35 = 0.297 < 0.30 \text{ Usual } M_u \text{ eq. \textbf{O.K.}}$$
$$a = A_{ps}f_{ps}/(0.85\, f'_c b) = 1\,000 \times 1\,105/0.85 \times 35 \times 250 = 149 \text{ mm}$$
$$M_n = A_{ps}f_{ps}(d - a/2) = 1\,000 \times 10^{-6} \times 1\,105(0.425 - \tfrac{1}{2} \times 0.149)$$
$$= 0.387 \text{ MN} \cdot \text{m} = 387 \text{ kN} \cdot \text{m}$$
$$M_u = \phi M_n = 0.9 \times 387 = 348 \text{ kN} \cdot \text{m}$$

Reqd. $M_u = 1.4\, M_d + 1.7\, M_\ell$
$$= 1.4 \times 90 + 1.7 \times 100 = 296 \text{ kN} \cdot \text{m} < 348 \text{ kN} \cdot \text{m} \qquad \textbf{O.K.}$$

---

*This ultimate and the initial wire stress of $f_{si} = 930$ MPa in Sec. 21.11 are both low for the currently available strand (having f_{pu} of 1 860 MPa) or the wire (having f_{pu} of 1 760 MPa) with yields about 80 percent of ultimate.

This comparison shows about 15% surplus strength, compared to 6% at service load.

For the post-tensioned beam of Sec. 21.12 and Fig. 21.10a with $f_{pu} = 1\,350$ MPa and the steel well grouted, the failure conditions of Fig. 21.10 k give:

$$\rho_p = 1\,000/250 \times 420 = 0.009\,52$$

$$f_{ps} = f_{pu}(1 - 0.5\,\rho_p f_{pu}/f_c')$$

$$= 1\,350(1 - \tfrac{1}{2} \times 0.009\,52 \times 1\,350/35) = 1\,102 \text{ MPa}$$

$$a = 1\,000 \times 1\,102/0.85 \times 35 \times 250 = 149 \text{ mm}$$

$$M_n = 1\,000 \times 10^{-6} \times 1\,102(0.420 - \tfrac{1}{2} \times 0.149) = 0.380 \text{ MN} \cdot \text{m}$$

$$= 380 \text{ kN} \cdot \text{m}$$

$$M_u = \phi M_n = 0.9 \times 380 = 342 \text{ kN} \cdot \text{m}$$

$$\text{Reqd. } M_u = 1.4 \times 80 + 1.7 \times 90 = 265 \text{ kN} \cdot \text{m} < 342 \text{ kN} \cdot \text{m}$$

The section provides about 23% more capacity than required.

If the beam of Sec. 21.11 were left ungrouted, it would be contrary to Code 18.9 and member strength would be lowered, because the entire stretch of the tendon tends to accumulate at the first flexural crack. Some tests by Baker indicated cracks 3 to 10 times as wide as for bonded conditions. Such cracks raise the neutral axis and bring premature compression crushing.

For the beam just investigated, if grouting is omitted (Fig. 21.10l), the calculations become:

$$\rho_p = 1\,000/250 \times 420 = 0.009\,52$$

$$f_{ps} = f_{se} + 70 + f_c'/(100\,\rho_p), \text{ in MPa}$$

$$f_{ps} = (930 - 170) + 70 + 35/0.952 = 867 \text{ MPa}$$

$$\rho_p f_{ps}/f_c' = 0.009\,52 \times 867/35 = 0.235 < 0.30. \text{ Usual } M_u \text{ eq. } \textbf{O.K.}$$

$$a = 1\,000 \times 867/0.85 \times 35 \times 250 = 116 \text{ mm}$$

$$M_n = 1\,000 \times 10^{-6} \times 867(0.420 - \tfrac{1}{2} \times 0.116) = 0.314 \text{ MN} \cdot \text{m} = 314 \text{ kN} \cdot \text{m}$$

$$M_u = \phi M_n = 0.9 \times 314 = 282 \text{ kN} \cdot \text{m}$$

$$\text{Reqd. } M_u = 1.4 \times 80 + 1.7 \times 90 = 265 \text{ kN} \cdot \text{m} < 282 \text{ kN} \cdot \text{m}$$

The beam has 6% excess strength at ultimate when ungrouted, which can be compared with the 23% excess strength when grouted. Note that the Code permits the unbonded tendons only when some bonded reinforcement is cast into the tension area; the yield strength of these bars would automatically be available for increasing the available M_u and would control the cracking.

If calculations for ungrouted sections are based on correlating strains, as suggested in the second paragraph of this section, it is not difficult to introduce a factor to represent relative crack width. The effective steel strain (beyond initial prestress) can simply be taken as about one-fifth of the strain indicated by the strain triangles. This reduced steel strain, when added to the initial prestress strain, gives the total to be used with the stress-strain curve.

21.14 TENDON EFFECT CONSIDERED AS A NEGATIVE LOADING

Since simple beam tendon eccentricity develops a negative moment that offsets much or all the dead load moment, this eccentricity must be reduced where dead load moment is small; otherwise a negative moment failure can occur as prestressing is applied. With post-tensioned steel the tendons can be formed to any desired shape. With pre-tensioned wires or strands practice is often using hold-down devices to give the effect of bent-up steel. Other pre-tensioned beams have been made with a curved soffit, the curved beam axis thus producing a varying eccentricity.

A complete analysis can be made by considering all critical sections with their respective eccentricities. The relieving moment is (almost) proportional to the eccentricity, since the slope angles are small and the horizontal component of the prestress is essentially a constant. Hence the relieving moment diagram has the same shape as the curve of plotted e values.

A more convenient way for some to visualize the effect of curvature is in terms of a negative or upward load on the beam. A curved tendon under heavy tension requires vertical load to deflect it from a nearly horizontal line. A concentrated load at the center will deflect the tendon as in Fig. 21.12a, loads at both third points as in Fig. 21.12b, and a uniform load as in Fig. 21.12c, a parabolic curve. When embedded in a concrete beam in a slot having one of these shapes, the tendon after prestressing exerts upward forces on the concrete opposite to those required to deform the free tension tendon into the same shape.

If the parabolic tendon shape is used, the free body of Fig. 21.12d indicates from ΣM that $Pe = (w_p\ell/2)(\ell/4) = w_p\ell^2/8$, or $w_p = 8\,Pe/\ell^2$. Hence the tendon can be visualized as contributing an axial load and an upward uniform load w_p, both acting on the plain concrete section. The tendon can be shaped in such a way as to offset, nearly completely, any dead load distribution that exists.

If the member is inclined instead of horizontal, the eccentricity is from the gravity axis of the member, not from a horizontal reference line. Likewise, if the member is of variable depth, the eccentricity is from the centroidal axis. The eccentricity may then be defined as the vertical distance between tendon and centroidal axis and Fig. 21.12c indicates the

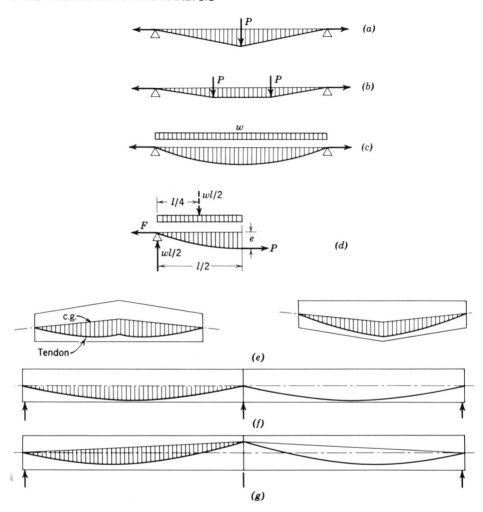

FIGURE 21.12   Relation between tendon curvature and load. Prestressed tendon exerts similar upward forces on (plain) concrete. (*c*) through (*g*) give uniform upward loads.

tendon shape that would correspond to the uniform upward load of Fig. 21.12*c* when the member is of varying depth. The shaded ordinates in (*e*) and (*c*) must be identical.

21.15 LOAD BALANCING

Lin[1] has expanded the negative loading concept to a design technique that balances with prestress any desired portion of the load, such as most or all of the dead load, or possibly under other conditions, dead load plus some portion of the live load.

The balanced load could ideally* provide no moment and no curvature or deflection. Hence, for continuous beams or slabs the unbalanced load alone may be used with ordinary elastic theory to compute the bending moments acting on the (nearly) uniform concrete section. This in the case of ungrouted tendons would be approximate in neglecting concrete displaced for ducts or tendons. Thus a continuous member might call for tendons, as in Fig. 21.12f or, what can be shown to be equivalent, the so-called "linearly transformed" tendon profile of Fig. 21.12g. The bend over the support obviously cannot be as sharp as shown and separate tendons for each span, possibly with inverted U strands over the reaction, or other modification may be needed. This is beyond the scope of this text.

21.16 SHEAR AND DIAGONAL TENSION

A tendon with varying eccentricity acts somewhat as a suspension cable, partially relieving the concrete not only of bending stress but also of shear stress. The shear thus carried by the tendon can be calculated either as the vertical component of the tendon pull or as the shear created by the equivalent upward loads discussed in Sec. 21.14. The remainder of the external load creates a shear which must be resisted by the concrete.

The axial compression from prestress reduces the diagonal tension stresses so long as the beam is uncracked from moment stress (Sec. 5.18). Cracking reduces this beneficial effect and inelastic deformations at loads near the ultimate largely eliminate it.

For a member having an effective prestress of at least 40% of the tensile strength of its reinforcement, the nominal shear resistance may be taken as

$$V_c = (0.05\sqrt{f_c'} + 5\ V_u d/M_u)b_w d \qquad \text{(Code Eq. 11-10)}$$

unless more detailed analyses (such as follows) are made. This V_c need not be taken lower than $\frac{1}{6}\sqrt{f_c'}$, and $0.4\sqrt{f_c'}$ is the upper limit. An exception is made where the transfer length of the prestressing tendons exceeds $h/2$, which calls for calculations using a reduced prestress force in the calculation for V_{cw} from Code Eq. 11-13.

For the more detailed methods the Code recognizes two types of shear failures.

The first failure type, web-shear cracking, starts from a web crack (principal diagonal tension stress of approximately $\frac{1}{3}\sqrt{f_c'}$) that develops

---

*Obviously changes in prestress from time effects of creep and relaxation modify the effective negative loading and upset the initial assumptions slightly.

into a failure without the presence of a moment crack at that section. This critical shear, called V_{cw}, is more apt to develop in thin web members such as I or T shapes. The equation from Code 11.4.2.2 is

$$V_{cw} = 0.3(\sqrt{f_c'} + f_{pc})b_w d + V_p \qquad \text{(Code Eq. 11-13)}$$

where $0.3\sqrt{f_c'}$ represents the shear strength considered as corresponding to a principal tensile stress of about $\frac{1}{3}\sqrt{f_c'}$ at the neutral axis and V_p is the component of prestress acting perpendicular to the axis of the member. Each of the last two terms increases the permissible external shear load, the longitudinal prestress by balancing out some of the potential diagonal tension and the normal component of prestress by reducing the vertical shear acting on the concrete.

An alternate calculation of V_{cw} is permitted as "the shear force corresponding to dead load plus live load that results in a principal stress of $\frac{1}{3}\sqrt{f_c'}$ at centroidal axis of member, or at intersection of flange and web" for a centroidal axis in flange.

The second failure type, related to ordinary reinforced concrete beam shear failures, is called flexure-shear cracking. A near vertical crack initiated by flexure becomes inclined as load increases (as in Fig. 5.2) and, unless it encounters a stirrup, goes on to a shear failure. For this case the Code 11.4.2.1 allows

$$V_{ci} = 0.05\sqrt{f_c'}b_w d + V_d + V_i M_{cr}/M_{max} \qquad \text{(Code Eq. 11-11)}$$

but $\qquad V_{ci} \not< 0.14\sqrt{f_c'}b_w d$

where V_d is the unfactored dead load shear at the section and M_{max} and V_i are computed from the load combination creating maximum moment* causing flexural cracking at the section and

$$M_{cr} = (I/y_t)(\tfrac{1}{2}\sqrt{f_c'} + f_{pe} - f_d) \qquad \text{(Code Eq. 11-12)}$$

In the above equations M_{cr} is the moment from externally applied loads causing flexural cracking at the section.

I = moment of inertia of the section

y_t = distance from centroid of gross section (neglecting reinforcement) to tension face of member

f_{pe} = compressive stress in concrete due to effective prestress (after losses) at extreme fiber where external loads cause tension

f_d = unfactored dead load stress at same extreme fiber as f_{pe}.

The smaller value, either V_{ci} or V_{cw}, limits the shear assigned to the concrete, although stirrups may be added to increase ultimate shear capacity.

---

*For heavy moving loads, maximum shear for stirrup design must control, instead of having the load produce maximum moment.

Minimum stirrups of $A_v = \frac{1}{3} b_w s / f_y$ are required wherever V_u exceeds half ϕV_c, assuming torsion is small. For effective prestress force not less than 40 percent of the tensile strength of flexural reinforcement, minimum shear reinforcement may be either that or

$$\text{Min. } A_v = \frac{A_{ps}}{80} \frac{f_{pu}}{f_y} \frac{s}{d} \sqrt{\frac{d}{b_w}}$$

where A_{ps} = area of prestressed reinforcement in tension zone

d = depth from compression face to centroid of longitudinal reinforcement $\not< 0.8\,h$

f_{pu} = ultimate strength of prestressing tendon

f_y = specified yield strength of non-prestressed reinforcement.

The maximum stirrup spacing is limited to $0.75\,h$ but not more than 600 mm.

When prestressed tendons are stopped (or left unbonded and unstressed) for a part of the member length, a problem in shear strength results very similar to that of bars cut off within a moment region of reinforced concrete. Until more is known about this phenomenon, excess stirrups should be used near these "cutoff" points.

21.17 DEVELOPMENT OF REINFORCEMENT

On pre-tensioned beams the self-anchorage of wires and strands at the ends of the beams leads to high bond stresses. This end anchorage requirement has received much attention and Lin's treatment[1] of the subject is quite complete.

For three- or seven-wire prestressing strand the Code (12.10.1) specifies a development length in mm

$$\ell_d = \tfrac{1}{7}(f_{ps} - 2\,f_{se}/3)d_b$$

where d_b is the nominal diameter in mm and the calculated stress f_{ps} and f_{se} (effective prestress after losses) are in MPa, but used here simply as constants. The specified length must be available between the end of the strand and the closest point where full strength is required, which should govern only in cantilever and short span members. The adequacy of this ℓ_d value is now under serious question.

Where bonding does not extend to the end of the member, the development length provided must be doubled.

In post-tensioned beams there is no problem of development because end anchorages are used. Even when flexural cracking occurs, the stresses adjacent to cracks when tendons are grouted are less important than in ordinary reinforced concrete where they are not separately computed.

21.18 DESIGN CONSIDERATIONS

Design considerations are treated *very* briefly. Allowable working load stresses have already been mentioned in analysis examples.

Since the optimum total prestress approximates half the allowable f_c multiplied by the total concrete area, economy is provided by reducing the concrete area near the centroid, where it is less effective in resisting moment, and by increasing the area near the extreme fibers. Hence I, T, double T, and ⌐¬ shapes are the general types that prove economical. Tables of section modulus values for such shapes are helpful in picking a size.

It is sometimes stated that the prestress carries the service dead load and that the size of the beam depends only on the live load. The negative moment $P_s e$ at transfer can be made to offset the dead load moment. Of course, as P_s decreases in time to P_{se}, part of this offsetting moment is lost. Furthermore, at ultimate strength the effect of prestress is largely lost and the beam must carry all the moment, essentially as an ordinary beam.

Referring to Fig. 21.6b, it will be noted that the top stress under prestress and dead load may be a tension as large as $\frac{1}{4}\sqrt{f_{ci}}$. Under the full load (Fig. 21.6c) this stress may be the full allowable compression $0.45 f_c'$. Thus, except for the effect of prestress losses, the full range of $\frac{1}{4}\sqrt{f_c'} + 0.45 f_c'$ is available to take care of M_ℓ. Likewise, on the bottom the stress range for M_ℓ can be from $0.60 f_{ci}$ to $\frac{1}{2}\sqrt{f_{ci}}$, except for the effect of prestress losses. Magnel[4] has thus worked out design requirements for a section modulus established from M_ℓ alone divided by the major part of this stress range. This approach is most suitable when M_ℓ is large compared to M_d. It indicates the saving sometimes possible in beam size compared to an ordinary beam which must be sized for $M_d + M_\ell$.

Lin has developed[1] design procedures for elastic designs based on visualizing the internal resisting couple. These are very straightforward in application. He has also devised a "load balancing" method for slabs and beams, already mentioned in Sec. 21.15.

21.19 CONTINUOUS BEAMS

Continuous beams cannot be considered here except to mention one special condition that often controls. Since an eccentrically placed prestressing tendon itself develops bending moment and curvature in a beam, the prestressing operation may well require reactions (up or down) to hold the member in contact with its supports. These reactions change the prestressing moment from the simple $P_s e$ value to something more involved. Cables can be so arranged as to cause zero external reactions, but this appears to be an unnecessary restriction. Lin[1] gives a complete discussion of this problem following basic methods originally suggested by Guyon[3] and has simplified it considerably in connection with his "load balancing" method.

SELECTED REFERENCES

1. T. Y. Lin, *Design of Prestressed Concrete Structures*, 2nd ed., John Wiley and Sons, New York, 1963.
2. James R. Libby, *Modern Prestressed Concrete*, Van Nostrand, Princeton, N.J., 1971.
3. Y. Guyon, *Prestressed Concrete*, John Wiley and Sons, New York, 1953.
4. Gustave Magnel, *Prestressed Concrete*, Concrete Publications, Ltd., London, 2nd ed., 1950.
5. N. Khachaturian and G. Gurfinkel, *Prestressed Concrete*, McGraw-Hill Co., New York, 1969.
6. *Standard Specifications for Highway Bridges*, AASHTO, Washington, 12th ed., 1977.
7. PCI Comm. on Prestress Losses, "Recommendations for Estimating Prestress Losses," *PCI Journal*, V. 20, No. 4, July–Aug. 1975, p. 44.

PROBLEMS

PROB. 21.1. The rectangular beam of Fig. 21.13 has pre-tensioned steel at two levels as shown with $f_{pi} = 1\,200$ MPa. Consider $f'_{ci} = f'_c = 350$ MPa, $n = 7$,

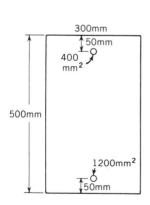

FIGURE 21.13 Beam for Prob. 21.1.

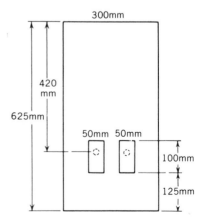

FIGURE 21.14 Beam for Probs. 21.2 and 21.3.

and a simple span of 12 m. Calculate concrete stresses at top and bottom and both steel stresses when prestressed tendons are first cut between units.

PROB. 21.2. The post-tensioned beam of Fig. 21.14 has a 12 m simple span. It has two holes as shown for tendons having a total area of $1\,400$ mm^2, which after tensioning will be 45 mm below the top of beam. $f'_c = 35$ MPa, $f_{so} = 1\,200$ MPa. Considering the holes ungrouted and without using the approximation of area as the gross area of the concrete, calculate:

(a) Concrete stresses initially under dead weight alone after post-tensioning.

(b) Concrete stresses after shrinkage and creep reduce the steel stress to $f_{se} = 1\,070$ MPa and a live load of 9 kN/m acts on the beam.

PROB. 21.3

(a) Recalculate condition (b) of Prob. 21.2 if each beam hole is well grouted.

(b) Calculate the ultimate moment capacity of this beam and its over-all factor of safety assuming it is well grouted and has $f_{pu} = 1\,750$ MPa.

PROB. 21.4.

(a) If the pre-tensioned beam of Fig. 21.9 is changed to $f'_c = 35$ MPa, $f_{si} = 1\,000$ MPa, $n = 7$, $A_s = 700$ mm^2 with M_d still 90 kN $\cdot$ m, calculate all stresses after the tendons are cut.

(b) If the tendons are straight and concrete stresses are limited to a (bottom) compression of 20 MPa and a top tension of $\frac{1}{4}\sqrt{f'_c}$, what is the maximum tendon eccentricity that can be used? [The student should note that M_d at the support would be zero. The change in c.g. and I from that used in part a may be ignored here and in (c), as a problem simplification.]

(c) If the tendons are draped such that the condition at the end of the span does not control, and if f_{so} decreases to f_{se} of 800 MPa, what service live load moment is permissible with the allowable f_c of 0.45 f'_c and allowable tension of $\frac{1}{2}\sqrt{f'_c}$? (Ultimate load check is also important, but not a part of this problem.)

22
COMPOSITE BEAMS

22.1 THE NATURE OF COMPOSITE BEAMS

Composite beam is the name given the combination of a steel beam with a concrete slab, or the combination of a precast concrete beam with a cast-in place concrete slab, when the two are so connected together that they act as a single unit in resisting flexure. Because every reinforced concrete slab or beam is really a composite member of steel and concrete, the basic theory for flexure requires almost no new concepts. Likewise the bonding together of the two units is theoretically a simple problem of horizontal shear. A roughened concrete surface has a considerable bonding strength if the slab is held tightly in place by stirrups extending into the slab. Shear lugs can be welded to steel beams to perform the same function.

Composite construction is economically important because the steel beams or precast concrete beams* can furnish supports for the slab forms and the dead weight of the concrete while, at the same time, the beams can be made lighter because the slab forms a stiff attached flange which helps in resisting live loads. The primary bending moment in a slab calls for slab steel transverse to the beam and hence neither this slab steel nor slab bending seriously complicates the analysis of the composite beam.

The engineering problems hinge around three considerations: flexural strength, horizontal shear strength, and deflection. The latter may call for shoring under the beams during construction since this makes the entire section available for resisting dead load as well as live load. The ultimate strength is not significantly influenced by shoring, shrinkage, or creep.

---

*Many pretensioned prestressed concrete beams are cast with stirrups extending out the top to bond into a cast-in-place slab to form a composite T-beam.

22.2 FLEXURAL STRENGTH WITH SLAB CAST ON STEEL BEAM

Flexural strength could be considered at three different stages of loading: under dead load prior to the time when the slab is effective for strength, under live load with the slab acting effectively, and at ultimate under overload conditions.

Consider first a symmetrical steel beam with the cast-in-place slab. If there is no shoring the beam acts to carry its own weight, the slab forms, the slab concrete, and all the superimposed construction load. As the concrete sets, the steel beam carries Mc/I stresses for the steady portion of this unfactored load, a very simple calculation, as indicated in Fig. 22.1a. After the concrete gains strength, the section becomes the transformed area of Fig. 22.1b for any further load, it being simplest to transform the concrete into equivalent steel by the relation A_c/n or b/n. This results in the addition of stresses as shown in Fig. 22.1c.

For long-time loads some multiple of n may be used to represent the effects of creep and shrinkage, up to $3n$.

Since neither creep nor shrinkage influences ultimate strength significantly, the ultimate design condition will first be discussed.* The neutral axis in deep beams will usually fall below the slab leading to the stress distribution shown in Fig. 22.2a. Viest, Fountain, and Singleton[2]

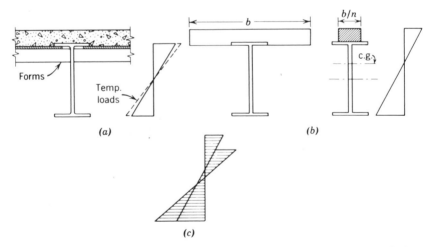

FIGURE 22.1   Stresses in an unshored composite beam. (a) Dead load alone. (b) Live load alone. (c) Total stresses.

---

*Although the AISC specification is written in terms of service loads, the specification is an adaptation from behavior at ultimate.

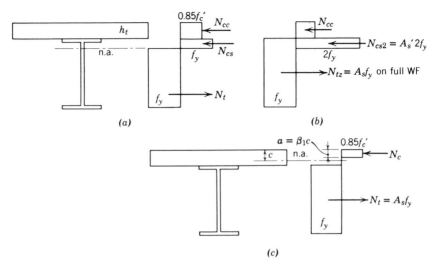

FIGURE 22.2 Ultimate stress conditions. (a) Neutral axis below slab. (b) Equivalent to (a) using N_t on full A_s. (c) Neutral axis within slab.

have pointed out that this is equivalent to Fig. 22.2b, which is considerably simpler to use.

The same authors point out, especially in buildings, that the neutral axis may fall in the slab, as in Fig. 22.2c, in which case the tension on the concrete below the neutral axis is ignored and the concrete compression above the neutral axis is handled as in any T-beam.

When the neutral axis appears close to the level of the top flange the lower level of strains nearby can be considered, at least roughly, as in the following design example.

The later discussion on deflection will point out that beams are frequently shored enough to avoid deflection and stresses of consequence from the dead weight of the slab until after the concrete slab is strong enough to act as a flange. In this case the dead load stresses of Fig. 22.1a are small, usually only from the weight of the beam itself, and the composite beam stresses of Fig. 22.1b become the major part. This shoring does not influence the ultimate moment capacity significantly.

22.3 EXAMPLE OF FLEXURAL CALCULATIONS

For a simple span of 7.8 m check the flexural capacity of an $f_y{}^* = 250\,\text{MPa}$ steel W 460 × 74 on 2.1 m center with a 100 mm slab to carry a service load of its own weight

*The AISC uses F_y.

and a live load of 20 kPa. Assume the beam is shored at the quarter points so as not to pick up the weight of the concrete until it reaches its intended f_c' of 20 MPa. The beam properties include $A = 9\,840 \text{ mm}^2$, $I = 334 \times 10^6 \text{ mm}^4$, $b_f = 190$ mm, and flange thickness $= 14.5$ mm. (The beam in such a case is usually designed by a service load procedure covered in the AISC Specifications.)

Solution

A composite beam based on a steel shape is usually designed under the AISC Specification,[3] which is really an ultimate design restated in terms of service load stresses. The effective flange width used here will be taken as that recommended by AISC, that is, the smallest of:

1. Spacing of beams $= 2.10$ m
2. Flange width of steel section plus 8 times slab thickness as the overhang on each side $= b_f + 16h_f = 0.19 + 1.60 = 1.79$ m
3. Beam span/4 $= 1.95$ m

The 1.79 m width controls.

This text section relates to flexural strength only. The necessary shear connectors are covered in Sec. 22.6 and deflections are discussed in Sec. 22.5.

Here the given section will first be checked by applying ACI Code strength concepts to the analysis of Fig. 22.2. It will then be rechecked by the AISC procedures.

Whether the beam is or is not initially shored, the ultimate condition includes all loads on the full span with the steel shape considered as yielding throughout its depth, or nearly so. This total tension fixes the depth of the compression block, as in a homogeneous section.

$$N_{nc} = N_{nt} = 250 \times 9\,480 \times 10^{-6} = 2.37 \text{ MN}$$

$$a = N_{nc}/(0.85f_c'b) = 2.37/0.85 \times 20 \times 1.79 = 0.078 \text{ m} = 78 \text{ mm}$$

$$c = a/0.85 = 78/0.85 = 92 \text{ mm}$$

The neutral axis is in the concrete, but the sketch in Fig. 22.3a indicates it is too low to put the upper steel flange at the full f_y of 250 MPa. Since N_{nt} is then less, try $c = 80$ mm and check whether the tension justifies this value. Tension is reduced

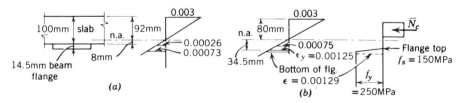

FIGURE 22.3   Ultimate analysis of Sec. 22.3. (*a*) First trial n.a. (*b*) Final n.a.

(Fig. 22.3b) by roughly $\frac{1}{2}(250 - 150) \times 0.19 \times 0.0145 = 0.14$ MN

$$N_{nt} = 2.37 - 0.14 = 2.23 \text{ MN}$$
$$c = 92 \times 2.23/2.37 = 86 \text{ mm} \qquad \textbf{O.K.}$$
$$M_n = 2.37(\tfrac{1}{2} \times 460 + 100 - 34.5) - 0.14(100 - 34.5 + \tfrac{1}{3} \times 14.5)$$
$$= 700 - 10 = 690 \text{ kN} \cdot \text{m}$$

The small influence of the reduction term confirms that its exact evaluation is not critical.

$$\text{Wt. beam} = 74 \times 9.8 \times 10^{-3} = 0.72 \text{ kN/m}$$
$$\text{Wt. slab} \quad = 2.1 \times 0.1 \times 24 = \underline{5.04}$$
$$\overline{5.76} \times 1.4 = 8.06 \text{ kN/m}$$
$$\text{L.L.} \qquad = 2.1 \times 20 = 42.00 \times 1.7 = \underline{71.40}$$
$$\text{Total} = \overline{79.46} \text{ kN/m}$$

$$M_u = \tfrac{1}{8} \times 79.46 \times 7.8^2 = 604.3 \text{ kN} \cdot \text{m}$$
$$M_n = M_u/\phi = 604.3/0.9 = 671 \text{ kN} \cdot \text{m} < 690 \text{ kN} \cdot \text{m} \qquad \textbf{O.K.}$$

The beam is safe at ultimate, by 3%

The beam will now be checked for strength under the service load procedure of the AISC Specification,[3] with allowable $f_s = 165$ MPa and allowable $f_c = 0.45 f'_c = 9$ MPa. Although using the service load format, this calculation derives its validity directly from ultimate strength calculations. Shrinkage and creep strains do not influence ultimate strength significantly and hence are not introduced in these strength calculations for either shored or unshored beams. (These strains do influence service load deflections; see Sec. 22.5.)

Since this is a shored* beam, the total moment applies directly to the composite section. For 20 MPa concrete, $n = 9$, which permits the flange to be considered as transformed into steel of width $1\,790/9 = 199$ mm as in Fig. 22.4. Moments about the bottom of the flange establish the neutral axis:

$$(199 \times 100)\tfrac{1}{2} \times 100 - 9\,840 \times \tfrac{1}{2} \times 460 = y(19\,900 + 9\,840)$$
$$\bar{y} = -42.6 \text{ mm}$$
$$I \text{ of slab} = \tfrac{1}{2} \times 199 \times 100^3 = \quad 16\,583\,000$$
$$+ (199 \times 100)(\tfrac{1}{2} \times 100 + 42.6)^2 = 170\,640\,000$$
$$I \text{ of } W = 334\,000\,000$$
$$+ 9\,480(\tfrac{1}{2} \times 460 - 42.6)^2 = 332\,926\,000$$
$$I = \overline{854\,149\,000} \text{ mm}^4 = 854 \times 10^{-6} \text{ m}^2$$

$$M = \tfrac{1}{8}(5.76 + 42.00) \times 7.8^2 = 363 \text{ kN} \cdot \text{m} = 0.363 \text{ MN} \cdot \text{m}$$
$$f_s = Mc_s/I = 0.363(0.460 - 0.040)/854 \times 10^{-6} = 178.5 \text{ MPa} > 165 \text{ MPa}$$

---

*If the beam had been unshored initially, the stresses would have to be combined from two calculations: first, those for the total dead load on the steel shape alone and, second, the live load on the total transformed area.

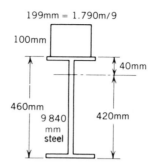

FIGURE 22.4 Transformed area for analysis in Sec. 22.3.

The section is overstressed by about 8% in tension. Compression is usually satisfactory:

$$f_c = (Mc_s/I)/n = 0.363(0.100 + 0.040)/9 \times 854 \times 10^{-6} = 6.6 \text{ MPa} < 9 \text{ MPa} \qquad \textbf{O.K.}$$

Attention is called to the AISC limitation on unshored beams expressed by limiting the section modulus S_{tr} (referred for a positive moment to the bottom flange stress, or for a negative moment to each flange) of the transformed composite section to

$$S_{tr} \leqslant (1.35 + 0.35 \ M_L/M_D)S_s$$

where M_L and M_D are live and dead load moments and S_s is the section modulus of the steel section alone. In this example:

Actual $S_{tr} = 854 \times 10^6/(460 - 40) = 2.033 \times 10^6 \text{ mm}^3$

Limiting $S_{tr} = (1.35 + 0.35 \ M_L/M_D)S_s = (1.35 + 0.35 \times 42/5.76)(334 \times 10^6/230)$

$$= 5.667 \times 10^6 \text{ mm}^3$$

Since $2.033 \ll 5.667$, this loading does not require shored beams. This provision is used by AISC to guard against excessive tensile stress at service loads, stresses not fully reflected in the nominal stress calculation above.

The AISC has many suggestions and rules[4] to cover composite construction. Our purpose here is to illustrate the calculations for a simple case.

In selection of steel shapes for composite slab construction a useful rule-of-thumb is that effective section modulus of the steel shape is increased about 50 percent by the composite slab. Thus, the required section modulus of such a steel beam will be approximately two-thirds that otherwise required.

22.4 FLEXURAL STRENGTH WITH SLAB CAST ON PRECAST CONCRETE BEAM

When the basic element is a precast beam, either with or without pre-stressing, the details of the calculations change, but not the general ideas. For example, without prestressing, the precast beam by itself must carry its own weight and this is a larger item than with a steel beam. If it is partially shored such that the slab weight is carried by shores until the slab reaches its design strength (say, $0.75f_c'$), the analysis proceeds as above, with the transformed area expressed as equivalent concrete instead of steel. In this case the concrete must be considered cracked wherever tension strains exist, but this is not a serious problem to handle.

The chief difference between using reinforced concrete and steel as the basic beam lies in the horizontal shear provisions discussed in Sec. 22.6.

22.5 DEFLECTIONS

For the uniform load and steel section of Sec. 22.3, deflections would be calculated at service loads by $(5/384)w\ell^4/EI$, using the transformed area (in steel) for the stress calculation. In this case, however, the long-term dead load deflection must be separately computed with n assigned at least twice the nominal value. The live load deflection would be based on the I with the usual value of n. The AISC suggests as a starting point in design a depth-span ratio of $1/22$ for $f_y = 250$ MPa and $1/16$ for $f_y = 350$ MPa, but deflection must still be calculated.

For composite construction involving precast reinforced concrete beams, the Code (9.5.5.1) provides that a composite member with shored con-struction may be considered as a cast-in-place member with the minimum thickness requirements of Code Table 9.5a* controlling unless deflections are computed. In unshored members the precast element existing before the slab is cast is the one to meet this thickness requirement; the composite member with slab may then be considered satisfactory† without deflection check. The precast portion may need investigation for long-time deflection occurring prior to the time the slab becomes effective.

The requirements of Code Table 9.5a may be waived if deflections are calculated and these are less than the values in Code Table 9.5b. Such calculations must involve differential shrinkage between precast and cast-

---

*See Table 3.2 in Sec. 3.8.

†This Table 9.5a does not apply when members are attached to or support partitions likely to be damaged by large deflections. For a short summary of Code Table 9.5b requirements governing in this case, see footnote near start of Sec. 8.5c.

in-place parts of the member. Otherwise the requirements and methods discussed in Chapter 8 are appropriate.

Composite beams with prestressed, precast sections are quite common, but are not covered here. The Code requires that deflections *always* be computed for these members. The 1971 Code Commentary suggested the calculation method of References 5, 6, and 7 as a suitable one.

22.6 HORIZONTAL SHEAR

(a) General

With steel beams all the vertical shear should be carried by the web of the beam. With reinforced concrete precast elements the vertical shear may be considered carried by the precast elements or by the entire composite member as in monolithic construction.

The special problem of composite beams is the horizontal shear between flange and beam below, whether it is a steel beam or a precast concrete beam. Steel beams may be used with either full or partial shear transfer; the reader is referred to the Steel Manual[3] for the partial shear transfer case. Full shear transfer is required by the ACI Code for reinforced concrete members.

Theoretically the horizontal shear varies along the beam length as the external shear varies and could be calculated on a $v = VQ/Ib$ basis. However, both in steel and concrete composite beam work it is recognized that other distributions of the horizontal shear resistance will also serve adequately, except in the case of large concentrated loads in a positive moment region near a point of inflection or near the end of a simple span.*

(b) Steel-Concrete Composite Beams

Although spirals or short transverse lengths of small steel channels welded to the flange make adequate shear connectors, the stud connector is most commonly used. The stud is round, automatically welded to the flange, and is topped with a hook or flange making a head that can engage the concrete slab and prevent any vertical separation that may tend to occur between slab and beam. The stud height must permit at least 25 mm of concrete cover over the head.

The AISC Manual[4] tabulates the shear load permitted on each stud or channel connector, for example, 50 kN for a 20 mm × 75 mm stud in 20 MPa concrete made from ordinary aggregate. This is a service load equivalent,

---

*The partial shear transfer with the steel beam is an extreme case of such reassignment of shears.

since AISC uses a service load computation (derived from ultimate strength values). The total (service load) horizontal shear between points of maximum and zero moment is taken as 0.5 of the ultimate strength of the concrete flange in compression, or 0.5 of the tensile yield strength of the entire steel section, whichever is smaller, that is, the smaller of

$$0.85\, f'_c bh_f/2 \qquad \text{or} \qquad A_s f_y/2$$

The shear is considered fully transferred if the connectors used have a total shear capacity of this amount. The connectors may be spaced uniformly from maximum moment point to zero moment point.*

(c) Reinforced Concrete Composite Beams

The designer may assume full shear transfer if four requirements given in Code 17.5.2 are met:

1. Contact surfaces are clean and intentionally roughened (full amplitude approximately 5 mm).
2. Minimum ties are provided extending into the slab at a spacing not more than four times the least dimension of the element or 600 mm $(A_v = \frac{1}{3}\, b_w s/f_y)$
3. Web below the slab is designed to resist the entire vertical shear.
4. The ties in item 2 are fully anchored in both elements.

In all other cases the horizontal design shear v_{nh} must be computed and satisfied.

The horizontal shear may be computed at any cross section from the relation

$$v_{nh} = V_u/\phi(b_v d)$$

where $\phi = 0.85$
  b_v = width being investigated
  d = effective depth (compression face to centroid of tension reinforcement) for the entire composite member.

Alternatively, the compression (or tension) in any segment may be computed, divided by 0.85, and the resulting force considered as the total horizontal shear to be transferred. The two alternates permit either a v_{nh} varying as the external shear or considered as uniformly distributed. In the latter case special consideration appears appropriate where large concentrated loads fall close to the end of a simple span or close to the point of inflection in a positive moment region of a continuous member.

---

*With some further check needed if a large moment closer to the zero moment point can occur.

For clean surfaces the resisting v_{nh} may be evaluated on the basis of the provision of minimum ties, or of an intentionally roughened surface, or both. The allowable v_{nh} is 0.5 MPa for *either* minimum ties or intentionally roughened surface *alone*, or 2.4 MPa for *both together*. These shears act over the contact area between slab and precast element where flexural strains might otherwise cause horizontal slip to occur. If the horizontal shear exceeds 2.4 MPa, the concepts of shear friction discussed in Sec. 5.20 should be used.

22.7 CONTINUOUS BEAMS—AASHTO SPECIFICATION

When beams are continuous over a support, the steel in the slab parallel to the beam may be considered a portion of the transformed area, with the slab concrete neglected because it is in tension. In such a case, the requirements for connectors are computed separately for positive and negative moment regions. The AASHTO Specification is an ultimate strength one insofar as connectors are concerned. Hence the connectors in the negative moment region are required to resist horizontal shear equal to the area of the longitudinal slab steel (within the effective width of the slab) times its yield stress. The AASHTO Specification limits the *total* effective slab width to $12h_f$, more restrictive than the AISC value of $16h_f$ added to the steel flange width.

Because there is some objection to welding connectors to the tension flange in the region of maximum tension, the AASHTO Specification provides that the pitch may be modified to avoid connectors at locations of high tension in the flanges. It allows closer spacings to be used away from the maximum negative moment point (closer to the point of inflection) to maintain the total number required while avoiding the highest flange stresses. Alternatively, the total negative moment may be assigned to the steel beam alone, but in such a case added connectors are required at the dead load point of inflection (within a length equal to one-third of the effective flange width, on either side of or centered on the *P.I.*), based on a fatigue requirement discussed in the next paragraph.

The basic design for connectors in the AASHTO Specification is based on fatigue and the ultimate strength design above is used as a check rather than as the chief consideration. The equation used (in AASHTO notation) is:

$$S_r = V_r Q / I$$

where S_r = range in horizontal shear stress per unit length at the connection
V_r = range in shear due to live load plus impact, that is, the difference between maximum and minimum shear envelopes, excluding dead load.

Q, I = statical area moment above the junction and moment of inertia of the total transformed area, both about the neutral axis

The reader is referred to that specification for the allowable shear on a connector, which is written in terms of length of channel or square of the diameter of stud multiplied in each case by factors which vary substantially with the design number of cycles.

It is noted that AASHTO requires 50 mm of concrete over the connectors and requires that they penetrate at least 50 mm into the slab.

SELECTED REFERENCES

1. ACI-ASCE Committee 333, "Tentative Recommendations for Design of Composite Beams and Girders for Buildings," *Jour. ACI*, 32, No. 6, Dec. 1960; *Proc. 57*, p. 609.
2. Ivan M. Viest, R. S. Fountain, and R. C. Singleton, *Composite Construction in Steel and Concrete*, McGraw-Hill Book Co., New York, 1958.
3. "Specifications for the Design of Structural Steel for Buildings," American Institute of Steel Construction, New York, 1969, with Supplements Nos. 1, 2, and 3, of 1970, 1971, and 1974.
4. *Manual of Steel Construction*, American Institute of Steel Construction, New York, 7th Edition, 1970, revised 1973.
5. Subcommittee 5, ACI Committee 435, "Deflections of Prestressed Concrete Members," *Jour. ACI*, 60, *No. 12*, Dec. 1963, p. 1697.
6. Subcommittee 2, ACI Committee 209, "Prediction of Creep, Shrinkage, and Temperature Effects in Concrete Structures," *Designing for the Effects of Creep, Shrinkage, and Temperature in Concrete Structures*, SP-27, American Concrete Institute, Detroit, 1971.
7. D. E. Branson, B. L. Meyers, and K. M. Kripanarayanan, "Time-Dependent Deformation of Noncomposite and Composite Prestressed Concrete Structures," Symposium on Concrete Deformation, *Highway Research Record 324*, Highway Research Board, Washington, D.C., 1970, p. 15.
8. Roger G. Slutter and John W. Fisher, "Fatigue Strength of Shear Connectors," *Highway Research Record 147*, Highway Research Board, Washington, D.C., 1966, p. 65.

PROBLEMS

PROB. 22.1.   Calculate the ultimate moment capacity of a structural steel W 610 × 113 beam spaced at 2.4 m on centers and adequately anchored to a 125 mm slab, assuming A36 steel (f_y = 250 MPa) and f'_c = 30 MPa.

PROB. 22.2. In Prob. 22.1, assume the beam is not shored. Calculate the AISC service load stresses for the dead load including 1 kPa for forms and a construction live load equal to 1.5 kPa. Simple span of 15 m.

PROB. 22.3. Calculate the AISC design horizontal shear to be developed between beam and slab in Prob. 22.1 under full live load.

23

DETAILING FOR
SEISMIC RESISTANCE

23.1 SEISMIC AREAS

A growing awareness exists among code authorities that seismic areas are not restricted to California and the West. Instead, some hazard exists over major portions of the United States, as indicated on the map of Fig. 23.1 taken from the Uniform Building Code of 1970. For the purpose here, the zone designations are more simply and less sharply characterized than in the Code.

> Zone 0—no damage
> Zone 1—minor damage
> Zone 2—moderate damage
> Zone 3—major damage

The Zone 0 area is quite limited, which means there must be a more general design acceptance of some earthquake hazard over most of the United States.

23.2 SEISMIC DESIGN—GENERAL OBJECTIVES

Under the present state of the art, earthquake hazard has to be treated as a matter of probability, based on past history of shocks, geologic structures, and evidence of previous faulting. The map of Fig. 23.1 does not consider frequency of occurrence.

Blume, Newmark, and Corning[1] consider several different design criteria appropriate in seismic design. First, earthquakes of an intensity that might be expected to occur several times during the life of the structure

631

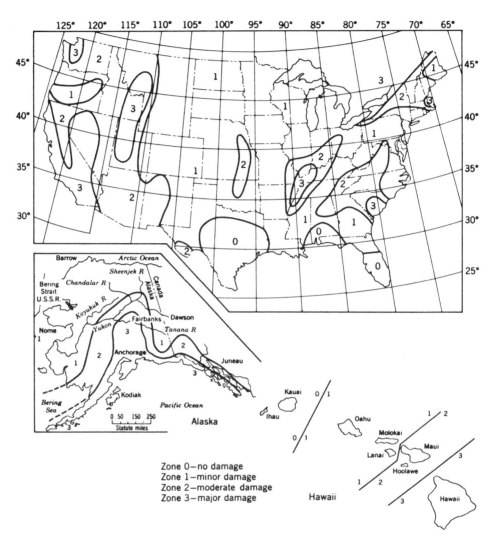

FIGURE 23.1   Seismic zone map of United States, from 1970 Uniform Building Code.

should be so covered by such a design that no damage will result, or at most that minor repairs will suffice if an unusual *number* of such earthquakes should occur.

Second, under the most severe *probable* earthquake, the structure should not collapse with possible loss of life, and severe damage should not occur. Design against this earthquake involves incorporating a reserve of strength in the structure beyond the elastic range.

Third and finally, even in an unusual earthquake—somewhat greater than that probable during the structure's life—the building must not collapse, even though under these conditions it may require major repair.

The uncertainty in any probability evaluation is such that the second and third conditions must be accepted as potentially possible. Also, it is not economically feasible to design to avoid all such damage. These conditions imply reversal of axial forces, reversal of shears, and reversal of moments. The ductility of a structure then becomes the governing consideration, ductility even in the presence of reversal of stresses.

It is not within the scope of this book to discuss the natural frequency of the structure or the methods used in defining design forces. But the basic requirements that beams, columns, joints, and walls must include for adequate service in a ductile moment-resisting space frame are summarized as Code Appendix A presents them. Appendix A applies where both (1) major damage has a high probability of occurring and (2) the total lateral seismic forces have been *reduced* because of using the "ductile moment-resisting space frame" with or without special shear walls. Seismic resistance is a complex area and Appendix A represents minimum requirements, not a guarantee of performance. (See Sec. 23.8.)

23.3 LIMITATIONS ON MATERIALS

Minimum concrete strength is specified as f'_c of 20 MPa. No maximum is set because ductility is primarily achieved by the use of members under-reinforced, a method that eliminates compression failure until ductility limits have been exceeded. The Uniform Building Code requires that connections and panel joints be designed to allow movement between stories of not less than twice the story drift from computed wind or seismic forces. Blume et al[1] recommend a ductility factor of at least 4 (measured as the ratio of maximum deflection to deflection at first yield) and discuss ductility ratios from 4 to 6 as reasonable in the context* of the then (1959) Structural Engineers' Association of California design code.

The maximum specified yield strength of reinforcement is limited to 400 MPa. If the design is based on Grade 300 bars, a higher grade such as Grade 400 may not be substituted.

23.4 REQUIREMENTS FOR FLEXURAL MEMBERS
(a) Primary Reinforcement Requirements

The maximum steel ratio ρ is limited to 0.5 ρ_b in order to insure a definite ductility, just as the nominal yield stress itself is limited.

---

*Any variation in code rules shifts the computed point at which first yield occurs; any earthquake code thus presents its own relatively arbitrary reference point.

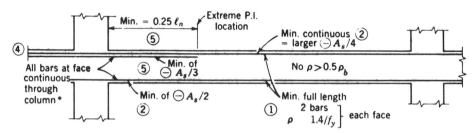

FIGURE 23.2 Minimum requirements for arrangement of reinforcement in beams. (*If impossible, see text.)

The various minimum requirements for A_s are summarized in Fig. 23.2. These minimums are:

1. Minimum top and bottom steel for the full length of beam, at least two bars top and bottom; on each face at least $\rho = 1.4/f_y$, which is 0.47% for Grade 300 and 0.35% for grade 400.
2. For top steel, a minimum of one-fourth of the larger negative moment A_s must be continuous all across the span.
3. At each support minimum bottom steel must equal half of the negative moment steel.
4. *All* bars at the face of column, top and bottom, shall extend *through the support into the adjacent beam*, if possible. Where, because of a change in member size this requirement is impossible for some bars, any bars stopped must be anchored into the column for their full f_y.
5. At least one-third of the negative moment tension bars must be carried to the extreme range of the *P.I.* and in no case less than 0.25 ℓ_n from the support.

(b) Bar Development and Splicing

Where a beam frames into a column with no beam continuing on the other side, all bars shall be extended to the far face of the confined region (Sec. 23.6) and anchored to develop f_y from the near face of column. "Every bar shall terminate with a standard 90 degree hook or with a hook and additional bar extension where needed to provide the required development length."* In the confined region the use of two-thirds the required development length ℓ_d of Sec. 7.10 (Code 12.2) is authorized; elsewhere the usual development length is required except that the minimum is set at 400 mm instead of the usual 300 mm.

---

*The effectiveness of extensions beyond the 90° hook is essentially nil, except possibly at the extreme state where the concrete at the start of the hook is seriously damaged.

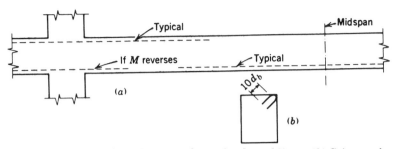

FIGURE 23.3 (*a*) Tensile zones shown by dotted lines. (*b*) Stirrup-tie.

No lap splices are permitted in a tensile zone or a zone of reversing stress (Fig. 23.3), unless stirrup-ties are provided at the lap as required around compression bars. This requirement is for stirrup-ties spaced at not more than 16 d_b or 300 mm. At all beam ends, such stirrup-ties must extend at least 2 d from the column face.

At all tension splices, at least two stirrup-ties shall be provided and splice lengths must be at least 24 d_b or 300 mm.

No welded splice may fall within an effective depth d of a plastic hinge.

(c) Shear Type Reinforcement for Flexure

Ordinary web reinforcement must be increased by designing for the sum of gravity load shears and the shears necessary to balance hinging moment at both ends of the beam, that is, an extra $V_h = 2\,M_h/\ell$ of either sign, since hinging moments M_h tend to be additive on the beam as a free body and may increase either positive or negative shear. The Code suggests that M_h may be taken the same as the usual M_u which would come from analysis of the cross section.

Wed reinforcement of at least #10 at a spacing of $d/2$ is required over the full length of the member.

Within the region near each end of the member for a length of 4 d, web reinforcement shall be at least

$$A_v d/s = 0.15\,A_s \text{ (or } 0.15\,A_s' \text{ if larger)}$$

and the spacing there shall not exceed $d/4$. The first stirrup shall be placed within 75 mm of the column face and the first two shall be stirrup-ties. Stirrup-ties (Fig. 23.3b) are closed stirrups that end in 135 degree bends (hooks) hooked around longitudinal bars and extended an additional 10 d_b past the bend.

Wherever inelastic deformation may cause the development of ultimate M_u *away* from the end of a member (as a concentrated load near midspan) this point is to be treated as a column face insofar as applying the requirements of the paragraph just above. In addition, the spacing of these stirrup-ties shall not exceed 16 d_b or 300 mm for at least 2 d from this point.

23.5 REQUIREMENTS FOR COLUMNS

(a) General

Joints between beams and columns are covered in Sec. 23.6.

Columns must carry from 1 to 6% of vertical reinforcement and no excess concrete may be disregarded in satisfying this minimum. Unless the strength of the column cores is sufficient to resist the design load, the sum of the moment strengths of the column at the joint (above and below) must exceed the sum of the beam moment strengths. This distribution of strength is intended to force hinging into the beams. A particular beam-column joint may be exempted if the others are able to handle the full shear without help from the exempted column.

Column detailing is divided into two sets of rules. If the maximum design load during an earthquake P_e is not more than 0.4 P_b, the member should be detailed as a flexural member.

(b) Confinement Reinforcement

When $P_e > 0.4 P_b$, confinement of the core is required, by spirals or hoop reinforcement, for a distance above and below the joint equal to the overall h of the column (the larger h if a rectangular column), but not less than one-sixth of the clear height or 450 mm. The requirements outlined in Sec. 23.6 in effect continue a similar reinforcement through the joint.

The Commentary mentions two objectives of the confinement reinforcement:

1. To care for hinging if it should occur in the column.
2. To compensate for spalling which occurs at overload.

In addition, the confinement steel provides some or all of the shear reinforcement over these lengths.

If a spiral is used, the ratio ρ_s is that normally used for a spiral column, where the spiral replaces the strength which was originally carried by the column shell. A second lower limit of not less than 0.12 f'_c/f_y is also set. If rectangular hoops, either separate or continuous, are used, the hoop is required to be

$$A_{sh} = \ell_h \rho_s s_h/2$$

where ℓ_h is the maximum unsupported length of rectangular hoop measured between perpendicular legs or crossties, ρ_s is the volumetric steel ratio required for a spiral when the area of the rectangular core area A_{ch} is substituted in the formula instead of A_c of the circular core. This relation is a semirational one presented in detail in the Commentary; some related tests in the Portland Cement Association laboratory have given good results.[2] The formula assumes the hoops about half as effective as a circular spiral. The hoop pitch or spacing center to center must not exceed 100 mm.

The concept of supplementary crossties to reduce the length of the straight hoop side is new, an attempt to increase the confinement of the

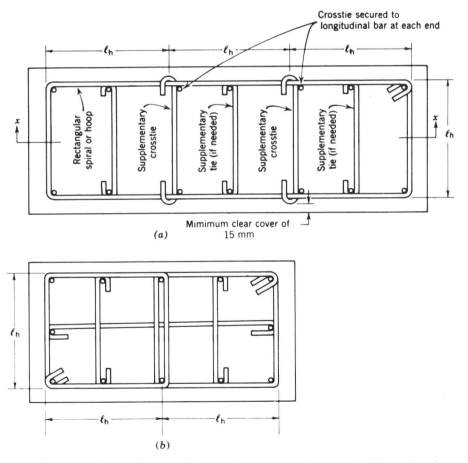

FIGURE 23.4 Two columns with supplementary ties. (*a*) With supplementary crossties also to reduce ℓ_h. (*b*) With overlapping hoops to reduce ℓ_h. (From Code Commentary.)

element. A long straight side of a hoop cannot be very effective in developing restraint on the hoop core. The supplementary crosstie must be of the same diameter as the basic hook and must have a standard semicircular hook at each end. These hooks must engage the hoop at each end and also must be fastened to a longitudinal bar at each end. The previous formula accepts a single supplementary tie as reducing ℓ_h to half the main hoop value, thus permitting A_{sh} half as large. Since there would then be three cross legs, each half as large, a 25% reduction in required steel area results. Two supplementary crossties double the number of legs but each is only one-third as heavy, a saving of 33%. Two details of column cross sections detailed under this concept are shown in Fig. 23.4.

Supplementary ties may be used in addition to crossties for shear resistance, these hooking around primary vertical bars rather than hoops.

Minimum cover over supplemental crossties may be reduced to 15 mm.

Where a wall or stiff partition is carried by columns, the Code requires that the entire column height have confinement reinforcement, because these often tend to pick up heavy loading from the wall as a cantilever. If it can be shown that the column loads are not significant, the confinement can be omitted, according to the Commentary.

(c) Shear Reinforcement

Lateral forces and joint displacement both contribute to shear forces on the columns. The shear the frame can deliver when hinging occurs should be considered. Figure 23.5 is from the Commentary, suggesting the possible combinations of hinges around the column joints. The hinging of the column itself produces the worst shear, if it is assumed possible, because of the relative strengths specified and already discussed in (a). It would always be safe to use shear reinforcement required by hinging moments at each end of the column.

The allowable shear on the concrete is that which includes the effect of axial tension or compression as may be appropriate (Sec. 5.18). Shear reinforcement shall not be spaced farther apart than $d/2$.

(d) Column Bar Splices

Splices must conform to the usual standard for compression bars, but a stricter minimum of 30 d_b or 400 mm is set for the seismic situation. Splices near midheight of the column seem desirable.

The Appendix also requires that mechanical or welded splices be staggered, not more than 25% of the bars spliced at any one level, and at least 300 mm of stagger between the levels used.

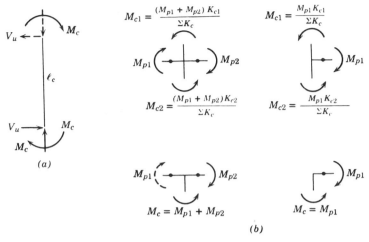

FIGURE 23.5 Column under seismic loading. (a) Maximum total column shear under seismic loading. (b) Column moments in joints where plastic hinges develop in beams. (From Code Commentary.)

23.6 BEAM-COLUMN CONNECTIONS

The column spiral or the hoop confinement must extend through the joint between beams and column, with consideration given to both axial load and shear. Usually the shear will be the critical one.

The joint receives a shear from the column above and below. Within the joint depth there will be moments from the adjoining beams, as shown on the joint in Fig. 23.6. If one assumes a hinging moment in each beam, there is on each side a tension pull of $A_s f_y$, to the right at the top (on the right of the figure) and to the left at the bottom. The total compression, although differently constituted in each case, matches the tension in amount, since axial force at most is small in comparison. These forces permit the shear diagram to be constructed and horizontal shear reinforcement to be computed. The necessary hoops or spiral demands may change as different sections are considered.

Where four beams frame in at a joint, the transverse reinforcement may be reduced 50% provided the beams have widths at least half that of the column and have heights such that the most shallow one is at least 75% the height of the deepest one.

The Code calls attention to the torsion, shears, and moments that occur when beams frame in to the column but off of its axis; these require consideration.

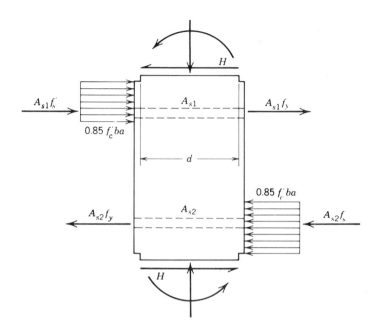

FIGURE 23.6 Horizontal shearing forces acting on a connection under seismic loading. (From Code Commentary.)

The recent ACI-ASCE Committee 352 recommendations on beam-column joints is significant.[5]

23.7 SPECIAL SHEAR WALLS

(a) General

Horizontal and vertical reinforcement in the wall shall provide at least $\rho = 0.0025$ on the gross area in each direction. Reduced horizontal force factors permitted for a ductile moment resisting space frame are *not* to be applied to the shear walls.

As in the case of columns, where $P_e \gtrless 0.4\,P_b$ the wall is designed primarily for flexure, but must be good for shear, axial load, and moment. Based on the wall (gross area) as an elastic homogeneous member, wherever the extreme fiber stress exceeds $0.15\,f_r$ (f_r being the modulus of rupture), the minimum area of vertical reinforcement must be

$$A_s = (1.4/f_y)hd$$

Here the reinforcement is concentrated near the ends of the wall, and d is the effective depth to the centroid of that steel and h is the member thickness.

(b) Walls with $P_e > 0.4\,P_b$

The Appendix suggests a design "based on accepted engineering principles" or the specific provisions of A.8.5.1 and A.8.5.2 in the Code. These sections call for vertical boundary elements designed to carry all vertical stresses resulting from the design wall load, tributary dead and live loads, and horizontal forces. The author envisions the wall thickened at the ends, or an I shape, with the end sections detailed as columns with confining reinforcement *for the full height* just as specified over a part of the height for *end* confinement on the columns.

23.8 GENERAL COMMENT

The general purpose of Appendix A is to add toughness or ductility to the frame, to provide for reversal of stresses, to guard against sudden shear failure, and to leave a confined core on columns even when some concrete has spalled off. The end result should be a greatly improved frame. But there appear to be traps for the careless designer. This system is much more complex than normal present-day construction; the designer must visualize the erection problem clearly. The author feels this paragraph from the Commentary is quite appropriate as a closing comment:

> Great care is needed to avoid impractical placement problems in seismic design. This is particularly true for the reinforcing steel in the connections of special ductile frames. If high percentages of reinforcement are used in beams and columns, it may be physically impossible to place the beam and column reinforcement and the required ties in the connections; or if the reinforcement is placeable, it may be impossible to place the concrete in the connection or to get a vibrator into it. It may prove economical to construct a full size model of the reinforcement in a typical connection to investigate its constructibility, unless the designer has had considerable experience in this type of work.

SELECTED REFERENCES

1. J. A. Blume, N. M. Newmark, and L. H. Corning, *Design of Multistory Reinforced Concrete Buildings for Earthquake Motions*, Portland Cement Association, Skokie, 1961.
2. N. W. Hanson and H. W. Conner, "Seismic Resistance of Reinforced Concrete Beam-Column Joints," ASCE, *Proc.*, 93, ST5, Oct. 1967, p. 533.

3. ACI Committee 315, "Seismic Details for Special Ductile Frames," *Jour. ACI.* 67, May 1970, p. 374.
4. *Recommended Lateral Force Requirements and Commentary*, Seismology Comm., Structural Engineers Assn. of California, San Francisco, California, 1975.
5. ACI-ASCE Committee 352, "Recommendations for Design of Beam-Column Joints in Monolithic Structures," *Jour. ACI*, 73, July 1976, p. 375.

SUMMARY TREATMENT OF SERVICE LOAD ANALYSIS AND DESIGN FOR FLEXURE

A.1 GENERAL FIELD OF USE

Elastic analysis and the transformed area concept for the purpose of calculating moments of inertia relating to deflections and shrinkage stresses were introduced in Sec. 8.4. In building design and analysis the old working stress design method is gradually being phased out. For slabs and beams the 1977 Code (Appendix B) classifies it as an alternate design method with a unity load factors.

For development and splicing of reinforcement and for torsion resistance, the method *almost* follows Code Chapters 11 and 12. Shear resistance also follows Code Chapter 11, with lower coefficients in the allowables. For columns it is so restricted as to be essentially the strength design of Chapters 18 and 19.

In the highway bridge field so much emphasis is necessary on service conditions that the 1977 AASHTO Specifications[1] still relate to what formerly was called working stress design, although now also presenting strength design as an alternate.

The presentation here is restricted to simple shapes, with and without compression steel, using AASHTO allowable stresses.

There is very little basic difference between calculating beam stresses resisting a given moment, for comparison with allowable stress values, and calculating an allowable moment based on these allowable stresses.

The *design* of beams for moment, on the other hand, is a process considerably different from analysis. This topic starts with Sec. A.8.

ANALYSIS OR REVIEW

A.2 COMPRESSION REINFORCEMENT

The elastic analysis of Sec. 8.4 accepted compression reinforcement as a portion of the elastic member with a steel stress $f'_s = nf_c$ and a transformed area nA'_s or effective addition of $(n-1)A'_s$ to the transformed area and $(n-1)f_c^*$ to the stress, after accounting for displaced concrete.

For strength, the creep of concrete when under load (or the flattening of the stress-strain curve for concrete under even short time high stresses) shortens the compressive concrete and steel more than an elastic analysis would suggest. Analysis or design by the working stress method recognizes this by using $2\,nA'_s$ as the transformed area and $2\,nf_c$ as the stress. It would be logical to use $(2\,n-1)A'_s$ and $(2\,n-1)f_c$ to recognize the displacement of useful concrete, but this will sometimes be simplified to $2\,n$ instead of $2\,n-1$ in recognition that the factor 2 applied to n is not really accurate enough to justify a 5 to 8% correction.

Under the ACI Code values of n should be based upon E_s/E_c from values in Code 8.5.

A.3 AASHTO ALLOWABLE STRESSES

In flexure AASHTO allows a maximum fiber stress in compression $f_c = 0.4\,f'_c$ with zero in tension (i.e., cracked over tension zone) except in plain concrete (as in some footings). On reinforcing bars it allows in tension 20 ksi (140 MPa) on Grade 40 ksi (Grade 300 MPa) and 24 ksi (165 MPa) on Grade 60 ksi (Grade 400 MPa) steel, and in compression $2\,n$ times the concrete stress at that level, but not greater than the allowable in tension. The AASHTO value of n is 10 for f'_c of 3 000 to 3 900 psi (20 to 27 MPa), 8 for f'_c of 4 000 to 4 900 psi (28 to 34 MPa), 6 for f'_c of 5 000 psi (35 MPa) or more; however, for deflections n is always 8.

A.4 RECTANGULAR BEAM ANALYSIS OR REVIEW

Since Sec. 8.4 used only the moment of inertia approach, this review will now be checked by looking directly at the internal stress system. This stress system is not an improvement for analysis but it is important as a better approach for design and for easy visualization. The data already used are $M = 130$ kN $\cdot$ m, $f'_c = 30$ MPa, Grade 600 steel, $c = 223$ mm (Sec. 8.4d) and the other dimensions of Fig. A.1, which is a copy of Fig. 8.4 with stress triangles added.

---

*Here f_c is at the level of the compression steel, a value less than the maximum stress on the extreme fiber.

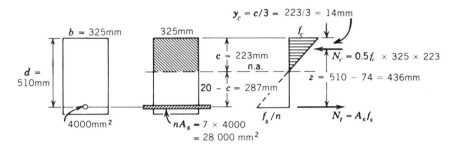

FIGURE A.1   Analysis of rectangular beam.

The problem in Section 8.4 was solved in accordance with the ACI Code 8.5.1 and 8.5.2.

$E_c = 5\,000\sqrt{f_c'}$ MPa, $E_s = 200\,000$ MPa, and $n = E_s/E_c = 40/\sqrt{f_c'} = 7.30$, or 7 to the nearest whole number. AASHTO specifies $n = 8$ for this grade of concrete, both for deflection and for strength calculations; but for consistency with the problem in Section 8.4, $n = 7$ is retained.

The resultant compression N_c and tension N_t form a couple with arm $z^* = d - \frac{1}{3}c = 436$ mm, as noted on the figure.

$$N_t = N_c = M/z = 110/0.436 = 252 \text{ kN}$$
$$N_t = A_s f_s$$
$$f_s = N_t/A_s = 252 \times 10^{-3}/4\,000 \times 10^{-6} = 63.0 \text{ MPa}$$

or

$$f_s = \frac{M}{A_s z} = \frac{110 \times 10^{-3}}{4\,000 \times 10^{-6} \times 0.436} = 63.0 \text{ MPa}$$

$$N_c = \tfrac{1}{2}f_c\, bc$$

$$252 \times 10^{-3} = \tfrac{1}{2}f_c \times 0.325 \times 0.223$$
$$f_c = 7.0 \text{ MPa}$$

Check: $$\frac{f_c}{f_s/n} = \frac{c}{d-c} \qquad f_c = \frac{63.0}{7} \times \frac{223}{510-223}$$
$$f_c = 7.0 \text{ MPa}$$

Under the AASHTO allowable stresses of $f_c = 0.4\,f_c' = 0.4 \times 30 = 12$ MPa and $f_s = 140$ MPa, the allowable moment on this beam may now be found by proportion from the stresses already established. Based on compression alone, f_c may be increased from 7 MPa to 12 MPa.

Allowable $M_c = 130 \times 12/7 = 223$ kN · m

---

*In the older notation, c was kd, z was jd, and y_c was sometimes designated as z.

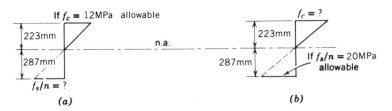

FIGURE A.2   Determination of limiting or governing stress.

Based only on tensile stress:

$$\text{Allowable } M_t = 130 \times 140/63 = 289 \text{ kN} \cdot \text{m}$$

The lower "allowable" is the real limit, in this case a limit on compression.

$$\text{Allowable } M = 223 \text{ kN} \cdot \text{m}$$

Economy would indicate the use of less A_s, with a deeper beam if the moment must be kept to this level.

If the calculated stresses had not already been available, the limiting stress could have been established as soon as the neutral axis was found. Consider *either* triangle in Fig. A.2 (considering both would be a waste of time). From the triangles in Fig. A.2a:

$$\text{If } f_c' = 12 \text{ MPa}, \quad f_s/n = 12 \times 287/223 = 15.44$$
$$n = 7, \quad f_s = 7 \times 15.44 = 108 \text{ MPa} < 140 \text{ MPa} \qquad \textbf{O.K.}$$

Or, from the triangles of Fig. A.2b:

$$\text{If } f_s/n = 140/7 = 20 \text{ MPa}, \quad f_c = 20 \times 223/287 = 15.5 \text{ MPa} > 12 \text{ MPa} \qquad \textbf{N.G.}$$

Either trial establishes that the moment is limited by f_c to:

$$\text{Allowable } M = N_c z = (\tfrac{1}{2} f_c bc)z = \tfrac{1}{2} \times 12 \times 0.325 \times 0.223 \times 0.436$$
$$= 0.190 \text{ MN} \cdot \text{m} = 190 \text{ kN} \cdot \text{m}$$

This example has covered the operating procedures for analysis. The details but not the principles change in the case of compression steel, T-beams, or irregular sections.

A.5 T-BEAM REVIEW

(a) Compared to a Rectangular Beam

The AASHTO specification for flange width is the same as in the ACI Code, as used here in Sec. 3.11, except for limiting the flange overhang to $6\,h_f$ instead of $8\,h_f$, where h_f is the flange thickness.

The flange area is usually more than adequate to care for compressive

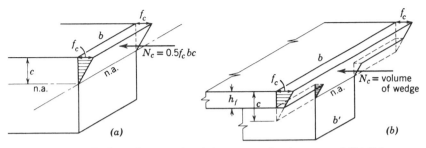

FIGURE A.3   Wedge of stress for (a) rectangular beam and (b) T-beam.

stresses; hence allowable moments are usually limited by the steel stress. Furthermore, the large compression area in the flange often pulls the centroid of the transformed area up into the flange itself. When this occurs, the analysis for moment is exactly that for a wide rectangular beam of width b, since the missing area below the flange would have been in tension and considered as cracked.

The neutral axis does more often fall below the flange with working stress analysis than in strength design, where it rarely does so. In this case the evaluation of N_c in terms of f_c and the location of N_c are no longer possible by simple inspection. The student will probably find it helpful to think in terms of the wedge of stress. For a rectangular beam, this wedge of stress is a simple triangular wedge (Fig. A.3a). The volume of the wedge gives the magnitude of N_c and the centroid of the wedge marks the location of the resultant N_c. When this concept is applied to the T-beam, as in Fig. A.3b, neither the volume nor the centroid can be determined except by a summation process, as in the following examples. The first example is calculated by the so-called exact analysis; the second example uses the approximate method that neglects the web area between neutral axis and flange. The approximate method fits much easier into a solution by formulas, charts, or curves, but has little advantage for a basic analysis such as is used here.

(b) Exact Analysis

Find f_c and f_s for the T-beam shown in Fig. A.4a under a moment of 270 kN · m, assuming $f'_c = 20$ MPa and Grade 400 steel, AASHTO specification.

Solution

The AASHTO specification for f'_c between 3 000 psi and 3 900 psi (20 MPa and 27 MPa) uses $n = 10$ for strength and $n = 8$ for deflection calculations, making here $nA_s = 10 \times 4\,000$ mm$^2 = 40\,000$ mm^2. Flange overhang of 375 mm is less than 6 $h_f = 600$ mm   O.K.

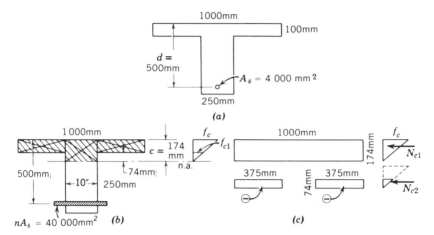

FIGURE A.4 Analysis of T-beam, exact method. (The 100 mm slab is too thin for highway use, but here it helps to emphasize the influence of $n.a.$ location.)

First make a quick check on where the neutral axis lies with respect to the bottom of the flange, using area moments about the bottom flange.

$$1\,000 \times 100 \times 50 = 5\,000\,000; \quad 400\,000(500 - 100) = 160\,000\,000$$

$$5\,000\,000 \ll 160\,000\,000; \text{ neutral axis is well below flange.}$$

Take area moments about the neutral axis using the area of Fig. A.4b.

$$250\,c \times \tfrac{1}{2}c + (1\,000 - 250) \times 100(c - \tfrac{1}{2} \times 100) = 40\,000(500 - c)$$

$$c^2 + 920\,c = 190\,000$$

$$c = 174 \text{ mm}$$

At the bottom of flange the stress triangle gives

$$f_{c1} = (74/174)f_c = 0.43\,f_c$$

The total compression can be calculated by subdividing the areas, and if necessary the stresses, into pieces that can be visualized simply (Fig. A.4b). In this case, the simplest pieces are N_{c1} on a rectangle $1\,000$ mm wide and extending all the way to the neutral axis and a negative N_{c2} deducting the surplus below the flanges. Sketches, as in Fig. A.4c, and a tabular form are desirable. To locate the resultant N_c, moments are taken about some convenient axis, by custom about the top of beam.

Converting the millimeter dimensions to meters:

$$N_{c1} = 1.0 \times 0.174 \times \tfrac{1}{2}f_c \qquad\qquad = +87.0 \times 10^{-3}\,f_c$$

$$N_{c2} = -0.75 \times 0.074 \times 0.43 \times \tfrac{1}{2}f_c = -11.9 \times 10^{-3}\,f_c$$

$$N_c = \overline{+75.1 \times 10^{-3}\,f_c}$$

Moments about the top of the beam due to:

$$N_{c1}: \quad +87.0 \times 10^{-3} f_c \times \tfrac{1}{3} \times 0.174 \qquad = +5.05 \times 10^{-3} f_c$$
$$N_{c2}: \quad -11.9 \times 10^{-3} f_c (0.1 \times \tfrac{1}{3} \times 0.074) = -1.49 \times 10^{-3} f_c$$
$$\overline{\phantom{N_{c2}: \quad} +3.56 \times 10^{-3} f_c}$$

$$y_c = 3.56 \times 10^{-3} f_c / 75.1 \times 10^{-3} f_c = 0.047 \text{ m} = 47 \text{ mm}$$

$$z = 500 - y_c = 453 \text{ mm} = 0.453 \text{ m}$$

$$f_s = \frac{M}{A_s z} = \frac{270 \times 10^{-3}}{400 \times 10^{-6} \times 0.453} = 149 \text{ MPa versus 165 MPa allowable} \qquad \textbf{O.K.}$$

$$N_c = \frac{M}{z} = \frac{270 \times 10^{-3}}{0.453} = 0.596 \text{ MN} = 75.1 \times 10^{-3} f_c \text{ (from foregoing)}$$

$$f_c = 7.94 \text{ MPa versus } 0.4 \times 20 = 8 \text{ MPa allowable} \qquad \textbf{O.K.}$$

(c) Approximate Analysis

Same beam as in (b) above, Fig. A.5a.

Solution

As before, the neutral axis falls below the flange. The approximate method neglects the small compressive area (and small unit stress) below the flange, Fig. A.5b. Then area moments about the neutral axis give:

$$(1\,000 \times 100)(c - 50) = 40\,000(500 - c)$$

$$140\,000\,c = 25\,000\,000$$

$$c = 179 \text{ mm}$$

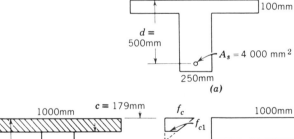

(a)

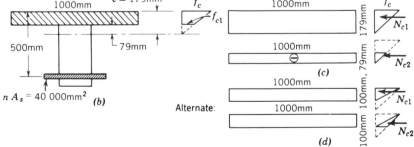

FIGURE A.5 Analysis of T-beam, approximate method.

At the bottom of the flange, $f_{c1} = f_c \times 79/179 = 0.44\, f_c$.
The areas and stresses shown in Fig. A.5c lead to:

$$N_{c1} = 1.0 \times 0.179 \times \tfrac{1}{2} f_c \qquad\qquad = +89.5 \times 10^{-3}\, f_c$$
$$N_{c2} = -1.0 \times 0.079 \times 0.44 \times \tfrac{1}{2} f_c = -17.4 \times 10^{-3}\, f_c$$
$$N_c = +72.1 \times 10^{-3}\, f_c$$

Moment about the top of the beam due to

$$N_{c1}: \quad +89.5 \times 10^{-3}\, f_c \times \tfrac{1}{3} \times 0.179 \qquad = +5.34 \times 10^{-3}\, f_c$$
$$N_{c2}: \quad -17.3 \times 10^{-3}\, f_c (0.1 + \tfrac{1}{3} \times 0.079) = -2.19 \times 10^{-3}\, f_c$$
$$+3.15 \times 10^{-3}\, f_c$$

$$y_c = 3.15 \times 10^{-3}\, f_c / 72.1 \times 10^{-3}\, f_c = 0.044\text{ m} = 44\text{ mm}$$

$$z = 500 - 44 = 456\text{ mm}$$

The student should see clearly that there are many different patterns for N_{c1} and N_{c2} which lead to identical values of N_c and m. For example, the pattern of Fig. A.5d where the unit stress picture has been subdivided rather than the area. It could have been subdivided into a rectangle and triangle of stress or the trapezoid could have been used undivided, provided its centroid was calculated as the location of N_c. The subdivisions to be used are a matter of convenience, but it should be noted that *in the exact solution* in (b) no other subdivision leads as directly to the desired result.

$$f_s = \frac{M}{A_s z} = \frac{270 \times 10^{-3}}{4\,000 \times 10^{-6} \times 0.456} = 148\text{ MPa vs. 165 MPa} \qquad \text{O.K.}$$

$$N_c = M/z = 270 \times 10^{-3}/0.456 = 0.592\text{ MN} = 72.1 \times 10^{-3}\, f_c$$

$f_c = 8.21$ MPa vs. 8.0 MPa allowable; but note that f_c is below the allowable stress in the exact analysis.

(d) Allowable Moment

Just as for the rectangular beam in Sec. A.4, the allowable moment may be found by proportion from stresses already calculated; or, the governing stress can be determined from the stress triangles and the allowable moment then calculated.

Although this particular beam with its narrow flange will be limited by compression, this is a rare case for a real T-beam. Hence an approximate allowable moment can usually be estimated as $A_s f_s$, with the stress at the allowable, times an estimated arm z which will usually be in the neighborhood of $0.9\, d$.

A.6 ANALYSIS OR REVIEW WITH COMPRESSION STEEL

(a) General

The use of $2\,n$ with A_s' as discussed in Sec. A.2 is direct and reasonable for design and fairly satisfactory (but occasionally awkward) for allowable moments. In the calculation of stresses from a given moment, the concepts become less significant and the stresses become only nominal ones.

(b) Analysis

Assume $1\,200\ \text{mm}^2$ of compression steel is added at 50 mm below the top of the rectangular beam of Sec. A.4a and Fig. A.1a, with $f'_c = 30\ \text{MPa}$, $n = 8$, Grade 300 steel. Find stresses for $M = 200\ \text{kN} \cdot \text{m}$. Also find the allowable moment.

Solution

The transformed area is shown in Fig. A.6b, using $(2\,n - 1)A'_s = 15\,000\ \text{mm}^2$ for the compression steel.

Area moments about neutral axis:

$$325\ c \times \tfrac{1}{2}c + 15\,000(c - 50) = 32\,000(500 - c)$$
$$c^2 + 289\ c = 103\,000$$
$$c = 207\ \text{mm}$$

Based on Fig. A.6c:

$$f'_{cs} = f_c \times 157/207 = 0.758\ f_c$$

Compressive forces are:

$$N_{c1} = 0.375 \times 0.207 \times \tfrac{1}{2} f_c = 38.81 \times 10^{-3}\ f_c$$
$$N_{c2} = 15 \times 10^{-3} \times 0.758\ f_c = 11.73 \times 10^{-3}\ f_c$$
$$N_c = \overline{50.54 \times 10^{-3}}\ f_c$$

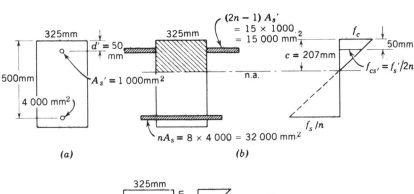

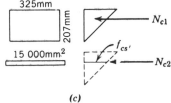

FIGURE A.6 Analysis of double-reinforced beam.

Moments about the top of the beam due to:

N_{c1}: $38.81 \times 10^{-3} f_c \times \frac{1}{3} \times 0.207$ $= 2.678 \times 10^{-3} f_c$

N_{c2}: $11.73 \times 10^{-3} f_c \times 0.050$ $= 0.587 \times 10^{-3} f_c$

$\overline{ 3.265 \times 10^{-3} f_c}$

$y_c = 3.265 \times 10^{-3} f_c / 50.54 \times 10^{-3} f_c = 0.064 \text{ m} = 64 \text{ mm}$

$z = 500 - 64 = 436 \text{ mm}$

$f = \dfrac{M}{A_s z} = \dfrac{200 \times 10^{-3}}{4\,000 \times 10^{-6} \times 0.436} = 115 \text{ MPa versus } 140 \text{ MPa allowable}$ **O.K.**

$N_c = M/z = 200 \times 10^{-3}/0.436 = 0.459 \text{ MN} = 50.54 \times 10^{-3} f_c$ (from above)

$f_c = 9.0 \text{ MPa versus } 0.40 \times 30 = 12 \text{ MPa allowable}$ **O.K.**

Check $f_c = \dfrac{115}{8} \times \dfrac{207}{293} = 10 \text{ MPa} < 12 \text{ MPa}$ **O.K.**

$f'_s = 2 \, n f'_{cs} = 2 \times 8 \times 0.758 \times 9.0 = 109 \text{ MPa versus } 140 \text{ MPa allowable}$ **O.K.**

For allowable moment: If $f_c = 12$ MPa, similar stress triangles give:

$$f_s = n f_c \, \frac{d - c}{c} = 8 \times 12 \times \frac{500 - 207}{207}$$

$= 136 \text{ MPa vs. } 140 \text{ MPa O.K. Compression controls.}$

$f'_s = 2 \, n f'_{cs} = 2 \times 8 \times 0.758 \times 12$

$= 145.5 \text{ MPa vs. } 140 \text{ MPa allowable; } 4\% \text{ too high.}$

With $M = A_s f_s z$ and $z = 436$ mm as above:

Allowable $M = 200 \times 140/145.5 = 192 \text{ kN} \cdot \text{m}$

A.7 IRREGULAR-SHAPED BEAMS—ANALYSIS OR REVIEW

Beams of irregular shape are often used for special purposes, such as curved-top or triangular-top shapes for railings, curbs over exterior bridge girders, beams with continuous side brackets, or beams with recesses for supporting a future slab or masonry. Where the shape is entirely bounded by vertical and horizontal surfaces, the location of the resultant N_c as for T-beams and double-reinforced beams, is a practical procedure. Where sloping faces or curved surfaces occur, the use of the moment of inertia and $Mc = fI$ is generally simpler.

It should be noted that many irregular-shaped beams are not symmetrical and are not loaded along their principal axes. Fortunately, most of these are restrained against lateral deflection by a monolithic slab. In such cases, lack of symmetry can be ignored and the bending axis will be essentially horizontal.

DESIGN

A.8 DESIGN VERSUS REVIEW

In review, whether for actual stresses or for allowable moments, the engineer deals with given beams, known both as to dimensions and steel. He has no control over the location of the neutral axis, which lies at the centroid of the transformed area.

In design, loads and allowable stresses are known and some or all the dimensions remain to be fixed. In this case designers have some control over the location of the neutral axis. They can shift it where they want it, to the extent that the change in dimensions can shift the centroid of the transformed area.

The student should understand clearly this fundamental difference between design and analysis or review problems.

A.9 RECTANGULAR BEAMS—MOMENT DESIGN

(a) The Balanced Beam in Design

A beam that reaches its allowable f_c and its allowable f_s under the same working moment is called a balanced beam.* Such a beam will often constitute the most economical construction; it usually represents the smallest desirable beam.

Although a balanced beam is frequently the design objective, the balanced depth will only occasionally be a practical dimension. Usually, over-all beam depth are specified in multiples of 25 mm. If the balanced depth (in mm) is adjusted to a more practical dimension, the beam will no longer be exactly balanced for a given moment. Hence practical design is concerned with the balanced beam only as an approach to practical beam dimensions. The design of the reinforcing steel should be based on the actual dimensions used.

Slightly deeper beams are in the direction of economy. Since less steel is required with the larger z, such a beam is termed an underreinforced beam. On the other hand, slightly shallower beams are also practical, but these require a considerable increase in steel area. Such overreinforced beams are somewhat more expensive and become impractical when the depth is reduced more than 5 or 10%. For larger reductions in depth compression steel is indicated.

---

*The student should note that a beam thus balanced for allowable working stresses will not be balanced at failure. The balance at failure calls for a much larger A_s.

(b) Balanced Beam Design

The design procedure for a beam exactly balanced *at working load* will be illustrated by the design of a rectangular beam for $M = 120 \text{ kN} \cdot \text{m}$, $f'_c = 20 \text{ MPa}$, Grade 40 steel.

Solution

Since the steel is still to be selected, the neutral axis can be made to fall at any convenient location. That location corresponding to balanced *stresses** is desired. In this example, the AASHTO specification establishes: allowable $f_c = 0.40 \times 20 = 8 \text{ MPa}$, $n = 10$, allowable $f_s = 140 \text{ MPa}$.

The beam, with dimensions still as symbols, with subscript b emphasizing balanced, can be sketched as in Fig. A.7 along with the unit stress triangles that must accompany a straight-line distribution of stress. When simultaneous values of allowable f_c and f_s are assigned to the maximum stress values, the similar triangles locate the proportionate depth c_b to the desired neutral axis. It is convenient to consider the large dashed triangle of height d and horizontal dimension $2000 + 1200$ as one of the similar triangles.

$$\frac{c_b}{d} = \frac{8}{14+8} = 0.364, \; c_b = 0.364$$

$$z_b = d - \tfrac{1}{3} c_b = d - \tfrac{1}{3} \times 0.364 \, d = 0.879 \, d$$

The resultant compression N_c can now be sketched in its proper location and evaluated in terms of b and c_b.

$$N_c = (b \times 0.364 \, d) \times \tfrac{1}{2} \times 8$$

$$M = N_c z_b = b \times 0.364 \, d \times \tfrac{1}{2} \times 8 \times 0.879 \, d = 1.28 \, bd^2 = k_b bd^2$$

where k_b represents the constant $197 = M/bd^2$ for these balanced stresses.

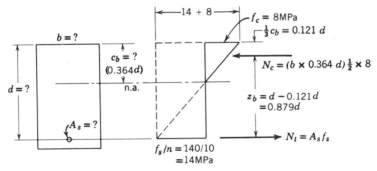

FIGURE A.7 Design of a balanced rectangular beam.

*Notice that the balanced stress condition at working load is unrelated to the strength concept of balanced beam at failure.

This constant, $k_b = 1.28$ MPa, is a function of the unit stresses and not of the loading. Hence, every balanced rectangular beam designed for these materials (or these unit stresses) will be concerned with this constant. One calculates it once on a given job and then uses it for many designs. It will usually be found tabulated in all handbooks.*

The size of the beam can be found by equating the maximum bending moment to this resisting moment, noting that units must be consistent.

$$M = 120 \times 10^{-3} = k_b bd^2 = 1.28\ bd^2$$
$$\text{Reqd. } bd^2 = 93.75 \times 10^{-3}\ \text{m}^3 = 93.75 \times 10^6\ \text{mm}^3$$

This requirement can be met by a wide range of sizes, for example:

If $b = 175$ mm $d = 732$ mm

200   685

225   646

250   613

275   584

300   559

325   537

350   518

375   500

$b = 400$ mm, $d = 484$ m

USE $b = 250$ mm, $d \geqslant 613$ mm.

The depth of 613 mm is probably not entirely practical; it is not a dimension one would turn over to a carpenter or steel worker. But either more or less depth than 613 mm, with this width, would *not* give a balanced beam. This theoretical balanced depth gives:

$$\text{Balanced } A_s = \frac{N_t}{f_s} = \frac{M}{f_s z} = \frac{120 \times 10^{-3}}{140 \times 0.879 \times 0.613} = 1\,591 \times 10^{-6}\ \text{m}^2 = 1\,591\ \text{mm}^2$$

The design should be changed to a practical depth before choosing bars.

(c) Beams Deeper Than a Balanced Section— Underreinforced

When dimensions must be modified to get practical sizes, the usual procedure is to add an allowance for depth from the center to the bottom of steel and for clear cover over the steel (including stirrups) and then to use the next larger practical over-all dimension. Beams are also made deeper because such a size may be required for shear or to match a level established by heavier loaded beams.

*Under older notations it would probably show as R or K.

Redesign the beam of (*b*) for $b = 250$ mm, $d = 700$ mm, in this case an arbitrary choice to show the method.

Solution

The steel area is the only element left to design. Since the beam is larger than the balanced section, f_c is less than allowable and f_s would also be less than allowable if the 1591 mm² steel area calculated above for the balanced beam were used. For economy, reduce A_s until f_s is the full allowable. The unit stress diagram in Fig. A.8 is thus shown with $f_s = 140$ MPa and with f_c unknown, but known (assumed) to be less than allowable. From similar triangles, the unknown f_c can be expressed in terms of the known f_s and unknown c. Then N_c and M can be calculated in terms of c.

$$f_c = \frac{f_s}{n} \frac{c}{d - c}$$

$$N_c = \tfrac{1}{2} f_c bc$$

$$z = d - \tfrac{1}{3}c$$

$$M = N_c z = \tfrac{1}{2} \frac{f_s}{n} \frac{c}{d - c} bc(d - c)$$

$$\frac{c/d}{1 - c/d} \cdot c/d \, (1 - \tfrac{1}{3} c/d) = \frac{2\,nM}{f_s b d^2} = \frac{2 \times 10 \times 120 \times 10^{-3}}{140 \times 0.25 \times 0.7^2} = 0.1400$$

This cubic equation is solved by trial. The required value of c must be slightly less than the balanced beam value $c_b = 0.364\,d$.

$$\text{For } c/d = 0.35: \quad \frac{(c/d)^2}{1 - c/d} (1 - \tfrac{1}{3} c/d) = \frac{0.35^2}{0.65} \times 0.883 = 0.166$$

$$0.33 \qquad\qquad \frac{0.33^2}{0.67} \times 0.890 = 0.145$$

$$0.325 \qquad\qquad \frac{0.325^2}{0.675} \times 0.892 = 0.139\,58$$

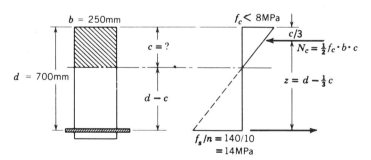

FIGURE A.8 Stress triangles for beam deeper than a balanced beam.

$$\text{USE } c = 0.325 \times 700 = 228 \text{ mm}$$
$$z = 700 - \tfrac{1}{3} \times 228 = 624 \text{ mm}$$
$$\text{Required } A_s = \frac{M}{f_s z} = \frac{120 \times 10^{-3}}{140 \times 0.624} = 1\,374 \times 10^{-6} \text{ m}^2 = 1\,374 \text{ mm}^2$$

Check by area moments about the neutral axis:

$$\tfrac{1}{2} \times 250 \times 228^2 = 10 \, A_s (700 - 228)$$
$$A_s = 1\,376 \text{ mm}^2 \qquad \textbf{O.K.}$$

Verify f_c by similar triangles:

$$f_c = \frac{140}{10} \times \frac{228}{700 - 228} = 6.8 \text{ MPa} < 8 \text{ MPa} \qquad \textbf{O.K.}$$

Since the cubic equation is awkward, an empirical approximation is usually substituted in the calculation for A_s, that is, the use of the balanced z for the given increased d.

$$\text{Reqd. } A_s = \frac{120 \times 10^3}{140 \times 0.879 \times 0.700} = 1\,393 \text{ mm}^2$$

Since less steel is used than would be required for a balanced beam, the neutral axis is higher and the real z value is always greater than the balanced z would be for the given increased d. Thus this approximation for A_s is always on the safe side. For small increases in d it is not wasteful and is recommended. When d is 20 to 50% greater than balanced d, some steel can be saved by using a more exact z.

(d) Beams Shallower Than a Balanced Section— Overreinforced

It is rarely feasible to reduce d below the balanced depth by more than possibly 10%, unless compression steel is used. A reduction in depth greatly increases the required steel and makes this type of construction uneconomical except for very small deficiencies in d.

Redesign the beam of Sec. A.9b for $b = 250$ mm, $d = 600$ mm.

Solution

In this case, the reduced depth tends to cause increased f_c and f_s. A reasonable addition to A_s would bring f_s within the allowable, but would leave f_c too large. The only way to remedy the overstress in compression (other than with compression steel) is to increase the compression area by lowering the neutral axis. Extra A_s will lower the neutral axis in this fashion, but a large amount is required; this operation tends to be inefficient since the steel has to work at a lowered unit stress. The unit stress diagram in Fig. A.9 is thus shown with f_c at the allowable and f_s below the allowable.

$$N_c = \tfrac{1}{2} \times 8 \times b \times c$$

$$M = 120 \times 10^{-3} \text{ MN} \cdot \text{m} = N_c z = \tfrac{1}{2} \times 8 \, bc(d - \tfrac{1}{3}c)$$

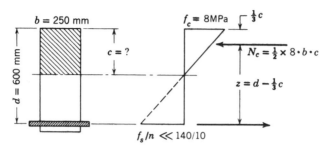

FIGURE A.9   Stress triangles for beam shallower than a balanced beam.

$$(c/d)^2 - 3(c/d) + 3\,M/4bd^2 = 0$$
$$c/d = +1.5 \pm \sqrt{1.5^2 - 3 \times 0.12/4 \times 0.25 \times 0.6^2} = 0.382;\; c = 0.382 \times 600 = 229\text{ mm}$$
$$z = 600 - \tfrac{1}{3} \times 229 = 524\text{ mm}$$

By similar triangles,

$$f_s/n = 8 \times \frac{600 - 229}{229} = 13.0\text{ MPa}$$

Limiting $f_s = 10 \times 13.0 = 130$ MPa (to protect against overstress in f_c)

$$\text{Reqd. } A_s = \frac{M}{f_s z} = \frac{120 \times 10^3}{130 \times 0.524} = 1\,762\text{ mm}^2$$

Check by area moments about the neutral axis:

$$250 \times 229 \times \tfrac{1}{2} \times 229 = 10\,A_s(600 - 229)$$
$$\text{Reqd. } A_s = 1\,767\text{ mm}^2 \qquad \textbf{O.K.}$$

Note that a 2% reduction in d has required a 10% increase in A_s compared to the balanced condition in Sec. A.9b. There is no accepted approximate solution for this case.

(e) Reinforcing Steel for Beam of Given Size

Design procedures for rectangular beams can be summarized as indicated in Table A.1. In this table c_b and z_b represent the balanced condition for the beam on which they appear, for instance, for the stresses of Sec. A.9b, always $c_b = 0.364\,d$ and $z_b = 0.879\,d$.

The first step after the beam size is known or has been established is to identify the beam as underreinforced, balanced, or overreinforced. As noted under "Identification," the actual moment may be compared to the

TABLE A.1 Rectangular Beams

Identification (use any one of the three checks):

$\text{Actual } M < (\text{bal. } k)bd^2 = k_b bd^2$	$\text{Actual } M = k_b bd^2 = \text{bal. } M$	$\text{Actual } M > (\text{bal. } k)bd^2 = k_b bd^2$
$\text{Actual } k = M/bd^2 < k_b$	$\text{Actual } k = M/bd^2 = k_b$	$\text{Actual } k = M/bd^2 > k_b$
$\text{Actual } d > \sqrt{M/k_b b}$	$d = \sqrt{M/k_b b} = \text{bal. } d$	$\text{Actual } d < \sqrt{M/k_b b}$

Designation:

Underreinforced	Balanced	Overreinforced (If $d \ll$ bal. d, use A'_s with balanced stresses.)

Design:

$f_c = \dfrac{f_s}{n}\dfrac{c}{d-c}$	c from similar triangles	c from $M = N_c z$ (quadratic eq. in c)
c from $M = N_c z$ (cubic eq. in c)	$M = k_b bd^2$	$\text{Reduced } f_s = nf_c\dfrac{d-c}{c}$
		$\ll \text{allow. } f_s$
$A_s = \dfrac{M}{(\text{allow.})f_s \times (d - c/3)}$	$A_s = \dfrac{M}{f_s z_b}$	$A_s = \dfrac{M}{(\text{reduced})f_s \times (d - c/3)}$
Approx. $A_s = $	*Alternate* for any of the three types:	
$\dfrac{M}{(\text{allow.})f_s \times z_b \text{ for this depth}}$	Find A_s by area moments about n.a. after c is established:	
	$nA_s(d - c) = bc^2/2$	

balanced moment $M = k_b b d^2$. An alternate equivalent calculation is to compare the actual k = actual M/bd^2 with the balanced value of k_b. The comparison can also be made between the actual d and the balanced $d = \sqrt{\text{actual } M/k_b b}$.*

If the actual d is less than 90 to 95% of the balanced d, it is more economical to design as a double-reinforced beam, as in the following section.

A.10 BEAMS WITH COMPRESSION STEEL—MOMENT DESIGN

(a) General

Although concrete is usually cheaper for carrying compression than steel, economy is often achieved by using compression steel over a short critical length. Compression steel increases the toughness of a beam, reducing the possibility of a sudden type of failure. It also reduces the creep and the creep-and-shrinkage deflection of a beam.

Design procedures are simple and are best shown by examples. In all cases the total design moment is subdivided into two parts, M_1 which is the allowable moment on the balanced beam without compression steel ($M_1 = k_b b d^2$), and M_2 which is the surplus moment to be resisted with compression steel and extra tension steel (Fig. A.10). The use of A'_s with $2n$ leads to a semielastic analysis where $f'_s \leq$ allowable f_s; if the strains should then indicate $f'_s >$ allowable f_s, complete validity requires that a nonelastic analysis be substituted.

(b) Semielastic Case

Redesign the steel for the beam of Sec. A.9d, using cover concrete $d' = 60$ mm.

Solution

Compression steel is to be valued at twice the stress an elastic analysis would indicate, provided this does not exceed the allowable tension of 140 MPa in this case. The requirement for ties around the compression steel must not be overlooked.

The procedure sketched in Fig. A.10 will be followed. As a balanced beam the moment capacity is $M_1 = k_b b d^2$. In Sec. A.9b, for $f_c = 8$ MPa, $f_s = 140$ MPa, $n = 10$, k_b has been established as 1.28 MPa, $c_b = 0.364\,d$, $z_b = 0.879\,d$.

$$M_1 = k_b b d^2 = 1.28 \times 0.25 \times 0.6^2 = 115.2 \times 10^{-3}\ \text{MN} \cdot \text{m}$$

$$A_{s1} = \frac{M_1}{f_s z_b} = \frac{115.2 \times 10^{-3}}{140 \times 0.879 \times 0.6} = 1\,560 \times 10^{-6}\ \text{m}^2$$

*In review problems (not design) the simplest identification is by the steel ratio ρ, below ρ_b being underreinforced, over ρ_b overreinforced.

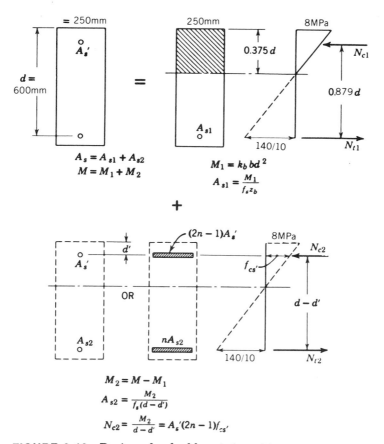

$$A_s = A_{s1} + A_{s2}$$
$$M = M_1 + M_2$$

$$M_1 = k_b bd^2$$
$$A_{s1} = \frac{M_1}{f_s z_b}$$

$$M_2 = M - M_1$$
$$A_{s2} = \frac{M_2}{f_s(d - d')}$$
$$N_{c2} = \frac{M_2}{d - d'} = A_s'(2n - 1)f_{cs'}$$

FIGURE A.10 Design of a double-reinforced beam.

$$M_2 = M - M_1 = (120 - 115.2) \times 10^{-3} = 4.8 \times 10^{-3} \text{ MN} \cdot \text{m}$$
$$A_{s2} = \frac{M_2}{f_s(d - d')} = \frac{4.8 \times 10^{-3}}{140(600 - 60) \times 10^{-3}} = 63 \times 10^{-6} \text{ m}^2$$
$$\text{Total } A_s = A_{s1} + A_{s2} = (1\,560 + 63) \times 10^{-6} \text{ m}^2 = 1\,623 \text{ mm}^2$$
$$N_{c2} = M_2/(d - d') = 4.8 \times 10^{-3}/0.540 = 8.9 \times 10^{-3} \text{ MN} = A_s'(2n - 1)f_{cs}$$
$$c = 0.364 = 0.364 \times 600 = 218 \text{ mm}$$
$$f_{cs}' = \frac{c - d'}{c} f_c = \frac{218 - 60}{218} \times 8 = 5.80 \text{ MPa}$$
$$f_s' = 2 nf_{cs}' = 2 \times 10 \times 5.80 = 116 \text{ MPa} < 140 \text{ MPa} \qquad \text{O.K.}$$
Effective $f_s'' = (2 n - 1)f_{cs}' = (20 - 1)5.80 = 110 \text{ MPa}$
$$A_s' = N_{c2}/110 = 8.9 \times 10^{-3}/110 = 81 \times 10^{-6} \text{ m}^2 = 81 \text{ mm}^2$$

This theory is identical with the semielastic analysis of Sec. A.6b. Check A_s' by area moments about the neutral axis with the effective transformed areas nA_{s2} and $(2n-1)A'_s$:

$$10 \times 63(600 - 218) = (2 \times 10 - 1)A'_s(218 - 60)$$
$$A'_s = 80 \text{ mm}^2$$

It might be noted that the total steel used in tension and compression is $1\,703$ mm^2 compared to $1\,767$ mm^2 when tension steel alone was used (Sec. A.9d). A double-reinforced beam is usually cheaper than the use of tension steel alone when the beam must be smaller than a balanced beam.

(c) Nonelastic Value of f'_s

Redesign the steel for the beam of (b) above, using $d' = 25$ mm. (This is a rather impractical assumption taken to illustrate a method.)

Solution

The values of M_1, M_2, A_{s1}, and c can be taken unchanged from (b) above:

$$A_{s2} = \frac{M_2}{f_s(d - d')} = \frac{4.8 \times 10^{-3}}{140(600 - 25) \times 10^{-3}} = 60 \times 10^{-6} \text{ m}^2$$

$$\text{Total } A_s = A_{s1} + A_{s2} = 1\,560 + 60 = 1\,620 \text{ mm}^2$$

$$N_c = 4.8 \times 10^{-3}/0.575 = 8.35 \times 10^{-3} \text{ MN} = A'_s(2n - 1)f'_{cs}$$

$$f'_{cs} = \frac{218 - 25}{218} \times 8 = 7.083 \text{ MPa}$$

$$f'_s = 2\,nf'_{cs} = 2 \times 10 \times 7.083 = 141.6 \text{ MPa} > 140 \text{ MPa}$$

$$\text{USE } f'_s = 140 \text{ MPa}$$

$$\text{Effective } f''_s = 140 - f'_{cs} = 140 - 7.083 = 133 \text{ MPa}$$

$$A'_s = N_{c2}/f''_s = 8.35 \times 10^{-3}/133 = 63 \times 10^{-6} \text{ m}^2 = 63 \text{ mm}^2$$

In this case the 140 MPa limit on f'_s has no relation to the elastic properties of the materials. Since total N_t = total N_c, there is no reason to shift the original neutral axis, but *it no longer falls at the centroid of the usual transformed areas.* Hence the check on A'_s based on area moments about the neutral axis as used in Sec. A.10b is invalid here.

A.11 T-BEAMS—MOMENT DESIGN

The choice of b_w (web width) and d is excluded since these are usually governed by factors other than flexure, although b_w is often modified to give more width for bars.

The flange ordinarily provides enough compression area to keep f_c low, and the check on this can be postponed until one of the last steps. The design for A_s is thus assumed to be for a beam much deeper than a balanced section. For a T-beam the algebraic approach used in Sec. A.9b for a similar rectangular beam becomes too involved for practical use. Instead, a cut-and-try procedure is recommended, based on the fact that the value of z can be estimated quite closely.

Example

Find the required area of Grade 400 steel ($f_y = 400$ MPa) for the T-beam of Fig. A.11 if $f'_c = 30$ MPa and $M = 270$ kN · m, AASHTO stresses.

Solution

Both the shape of the compression area and the high neutral axis resulting from a low f_v tend to locate the resultant N_c high in the beam. On this basis, an average value of z could be taken as $0.9\,d$. If the small amount of compression in the web below the flange is neglected, the trapezoidal stress distribution locates the resultant N_c higher than the middle of the flange. This means that z is greater than $d - h_f/2$, where h_f is the flange thickness. For the initial approximation, it is recommended that z be taken as the larger of $0.9\,d$ or $d - h_f/2$, this giving a value probably within 1 to 3% of the exact one.

$$0.9\,d = 0.9 \times 600 = 540 \text{ mm}$$
$$d - h_f/2 = 600 - 50 = 550 \text{ mm} \quad \text{USE for trial } z$$

Allowable $f_s = 165$ MPa for steel with $f_y = 400$ MPa

$$\text{Approx. } A_s = \frac{M}{z} = \frac{270 \times 10^{-3}}{165 \times 0.550} = 2\,975 \times 10^{-6}\,\text{m}^2 = 2\,975\,\text{mm}^2$$

This approximate A_s will be used to establish c and a better (almost exact) value of z. The "exact" method for T-beams, Sec. A.5b, will be used (an arbitrary choice), as indicated in Fig. A.12a.

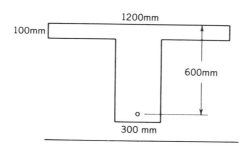

FIGURE A.11 T-beam for calculation of A_s.

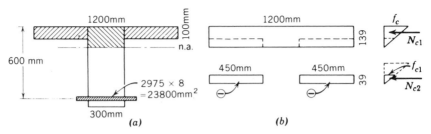

FIGURE A.12 Areas used in T-beam calculations. (a) For location of neutral axis. (b) For location of N_c.

Area moments about the neutral axis:

$$300\,c \times \tfrac{1}{2}c + 2 \times 450 \times 100(c - \tfrac{1}{2} \times 100) = 2\,975 \times 8(600 - c)$$
$$c^2 + 759\,c = 125\,200$$
$$c = 139\ \text{mm}$$
$$f_{c1}\ \text{at bottom of flange} = (39/139)\,f_c = 0.281\,f_c$$

Compressive forces shown in Fig. A.12b:

$$N_{c1} = 1.20 \times 0.139 \times \tfrac{1}{2}\,f_c \qquad\qquad = +83.40 \times 10^{-3}\,f_c$$
$$N_{c2} = -2 \times 0.45 \times 0.039 \times 0.281 \times \tfrac{1}{2}\,f_c = -\ 4.93 \times 10^{-3}\,f_c$$
$$N_c = +78.47 \times 10^{-3}\,f_c$$

Moments about the top of the beam due to:

$$N_{c1}:\quad +83.40 \times 10^{-3}\,f_c \times \tfrac{1}{3} \times 0.139 \quad = +3.864 \times 10^{-3}\,f_c$$
$$N_{c2}:\quad -4.93 \times 10^{-3}\,f_c(0.1 + \tfrac{1}{3} \times 0.039) = -0.557 \times 10^{-3}\,f_c$$
$$+3.307 \times 10^{-3}\,f_c$$
$$\bullet\ y_c = 3.307 \times 10^{-3}\,f_c/78.47 \times 10^{-3}\,f_c = 0.042\ \text{m} = 42\ \text{mm}$$
$$z = 600 - 42 = 558\ \text{mm}$$
$$A_s = 0.270 \times 10^6/165 \times 0.558 = 2\,933\ \text{mm}^2$$

The calculation of y_c above used $N_c = 78.47 \times 10^{-3}\,f_c$

$$f_c = N_c/78.47 \times 10^{-3} = M/z \times 78.47 \times 10^{-3}$$
$$= 270 \times 10^{-3}/0.558 \times 78.47 \times 10^{-3} = 6.16\ \text{MPa}$$

This is less than the allowable stress of $0.4 \times 40 = 16$ MPa   **O.K.**

An estimate of f_c based on approximate theory requires less preliminary calculation and is often adequate. A trapezoidal stress of the flange gives:

$$f_c < 2 \times \text{average } f_c, \text{ since average } f_c > \tfrac{1}{2}f_c$$
$$f_c < 2\,M/bh_f z = 2 \times 0.270/1.2 \times 0.1 \times 0.558 = 8.06\ \text{MPa}$$

This is less than the allowable stress of 12 MPa, although higher than the stress calculated by the precise method.   **O.K.**

Twice the average f_c in this approximate calculation could exceed the allowable stress and still give a satisfactory answer by the precise method.

SELECTED REFERENCE

1. *Standard Specifications for Highway Bridges*, AASHTO, Washington, 11 ed., 1973.

PROBLEMS

Note: Sec. A.3 states the AASHTO allowable stresses. When stresses are calculated for a given moment, compare these with the allowables.

Designs in this series of problems are to be partial designs, for flexure alone; bond, shear, and bar spacing are not a part of these problems.

When the actual moment exceeds the balanced beam moment, the student should decide whether excess tension steel or compression steel is preferable.

The instructor should indicate whether the student is to work with or without the use of curves.

PROB. A.1.
(a) For the rectangular beam of Fig. A.13, $f'_c = 30$ MPa, Grade 300 steel, calculate f_c and f_s from the internal couple when $M = 180$ kN · m.
(b) Calculate f_c and f_s from $Mc = fI$. (See Sec. 8.4d)
(c) What is allowable (working) moment on this beam under AASHTO Specification if allowable $f_c = 12$ MPa, $f_s = 140$ MPa?

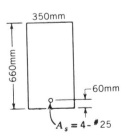

FIGURE   A.13
Beam of Prob.
A.1.

PROB. A2.   Same as Prob. A.1 except A_s is 4-#30 bars of Grade 400 steel.

PROB. A. 3. Same as Prob. A.1 except A_s is 4-#30 bars and A'_s of 2-#20 bars has been added, centered 60 mm below the top, and steel is Grade 400 steel.

PROB. A.4.

(a) For the T-beam of Fig. A.14, $f'_c = 20$ MPa, A_s of 4-#30 bars, Grade 300 steel, calculate f_c and f_s from the internal couple when $M = 135$ kN · m. Use the approximate method (Sec. A.5c).

(b) Calculate f_c and f_s from $Mc = fI$.

(c) What is the allowable moment?

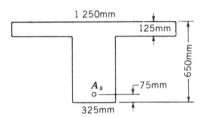

FIGURE A.14 T-beam for Prob. A.4.

PROB. A.5. Repeat Prob. A.4 except change A_s to 4-#35 bars and use the exact method of analysis.

PROB. A.6. Find the allowable moment on a 1-m strip of the slab of Fig. 3.9 if the bars are made #20 at 150 mm on centers and $d = 150$ mm, $f'_c = 30$ MPa, Grade 400 steel.

PROB. A.7. Find the allowable moment about the horizontal axis of the regular hexagonal pile of Fig. A.14 in the position shown, if $f'_c = 30$ MPa and steel is Grade 400. (*Suggestion*: Use $M_c = fI$).

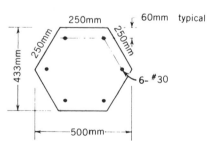

FIGURE A.15 Hexagonal pile for Probs. A.7 and A.8.

PROB. A.8. If the hexagonal pile of Fig. A.15 and Prob. A.7 is rotated 90°, find the allowable moment.

PROB. A.9. Neglecting the lack of symmetry (because the flange is part of a continuing slab), find f_c and f_s in the beam of Fig. A.16 if $f'_c = 20$ MPa and $A_s = 2$-#30 bars of Grade 400 steel. $M = 95$ kN · m.

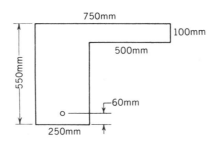

FIGURE A.16 Spandrel T-beam for Prob. A.9.

PROB. A.10.
(a) Design a rectangular beam exactly balanced for a total $M = 265$ kN · m using $f'_c = 30$ MPa, Grade 400 steel. After the required bd^2 has been established, use $b = 375$ mm.
(b) Redesign steel if d is made 650 mm and b is maintained at 375 mm.
(c) Redesign steel if d is made 575 mm and b is maintained at 375 mm.

PROB. A.11. A rectangular beam 300 mm wide by 500 mm deep to center of steel must carry a total moment of 135 kN · m with $f'_c = 20$ MPa and Grade 300 steel. Find the required steel.

PROB. A.12. A slab 200 mm thick with 40 mm cover to center of steel must carry a total moment of 27 kN · m/m width. Find the required A_s per foot if $f'_c = 20$ MPa and steel is Grade 400.

PROB. A.13. A rectangular beam 350 mm wide by 625 mm deep over-all with 75 mm cover to center of steel must carry a total moment of 300 kN · m. Find the necessary steel if $f'_c = 30$ MPa and steel is Grade 400.

PROB. A.14. From basic principles establish the value of the balanced steel ratio ρ_b for a rectangular beam with $f'_c = 30$ MPa and $f_s = 140$ MPa.

DESIGN TABLES AND CURVES

TABLE B.1 Flexural Resistance Coefficients k_m for Rectangular Sections without A'_s (Modified from ACI *Ultimate Strength Design Handbook*). $M_u = k_m bs^2$; $M_n = (k_m/0.9)bd^2$

$\omega = \rho \dfrac{f_y}{f'_c}$	$f'_c = 25$ MPa			$f'_c = 30$ MPa			$f'_c = 35$ MPa			Any $f'_c \leqslant 35$ MPa		
	f_y —	300 MPa	400 MPa	—	300 MPa	400 MPa	—	300 MPa	400 MPa			
	k_m	ρ	ρ	k_m	ρ	ρ	k_m	ρ	ρ	c/d^a	a/d	z/d
0.020	0.44	0.0017	0.0013	0.53	0.0020	0.0015	0.62	0.0023	0.0018	0.028	0.024	0.988
0.030	0.66	0.0025	0.0019	0.80	0.0030	0.0023	0.93	0.0035	0.0026^b	0.042	0.035	0.982
0.040	0.88	0.0033	0.0025	1.05	0.0040	0.0030^b	1.23	0.0047	0.0035	0.055	0.047	0.976
0.050	1.09	0.0042	0.0031^b	1.31	0.0050	0.0038	1.53	0.0058	0.0044	0.069	0.059	0.971
0.060	1.30	0.0050	0.0038	1.56	0.0060	0.0045	1.82	0.0070	0.0053	0.083	0.071	0.965
0.070	1.51	0.0058	0.0044	1.81	0.0070	0.0053	2.11	0.0082	0.0061	0.097	0.082	0.959
0.080	1.72	0.0067	0.0050	2.06	0.0080	0.0060	2.40	0.0093	0.0070	0.111	0.094	0.953
0.090	1.92	0.0075	0.0056	2.30	0.0090	0.0068	2.68	0.0105	0.0079	0.125	0.106	0.947
0.100	2.12	0.0083	0.0063	2.54	0.0100	0.0075	2.96	0.0117	0.0088	0.138	0.118	0.941
0.110	2.31	0.0092	0.0069	2.78	0.0110	0.0083	3.24	0.0128	0.0096	0.152	0.129	0.935
0.120	2.51	0.0100	0.0075	3.01	0.0120	0.0090	3.51	0.0140	0.0105	0.166	0.141	0.929
0.130	2.70	0.0108	0.0081	3.24	0.0130	0.0098	3.77	0.0152	0.0114	0.180	0.153	0.923
0.140	2.89	0.0117	0.0088	3.47	0.0140	0.0105	4.04	0.0163	0.0123	0.194	0.165	0.917
0.150	3.08	0.0125	0.0094	3.69	0.0150	0.0113	4.30	0.0175	0.0131	0.208	0.176	0.912
0.160	3.26	0.0133	0.0100	3.91	0.0160	0.0120	4.55	0.0187	0.0140	0.221	0.188	0.906

c/d												
0.170	3.44	0.0142	0.0106	4.13	0.0170	0.0128	4.81	0.0198	0.0149	0.235	0.200	0.900
0.180	3.62	0.0150	0.0113	4.34	0.0180	0.0135	5.06	0.0210	0.0158	0.249	0.212	0.894
0.190	3.80	0.0158	0.0119	4.55	0.0190	0.0143	5.30	0.0222	0.0166	0.263	0.224	0.888
0.200	3.97	0.0167	0.0125	4.76	0.0200	0.0150	5.54	0.0233	0.0175	0.277	0.235	0.882
0.210	4.14	0.0175	0.0131	4.97	0.0210	0.0158	5.78	0.0245	0.0184	0.291	0.247	0.876
0.220	4.31	0.0183	0.0138	5.17	0.0220	0.0165	6.01	0.0257	0.0193	0.304	0.259	0.870
0.230	4.47	0.0192	0.0144	5.37	0.0230	0.0173	6.24	0.0268	0.0201	0.318	0.271	0.864
0.240	4.64	0.0200	0.0150	5.56	0.0240	0.0180	6.47	0.0280	0.0210	0.332	0.282	0.858
0.250	4.80	0.0208	0.0156	5.75	0.0250	0.0188	6.69	0.0292	0.0219	0.346	0.294	0.853
0.260	4.95	0.0217	0.0163	5.94	0.0260	0.0195	6.91	0.0303	0.0228	0.360	0.306	0.847
0.270	5.11	0.0225	0.0169	6.13	0.0270	0.0203	7.12	0.0315	0.0236	0.374	0.318	0.841
0.280	5.26	0.0233	0.0175	6.31	0.0280	0.0210	7.33	0.0327	0.0245	0.388	0.329	0.835
0.290	5.41	0.0242	0.0181	6.49	0.0290	0.0218	7.54	0.0338	0.0254	0.401	0.341	0.829
0.300	5.56	0.0250	0.0188	6.67	0.0300	0.0225	7.74	0.0350	0.0263	0.415	0.353	0.823
0.310	5.70	0.0258	0.0194	6.84	0.0310	0.0233	7.94	0.0362	0.0271	0.429	0.365	0.817
0.320	5.84	0.0267	0.0200	7.01	0.0320	0.0240	8.14	0.0373	>0.75 ρ_b	0.443	0.376	0.811
0.330	5.98	0.0275	>0.75 ρ_b	7.18	0.0330	>0.75 ρ_b	8.33	0.0385		0.457	0.388	0.805
0.340	6.12	0.0283		7.34	0.0340		8.52	0.0397		0.471	0.400	0.799
0.350	6.25	0.0292		7.50	0.0350		8.70	>0.75 ρ_b		0.484	0.412	0.794
0.360	6.38	0.0300		7.66	0.0360		8.88			0.498	0.424	0.788
0.370	6.51	>0.75 ρ_b		7.81	>0.75 ρ_b		9.06			0.512	0.435	0.782

[a] For $f'_c > 30$ MPa the value of c/d must be adjusted (see Section 3.3).
For $f'_c = 35$ MPa c/d must be multiplied by $0.85/0.81 = 1.05$.
[b] Above these heavy lines $\rho < 1.4/f_y$, f_y in MPa.

TABLE B.2 Ultimate Moment Coefficients k_n for Rectangular Sections (Strength Design) for $f'_c \leq 30$ MPa

$$\text{Coef.} = \frac{M_n}{f'_c bd^2} = \omega(1 - 0.59\,\omega), \quad k_n = f'_c(\text{coef.}), \quad k_m = 0.9\, f'_c(\text{coef.})$$

$$\omega = \frac{\rho f_y}{f'_c} \text{ or } \rho = \frac{\omega f'_c}{f_y}$$

ω	.000	.001	.002	.003	.004	.005	.006	.007	.008	.009
0.0	0	0.0010	0.0020	0.0030	0.0040	0.0050	0.0060	0.0070	0.0080	0.0090
.01	0.0099	.0109	.0119	.0129	.0139	.0149	.0159	.0168	.0178	.0188
.02	.0197	.0207	.0217	.0226	.0236	.0246	.0256	.0266	.0275	.0285
.03	.0295	.0304	.0314	.0324	.0333	.0343	.0352	.0362	.0372	.0381
.04	.0391	.0400	.0410	.0420	.0429	.0438	.0448	.0457	.0467	.0476
.05	.0485	.0495	.0504	.0513	.0523	.0532	.0541	.0551	.0560	.0569
.06	.0579	.0588	.0597	.0607	.0616	.0625	.0634	.0643	.0653	.0662
.07	.0671	.0680	.0689	.0699	.0708	.0717	.0726	.0735	.0744	.0753
.08	.0762	.0771	.0780	.0789	.0798	.0807	.0816	.0825	.0834	.0843
.09	.0852	.0861	.0870	.0879	.0888	.0897	.0906	.0915	.0923	.0932
.10	.0941	.0950	.0959	.0967	.0976	.0985	.0994	.1002	.1011	.1020
.11	.1029	.1037	.1046	.1055	.1063	.1072	.1081	.1089	.1098	.1106
.12	.1115	.1124	.1133	.1141	.1149	.1158	.1166	.1175	.1183	.1192
.13	.1200	.1209	.1217	.1226	.1234	.1243	.1251	.1259	.1268	.1276
.14	.1284	.1293	.1301	.1309	.1318	.1326	.1334	.1342	.1351	.1359
.15	.1367	.1375	.1384	.1392	.1400	.1408	.1416	.1425	.1433	.1441
.16	.1449	.1457	.1465	.1473	.1481	.1489	.1497	.1506	.1514	.1522

ω										
.17	.1529	.1537	.1545	.1553	.1561	.1569	.1577	.1585	.1595	.1601
.18	.1609	.1617	.1624	.1632	.1640	.1648	.1656	.1664	.1671	.1679
.19	.1687	.1695	.1703	.1710	.1718	.1726	.1733	.1741	.1749	.1756
.20	.1764	.1772	.1779	.1787	.1794	.1802	.1810	.1817	.1825	.1832
.21	.1840	.1847	.1855	.1862	.1870	.1877	.1885	.1892	.1900	.1907
.22	.1914	.1922	.1929	.1937	.1944	.1951	.1959	.1966	.1973	.1981
.23	.1988	.1995	.2002	.2010	.2017	.2024	.2031	.2039	.2046	.2053
.24	.2060	.2067	.2075	.2082	.2089	.2096	.2103	.2110	.2117	.2124
.25	.2131	.2138	.2145	.2152	.2159	.2166	.2173	.2180	.2187	.2194
.26	.2201	.2208	.2215	.2222	.2229	.2236	.2243	.2249	.2256	.2263
.27	.2270	.2277	.2284	.2290	.2297	.2304	.2311	.2317	.2324	.2331
.28	.2337	.2344	.2351	.2357	.2364	.2371	.2377	.2384	.2391	.2397
.29	.2404	.2410	.2417	.2423	.2430	.2437	.2443	.2450	.2456	.2463
.30	.2469	.2475	.2482	.2488	.2495	.2501	.2508	.2514	.2520	.2527
.31	.2533	.2539	.2546	.2552	.2558	.2565	.2571	.2577	.2583	.2590
.32	.2596	.2602	.2608	.2614	.1621	.2627	.2633	.2639	.2645	.2651
.33	.2657	.2664	.2670	.2676	.2682	.2688	.2694	.2700	.2706	.2712
.34	.2718	.2724	.2730	.2736	.2742	.2748	.2754	.2760	.2766	.2771
.35	.2777	.2783	.2789	.2795	.2801	.2807	.2812	.2818	.2824	.2830
.36	.2835	.2841	.2847	.2853	.2858	.2864	.2870	.2875	.2881	.2887
.37	.2892	.2898	.2904	.2909	.2915	.2920	.2926	.2931	.2937	.2943
.38	.2948	.2954	.2959	.2965	.2970	.2975	.2981	.2986	.2992	.2997
.39	.3003	.3008	.3013	.3019	.3024	.3029	.3035	.3040	.3045	.3051
.40	.3056									

Read approximate ω at left and move to right to read k_n for interpolated values of ω (increments shown at top).

TABLE B.3 Tension Bar Development Lengths, mm

Tension bar ℓ_d (mm)-Ordinary concrete-$ft = 20$ MPA^a

Bar	← $f_y = 400$ MPa →				← $f_y = 300$ MPa →				Hooks, $f_y = 400$ MPa	
	$s < 150$ mm		$s \geqq 150$ mm		$s < 150$ mm		$s \geqq 150$ mm		f_h in MPa	
	Other	Top	Other	Top	Other	Top	Other	Top	Other	Top
#10	300	325	300	300	300	300	300	300	192	192
#15	348	487	300	390	300	365	300	300	192	192
#20	510	714	408	571	383	536	306	428	165	192
#25	850	1190	680	952	638	893	510	714	134	192
#30	1190	1666	952	1333	893	1250	714	1000	134	174
#35	1700	2380	1360	1904	1275	1785	1020	1428	134	152
#45	2326	3256	1860	2605	1745	2442	1395	1954	121	121
#55	3041	4257	2433	3406	2281	3193	1825	2555	80	80

^{a}For other f_c' multiply by $\sqrt{20/f_c'}$ with min. ℓ_d still 300 mm except for splice calculations including stirrup splices.

Tension bar ℓ_d (mm)-Lightweight concrete-$f_c' = 20$ MPa^a − f_{ct} not specified

Bar	← $f_y = 400$ MPa →				← $f_y = 300$ MPa →				
	$s < 150$ mm		$s > 150$ mm		$s < 150$ mm		$s > 150$ mm		
	Other	Top	Other	Top	Other	Top	Other	Top	
#10	308	432	300	346	300	324	300	300	For sand-
#15	463	648	370	519	347	485	300	388	lightweight
#20	678	950	543	759	509	713	407	569	multiply table
#25	1131	1583	904	1266	849	1188	678	950	values by
#30	1583	2216	1266	1773	1188	1663	950	1333	factor
#35	2261	3165	1809	2531	1696	2374	1357	1899	1.18/1.33 but
#45	3094	4330	2474	3465	2321	3248	1855	2599	min. ℓ_d still
#55	4045	5562	3236	4530	3034	4247	2427	3398	300 mm

^{a}For other f_c' multiply by $\sqrt{20/f_c'}$ with min. ℓ_d still 300 mm except for splice calculations including stirrup splices.

TABLE B.4 Compression Bar Development Lengths, mm

Bar	$f_y = 400$ MPa		$f_y = 300$ MPa		
	$\ell_d, f_c' = 20$	Min. ℓ_d	$\ell_d, f_c' = 20$	Min. ℓ_d	
#10	214	200	200	200	For $f_c' \neq 20$ MPa
#15	322	264	241	200	multiply ℓ_d values
#20	429	350	322	263	by $\sqrt{20/f_c'}$, but
#25	537	440	403	330	ℓ_d must not be below
#30	644	528	483	396	minimum tabulated.
#35	751	616	563	462	
#45	996	792	747	594	
#55	1181	968	886	726	

TABLE B.5 Limiting Constants for Rectangular Beams (from Sec. 3.7)

Steel:		$f_y = 300$ MPa				$f_y = 400$ MPa			
Concrete Gr. f_c', MPa		k_m MPa	k_n MPa	$100\,\rho$	a/d	k_m MPa	k_n MPa	$100\,\rho$	a/d
C20	20	5.11	5.68	2.40	0.43	4.73	5.25	1.62	0.38
C25	25	6.40	7.11	3.01	0.43	5.92	6.58	2.03	0.38
C30	30	7.68	8.53	3.61	0.43	7.10	7.89	2.44	0.38
C35	35	8.65	9.61	4.01	0.41	7.98	8.87	2.71	0.36
C40	40	9.51	10.56	4.36	0.39	8.77	9.74	2.94	0.35
$\rho = 0.18\dfrac{f_c'}{f_y}$		$0.144 f_c'$	$0.160 f_c'$	$0.18\dfrac{f_c'}{f_y}$	0.21	$0.144 f_c'$	$0.160 f_c'$	$0.18\dfrac{f_c'}{f_y}$	0.21

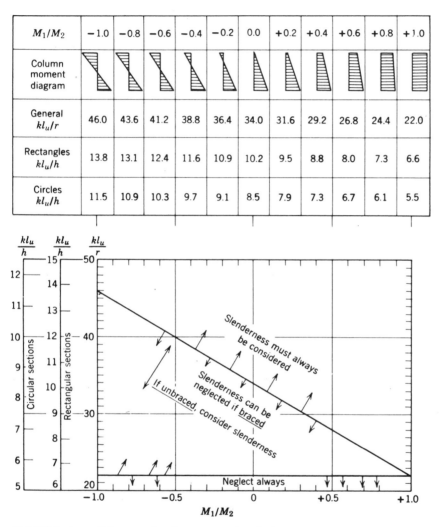

M_1/M_2	-1.0	-0.8	-0.6	-0.4	-0.2	0.0	$+0.2$	$+0.4$	$+0.6$	$+0.8$	$+1.0$
Column moment diagram											
General kl_u/r	46.0	43.6	41.2	38.8	36.4	34.0	31.6	29.2	26.8	24.4	22.0
Rectangles kl_u/h	13.8	13.1	12.4	11.6	10.9	10.2	9.5	8.8	8.0	7.3	6.6
Circles kl_u/h	11.5	10.9	10.3	9.7	9.1	8.5	7.9	7.3	6.7	6.1	5.5

FIGURE B.1 Column slenderness ratios below which the effects of slenderness may be neglected. (Basic chart by Dr. R. W. Furlong.)

674

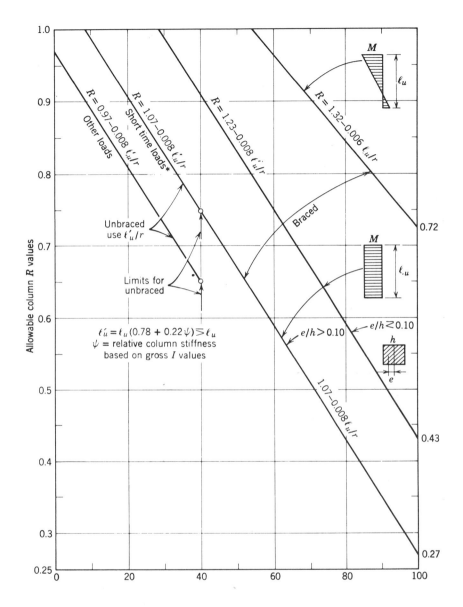

FIGURE B.2 Column R values for alternate column design method. Design $M = M_u/R$ and $P = P_u/R$; e unchanged at $M_u/P_u = M/P$. (*Min. beam $\rho = 0.01$ for this R eq.)

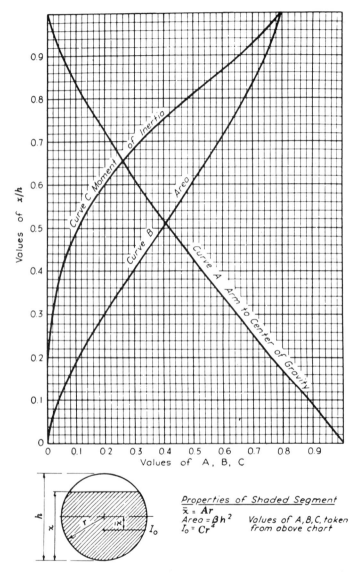

FIGURE B.3 Constants for properties of circular elements.
(Courtesy Prof. J. R. Shank, Ohio State University.)

676

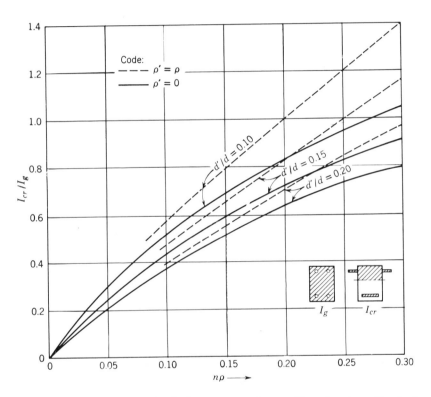

FIGURE B.4 Ratio of I_{cr}/I_g for rectangular sections having $d' = h\text{-}d$.

C

TABLES OF REINFORCING BAR AREAS

TABLE C.1 Cross-sectional Area (mm²) of Combinations of Metric Bars of the Same Size

Number of Bars	Bar No. Area	10 mm²	15 mm²	20 mm²	25 mm²	30 mm²	35 mm²	45 mm²	55 mm²
1		100	200	300	500	700	1 000	1 500	2 500
2		200	400	600	1 000	1 400	2 000	3 000	5 000
3		300	600	900	1 500	2 100	3 000	4 500	7 500
4		400	800	1 200	2 000	2 800	4 000	6 000	10 000
5		500	1 000	1 500	2 500	3 500	5 000	7 500	12 500
6		600	1 200	1 800	3 000	4 200	6 000	9 000	15 000
7		700	1 400	2 100	3 500	4 900	7 000	10 500	17 500
8		800	1 600	2 400	4 000	5 600	8 000	12 000	20 000
9		900	1 800	2 700	4 500	6 300	9 000	13 500	22 500
10		1 000	2 000	3 000	5 000	7 000	10 000	15 000	25 000
11		1 100	2 200	3 300	5 500	7 700	11 000	16 500	27 500
12		1 200	2 400	3 600	6 000	8 400	12 000	18 000	30 000
13		1 300	2 600	3 900	6 500	9 100	13 000	19 500	32 500
14		1 400	2 800	4 200	7 000	9 800	14 000	21 000	35 000
15		1 500	3 000	4 500	7 500	10 500	15 000	22 500	37 500
16		1 600	3 200	4 800	8 000	11 200	16 000	24 000	40 000
17		1 700	3 400	5 100	8 500	11 900	17 000	25 500	42 500
18		1 800	3 600	5 400	9 000	12 600	18 000	27 000	45 000
19		1 900	3 800	5 700	9 500	13 300	19 000	28 500	47 500
20		2 000	4 000	6 000	10 000	14 000	20 000	30 000	50 000

TABLE C.2 Cross-sectional Area per Meter Width (mm²/m) of Metric Bars of the Same Size

Bar Spacing mm	Bar No. 10 Area mm²/m	15 mm²/m	20 mm²/m	25 mm²/m	30 mm²/m	35 mm²/m	45 mm²/m	55 mm²/m
50	2 000	4 000	6 000	10 000	14 000	20 000	30 000	50 000
75	1 333	2 666	4 000	6 667	9 333	13 333	20 000	33 333
100	1 000	2 000	3 000	5 000	7 000	10 000	15 000	25 000
125	800	1 600	2 400	4 000	5 600	8 000	12 000	20 000
150	667	1 333	2 000	3 333	4 667	6 667	10 000	16 667
175	571	1 143	1 714	2 857	4 000	5 714	8 571	14 286
200	500	1 000	1 500	2 500	3 500	5 000	7 500	12 500
250	400	800	1 200	2 000	2 800	4 000	6 000	10 000
300	333	667	1 000	1 667	2 333	3 333	5 000	8 333
350	286	571	857	1 429	2 000	2 857	4 286	7 143
400	250	500	750	1 250	1 750	2 500	3 750	6 250
500	200	400	600	1 000	1 400	2 000	3 000	5 000
600	167	333	500	833	1 167	1 667	2 500	4 167
1000	100	200	300	500	700	1 000	1 500	2 500

APPENDIX
D
LIST OF SYMBOLS

a depth of equivalent stress block of concrete; arbitrary extension of reinforcing bar

a_b depth of equivalent rectangular stress block that produces balanced strain conditions

A area

A_1 bearing area

A_2 supporting surface that is geometrically similar to A_1

A_b cross-sectional area of an individual reinforcing bar

A_g gross area of cross section

A total area of longitudinal reinforcement to resist torsion

A_n net cross-sectional area of concrete $= A_g - A_{st}$

A_s area of tension reinforcement

A_s' area of compression reinforcement

A_{st} area of vertical column steel

A_t area of one leg of a closed stirrup resisting torsion within a distance s

A_v area of shear reinforcement within a distance s

b width of compression face of member

b_0 perimeter of critical section for shear in slabs and footings

b_w width of web

c distance from extreme compression fiber to neutral axis; unit diagonal compression; diameter of column capital

c' clear cover over the splice

c_b distance of extreme compression fiber to neutral axis that produces balanced strain conditions

C compressive force

C_t torsion constant for sections composed of rectangles

d effective depth

d' distance of extreme compression fiber to centroid of compression reinforcement

d'' distance of centroid of tension steel from plastic centroid of column

d_b nominal diameter of reinforcing bar

d_s distance of centroid of tension steel from tension face

D (unfactored) dead load

e eccentricity of load parallel to axis of member, measured from centroid of gross cross section; base of Naperian logarithm

e'　eccentricity of load parallel to axis of member, measured from centroid of tension reinforcement

e_b　column eccentricity for balanced design

E_c　modulus of elasticity of concrete

E_s　modulus of elasticity of steel

f　direct stress

f_c　actual compressive stress in concrete due to actual (unfactored) load

f'_c　specified compressive strength of concrete

f_{ct}　tensile strength of concrete

f_h　tensile stress developed by a standard hook

f_{pi}　initial steel stress in prestressed concrete

f_s　actual stress in tension steel

f'_s　actual stress in compression steel

f_{si}　initial steel stress in prestressed concrete

f_t　tensile stress in concrete

f_u　ultimate strength of reinforcing steel

f_y　specified yield strength of reinforcement

F_y　force in y (vertical) direction

h　overall depth; minimum depth of beams and slabs; height

h_f　slab (flange) thickness

H　horizontal force

I　moment of inertia

I_g　moment of inertia of gross concrete section about centroidal axis

J　polar moment of inertia

k　effective length factor for compression members

k_b　neutral axis depth ratio for balanced design (ultimate load design); design constant for service load analysis, in MPa

k_m　fully factored design constant for rectangular beams and slabs (ultimate load design), in MPa

k_n　nominal design constant for rectangular beams and slabs (ultimate load design), in MPa

ℓ　span

ℓ_a　additional embedment length at support or a point of inflection

ℓ_d　development length

ℓ_n　clear span

ℓ_s　length of splice

ℓ_u　unsupported length of compression member

L　(unfactored) live load

m　moment per unit length in yield-line theory

M　bending moment

M_b　bending moment for balanced design

M_d　moment due to dead load

M_ℓ　moment due to live load

M_n　nominal moment

M_p plastic moment

M_s moment due to service load

M_u fully factored bending moment

n modular ratio $= E_s/E_c$

N_c resultant compressive force in concrete

N_t resultant tensile force

N_u factored load axial to cross section occurring simultaneously with V_u, to be taken positive for compression and negative for tension

p pressure

P_b column load for balanced design

P_c critical column load

P_i initial prestress

P_n nominal axial load strength

P_{se} effective prestress

P_u fully factored axial load

q_{net} average allowable soil pressure

r radius; radius of gyration

s stirrup spacing

s' clear spacing between splices

S section modulus

S_s section modulus of structural steel section in composite construction

S_r range of horizontal shear stress for composite construction (AASHTO notation)

S_{tr} limiting section modulus of steel section in composite construction

t unit diagonal tension

T tensile force; torsion

T_c nominal torsional strength provided by concrete

T_n nominal torsional moment strength

T_u fully factored torsional moment at section

u bond stress

U required strength to resist factored load

v unit shear stress

v_c shear stress carried by concrete

v_s contribution of shear reinforcement to unit shear capacity

v_t torsional shear strength of concrete

v_{tc} torsional shear stress carried by concrete

v_{tn} shear stress due to torsion

V shear force

V_c nominal shear strength provided by concrete

V_n nominal shear strength

V_s nominal shear strength provided by steel

V_u fully factored shear force

w load per unit length, or per unit area

w_f equivalent fluid weight of soil

W (unfactored) wind load

x distance parallel to span; shorter dimension of rectangle in torsional analysis

y longer dimension of rectangle in torsional analysis

$\bar{y}$ depth of center of gravity

y_c distance of extreme compression fiber from centroid

z arm of resistance moment; quantity in Section 8.13 for limiting crack width

α_m ratio of average stiffness of the four beams supporting a two-way slab to that of the slab

β ratio of longer to shorter clear span of a two-way slab

β_1 factor used in the equivalent rectangular stress diagram for concrete at the ultimate load. It is 0.85 for f'_c up to 30 MPa. For concrete strength above 30 MPa it is reduced at the rate of 0.08 for each 10 MPa of strength in excess of 30 MPa

β_c ratio of long side to short side in two-way slabs and footings

β_d ratio of maximum factored dead load moment to maximum factored total moment

β_s ratio of the sum of the lengths of the continuous edges of a two-way slab to the total perimeter of the slab panel

γ unit weight of soil; load factor in limit design; ratio of distance between tension and compression reinforcement to overall thickness of column $= (h - d' - d_s)/h$

δ moment magnifier

Δ elastically computed lateral deflection

ϵ compressive strain in concrete

ϵ_s strain in tension steel

ϵ'_s strain in compression steel

ϵ_u ultimate concrete strain $= 0.003$

ϵ_y yield strain of steel $= f_y/E_s$

θ angle

ρ ratio of tension reinforcement to effective concrete section

ρ' ratio of compression reinforcement to effective concrete section

ρ_b reinforcement ratio producing balanced strain conditions

ρ_n ratio of vertical shear reinforcement to gross concrete area of horizontal section

ρ_w $= A_s/b_w d$

Σ_0 total perimeter of reinforcing bars

ϕ strength reduction factor; curvature; angle of friction; ratio of the sum of the stiffnesses of the compression members to that of the flexural members in a plane at the end of a compression member; unit rotation

ω shorthand notation for $\rho f_y/f'_c$

E

CONVERSION FROM ENGLISH TO SI UNITS

Length
1 ft   = 0.304 8 m = 304.8 mm
1 in.   = 25.40 mm

Area
1 ft^2 = 0.092 903 m^2 = 92 903 mm^2
1 in.^{2} = 645.16 mm^2

Volume, Section Modulus
1 ft^3 = 0.028 316 m^3 = 28 316 000 mm^3
1 in.^{3} = 16 387 mm^3

Moment of Inertia
1 ft^4 = 0.008 631 m^4 = 8 631 000 000 mm^4
1 in.^{4} = 416 231 mm^4

Mass
1 lb   = 0.453 59 kg

Weight
1 k   = 4.448 22 kN
1 lb   = 4.448 22 N

Force
1 k   = 4.448 22 kN
1 lb   = 4.448 22 N

Force per Unit Length
1 klf   = 14.594 kN/m
1 plf   = 14.594 N/m

Force per Unit Area
1 psf   = 47.880 Pa
1 psi   = 6.894 76 kPa

Moment
1 k-ft   = 1.355 82 kN · m
1 lb-ft   = 1.355 82 N · m
1 k-in.   = 112.985 N · m
1 lb-in.   = 0.112 985 N · m

Moment per Unit Width
1 k-ft per foot   = 4.448 22 kN · m/m
1 lb-ft per foot   = 4.448 22 N · m/m

Stress, Pressure

1 psf	= 47.880 Pa
1 ksi	= 6.894 76 MPa
1 psi	= 6.894 76 kPa

CONVERSION FROM SI TO ENGLISH UNITS

Length

1 m	= 3.280 833 ft = 39.370 in.
1 mm	= 0.039 370 in.

Area

1 m^2	= 10.764 ft^2 = 1 550.003 in.^{2}
1 mm^2	= 0.001 550 in.^{2}

Volume, Section Modulus

1 m^3	= 35.315 ft^3
1 mm^3	= 0.000 061 024 in.^{3}

Moment of Inertia

1 m^4	= 2 402 500 in.^{4}
1 mm^4	= 0.000 002 403 in.^{4}

Mass

1 kg	= 2.204 62 lb

Weight

1 MN	= 224.809 k
1 kN	= 224.809 lb

Force

1 MN	= 224.809 k
1 kN	= 224.809 lb

Force per Unit Length

1 MN/m	= 68.522 klf
1 kN/m	= 68.522 plf

Force per Unit Area

1 MPa	= 20.885 434 ksf = 145.037 738 psi
1 kPa	= 20.885 434 psf = 0.145 038 psi

Moment

1 MN · m	= 737.562 k-ft
1 kN · m	= 737.562 lb-ft = 8.850 732 k-in.

Moment per Unit Width

1 MN · m/m	= 224.809 k-ft per foot
1 kN · m/m	= 224.809 lb-ft per foot

Stress, Pressure

1 MPa	$= 20.885\,434$ ksf $= 145.037\,738$ psi
1 kPa	$= 20.885\,434$ psf $= 0.145\,038$ psi

USEFUL DATA

Acceleration due to Gravity
(Multiply SI mass to obtain SI weight by)

g $\qquad = 9.806\,65$ m/s^2

Mass of concrete and Weight of Concrete
A mass of 150 pcf (lb/ft^3) corresponds almost exactly to a mass of 2 400 kg/m^3. A weight of 150 pcf corresponds approximately to 24 kN/m^3.

INDEX